Basic Electrical and Electronics Engineering

Basic Electrical and Electronics Engineering

Dr. V. Jegathesan
Assistant Professor (SG)
Department of Electrical and Electronics Engineering
School of Electrical Sciences
Karunya University
Karunya Nagar, Coimbatore
Tamil Nadu, India

Mr. K. Vinoth Kumar
Assistant Professor
Department of Electrical and Electronics Engineering
School of Electrical Sciences
Karunya University
Karunya Nagar, Coimbatore
Tamil Nadu, India

Dr. R. Saravanakumar
Division Leader – Associate Professor
Control System Division
School of Electrical Engineering
VIT University, Vellore
Tamil Nadu, India

WILEY-INDIA

Basic Electrical and Electronics Engineering

Other Wiley Editorial Offices:

John Wiley & Sons, Inc. 111 River Street, Hoboken, NJ 07030, USA

Wiley-VCH Verlag GmbH, Pappellaee 3, D-69469 Weinheim, Germany

John Wiley & Sons Australia Ltd, 42 McDougall Street, Milton, Queensland 4064, Australia

John Wiley & Sons (Asia) Pte Ltd, 2 Clementi Loop #02-01, Jin Xing Distripark, Singapore 129809

John Wiley & Sons Canada Ltd, 22 Worcester Road, Etobicoke, Ontario, Canada, M9W 1L1

First Edition: 2011

ISBN: 978-81-265-2951-3

www.wileyindia.com

Printed at: ***Ar Emm International, New Delhi***

Preface

About the Book

This book is meant for a wide range of readers who wish to learn the basic concepts of Basic Electrical and Electronics Engineering. This book addresses the basic requirements of B.E, B.Tech, B.Sc. Electrical and Electronics Engineering, Electronics and Communication Engineering, Aerospace, Marine Engineering and of other disciplines. Fundamental concepts are presented in a clear and lucid manner. The entire book is written in a style to enable the students to understand the subject matter without much difficulty. The worked-out examples provide support in explaining the hidden theory, and the step-by-step procedure adopted in solving the problems will induce self-confidence in the minds of readers to solve any complicated networking problem. In short, the book is student-friendly and it is hoped that it will be well-received by the student community. The main aim of developing this book is to provide students with the basics of Electrical and Electronics Engineering in a self-learning style and further motivate them to go deep into the subject by referring to books of higher levels.

Salient Features of the Book

The salient features of this book are:

- Easy and clear presentation of each chapter.
- Detailed explanation of basics of electrical quantities with SI units.
- Study on DC circuits along with the combination of series and parallel circuits.
- Study on principles of electrical machines and also application of AC in the electric circuits.
- Exhaustive coverage of polyphase systems.
- Introduction to electrical machines and its types in AC and DC machines with discussion on principle and operation.
- Discussion on fundamentals of measuring instruments and domestic wiring.
- Explanation of fundamentals of power system and its types.
- Study on fundamentals of passive electronic components and semiconductor devices.
- Detailed description of various types of transducers.
- Study on linear and digital integrated circuits.
- Study on fundamental concept of computer and its communication.

Organization of the Book

The book contains 17 chapters. It starts with the basics of Electrical Engineering. The chapters are organized as follows:

Chapter 1 gives the basic of electrical quantities along with their SI units as well as basic definition related to the electrical quantities, terms such as current, voltage, power, etc.

Chapter 2 enhances the governing equations of the given circuit based on Ohm's law and Kirchhoff's law. Elementary laws and reduction of simple circuits consisting of resistances in series–parallel combination are analysed.

Chapter 3 discusses the basic fundamentals of magnetic circuit. There is a lot of similarity between the magnetic and electric circuits. A comparative study is done for both the circuits. Parallel and composite magnetic circuits with derivation are also mentioned. Magnetic hysteresis, Steinmetz law and Eddy current loss are also discussed.

Chapter 4 explains permanent magnets with their types. It also covers self-inductance, mutual inductance and coefficient of coupling, energy stored in the magnetic field, stored energy per unit volume of magnetic field, electro-mechanical energy conversion and force on current-carrying conductor. Problems related to self-inductance as well as electromechanical energy conversion are discussed.

Chapter 5 discusses the concept of alternating current in the circuits, requirements as well basic terms related to the alternating current waveforms. Phasor representations related to resistance, inductance and capacitance are also discussed in this chapter.

Chapter 6 provides information related to three-phase system. In this chapter, phase sequence, balanced and unbalanced systems for star and delta connected load along with their comparison are also discussed. Measurements of power using two- and three-wattmeter method are also explained in detail.

Chapter 7 discusses types of electrical machines, that is, DC generator, induction motor with single and three phase, synchronous motor, DC motor, transformer, alternator with their construction and principle of operation. In this chapter applications of individual electrical machines both AC and DC with advantages and disadvantages are explained neatly.

Chapter 8 gives the classification of measuring instruments, deflecting torque, controlling torque, damping torque, permanent magnet moving iron and moving coil instrument. In this chapter megger construction and operation are also discussed.

Chapter 9 provides information about domestic wiring with specifications. Types of wiring and also simple domestic layout, earthing and fuses are also discussed.

Chapter 10 discusses the generation of power in thermal, hydro and nuclear power stations with operation as well as layout. In this chapter transmission and distribution systems are also discussed along with overhead and underground cables.

Chapter 11 explains passive electronics components, that is, resistors, inductors and capacitors.

Chapter 12 discusses classification of materials, energy band structures of conductor, insulator and semiconductor, properties and classification of semiconductors, PN-junction, rectifiers, *V–I* characteristics, Zener diode as voltage regulator, filters, bipolar transistors, transistor configurations, CE configuration, current amplification factors, field effect transistors (FET), metal oxide semiconductor FET (MOSFET), enhancement MOSFET and depletion MOSFET.

Chapter 13 discusses the silicon-controlled rectifier construction, operation and characteristics, TRIAC, unijunction transistor (UJT), application of UJT as a relaxation oscillator, opto-electronic device, photo-conductive cell (PCC), photodiode, photo-transistor, light-emitting diode (LED).

Chapter 14 provides the information of classification of transducers, basic requirements of transducers, displacement transducer, electrical strain gauge, capacitance transducers, linear variable differential transformer (LVDT), temperature transducer, resistance temperature detectors (RTD), piezoelectric transducers, Hall effect transducers with advantages and disadvantages.

Chapter 15 explains the concepts of advantages and limitations of ICs, classification of ICs, monolithic ICs, linear ICs, operational amplifier, IC timer, digital ICs. In this chapter Boolean algebra, logic gates, universal gates, flip-flop with some problems.

Chapter 16 explains the fundamental concepts of central processing unit, output devices, memory devices, CPU–memory connection, ROM, flash memory, semiconductor, magnetic tape storage, optical disk data storage, printers, personal computers, microprocessors and internal Architecture.

Chapter 17 provides the information about Integrated Services Digital Network (ISDN) Analog and digital signals, telecommunication services, basic principle of modulation, digital data transmission, TV communication system, satellite communication.

Dr. V. Jegathesan
Mr. K. Vinoth Kumar
Dr. R. Saravanakumar
February 2011

About the Authors

Dr. V. Jegathesan received his B.E. degree in Electrical and Electronics Engineering from Bharathiar University, Coimbatore, Tamil Nadu, India in 1999. He has obtained M.E. from Bharathiar University, Coimbatore, Tamil Nadu, India in 2002 and his doctoral degree from Anna University, Chennai, Tamil Nadu, India in 2010. Presently, he is working as an Assistant Professor (SG) in Department of Electrical and Electronics Engineering in Karunya Institute of Technology and Sciences (Karunya University), Coimbatore, Tamil Nadu, India. Dr. Jegathesan has 12 years of teaching experience from Karunya University. His present research interests are Electric Circuits and Networks, Development of Heuristic Algorithm for Power Electronics Applications, Application of Non-traditional Methods to Power Electronics, Power Electronics and Drives. He has already published various research papers in international journals and also in international conferences. He is a member of IEEE (USA), Indian Society for Technical Education, International Association of Engineers (IAENG), International Association of Computer Science and Information Technology (IACSIT).

Mr. K. Vinoth Kumar received his B.E. degree in Electrical and Electronics Engineering from Anna University, Chennai, Tamil Nadu, India in 2006. He obtained M.Tech in Power Electronics and Drives from VIT University, Vellore, Tamil Nadu, India in 2008. Presently he is working as an Assistant Professor in the School of Electrical Science, Karunya Institute of Technology and Sciences (Karunya University), Coimbatore, Tamil Nadu, India. He is pursuing PhD degree in Karunya University, Coimbatore, India. His present research interests are Condition Monitoring of Industrial Drives, Neural Networks and Fuzzy Logic, Special Machines, Application of Soft Computing Technique. He has published various papers in international journals and conferences and also published four textbooks. He is a member of IEEE (USA), MISTE and International Association of Electrical Engineers (IAENG).

Dr. R. Saravanakumar is currently working as an Associate Professor in the School of Electrical Engineering, Vellore Institute of Technology University (VIT). He has been teaching Power Electronics and Special Electrical Machines since 1999 to B.Tech students and has also been teaching students of M.Tech Power Electronics and Drives since 2002. He has a total teaching experience of 12 years and a research experience of 7 years. He received his postgraduate degree in Power Electronics and Drives from the College of Engineering, Guindy, in 1998 and PhD degree from VIT University, Vellore, India. He has published articles in both international and national conferences, journals and has also published papers in IEEE Xplore. He has published three textbooks. His areas of interest are Power Electronics, Modeling and Simulation of Power Electronic Systems, and Special Electrical Machines and Control.

Contents

1 Basics of Electrical Engineering

LEARNING OBJECTIVES

After reading this chapter, you will be able to:

- Understand the basic SI units.
- Understand derived SI units.
- Understand prefixes denoting multiplication and division.
- Learn the units of charge, force, work and power, and perform simple calculations involving these units.

1.1 Electrical Quantities

Electricity is a form of energy that has been known to mankind for many centuries. However, the laws and relations that govern its fundamentals, uses and conversions to other forms of energy, such as heat, light, sound or mechanical energies, were discovered and formulated in due course of time.

Electrical energy is concerned with transmission and utilization of mainly two things – energy and information. Usually electrical energy is generated by four methods:

1. Hydro power station is used to generate electric power from water.
2. Thermal power station is used to generate electric energy from coal.
3. Nuclear power station is used to generate electric energy from atomic fusion.
4. Electric energy is also generated from non-conventional sources.

The purpose of this chapter is to obtain a relationship between the energy delivered to a device and the integrated effects of the electric field intensity (voltage) and current density (current). A battery delivers energy to a circuit element appearing between terminals whether it is a resistance or an inductance. The energy is delivered to the device by the charge carriers. There are two methods, namely series and parallel, used for calculating voltage, current, power, etc. from electric circuits.

There are two types of electric currents, direct current (DC) and alternating current (AC).

1. AC is a specific type of electric current in which the direction of current flow is reversed or alternated periodically.
2. Electrically, DC is no different from AC except for the fact that it flows in the same direction at all times.

Electrical devices that convert electricity directly into other forms of energy can operate with both AC and DC. In this chapter, we provide the definitions of various quantities related to electrical circuits.

1.2 Definitions About Various Quantities

1.2.1 Voltage

Voltage difference, also called potential difference, is used to find voltage values of passive elements (resistors, inductors and capacitors) in series circuits because in series circuits, voltage is different across different elements, while in parallel circuits the voltage value is same across any element. Normally, potential difference is designated by the notation "V" and its unit is volt (V). The device used to measure voltage value or potential difference in an electrical circuit is known as voltmeter.

Definition The voltage or potential difference between two points in an electric circuit is the amount of energy or work required to move a unit charge between these two points. In mathematical terms,

$$V \text{ or } E \text{ (potential difference)} = W \text{ (work or energy)}/Q \text{ (charge)}$$

When the work done in order to transfer 1 C of charge between two points is 1 J, the voltage between the two points is 1 V; that is,

$$1\text{ V} = 1\text{ J}/1\text{ C}$$

Batteries can store only a specified amount of energy. This energy is provided by the chemical reaction to sustain a constant potential energy difference across the battery terminals. When this charge capacity "dries up" the source can no longer sustain the specified voltage across its terminals. Hence the charge Q can be expressed as

$$Q = I \times t\text{ C}$$

The unit of charge is coulomb or ampere-hour. There are many types of electrical energy sources. They are broadly classified as follows:

1. **Voltage sources:** It has a fixed voltage across its terminals, independent of the load connected to it.
2. **Current sources:** It has a fixed specified current flow between its terminals, independent of the load connected to it.

Sources are also divided into two types as shown in Fig. 1:

1. **DC sources:** They produce an output independent of time.
2. **AC sources:** They produce an output that varies continuously as a function of time. Polarities of AC sources also reverse continuously after each specified interval of time.

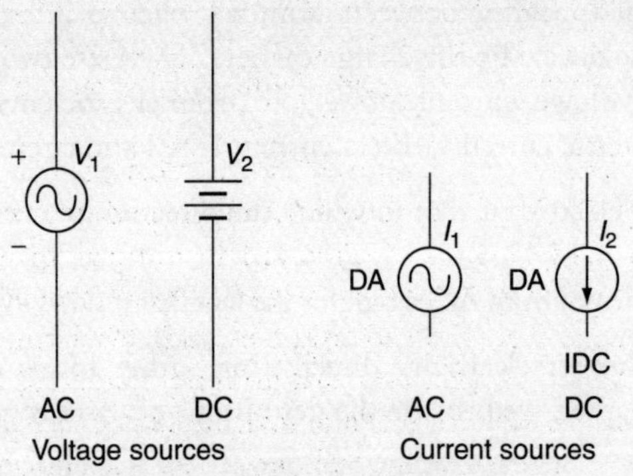

Figure 1 Voltage and current sources.

Example 1

If a dry cell chemical reaction needed 30 J of work to produce 15 C of charge separation between the battery terminals, what is the terminal voltage of this battery?
Given data: $W = 30$ J, $Q = 15$ C

Solution

Based on the definition of voltage or potential difference, we have

$$V = \frac{W}{Q} = \frac{30}{15} = 2\ \text{V}$$

Example 2

A current of 0.2 A flows through a filament and releases 8 J of heat energy in 2 s. Find the potential difference across the element.
Given data: $I = 0.2$ A, $W = 8$ J, $t = 2$ s

Solution

Here the work done by the charge as it flows in the element is known, but the amount of energy is unknown. The current and time are given. We have

$$I = \frac{Q}{t}$$

Therefore,

$$Q = I \times t = 0.2 \times 2 = 4\text{C}$$

We already know $V = W/Q = 8/4 = 2$ V.

1.2.2 Current

The flow of current carriers (electrons) through a conductor such as a copper wire is called the electric current or simply the current shown in Fig. 2. The symbol used for the magnitude of the current is I or i, originally derived from the French word "intensite". There are two types of currents, namely direct current (DC) and alternating current (AC).

When the magnitude and direction of current flowing in an element of an electric circuit do not vary with time then it is called direct current or DC.

Conversely, if the current in a certain electric circuit has a continuously varying magnitude with time and changes its direction of flow it is called alternating current or AC.

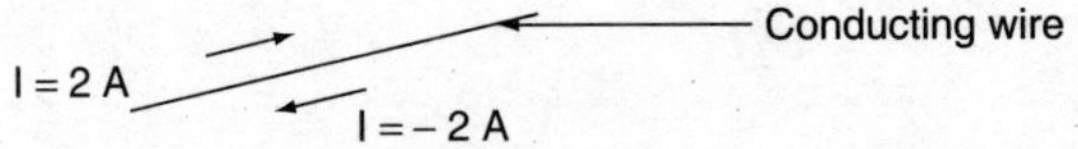

Figure 2 Representation of current direction.

Current can also be defined as the rate of flow of charge. Rate refers to variation with respect to time. Clearly, if a certain amount of charge Q flows through a cross-section of a conductor in time t, then the smaller t is, the faster is the motion (i.e., the higher must be the current). Mathematically, current I is expressed as

$$I = \frac{Q}{t} \tag{1.1}$$

When 1 C of charge flows through a cross-section of a conductor in 1 s, then one unit of current, called 1 ampere or 1 A, results. Thus,

$$1\text{ A} = 1\text{ C/s}$$

1.2.3 Electrical Power

When a person raises a certain mass to a fixed distance from the ground, he/she would feel more tired if the mass were lifted very fast than if it were raised slowly. Even though the same amount of **work** is done in both cases, in one case the time taken to do this work is short, while in the other case it is longer. The rate of spending energy is important in all practical applications. This concept is called power. The symbol of power is P or p.

Definition Power is the rate of doing work. The word rate implies "with respect to time". Mathematically,

$$P = \frac{W}{t}$$

The unit of power is watt (W) named in honor of James Watt, who invented the steam engine. Thus,

$$1\text{ W} = 1\text{ J/s}$$

Since work and time have been identified as influencing the definitions of electric voltage and current, power can be related to these quantities as shown below.

$$P = \frac{W}{t} = \left(\frac{W}{Q}\right) \times \left(\frac{Q}{t}\right) = V \times I \text{ Watt}$$

Thus,

$$1\text{ W} = 1\text{ V} \times 1\text{ A}$$

For practical applications, watt is usually a small unit. One practical unit of power that is still being used extensively is the horse power (hp) defined as

$$1\text{ hp} = 746\text{ W}$$

Another practical unit of power used extensively for appliances and industrial machinery is the kilowatt (kW) defined as

$$1 \text{ kW} = 1000 \text{ W}$$

Example 3

What is the power delivered from a 20 V source if the current supplied by this source is 1 A?
Given data: $V = 20$ V, $I = 1$ A

Solution

The power delivered by the source is

$$P = V \times I = 20 \times 1 = 20 \text{ W}$$

Example 4

Find the current flowing in a 20 W, 60 V lamp?
Given data: $P = 20$ W, $V = 60$ V

Solution

The lamp is a load, absorbing energy at the rate of 20 W. We have

$$P = V \times I$$

Therefore,

$$I = \frac{P}{V}$$

$$I = \frac{20}{60} = 0.33 \text{ A}$$

1.2.4 Electrical Energy

Another unit of energy which is used commercially is the kilowatt hour (kWh). It represents the work done by working at the rate of 1 kW for a period of 1 h. Once known as the board of trade unit, it is still widely referred to, especially by electricity suppliers, as the unit.

$$1 \text{ kWh} = 1000 \text{ Watt hours}$$

$$= 1000 \times 3600 \text{ Watt seconds or joules}$$

$$= 3{,}600{,}000 \text{ J} = 3.6 \text{ MJ}$$

Example 5

An electric motor is developing 10 kW at a speed of 900 revolution/min. Calculate the torque available at the shaft.

Solution

$$\text{Speed} = 900/60 = 15 \text{ revolution/second}$$
$$P = 2 \times \pi NrM/60$$
$$10000 = T \times 2\pi \times 15$$
$$T = 106 \text{ N m}$$

Example 6

A generating station has a daily output of 280 MWh and uses 500 tonnes of coal in the process. The coal releases 7 MJ/kg when burnt. Calculate the overall efficiency of the station.

Solution

Input energy per day is

$$W_{in} = 7 \times 10^6 \times 500 \times 1{,}000 = 35.0 \times 10^{11} \text{ J}$$

Output energy per day is

$$W_o = 280 \text{ MWh} = 280 \times 10^6 \times 3.6 \times 10^3 = 10.1 \times 10^{11} \text{ J}$$

Efficiency is

$$\eta = W_o / W_{in} = 0.288$$

1.2.5 Resistance

A resistance is an electrical load. The conventional associated reference directions of current flow and voltage drop for an electrical resistance are shown in Fig. 3. Current flows from the higher potential terminal toward the lower potential one, whereas the voltage arrow points toward the higher potential terminal (which is marked with a + sign).

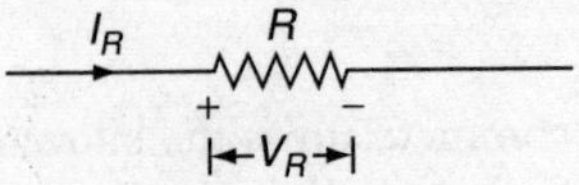

(a) Fixed resistance

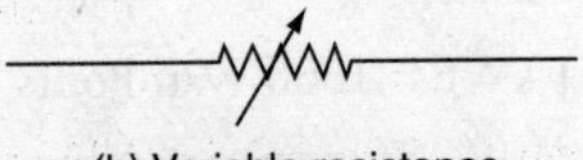
(b) Variable resistance

Figure 3 Resistance representation.

For $R = 0$, that is a short ciruit or connecting wire, with any finite amount of current flow,

$$V = I \times R = I \times 0 = 0 \text{ V}$$

For $R =$ infinity, that is an open circuit or open switch, with any finite voltage across the terminals,

$$I = \frac{V}{R} = \frac{V}{\text{Infinity}} = 0 \text{ A}$$

To measure the current flowing in a resistance, an ammeter is connected to the resistance. The ammeter monitors the current flowing through it, which is same as the current flowing through R if the voltmeter is not connected, or if the voltmeter is ideal, because then no current will be flowing in the voltmeter. The ammeter is a polarized instrument. This means that the current must enter the instrument from the positive side to obtain an upscale deflection of the pointer. The value of resistance is $R = V/I$. The unit of resistance is ohm (Ω).

Example 7

What is the resistance of a lamp if a 6 V battery results in a 100 mA current flow in it?
Given data: $V = 6$ V, $I = 100$ mA

Solution

$$R = \frac{V}{I} = \frac{6}{100 \times 10^{-3}} = 60\ \Omega$$

Example 8

Using Pspice, find the current through the lamp if a 10 V battery is connected in series with a 20 Ω resistor.

Solution

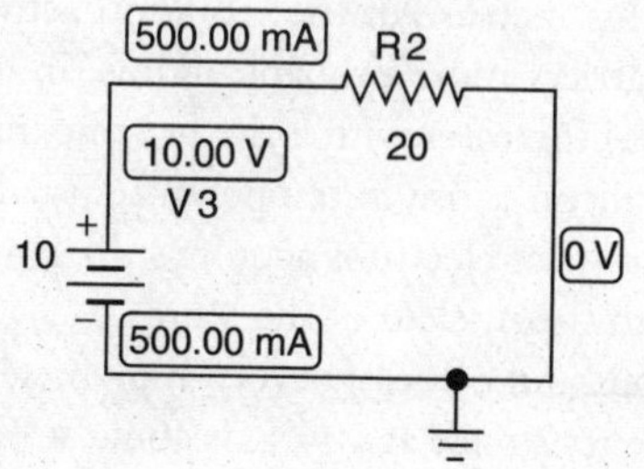

```
Output file: From Pspice simulator
VOLTAGE SOURCE CURRENTS
NAME      CURRENT
V_V3      -5.000E - 01
TOTAL POWER DISSIPATION 5.00E + 00 WATTS
********** Evaluation PSpice *************
```

The rate of energy absorption is the electrical power (P) dissipated in a resistance (R). Power is defined as

$$P = V \times I$$

From Ohm's law, $I = V/R$. Therefore,

$$P = V \times \frac{V}{R} = \frac{V^2}{R}$$

or

$$P = I^2 R$$

All these forms of power are equivalent to each other.

The cross-sectional dimension of the conductor with area and length of the resistance is given by

$$R = \rho \frac{l}{a} \tag{1.2}$$

where l is the length of the conductor in meter, a is the area of cross-section of the conductor in mm^2 or m^2 and ρ is the resistivity or specific resistance of the conductor material expressed in $\Omega\ mm^2/m$ or in Ω m.

The reciprocal of resistance is called conductance. It is represented by the symbol G and its unit is mho, or siemens (S).

$$G = \frac{l}{R} \tag{1.3}$$

1.2.6 Temperature Coefficient of Resistance

As the temperature of a metallic conductor is increased, the parent atoms constituting the crystalline structure of the material will acquire a larger amount of energy, exhibited in the form of vibrations around their average positions. This increase in the vibration activity of the parent atoms makes the collision process of the current carriers more probable. Thus, in general, the resistance of a metallic conductor (copper, aluminum, etc.) increases with a rise in temperature.

For carbon, germanium and silicon a rise in temperature usually results in the liberation of more electrons from their parent atom as these electrons acquire sufficient amounts of energy to enable them to overcome the binding force with the nucleus of the atom.

The resistance of all pure metals and of nearly every alloy increases with a rise in temperature. For example, in the case of copper, the resistance at 100°C is about 43% greater than the resistance at 0°C, and in this temperature range, the resistance–temperature curve is nearly linear.

The specific resistance of a metal at t °C (ρ_t) is given by the following expression:

$$\rho_t = \rho_0 (1 + \alpha t + \beta t^2)$$

where ρ_0 is the specific resistance at 0°C and α and β are constants for a given metal.

The value of β is small so that except when t is high, such as in the case of platinum-resistance thermometers, it is sufficiently accurate to use only the first two terms of the above expression. Then,

$$\rho_t = \rho_0 (1 + \alpha t)$$

and

$$R_t = R_0(1 + \alpha t)$$

where R_t and R_0 are the resistances corresponding to ρ_t and ρ_0, respectively.

It follows that $\alpha = (R_t - R_0)/R_0 t$ where α is called the temperature coefficient of increase in resistance and has the units of ohm/ohm per °C or simply just °C.

Certain alloys have very small values of α and high values of ρ and so they are valuable for constructing precision resistors. Carbon resistors and most thermistors have negative temperature coefficients. Thermistors are commonly made by embedding oxides of nickel, iron, etc. in ceramic binders and have large temperature coefficients; their resistance variation with temperature is less linear compared to metals.

Temperature coefficient for some of the conducting materials are given in Table 1.

Table 1 Temperature coefficients of some conducting materials

Material	*Temperature Coefficient Per °C at 0°C*
Copper	0.00428
German silver	0.00044
Platinum silver	0.00027
Manganin	0.00001

Example 9

Find the energy used by a 20 Ω heater element operating from a 120 V source for 2 h.

Solution

Heater power = $V^2/R = 120^2/20 = 0.72$ kW

Therefore energy used by the heater is

$$P \times t = 0.72 \times 2 = 1.44 \text{ kWh}$$

Example 10

The field current of a motor at 18°C was 1.75 A and after running for 5 h at full load it had fallen to 1.55 A. If the voltage applied to the field was 230 V and the temperature coefficient of the material is 0.00428°C^{-1}, calculate the final temperature of the coils.

Solution

At the start of the run the resistance $R_1 = (230/1.75)\ \Omega$

At the end of the run the resistance $R_2 = (230/1.55)\ \Omega$

Now

$$R_1 = R_0\,(1 + 18\alpha)$$

$$R_2 = R_0\,(1 + \alpha t)$$

where t is the final temperature. Therefore,

$$\frac{R_1}{R_2} = \frac{1.55}{1.75} = \frac{1+18\alpha}{1+\alpha t}$$

But $\alpha = 0.0428/°\text{C}$. So $t = 50.5°\text{C}$.

Example 11

The resistance of the armature winding of DC machine is 1.5 Ω at 27°C. What will be its resistance at 40°C? The resistance temperature coefficient of copper is 0.0043 / °C at 0°C.

Solution

$$R_2 = \frac{R_1(1+\alpha_0 t_2)}{1+\alpha_0 t_1} = \frac{1.5(1+0.0043\times 40)}{1+0.0043\times 27} = 1.575\ \Omega$$

Therefore, the resistance of the armature winding at 40°C is 1.575 Ω.

1.2.7 Inductance

A wire of certain length, when twisted into a coil, becomes a basic inductor. Inductance is defined as the energy stored in a magnetic field. If the current is passed through an inductor, an electromagnetic field is formed. A change in the magnitude of the current changes the electromagnetic field. If current increases, the field will expand otherwise it will not expand. Therefore the change in current produces a change in the electromagnetic field, which induces a voltage across the coil according to Faraday's law of electromagnetic induction. The unit of inductance is Henry and is denoted by H. The current–voltage relation is given by $v = L\,\mathrm{d}i/\mathrm{d}t$, where v is the voltage across the inductor in volts and i is the current through the inductor in amperes. We can write the above equation as

$$\mathrm{d}i = (1/L)\,v\mathrm{d}t$$

Integrating both sides, we get

$$\int_0^t \mathrm{d}i = \frac{1}{L}\int_0^t v\mathrm{d}t$$

$$i(t) - i(0) = \frac{1}{L}\int_0^t v\mathrm{d}t$$

$$i(t) = \frac{1}{L}\int_0^t v\mathrm{d}t + i(0)$$

The power absorbed by the inductor is

$$p = vi = Li\frac{\mathrm{d}i}{\mathrm{d}t} \text{ Watt}$$

The energy stored by the inductor is

$$W = \int_0^t p\mathrm{d}t$$

$$= \int_0^t Li\frac{\mathrm{d}i}{\mathrm{d}t}\mathrm{d}t = \frac{Li^2}{2}$$

Figure 4 shows the schematic representation of an inductance.

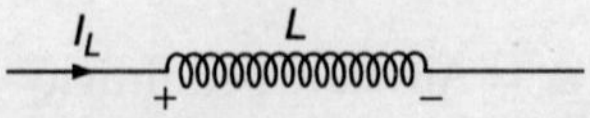

Figure 4 Inductance representation.

Example 12

The current in a 2 H inductor varies at a rate of 2 A/s. Find the voltage across the inductor and the energy stored in the magnetic field after 2 s.

Solution

$$v = L\frac{\mathrm{d}i}{\mathrm{d}t}$$

$$= 2 \times 4 = 8 \text{ V}$$

$$W = \int_0^t p\mathrm{d}t$$

$$= \int_0^t Li\frac{\mathrm{d}i}{\mathrm{d}t} = \frac{Li^2}{2}$$

$$= (1/2) \times 2 \times 4^2 = 16 \text{ J}$$

1.2.8 Capacitance

Any two conducting surfaces separated by an insulating medium exhibit the property of a capacitor. The conducting surfaces are called electrodes, and the insulating medium is called dielectric. A capacitor stores energy in the form of an electric field that is established by the opposite charges on the two electrodes. The unit of capacitance is farad and is denoted by F. Figure 5 shows the schematic representation of capacitance.

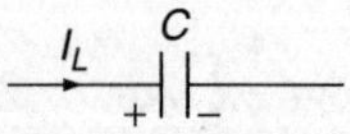

Figure 5 Capacitance representation.

Another definition of capacitance is stored energy in an electric field. Mathematically,

$$C = \frac{Q}{V} \tag{1.4}$$

If a dielectric of relative permittivity ε_r is placed in-between two conducting plates of area A each in m^2 and separated by a distance of d meter, then

$$C = \varepsilon_0 \varepsilon_r \frac{A}{d} \tag{1.5}$$

where

$$\begin{aligned} \varepsilon_0 &= \text{Absolute permittivity} \\ &= 8.854 \times 10^{-12}\ \text{H/m} \end{aligned}$$

We can write the capacitance equation in terms of current as mentioned below:

$$i = C\frac{dv}{dt}$$

where v is the voltage across capacitor and i is the current through the capacitor. So,

$$dv = \frac{1}{C} i dt$$

Integrating on both sides, we get

$$\int_0^t dv = \frac{1}{C}\int_0^t i dt$$

$$v(t) - v(0) = \frac{1}{C}\int_0^t i dt$$

$$v(t) = \frac{1}{C}\int_0^t i dt + v(0)$$

where $v(0)$ indicates the initial voltage across the capacitor. From the above equation, we can calculate the power absorbed by the capacitor as

$$p = vi = vC\frac{dv}{dt}$$

The energy stored by the capacitor in an electric field is

$$W = \int_0^t p\,dt = \int_0^t vC\frac{dv}{dt}\,dt$$

$$W = \frac{1}{2}Cv^2$$

From the above derivation, the capacitor can store a finite amount of energy, even if the current through it is zero and also the capacitor never dissipates energy, but only stores it.

Example 13

A capacitor having a capacitance of 2 μF is charged to a voltage of 1000 V. Calculate the stored energy in joules.

Solution

$$W = \frac{1}{2}Cv^2$$

$$= (1/2)\times 2\times 10^{-6}\times (1000)^2 = 1\ \text{J}$$

1.3 SI Units

Engineering is an applied science dealing with physical quantities. A unit is a physical quantity of standard of measurement regarded as an undivided entity. All other quantities can be expressed in terms of the three arbitrary units by means of the experimental, derived and defined relationships between the physical quantities. The three quantities of concern are called the fundamental quantities.

The adoption of International System of units (SI) is intended to allow the scientists in all the fields of physics and engineering to publish the results of their investigations in a manner that can be understood, verified and made use of by other workers in all countries. The SI system is based on the meter–kilogram–second (MKS) units for distance, weight and time rather than the centimeter–gram–second (CGS) units or the British system.

1. **International ohm:** It is defined as the resistance offered to the flow of current by a column of mercury of length 1063.3 cm and mass 14.452 g with uniform cross-section at 0°C.

2. **International ampere:** It is defined as the value of steady current which, when flowing through a solution of silver nitrate in water, deposits silver at the cathode at a rate of 1.118 mg/s.
3. **International volt:** It is defined as that amount of potential difference which, when applied across a conductor of one international ohm resistance, produces a current of one international ampere.

A summary of the most commonly used prefixes of the SI system is given in Table 2.

Table 2 SI sytem specifications

Power of 10	*Prefix Name*	*Symbol*
10^{12}	Tera	T
10^{9}	Giga	G
10^{6}	Mega	M
10^{3}	Kilo	K
10^{-2}	Centi	C
10^{-3}	Milli	M
10^{-6}	Micro	μ
10^{-9}	Nano	N
10^{-12}	Pico	P
10^{-15}	Femto	F

Summary

In this chapter, elementary concepts of electrical quantities of an electrical circuit are discussed. Definitions of various electrical terms with their units are also discussed and are explained in detail.

1. Potential difference, voltage and electromotive force are all names for the electrical pressure that tends to cause current flow.
2. Electrical current is a movement of electrons from a negative potential to a positive potential.
3. Electrical resistance is opposition to current flow.
4. Conventional current direction is from positive to negative.
5. Direct current is a current that constantly flows in a single direction.
6. Alternating current is a current that switches its direction of flow periodically.

Solved Problems

Problem 1

Convert the following

(i) 120 MW into Watt
(ii) 33 kV into volt
(iii) 12 μF into farad

Solution

(i) $120\text{ MW} = 120 \times 10^6\text{ W}$
(ii) $33\text{ kV} = 33 \times 10^3\text{ V}$
(iii) $12\ \mu\text{F} = 12 \times 10^{-6}\text{ F}$

Problem 2

A charge of 2 C is flowing between points A and B of a circuit. If potential difference between A and B is 4 V, how many joules will be released by this part of the circuit?

Solution

The potential difference of 4 V between points A and B means that each coulomb of charge will give up of 2 J of energy in moving from A to B. As the charge flowing is 2 C, therefore total energy released by part AB of the circuit is

$$4 \times 2 = 8\text{ J}$$

Problem 3

3 C of charge flow passed through a given wire in 1 second. How many amperes of current are flowing?

Solution

Given data

$$Q = 3\text{ C}$$

$$t = 1\text{ s}$$

$$I = \frac{Q}{t} = \frac{3}{1} = 3\text{ A}$$

Problem 4

What is the power in Watts if 20 J of energy is used in 2 s?

Solution

$$P = \frac{\text{Energy}}{\text{time}} = \frac{20}{2} = 10 \text{ W}$$

Problem 5

A conductor carries a constant current of 8 A. How many electrons pass a fixed point on the conductor in 3 min?

Solution

$$8 \text{ A} = 8 \times \frac{\text{C}}{\text{s}} = \frac{8 \text{ C}}{\text{s}} \times \frac{60 \text{ s} \times 3}{\text{min}}$$
$$= 3 \times 480 \text{ C/min}$$
$$e^- = 1.602 \times 10^{-19} \text{ C}$$
$$\frac{480 e^- \times 3 \text{ C/min}}{1.602 \times 10^{-19}} = 8.988 \times 10^{21} \text{ electrons/min}$$

Problem 6

In an electric circuit, 18 μJ of energy is required to transport 1 μC from point A to B. What electric potential difference exists between the two points?

Solution

$$V = \frac{W}{Q} = \frac{18 \text{ μJ}}{1 \text{ μC}} = 18 \text{ V}$$

Problem 7

A resistor has a potential difference of 50 V across its terminals and 160 C of charge/min passes a fixed point. Under these conditions, at what rate is electrical energy converted to heat?

Solution

Given

$$V = 50 \text{ V}$$
$$W = VQ$$
$$Q = \frac{160 \text{ C/min}}{60 \text{ s/min}} = 2.67 \text{ A}$$
$$\text{Power, } P = I \times V$$
$$= 2.67 \times 50$$
$$= 133.33 \text{ W}$$

Problem 8

Electrical energy is converted to heat at rate of 7.56 kJ/min in a resistor which has 290 C/min passing through it. What is the voltage across the resistor terminals?

Solution

Given

$$Q = 290 \text{ C/min}$$
$$W = 7.56 \times 1000 \text{ J}$$
$$V = \frac{W}{Q} = \frac{7.56 \times 1000}{290} = 26.06 \text{ V}$$

Problem 9

A resistor having a current of 3 A, converts 100 J of electric energy to heat energy in 22 s. What is the voltage across the resistor?

Solution

$$V = \frac{W}{Q}$$
$$I = 3 \text{ A}$$
$$W = 100 \text{ J}$$
$$t = 22 \text{ s}$$
$$Q = It = 3 \times 22 = 66 \text{ C}$$
$$V = \frac{W}{Q} = \frac{100 \text{ J}}{66}$$
$$V = 1.515 \text{ V}$$

Problem 10

If an oven has a 120 V heating element with resistor of 12 Ω, what is the minimum rating of a fuse that can be used in line to the heating element?

Solution

Voltage $V = 120\,\text{volts}$

$$R = 12\ \Omega$$

$$I = \frac{V}{R} = \frac{120}{12} = 10\ \text{A}$$

Multiple Choice Questions

1. A resistance of 50 kΩ has a conductance of
   (a) 20 S
   (b) 0.02 S
   (c) 0.02 mS
   (d) 20 kS

2. Which of the following statements is incorrect?
   (a) $1\ \text{N} = 1\ \text{kg m/s}^2$
   (b) $1\ \text{V} = 1\ \text{J/C}$
   (c) $30\ \text{mA} = 0.03\ \text{A}$
   (d) $1\text{J} = 1\ \text{N/m}$

3. The power dissipated by a resistor of 10 Ω when a current of 2 A passes through it is
   (a) 0.4 W
   (b) 20 W
   (c) 40 W
   (d) 200 W

4. A mass of 1200 g is accelerated at 200 cm/s^2 by a force. The value of the force required is
   (a) 2.4 N
   (b) 2400 N
   (c) 240 kN
   (d) 0.24 N

5. A charge of 240 C is transferred in 2 min. The current flowing is
   (a) 120 A
   (b) 480 A
   (c) 2 A
   (d) 8 A

6. The unit of quantity of electricity is the
   (a) Volt
   (b) Coulomb
   (c) Ohm
   (d) Joule

7. A current of 2 A flows for 10 h through a 100 Ω resistor. The energy consumed by the resistor is
   (a) 0.5 kWh
   (b) 4 kWh
   (c) 2 kWh
   (d) 0.02 kWh

8. Electromotive force is provided by
   (a) the resistance
   (b) a conducting path
   (c) an electric current
   (d) an electrical supply source

9. The coulomb is a unit of
   (a) power
   (b) voltage
   (c) energy
   (d) quantity of electricity

10. The ohm is the unit of
   (a) charge
   (b) resistance
   (c) power
   (d) current

Review Questions

1. Name the fundamental units in SI system.
2. Why is the SI system preferred?
3. How does electric current flow? How do you use sign notation for voltages?
4. Define inductance.
5. Define capacitance.
6. Write the differences between the passive elements.
7. Derive the expression of inductive parameter.
8. What is the difference between power and energy?
9. What is resistivity? How it is expressed?
10. Define electric current.
11. Define potential difference.
12. Derive an expression for energy stored in a capacitor.
13. Define the temperature coefficient of resistance.
14. Define conductance.
15. Derive an expression for energy stored in magnetic field.
16. Define energy.

Problems

1. If a current of 5 A flows for 2 min, find the quantity of electricity transferred.
2. A mass of 5000 g is accelerated at 2 m/s^2 by a force. Determine the force.
3. Find the force acting vertically downwards on a mass of 200 g attached to a wire.
4. A force of 10 N acts on a mass of 100 kg. Calculate the acceleration.
5. An electric motor develops a power of 20 kW at speed of 1000 rpm. Calculate the torque available at shaft.
6. A lift of 150 kg mass is raised with velocity of 5 m/s. If the driving motor has an efficiency of 75%, calculate the input power to the motor.
7. Find the conductance of a conductor of resistance: **(a)** 10 Ω, **(b)** 5 kΩ, **(c)** 100 mΩ.
8. A source emf of 5 V supplies a current of 3 A for 10 min. How much energy is provided in this time?
9. An electric heater consumes 1.8 MJ when connected to a 250 V supply for 30 min.

Find the **(a)** power rating of heater and **(b)** the current taken from the supply.

10. What is the amount of charge that passes by an arbitrary point in a wire in 2.5 ms if the current was measured to be 40 μA?

11. What is the power absorbed by an element if 3 C of charge passes through it in 6 s and the voltage across the element is 12 V?

12. Find the amount of energy gained by 10 C of charge, if the terminal voltage of the battery is 9 V.

13. An element carrying a current of 10 mA dissipates 5 J of energy in 20 s. What is the voltage across the element?

14. A lift of 250 kg mass is raised with velocity of 5 m/s. If the driving motor has an efficiency of 85%, calculate the input power to motor.

15. A battery can store 1.2 MJ of chemical energy. It has terminal of 6 V. If it is to deliver 4 A continuously, what is the charge delivered?

16. Find the resistance of 120 m of copper wire which has circular cross-section with a diameter 2.5 cm, assuming that this conductor is at room temperature.

17. The power dissipated in a 7.5 kΩ resistor is 3 W.

(a) What is the current through this element?

(b) What is voltage across the resistor

18. Find the voltage drop across a 560 Ω resistor when the current flowing in it is 20 mA?

19. If the voltage drop across 2 kΩ resistor is 8 V, find value of current flowing in the resistor.

20. A cube has resistivity of its material as 1.1×10^{-6} Ωm. If it has all sides of length 2 cm each, determine the resistance of cube between any two faces.

21. A copper conductor of circular cross-section has length 10 m and diameter 2 mm. Calculate its resistance if resistivity of copper is given as 1.72×10^{-8} Ωm.

22. Find (a) potential difference across a resistance dissipating 50 W of power while the strength of current is 5 A and (b) value of the resistance?

23. A current of 10 A flows in resistor for 10 s, how many coulombs of charge pass within this time?

24. An electric heater consumes 1 kWh of energy in 30 min at 220 V. What is the current rating of the heating element?

25. An energy source supplies 500 J of energy at 100 V for a certain period of time. Determine the quantity of charge that has passed through.

26. A given conductor has resistor of 5 Ω. What is resistance of another conductor of the same material, which has one-half the diameter and five times the length of the given conductor?

27. Determine the force between two charges each 20 μC when they are separated by a distance of 15 cm:

(a) In air.

(b) In a medium of permittivity 2.08.

28. A 10 mA DC charges a capacitor for 10 s. Find the charge stored in the capacitor.

29. A 100 mA DC charges capacitor of 47 μF for 100 ms. Calculate the voltage developed across the capacitor.

30. Two equal charges are placed 0.6 m apart in air. They repel each other with force of 200 μN. What is the strength of each charge?

31. A 20 μF, 200 V capacitor is charged to its rated voltage and then connected to a load requiring only 75 V for 30 m. Find the **(a)** remaining charge on the capacitor and **(b)** the power delivered.

32. A portable machine requires a force of 200 N to move it. Find

(a) How much work is done if the machine is moved 20 m

(b) What average power is utilized if the movement takes 25 s

33. A mass of 1000 kg is raised to a height of 10 m in 20 s. What is **(a)** the work done and **(b)** the power developed? Assume g = 9.81 m/s^2

Answers

Multiple Choice Questions

1. (c)	**3.** (c)	**5.** (c)	**7.** (b)	**9.** (d)
2. (d)	**4.** (a)	**6.** (b)	**8.** (c)	**10.** (b)

Problems

1. 600 C

2. 10 N

3. 1.962 N

4. 1 m/s^2

5. 191.75 Nm

6. 9810 W

7. **(a)** 0.1 S, **(b)** 0.2 mS, **(c)** 10 S

8. 9 kJ

9. **(a)** 1000 W, **(b)** 4 A

10. 0.1 μC

11. 6 W

12. 90 J

13. 25 V

14. 14.426 kW

15. 0.2×10^6 W

16. 0.1025 Ω

17. **(a)** 20 mA, **(b)** 150 V

18. 11.2 V

19. 4×10^{-3} A

20. 55 μΩ

21. 0.0548 Ω

22. **(a)** 10 V, **(b)** 2 Ω

23. 100 C

24. 9.09 mA

25. 5 C
26. 100 Ω
27. **(a)** 24, **(b)** 11.53
28. 100 mC
29. 212.76 V
30. 4995.72×10^3 C
31. **(a)** 1.5×10^{-3} C, **(b)** 31.25×10^6 W
32. **(a)** 4 kJ, **(b)** 160 W
33. **(a)** 98.1 kJ, **(b)** 4.905 kW

2 DC Circuits

LEARNING OBJECTIVES

After reading this chapter, you will be able to:

- Understand the Kirchhoff's voltage and current laws to determine unknown voltages and currents in DC circuits.
- Understand and study the Ohm's laws.
- Learn to calculate the unknown voltages, currents and resistances in a series circuit.
- Understand voltage division in a series circuit.
- Learn to calculate the unknown voltages, currents and resistances in a parallel network.
- Learn to calculate the unknown voltages, currents and resistances in series–parallel networks.

2.1 Introduction

A direct current (DC) circuit essentially consists of a DC power (e.g., battery, DC generator, etc.), the conductors used to carry the current and the load. The source supplies electrical energy to the load which converts it into heat or other forms of energy. Resistance is the load usually used in DC circuits. This load may be connected in series, parallel or series–parallel. The closed path followed by the direct current is called a DC circuit.

2.2 Circuit Laws

2.2.1 Ohm's Law

George Simon Ohm (1787–1854) was a Professor of mathematics and physics who devoted considerable time and effort to experiments dealing with voltaic cells and conductors. These experiments included the effects of temperature on the resistivity of various metals. Ohm was able to show that the current flow in a circuit composed of a battery and conductors can be expressed as

$$I = \frac{A}{\rho}\frac{\mathrm{d}v}{\mathrm{d}l} \tag{2.1}$$

where the derivative term denotes the electric field gradient. Here v is called the electroscopic force by way of representing the volume density of electricity at a point in the conductor. In the case where a conductor of uniform cross-sectional area is used, Eq. (2.1) may be written as

$$I = \frac{A}{\rho}\frac{V}{L} = \frac{V}{R} \text{ Amperes} \tag{2.2}$$

where V is the potential difference in volts appearing across the conductor of length L, A is the area of cross-section, ρ is a property of the material called specific resistivity and R is the resistance of the conductor in ohms. Equation (2.2) is a mathematical expression of Ohm's law.

Statement: Ohm's law states that the strength of the current in a wire is proportional to the potential difference between its ends.

For using Eq. (2.2) it is necessary to maintain a constant temperature; otherwise the results change accordingly.

Ohm's law may be alternatively expressed as

$$V = IR \tag{2.3}$$

In this form it states that for any given potential difference, the amount of current produced is proportional to the resistance, which in turn is dependent upon the composition of the wire.

Ohm's law can be expressed in differential form also:

$$\mathrm{d}v = R\mathrm{d}i \tag{2.4}$$

This concept of the Ohm's law can be applied to many materials like copper, aluminum, iron, brass, bronze, tungsten, etc. However, it does not apply to nonlinear devices involving step discontinuities, such as Zener diodes and voltage regulator tubes. From Eq. (2.3), we have

$$I = \frac{V}{R} \tag{2.5}$$

or

$$R = \frac{V}{I} \tag{2.6}$$

The general power equation is $P = VI$. Substituting Eq. (2.3) in this expression, we get

$$P = I \times R \times I = I^2 R \tag{2.7}$$

The general energy equations is $E = P \times t$. Substituting Eq. (2.7) in this energy expression, we get

$$E = I^2 \times R \times t \text{ Watt-s or Joules} \tag{2.8}$$

2.2.2 Kirchhoff's Law

Gustav Robert Kirchhoff (1824–1887) published the first systematic formulation of the principles governing the behavior of electric circuits. He advanced no new experimental facts or concepts but merely restated familiar principles. His work was embodied in two laws, namely a current law and a voltage law, which together are known as Kirchhoff's Laws. It is upon these laws that all electric circuit theory is based.

2.2.3 Kirchhoff's Current Law

Statement: Kirchhoff's current law states that the sum of the currents entering or leaving a junction point at any instant is equal to zero.

A junction point is that place in a circuit where two or more circuit elements are joined together. It is often called a node. The current law may be expressed mathematically as

$$\sum_{j=1}^{k} I_j = 0 \tag{2.9}$$

where k denotes the number of circuit elements connected to the node and Σ is the Greek symbol used to indicate summation. For the circuit shown in Fig. 1 there are five circuit elements joined at node n – four passive elements (the resistors) and one active element (DC voltage source). Applying Eq. (2.9) to node n for the situation where the currents are all assumed to be entering the node, we have

$$I_S - I_1 - I_2 - I_3 - I_4 = 0 \tag{2.10}$$

The minus signs are used because these currents are assumed to be leaving rather than entering the node. Equation (2.10) may be rewritten as

$$I_S = I_1 + I_2 + I_3 + I_4 \tag{2.11}$$

In mathematical terms it states that the sum of the currents entering the node is equal to the sum of the currents leaving it.

2.2.4 Kirchhoff's Voltage Law

Statement: Kirchhoff's voltage law states that at any instant of time the sum of voltages in a closed circuit is zero. Essentially this law is a restatement of the law of conservation of energy.

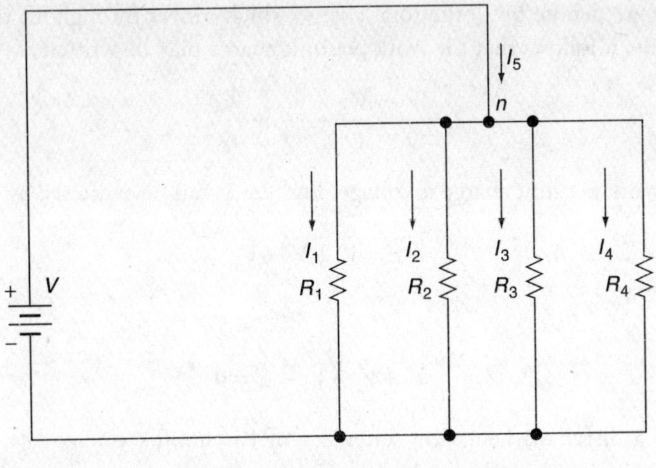

Figure 1 Kirchhoff's current law.

Figure 2 shows the illustration of Kirchhoff's voltage law; it represents a 12 V battery supplying energy to three series-connected resistors representing a total resistance of 12 Ω. By Ohm's law there exists a current of 1 A, which means that charge flows throughout the circuit at the rate of 1 C/s. This current is denoted by I. The positive direction of current flow is *a–b–c–d–a*. The electrons move in the reverse sense.

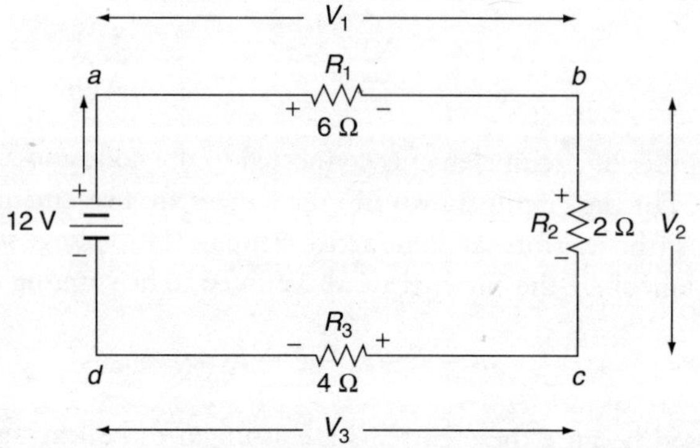

Figure 2 Kirchhoff's voltage law.

The constant current which flows in the circuit means that the average rate of movement of charge is constant. However, the electrons traverse the circuit elements R_1, R_2 and R_3, and they collide with the atoms in the space lattice configurations of these elements, thereby dissipating energy in the form of heat. Since this energy must be supplied by the battery, the law of conservation of energy permits us to write

$$W = W_1 + W_2 + W_3 \tag{2.12}$$

where W denotes the energy supplied by the battery in a specified time interval, W_1 denotes the energy dissipated in R_1, W_2 denotes the energy dissipated in R_2 and W_3 denotes the energy dissipated in R_3 in the same interval of time.

Furthermore, if we denote by Q the total charge which moves throughout the circuit during this same interval of time, it follows that the work per unit charge may be written as

$$\frac{W}{Q} = \frac{W_1}{Q} + \frac{W_2}{Q} + \frac{W_3}{Q} \tag{2.13}$$

However, because work per unit charge is voltage, Eq. (2.13) can be expressed as

$$E = V_1 + V_2 + V_3 \tag{2.14}$$

or

$$V_1 + V_2 + V_3 - E = 0 \tag{2.15}$$

Equation (2.15) is a direct mathematical statement of Kirchhoff's voltage law. Here all the voltage drops are treated as positive quantities. Conversely, all voltage rises encountered in the loop must be prefixed with a negative sign.

A generalized formulation of Kirchhoff's voltage law is rewritten as

$$\sum_{j=1}^{k} V_j = 0 \tag{2.16}$$

where V_j represents the voltage drop of the jth element in any given closed circuit which is assumed to have k elements.

Figure 3 shows that the source voltage was assumed to be a battery of fixed magnitude, thus producing a constant current. However, Kirchhoff's voltage law is general enough so that it applies equally well when the source voltage is a time-varying quantity producing a time-varying current. Referring to Fig. 3, the time-varying source voltage appears across d–a. Between points a and b there is a resistor R similar to those used by Ohm in his experiments. If the instantaneous value of the current is denoted as I then the potential difference between the ends of the resistor is IR. In between points b and c, appears a coil of the type used by Faraday across which an emf of self-induction $L(\mathrm{d}i/\mathrm{d}t)$ is developed. Finally, between points c and d, a capacitor appears. The voltage drop across the capacitor is $(1/C)\int i\mathrm{d}t$. Then by Kirchhoff's voltage law, we can write

$$e = iR + L\frac{\mathrm{d}i}{\mathrm{d}t} + \frac{1}{C}\int i\mathrm{d}t \tag{2.17}$$

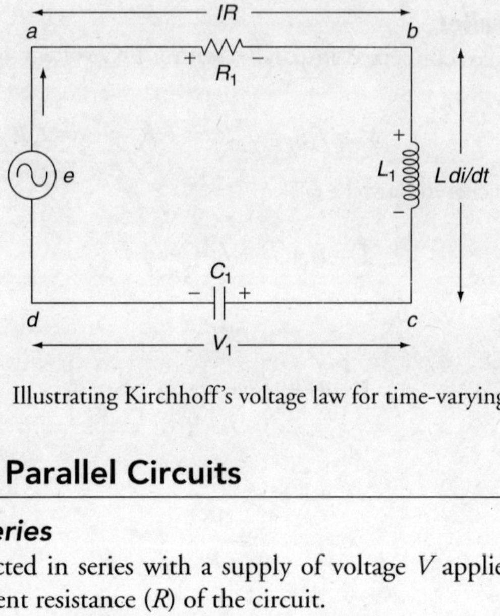

Figure 3 Illustrating Kirchhoff's voltage law for time-varying source.

2.3 Series and Parallel Circuits

2.3.1 Resistance in Series

Consider n resistors connected in series with a supply of voltage V applied as shown in Fig. 4. It is required to find the equivalent resistance (R) of the circuit.

Denote the voltages across the resistors by $V_1, V_2, \ldots, V_n$. Applying Ohm's law to each resistor, we have

$$V_1 = IR_1, \quad V_2 = IR_2, \quad V_3 = IR_3, \ldots \tag{2.18}$$

$$V = V_1 + V_2 + \cdots + V_n = I(R_1 + R_2 + \cdots + R_n) \tag{2.19}$$

However, from Fig. 4(b),

$$V = RI$$

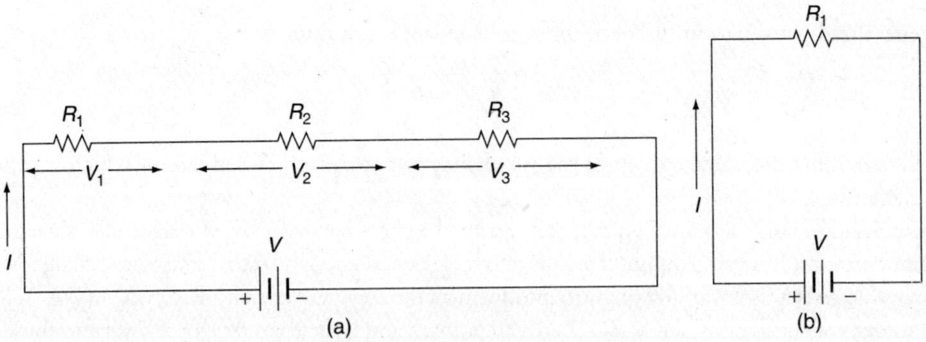

Figure 4 Resistors in series.

Therefore,

$$R = R_1 + R_2 + R_3 \tag{2.20}$$

That is, the total resistance of the circuit is the sum of the resistances connected in series.

2.3.2 Resistors in Parallel

Consider the three resistances connected in parallel with a DC voltage source as shown in Fig. 5. In this case,

$$V = I_1R_1 = I_2R_2 = I_3R_3 = \cdots = I_nR_n \tag{2.21}$$

However, from Kirchhoff's current law,

$$I = I_1 + I_2 + I_3 + \cdots + I_n \tag{2.22}$$

Therefore,

$$I = V\left(\frac{1}{R_1} + \frac{1}{R_2} + \frac{1}{R_3} + \cdots + \frac{1}{R_n}\right) \tag{2.23}$$

Also, from Fig. 5(b),

$$I = \frac{V}{R}$$

Therefore,

$$\frac{1}{R} = \frac{1}{R_1} + \frac{1}{R_2} + \frac{1}{R_3} + \cdots + \frac{1}{R_n} \tag{2.24}$$

That is, the reciprocal of the equivalent resistance of the circuit is the sum of the reciprocals of the separate resistances.

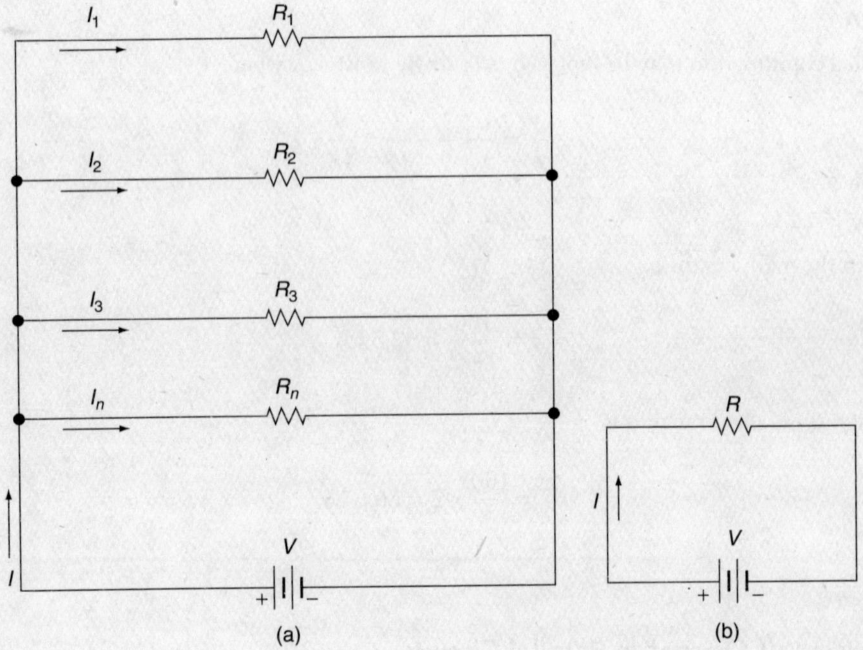

Figure 5 Resistors in parallel.

2.3.3 Series–Parallel Circuits

Example 1

Calculate the voltage across the parallel branches of the circuit arrangement shown in Fig. 6 and the current in the main circuit.

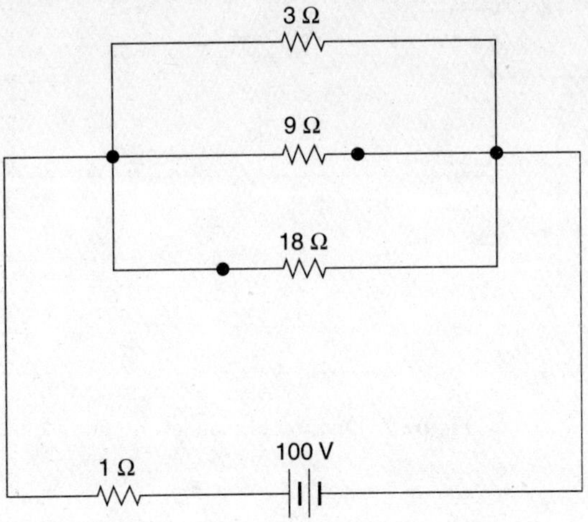

Figure 6 Example 1.

Solution

The parallel combination may be replaced by a single resistor R where

$$\frac{1}{R} = \frac{1}{3} + \frac{1}{9} + \frac{1}{18} = \frac{9}{18}$$

Therefore $R = 2\ \Omega$.

Current in the main circuit is

$$I = \frac{100}{1+2} = 33.33\ \text{A}$$

Voltage across parallel branches is

$$V = \frac{100}{3} \times 2 = 66.67\ \text{V}$$

2.3.4 Division of Current in Parallel Circuits

Consider the parallel combination of two resistors arranged in Fig. 7. Let 'R' be the equivalent resistance.

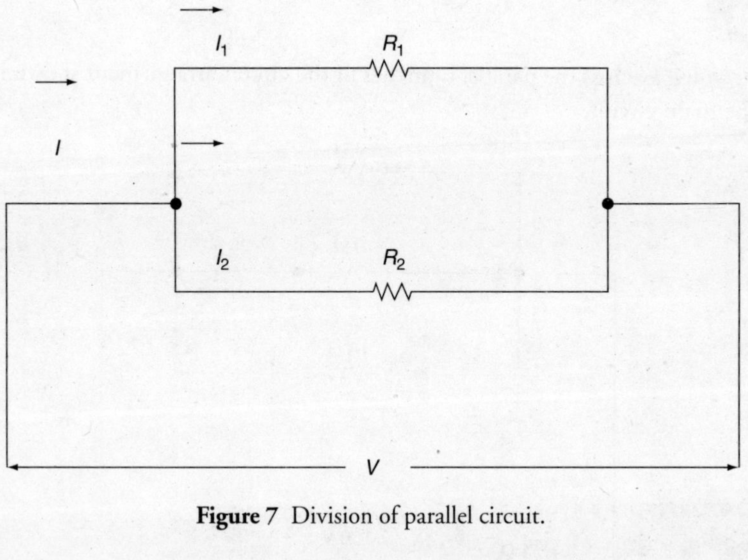

Figure 7 Division of parallel circuit.

Then according to Ohm's law,

$$I = \frac{V}{R} \Rightarrow V = IR \tag{2.25}$$

$$I_1 = \frac{V}{R_1} \Rightarrow V = I_1 R_1 \tag{2.26}$$

$$I_2 = \frac{V}{R_2} \Rightarrow V = I_2 R_2 \tag{2.27}$$

From Eqs. (2.25) and (2.26), we have

$$I_1 R_1 = IR$$

$$I_1 = \frac{IR}{R_1}$$

$$I_1 = \frac{I(R_1 R_2)}{R_1(R_1 + R_2)}$$

$$I_1 = \frac{IR_2}{R_1 + R_2} \tag{2.28}$$

From Eqs. (2.25) and (2.27), we have

$$I_2 R_2 = IR$$

$$I_2 = \frac{IR}{R_2}$$

$$R_2 = \frac{I}{R_2}\left(\frac{R_1 R_2}{R_1 + R_2}\right)$$

$$I_2 = \frac{IR_1}{R_1 + R_2} \tag{2.29}$$

2.3.5 Comparison of Series and Parallel Circuits

A comparison of series and parallel circuits is given in Table 1.

Table 1 Comparison between series and parallel circuits

Sr. No.	*Series Circuits*	*Parallel Circuits*
1	The current passing through all the connected elements is the same.	Voltage across each element connected in parallel is the same.
2	There is only one path for flow of current.	The paths for flow of current are more than one.
3	The total voltage applied across series circuit is equal to the sum of the voltage drops across all the elements connected in series.	The total current flowing through the parallel combination is equal to sum of all the currents flowing through all the elements connected in parallel.
4	The equivalent resistance is greater than the greatest resistance connected in series circuit.	The equivalent resistance is less than the least resistance connected in the parallel circuit.

Example 2

Consider the three resistances 5 Ω, 10 Ω, 15 Ω connected in series across a 120 V supply. Calculate the total resistance, total current and voltage across the element.
Given data: $R_1 = 5\ \Omega$, $R_2 = 10\ \Omega$, $R_3 = 15\ \Omega$, $V = 120$ V

Solution

The equivalent resistance is

$$R = R_1 + R_2 + R_3$$
$$= 5 + 15 + 15$$
$$= 30\ \Omega$$

Total current is

$$I = \frac{V}{R} = \frac{120}{30} = 4\ \text{A}$$

Voltage across R_1 is

$$V_1 = IR_1 = 4 \times 5 = 20\ \text{V}$$

Voltage across R_2 is

$$V_2 = IR_2 = 4 \times 10 = 40\ \text{V}$$

Voltage across R_3 is

$$V_3 = IR_3 = 4 \times 15 = 60\ \text{V}$$

Example 3

Find the equivalent resistance of thee series–parallel combination shown in Fig. 8.

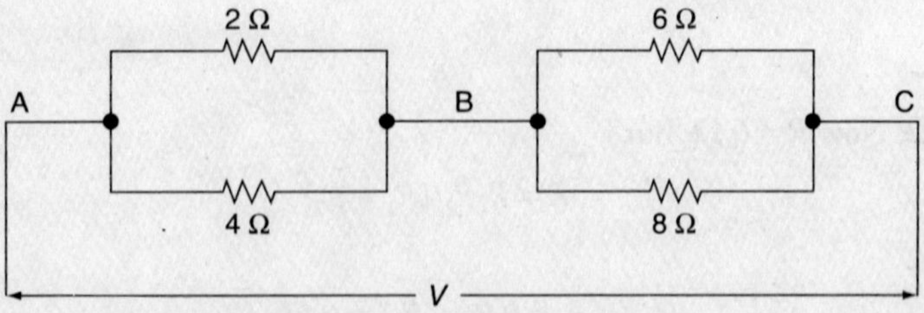

Figure 8 Equivalent circuit of series and parallel combination for Example 3.

Solution

The resistance across AB,

$$R_{AB} = (2 \times 4)/(2 + 4) = 1.33\ \Omega$$

The resistance across BC,

$$R_{BC} = (6\times8)/(6+8) = 3.42\ \Omega$$

The resistance across AC,

$$R_{AC} = R_{AB} + R_{BC} = 1.33 + 3.42 = 4.75\ \Omega$$

Example 4

Find the effective resistance of the network shown in the Fig. 9.

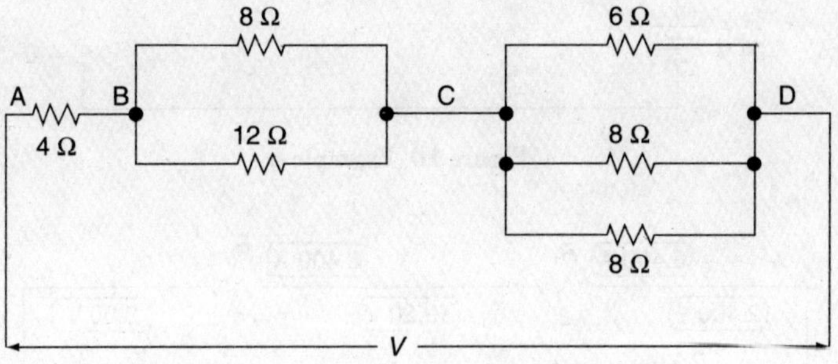

Figure 9 Effective resistance for Example 4.

Solution

Resistance across AB, $R_{AB} = 4\ \Omega$

Resistance across BC, $R_{BC} = (8 \times 12)/(8 + 12) = 4.8\ \Omega$

Resistance across CD, $R_{CD} = (6 \times 6 \times 6)/(6 \times 6 + 6 \times 6 + 6 \times 6) = 2\ \Omega$

Therefore, resistance across AD,

$$R_{AD} = R_{AB} + R_{BC} + R_{CD} = 4 + 4.8 + 2 = 10.8\ \Omega$$

Example 5

What resistance must be connected in parallel with a 8 Ω resistance to give an equivalent resistance of 6 Ω.

Solution

Given: $R_1 = 8\ \Omega$. Now $R = 6\ \Omega$. But

$$R = R_1R_2/(R_1 + R_2)$$

Therefore,

$$R_1R_2/(R_1 + R_2) = 6\ \Omega$$

$$(8 \times R_2)/(8 + R_2) = 6$$

$$8\ R_2 = (8 + R_2) \times 6$$

$$R_2 = 24\ \Omega$$

Example 6

A series circuit is shown in Fig. 10. Calculate current and voltage across each resistance using Pspice simulation.

Figure 10 Example 6.

Solution

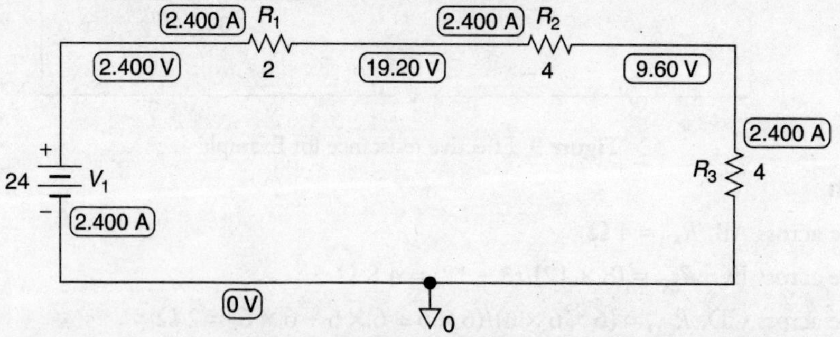

Figure 11 Solution of Example 6.

Example 7

Determine the value of the unknown current of the circuit as shown in Fig. 12.

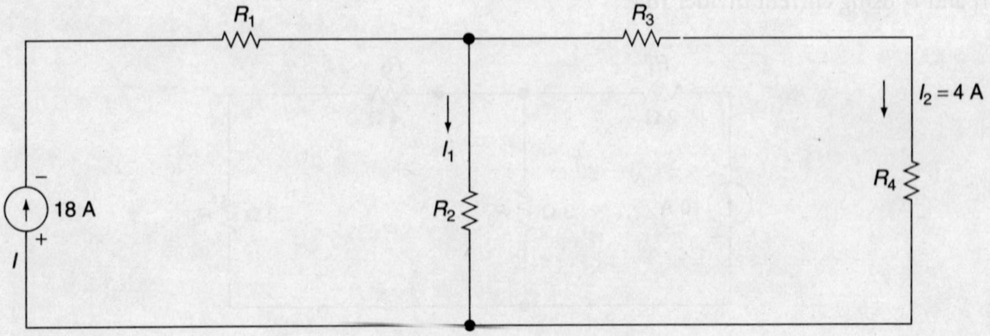

Figure 12 Example 7.

Solution

Applying Kirchhoff's current law at the node, we have

$$18 = I_1 + I_2$$

$$I_1 = 18 - I2 = 18 - 4 = 14 \text{ A}$$

Example 8

Calculate the unknown voltage across R_3 resistor of the circuit as shown in Fig. 13.

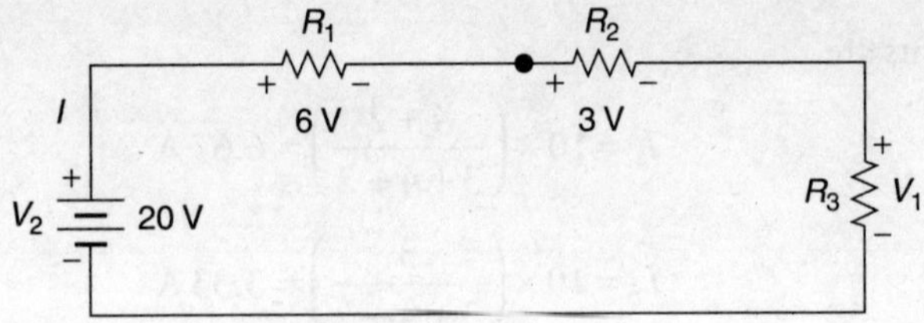

Figure 13 Example 8.

Solution

Applying Kirchhoff's voltage law to the circuit in Fig. 13, we have

$$-20 + 6 + 3 + V_1 = 0$$

$$V_1 = 11 \text{ V}$$

Example 9

Simulate the series–parallel circuit shown in Fig. 14 using Pspice. Calculate the branch currents I_1 and I_2 using current divider rule.

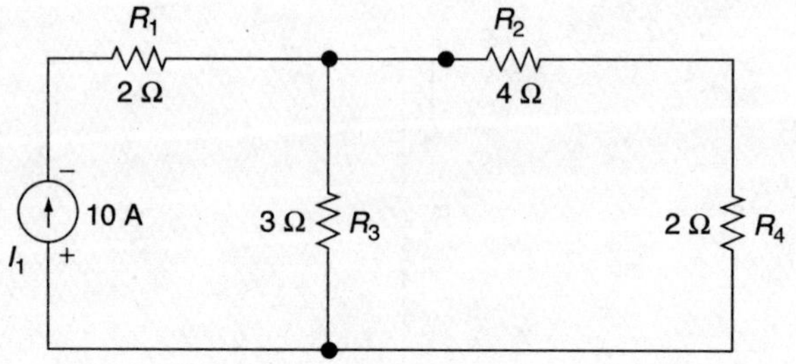

Figure 14 Example 9.

Solution

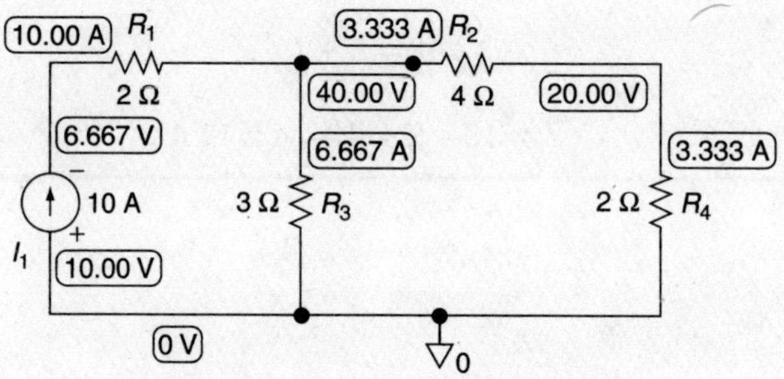

Figure 15 Solution of Example 9.

The branch currents are

$$I_1 = 10 \times \left(\frac{4+2}{3+4+2} \right) = 6.67\ \text{A}$$

$$I_2 = 10 \times \left(\frac{3}{3+4+2} \right) = 3.33\ \text{A}$$

Example 10

Calculate the currents I_1 and I_2 and the voltages V_1 and V_2 using Pspice simulation as well as manual calculation for the series–parallel circuit shown in Fig. 16.

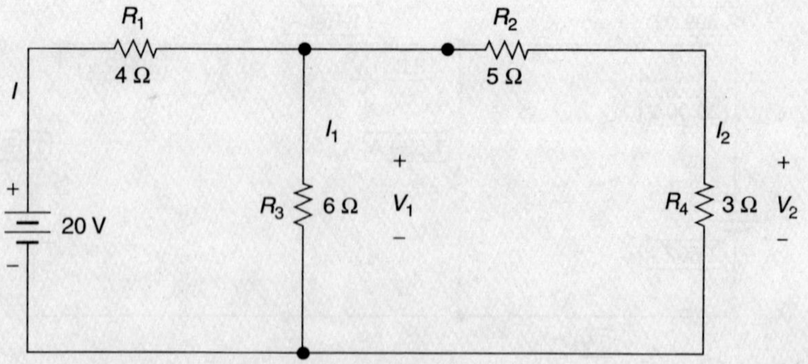

Figure 16 Example 10.

Solution

Resistors 5 Ω and 3 Ω are connected in series and their equivalent resistance is

$$R_s = 5 + 3 = 8\ \Omega$$

The 8 Ω resistor is in parallel with the 6 Ω resistor and the equivalent resistance is

$$R_p = \frac{8 \times 6}{8 + 6} = 3.42\,\Omega$$

Therefore, total resistance is

$$R_t = 4 + 3.42 = 7.52\ \Omega$$

The source current is

$$I = \frac{V}{R_t} = \frac{20}{7.42} = 2.69\,\text{A}$$

The branch currents are

$$I_1 = 2.69 \times \left(\frac{8}{8+6}\right) = 1.54\,\text{A}$$

$$I_2 = 2.69 \times \left(\frac{6}{8+6}\right) = 1.15\,\text{A}$$

The voltage drops are

$$V_1 = I_1 \times 6 = 1.54 \times 6 = 9.24\ \text{V}$$
$$V_2 = I_2 \times 3 = 1.15 \times 3 = 3.45\ \text{V}$$

The Pspice simulation results are shown in Fig. 17.

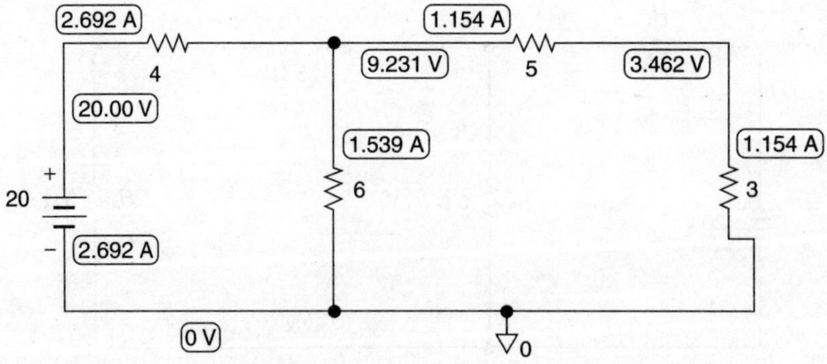

Figure 17 Solution of Example 10.

Summary

In this chapter we have discussed the basics of DC circuit and its growth.

1. Ohm's law statement: temperature remaining constant, the current through a passive element in the circuit is directly propositional to the voltage across the element.
2. Kirchhoff's current law: algebraic sum of the current meeting at any junction or node is zero.
3. Kirchhoff's voltage law: algebraic sum of all voltages around a closed path or closed loop at any instant is zero.
4. For the problems in this chapter, the main objective is normally achieved in two steps. The first step is to write the governing equations of the given circuit based on some well-known laws and network theorems. The second step involves solving these equations for the unknowns.

Solved Problems

Problem 1

Find the resistance at 30°C of the following annealed standard copper wires:
(i) 2 mm^2 cross-section and 10 m long.
(ii) 20 mm^2 cross-section 100 m long.

Solution

At 30°C resistance of copper wire 1 m long and 1 mm^2 in cross-section is 1/58 ohms. This is ρ.

(i) The resistance

$$R = \rho\frac{L}{A}$$

$$R = \frac{1}{58}\frac{\text{(length in m)}}{A \text{ in mm}^3} = \frac{1}{58} \times \frac{10\text{ m}}{2\text{ mm}^2} = 0.086\ \Omega$$

(ii) $R = \dfrac{1}{58} \times \dfrac{1200\text{ m}}{20\text{ mm}^2} = 0.086\ \Omega$

Problem 2

A rectangular metal strip has the dimension x = 10 cm, y = 0.5 cm and z = 0.2 cm. Determine the ratio of the resistance R_x, R_y and R_z between the respective pairs of opposite faces.

Solution

$$R_x : R_y : R_z = \frac{\rho x}{yz} : \frac{\rho y}{zx} : \frac{\rho z}{xy}$$

$$= \frac{10}{0.5 \times 0.2} : \frac{0.5}{0.2 \times 10} : \frac{0.2}{10 \times 0.5}$$

$$= \frac{10}{0.1} : \frac{1}{4} : 0.04 = 2500 : 6.25 : 1$$

Problem 3

Determine the resistance of a metal tube in terms of the external diameter D, the internal diameter d, the length L and the resistivity ρ. Calculate the resistance of a copper tube 0.5 cm thick and 2 m long, and the external diameter of 10 cm.

Solution

$\rho = 1\ \text{mm}^2$ cross-section of copper tube = 1/58 ohms
External diameter $D = 10 \times 10^2 = 10 \times 100\ \text{mm}^2$
Area of copper tube (metal tube) $A = (\pi/4)(D^2 - d^2)$
Length of copper tube $l = 2$ m

$$R = \frac{\rho l}{A} = \frac{\rho l}{(\pi/4)(D^2 - d^2)}$$

$$= \frac{1}{58} \times \frac{2\ \text{m}}{(\pi/4)(10^2 - 9^2) \times 100\ \text{mm}^2}$$

$$= 23.14\ \mu\Omega$$

Problem 4

Calculate the loss in watts per kg of a current-carrying conductor in terms of the resistivity ρμ cm, the current density σ A per mm² and the specific gravity S.

Solution

$$\text{Loss in watts} = RI^2 = \frac{\rho L}{A} I^2$$

$$= \frac{\rho L}{A} \cdot A^2 \cdot \left(\frac{I}{A}\right)^2$$

$$= \rho \left(\frac{I}{A}\right)^2 AL$$

Therefore, the loss in watts per $\text{cm}^3 = \rho\left(\frac{I}{A}\right)^2$

The mass of 1 $\text{cm}^3 = S\text{g}$

The loss in watts for $S\text{g} = \rho\left(\frac{I}{A}\right)^2$

Therefore, the loss in watts

$$= \rho\left(\frac{I}{A}\right)^2 \times \frac{1}{S} \text{ per gm}$$

$$= \frac{\rho \times 10^{-6}(\sigma \times 100)^2 \times 1000}{S} \text{ per kg}$$

$$= \frac{10\sigma^2\rho}{S}$$

Problem 5

Determine the length L and diameter d of a cylinder of copper in terms of the volume, x; resistivity, ρ and the resistance between opposite ends r.

Solution

$$r = \frac{\rho L}{A} = \frac{\rho L^2}{(\pi d^2/4)L} = \frac{\rho L^2}{x}$$

This implies

$$L = \left(\frac{rx}{\rho}\right)^{1/2}$$

Also

$$r = \frac{\rho L}{(\pi d^2/4)} = \frac{\rho L(\pi d^2/4)}{(\pi^2 d^4/16)} = \frac{\rho x}{(\pi^2 d^4/16)} = \frac{16\rho x}{\pi^3 d^4}$$

$$d = \left(\frac{16x\rho}{\pi^2 x}\right)^{1/4}$$

Problem 6

Find the resistance of the semi-circular copper conductor between equipotential faces A and B. Inner radius 5 cm, radial thickness 3 cm and axial thickness 3 cm.

Solution

Mean radius (r) = 5 + 1.5 = 6.5 cm

Mean length is given by

$$L = \frac{1}{2} \times 2\pi r = \frac{1}{2} \times 2\pi \times 6.5$$
$$= 6.5\pi \text{ cm}$$
$$= 0.065\pi \text{ m}$$

Area is given by

$$A = 3 \times 3 = 9 \text{ cm}^2 = 900 \text{ mm}^2$$

Therefore the resistance equals

$$\frac{1}{58} \frac{L \text{ in m}}{A \text{ in mm}^2} = \frac{1}{58} \times \frac{0.065\pi}{900} \times 10^6 \ \mu\Omega$$
$$= 3.91 \ \mu\Omega$$

Problem 7

A liquid resistance with parallel plate electrodes has to absorb 40 kW at 400 V. The 5% soda-solution used has a resistivity of 22 Ω cm. Allowing a current density on the electrodes of 0.25 A per cm², find the distance between the electrodes.

Solution

The current

$$I = \frac{\text{Watts}}{\text{Volts}} = \frac{40 \times 1000}{400} = 100 \text{ A}$$

Therefore, the area of electrode is

$$\frac{I}{\text{Current density}} = \frac{100}{0.25} = 400 \text{ cm}^2$$

Now

$$R = \frac{V}{I} = \frac{400}{100} = 4$$

But

$$R = \frac{\rho L}{A}$$

This implies

$$\frac{\rho L}{A} = 4 \Rightarrow \frac{22 \times L}{400} = 4 \Rightarrow L = \frac{1600}{22} = 73 \text{ cm}$$

Problem 8

In the arrangement shown in Fig. 18, calculate the impedance AB and the phase angle between voltage and current.

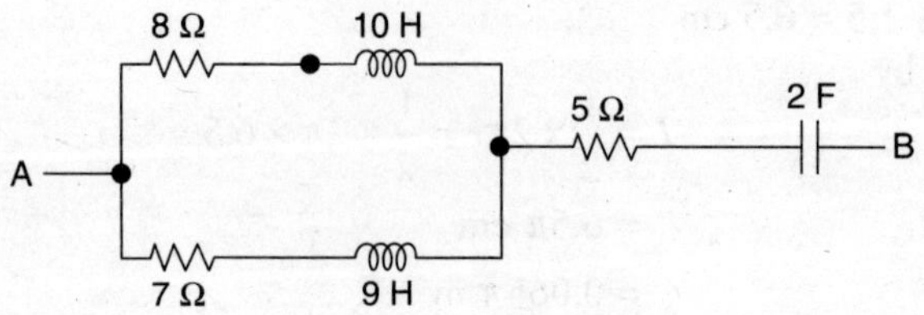

Figure 18 Problem 8.

Solution

$$
\begin{aligned}
Z_{AB} &= \frac{(8+j10)(7+j9)}{(8+j10)+(7+j9)} + 5 - j2 \\
&= \frac{56-90+j(70+72)}{15+j19} + 5 - j2 \\
&= \frac{-34+j42}{15+j19} + 5 - j2 \\
&= \frac{146\angle 103.5^\circ}{24.2\angle 51.7^\circ} + 5 - j2 \\
&= 6.03\angle 51.8^\circ + 5 - j2 \\
&= 8.73 + j2.74 \\
&= 9.15\angle 17.2^\circ
\end{aligned}
$$

That is, the impedance is 9.15 Ω and the angle 17.2° lag (the circuit is inductive).

Problem 9

A capacitor of 50 μF, shunted across a non-inductive resistor of 100 Ω is connected in series with a resistor of 50 Ω to a 200 V, 50 Hz supply as in Fig. 19. Calculate

(i) The current in the capacitor.
(ii) The current in the shunted resistor.
(iii) The total current.

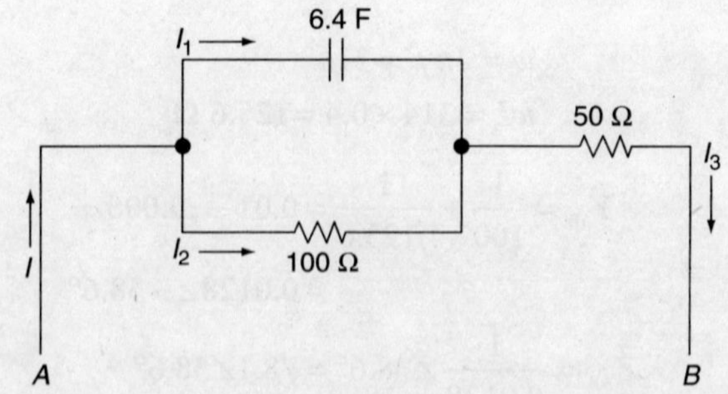

Figure 19 Problem 9.

Solution

$$\bar{Y}_{AB} = \frac{1}{100} + j\omega C = 0.01 + j314 \times 50 \times 10^{-6}$$
$$= 0.01 + j0.0157 = 0.0116\angle 57.5°$$

Now

$$\bar{Z}_{AB} = \frac{1}{\bar{Y}_{AB}} = \frac{1}{0.0186}\angle -57.5°$$
$$= 53.76\angle -57.5° = 28.89 - j45.36$$

Therefore,

$$\bar{Z}_{AC} = 78.89 - j45.36 = 91\angle -29.9°$$

The current I_3 is

$$I_3 = \frac{200}{91} = 2.2\text{ A} = \text{Total current}$$

The capacitor current equals

$$I_3 \times \frac{\text{Capacitive admittance}}{Y_{AB}} = I_3 \times \frac{0.0157}{0.0186} = 1.85\text{ A}$$

Current through 100 Ω is

$$I_2 = 2.2 \times \frac{0.01}{0.0186} = 1.18\text{ A}$$

Problem 10

A 100 Ω resistor is shunted by a 0.4 H inductor in series with a capacitor C. A voltage of 250 V at 50 Hz is applied to the circuit. Find

(i) The value of C to give unity.
(ii) The total current.
(iii) The current in the inductive branch. Draw the vector diagram.

Solution

$$w = 2\pi f = 314$$

$$wL = 314 \times 0.4 = 125.6\ \Omega$$

$$\bar{Y}_{AB} = \frac{1}{100} + \frac{1}{j125.6} = 0.01 - j0.008$$

$$= 0.0128\angle -38.6^\circ$$

$$Z_{AB} = \frac{1}{0.0128}\angle 38.6^\circ = 78.1\angle 38.6^\circ$$

$$= 61.06 + j48.7$$

The capacitive reactance required equals 48.7 Ω. Therefore

$$\frac{1}{wC} = 48.7$$

$$C \text{ in } \mu\text{F} = \frac{10^6}{314 \times 48.7} = 64.9\ \mu\text{F}$$

$$Z_{AO} = 61.06 + j48.7 - j48.7 = 61.06\ \Omega$$

The total current is

$$\frac{250}{61.06} = 4.09\ \text{A}$$

Current through the inductor branch equals

$$\text{Total current} \times \frac{\text{Admittance of inductor}}{Y_{AB}}$$

$$= 4.09 \times \frac{0.008}{0.0128}$$

$$= 2.56\ \text{A}$$

The vector diagram is

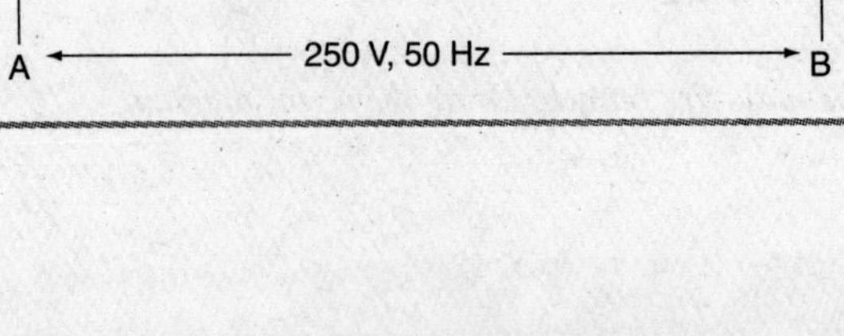

Problem 11

In the circuit shown in Fig. 20, determine the characteristic of the impedance between points B and D.

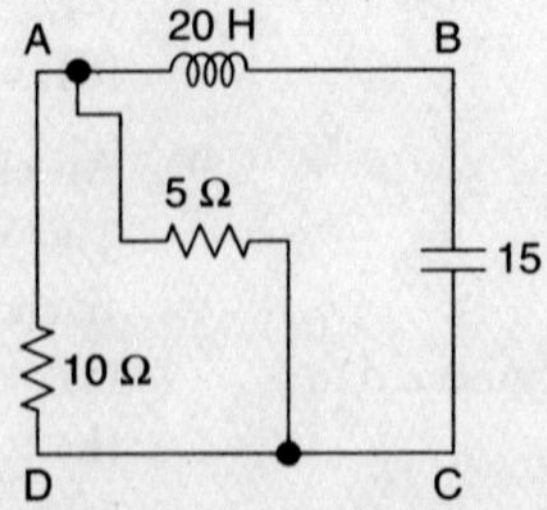

Figure 20 Problem 11.

Solution

The points C and D coincide 5 Ω in parallel with 10 Ω therefore

$$\frac{5\times 10}{15} = 3.33\ \Omega$$

Impedance between points B and D is

$$Z_{BC} = \text{Parallel combination of } (3.33 + j20) \text{ and} (-j15)$$

$$= \frac{(3.33 + j20)(-j15)}{3.33 + j5}$$

$$= \frac{304.1\angle -9.5^\circ}{6\angle 56.3^\circ} = 50.7\angle -6.58^\circ\ \Omega$$

Multiple Choice Questions

1. If two 4 Ω resistors are connected in series the effective resistance of the circuit is
   (a) 8 Ω
   (b) 4 Ω
   (c) 2 Ω
   (d) 1 Ω

2. If two 4 Ω resistors are connected in parallel, then the effective resistance of the circuit is
   (a) 8 Ω
   (b) 4 Ω
   (c) 2 Ω
   (d) 1 Ω

3. The effect of connecting an additional parallel load to an electrical supply source is to increase the
   (a) resistance to the load
   (b) voltage of the sources
   (c) current taken from the source
   (d) potential difference across the load

4. The equivalent resistance when a resistor of 1/3 Ω is connected in parallel with a 1/4 Ω resistance is
   (a) 1/7 Ω
   (b) 7 Ω
   (c) 1/2 Ω
   (d) 3/4 Ω

5. A 10 Ω resistor is connected in parallel with a 15 Ω resistor and the combination is in series with a 12 Ω resistor. The equivalent resistance of the circuit is
   (a) 37 Ω
   (b) 18 Ω
   (c) 27 Ω
   (d) 4 Ω
6. When three 3 Ω resistances are connected in parallel, the total resistances is
   (a) 3 Ω
   (b) 9 Ω
   (c) 1 Ω
   (d) 0.33 Ω
7. A DC circuit usually has ________ as the load.
   (a) Resistance
   (b) Inductance
   (c) Capacitance
   (d) Both inductance and capacitance
8. When cells are arranged in parallel
   (a) the current capacity increases
   (b) the current capacity decreases
   (c) the emf increases
   (d) the emf decreases
9. An electric fan and a heater are marked as 100 W, 220 V and 1000 W, 220 V, respectively. The resistance of the heater is
   (a) zero
   (b) greater than that of fan
   (c) less than that of fan
   (d) equal to that of fan
10. A wire has a resistance of 12 Ω. It is bent in the form of a circle. The effective resistance between the two points on any diameter of the circle is
   (a) 6 Ω
   (b) 3 Ω
   (c) 12 Ω
   (d) 24 Ω

Review Questions

1. State Kirchhoff's current and voltage laws.
2. Briefly describe the concept of Ohm's law.
3. Derive an expression for the equivalent resistance of a number or resistors connected in series.
4. Derive the equation for equivalent resistance of a number of resistors connected in parallel.
5. Write the difference between series and parallel circuits.
6. How will you prove the validity of Kirchoff's laws?

Problems

1. For the circuit shown in Fig. 21, determine
   (a) The battery voltage V.
   (b) The total resistance of the circuit.
   (c) The values of resistors R_1, R_2, R_3 given that potential difference across R_1, R_2 and R_3 are 5 V, 2 V and 6 V, respectively.

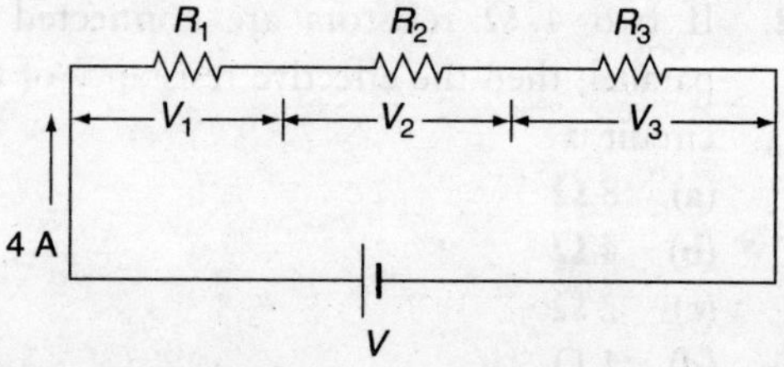

Figure 21 Problem 1.

2. For the circuit shown in Fig. 22,
   (a) Determine the potential difference across resistor R_3
   (b) Find the value of resistor R_3, if total resistance is 100 Ω.

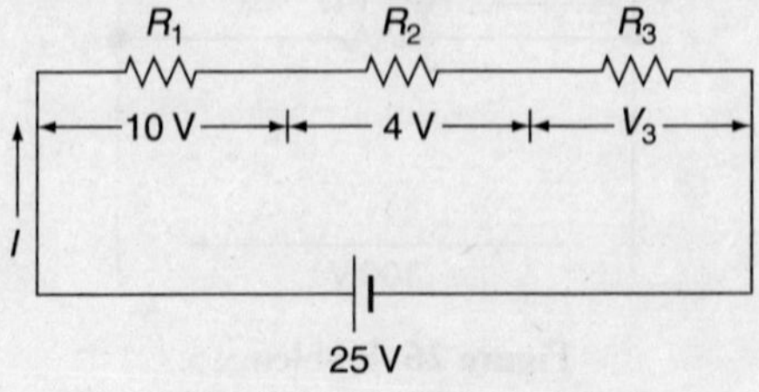

Figure 22 Problem 2.

3. A 12 V battery is connected in a circuit having three series-connected resistors having resistances of 4 Ω, 9 Ω and 11 Ω.
   (a) Determine the current flowing through the 9 Ω resistor.
   (b) Determine the potential difference across it.
   (c) Find the power dissipated in the 11 Ω resistor.

4. Two resistors, R_1 and R_2, are connected in series across a 24 V supply and a current of 3 A flows in the circuit. If R_1 has a resistance of 2 Ω, determine
   (a) The value of resistance R_2.
   (b) The potential difference across the 2 Ω resistor.

   How much electrical energy is used if the circuit is conncected for 50 h?

5. Two resistors of resistance 3 Ω and 6 Ω are connected in parallel across a battery having a voltage of 12 V. Determine
   (a) The total circuit resistance.
   (b) The current flowing in the 3 Ω resistor.

6. Resistances of 10 Ω, 20 Ω and 30 Ω are connected **(a)** in series and **(b)** in parallel to a 240 V supply. Calculate the supply current in each case.

7. For the series–parallel arrangement shown in Fig. 23, find
   (a) The supply current.
   (b) The current flowing through each resistor
   (c) The potential difference across each resistor.

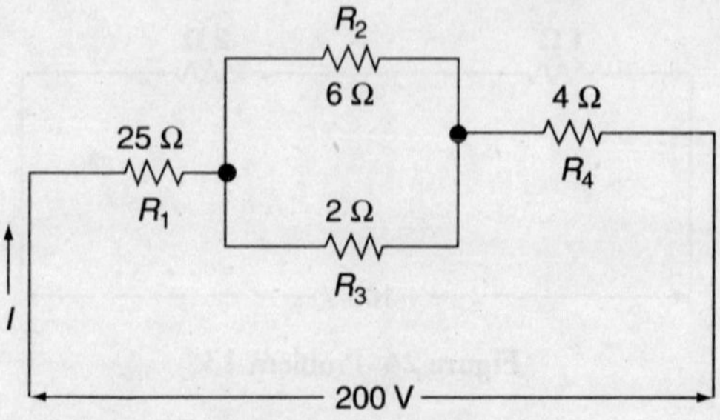

Figure 23 Problem 7.

8. Three resistors 4 Ω, 12 Ω and 6 Ω are connected in parallel. If the total current taken is 12 A, find the current through each resistor.

9. A 100 W, 250 V lamp is connected in parallel with an unknown resistance R across a 250 V supply. The total power dissipated in the circuit is 1100 W. Find the value of the unknown resistance. Assume that the resistance of lamp remains unaltered.

10. Two batteries E_1 and E_2 having emf of 6 V and 2 V and internal resistances of 2 Ω and 3 Ω, respectively, are connected in parallel across a 5 Ω resistor. Calculate
   (a) The current through each battery.
   (b) Terminal voltage.
   (c) Energy dissipated in the 5 Ω resistor in 10 min.

11. A 200 W and a 100 W bulb both meant for operation at 220 V are connected in series. When they are connected to a 220 V supply, find the power consumption.

12. The internal resistance of a cell of emf 2.0 V is 0.1 Ω. It is connected to a resistance of 3.9 Ω. Find the voltage across the cell.

13. Calculate the current flowing into circuit as shown in Fig. 24.

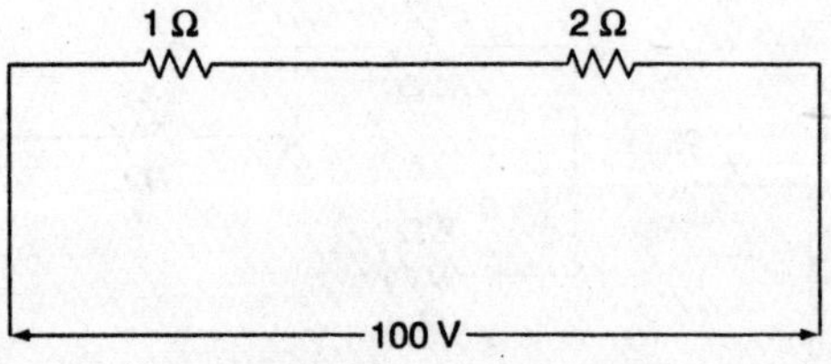

Figure 24 Problem 13.

14. Find voltage across each resistor as shown in Fig. 25.

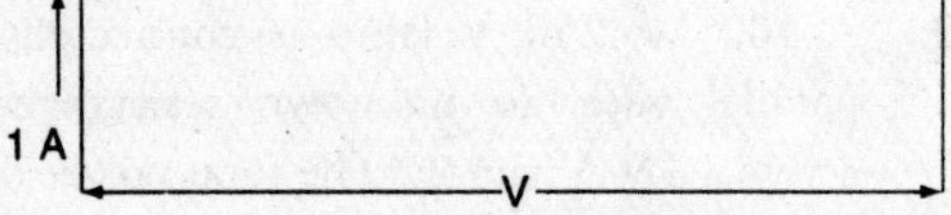

Figure 25 Problem 14.

15. Calculate the supplied current as shown in Fig. 26.

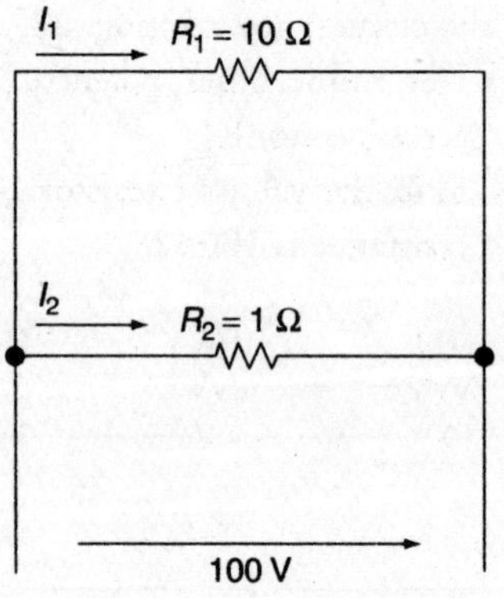

Figure 26 Problem 15.

16. A 50 Ω relay is connected is series with a resistance of 30 Ω and a small lamp of resistance 3 Ω in Fig. 27.

(a) Determine current in the circuit if the applied voltage to the circuit is 230 V.

(b) Find voltage across three elements.

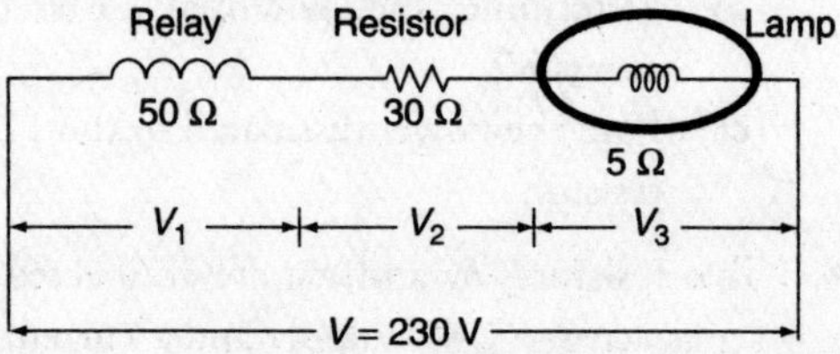

Figure 27 Problem 16.

17. In the circuit of Fig. 28 I_1 = 2.5 A and I_2 = −1.5 A. Calculate current I_3.

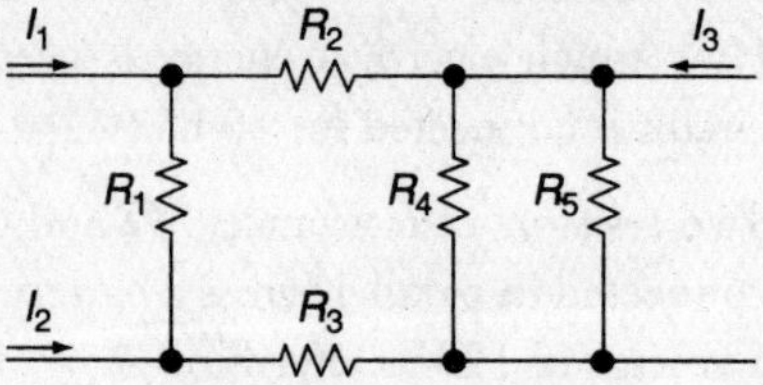

Figure 28 Problem 17.

18. Find current I_1 when $R_1 = 1\ \Omega$, $R_2 = 2\ \Omega$, $R_3 = 3\ \Omega$ and $V = 100$ V shown in Fig. 29.

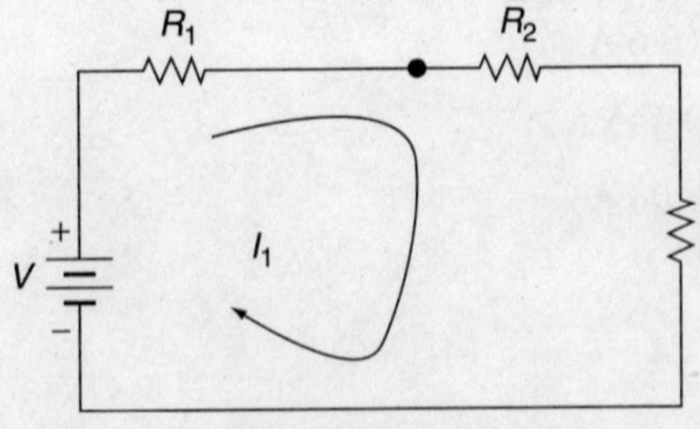

Figure 29 Problem 18.

19. Find current through R_2 shown in Fig. 30.

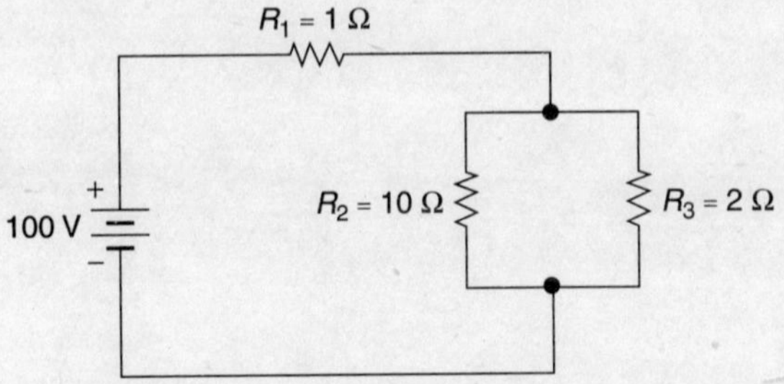

Figure 30 Problem 19.

20. Find current through 4 Ω resistor, that is, through R_4 shown in Fig. 31.

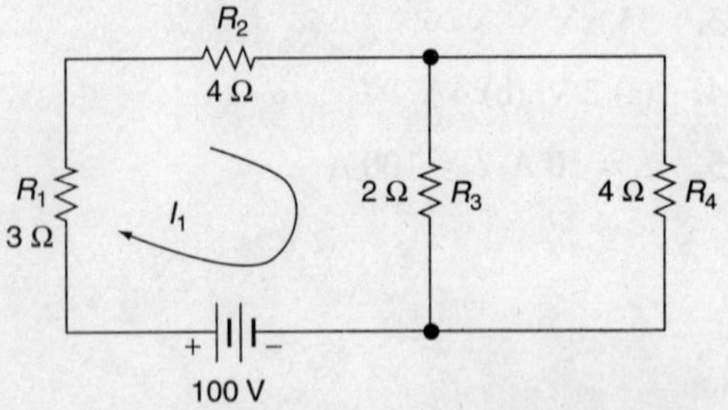

Figure 31 Problem 20.

Answers

Multiple Choice Questions

1. (a)	**3.** (c)	**5.** (b)	**7.** (a)	**9.** (c)
2. (c)	**4.** (a)	**6.** (c)	**8.** (a)	**10.** (b)

Problems

1. **(a)** 5/4 Ω, **(b)** 1/2 Ω, **(c)** 13 V
2. **(a)** 11 V, **(b)** 56 Ω
3. **(a)** 0.5A, **(b)** 4.5V, **(c)** 2.75W
4. **(a)** 6 Ω, **(b)** 6 V, 3600 J
5. **(a)** 2 Ω, **(b)** 4 A
6. **(a)** 4 A, **(b)** 44 A
7. **(a)** 6.55 A, **(b)** 6.55 A, **(c)** 26.2 V
8. 4 Ω
9. 62.5 Ω
10. **(a)** 0.8 A, **(b)** 4 V, **(c)** 1920 J

11. 66 W

12. 1.95 V

13. 33.3 V

14. **(a)** 2 V, **(b)** 4 V

15. $I_1 = 10$ A, $I_2 = 100$ A

16. **(a)** 2.706 A, **(b)** 13.5 V

17. –1.0 A

18. 16.6 A

19. 0.445 A

20. 4.04 A

3 Magnetic Circuits

LEARNING OBJECTIVES

After reading this chapter, you will be able to:

- Understand different types of magnets.
- Describe the magnetic field around a permanent magnet.
- Understand the laws of magnetic attraction and repulsion for two magnets in close proximity.
- Understand magnetic flux, magnetic flux density and their units.
- Perform simple calculations involving $B = \phi/A$.
- Understand permeability and distinguish between μ_0, μ_r and μ.
- Understand magnetomotive force and magnetic field strength.
- Understand simple magnetic circuit.
- Understand the derivation of composite and parallel magnetic circuits.
- Understand magnetic hysteresis, retentivity, magnetic leakage and Steinmetz law.
- Compare electric and magnetic circuits.

3.1 Definitions of Magnetic Quantities

A magnetic circuit provides a path for magnetic flux, just as an electric circuit provides a path for the flow of current. The knowledge of magnetic circuit concepts is useful in the analysis, design and application of electromagnetic devices such as rotating machines and transformers.

3.1.1 Introduction about Magnet

A magnet is a substance that attracts iron filling and points in the north–south direction when suspended freely.

3.1.2 Types of Magnet

There are two types of magnet:

1. **Permanent magnet:** It is made up of cobalt, steel or tungsten steel. It is used in moving coil instruments, energy meters, loud speakers, microphones, etc.
2. **Temporary magnet:** Temporary magnet, also called an electromagnet, is made up of soft iron or silicon steel. A soft iron piece wound with a coil acts as a magnet as long as current flows through the coil. It is used in electrical machines such as motor or generator.

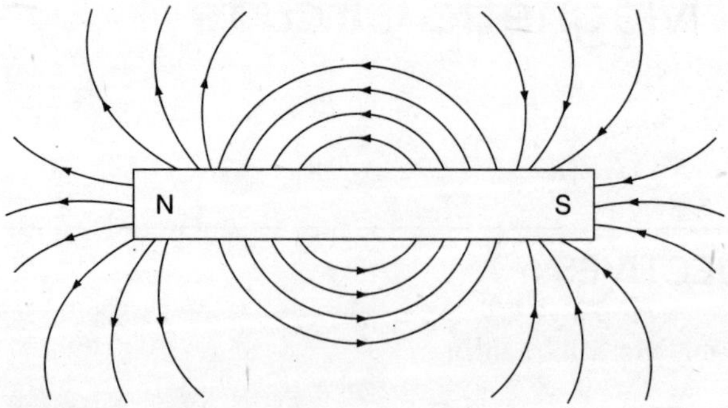

Figure 1 Magnetic lines of force.

3.1.3 Magnetic Lines of Force

The imaginary magnetic lines that travel from the north pole to the south pole outside the magnet and from the south pole to the north pole inside the magnet are called magnetic lines of force (Fig. 1).

3.1.4 Magnetic Field

The region around which the magnetic lines of force act is called magnetic field.

3.1.5 Magnetic Flux

The total number of lines of force in the magnetic field is called magnetic flux. It is denoted by ϕ and its unit is Weber.

3.1.6 Magnetic Flux Density (B)

The magnetic flux passing through a unit cross-section is called flux density. It is denoted by B. If ϕ is magnetic flux in webers, B is flux density in Wb/m^2 and A is area of cross-section in m^2, then

$$B = \frac{\phi}{A} \text{ Wb/m}^2$$

3.1.7 Permeability

The magnetic conductivity of a material as compared to that of air is called relative permeability of that material. The absolute permeability μ of a medium is given by

$$\mu = \frac{B}{H}$$

where B is flux density in Wb/m^2 and H is magnetic field strength in N/Wb. μ is also given by the equation

$$\mu = \mu_0 \mu_r$$

where $\mu_0 = 4\pi \times 10^7$ H/m and μ_r is relative permeability. The relative permeability μ_r of a medium is given by

$$\mu_r = \frac{B}{B_0}$$

where B_0 is the flux density in vacuum and B is the flux density in the medium under construction in units of Wb/m².

3.1.8 Magnetic Field Strength (H)

Magnetic field strength at a point is given by the force experienced by a unit north pole placed at that point. It is denoted by *H*. If ϕ is the flux in weber, *F* is the force in Newton (N) and *H* is the field strength in N/Wb, then

$$H = \frac{F}{\phi} \text{ N/Wb} \tag{3.1}$$

H is also given by

$$H = \frac{B}{\mu_0 \mu_r} \text{ N/Wb} \tag{3.2}$$

3.1.9 Magnetomotive Force

Magnetomotive force is given by the work done in moving a unit north pole once around the entire magnetic circuit. It is also the driving force required to drive the magnetic flux through a magnetic circuit. It is given by *NI* and its unit is ampere turns. *N* is the number of turns and *I* is the current.

3.1.10 Reluctance and Permeance

The property of a magnetic material by which it opposes the establishment of magnetic flux is called reluctance. Reluctance is denoted by "*S*" and its unit is ampere turns/weber.

Permeance is defined as the reciprocal of reluctance and is the readiness with which magnetic flux is developed. It is analogous to conductance in an electric circuit. Its unit is weber per ampere turns (AT).

3.2 Analysis of Magnetic Circuit

3.2.1 Simple Magnetic Circuit

Consider a closed iron ring wound with a magnetizing coil. Here, the coil produces the magnetic flux. Thus, it acts as a source of magnetomotive force (m.m.f.), and the iron ring offers reluctance to the establishment of magnetic flux.

3.2.1.1 Ohm's Law for Magnetism

Consider a simple magnetic circuit (Fig. 2) wound with a coil of "*N*" turns. Let *I* amperes be the current carried by the coil, "*A*" be the area of cross-section of the ring in m² and *l* be length of the

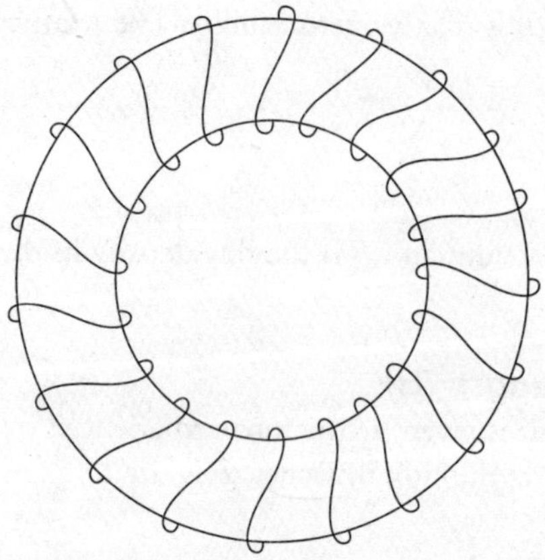

Figure 2 Simple magnetic circuit.

magnetic path in meters. Let ϕ be flux in weber established in the iron ring. Then the magnetic flux density is

$$B = \frac{\phi}{A}\,\text{Wb/m}^2$$

Also,

$$B = \mu_0\, \mu_r\, H\,\text{Wb/m}^2$$

where μ_r is the relative permeability of the ring and

$$\mu_0 = 4\pi \times 10^{-7}\ \text{H/m}$$

where H is the magnetic field strength in AT/m

$$H = \frac{NI}{l}\,\text{AT/m} \tag{3.3}$$

Using Eq. (3.3) we can write B as

$$B = \mu_0 \mu_r \frac{NI}{l}$$

which implies

$$\frac{\phi}{A} = \mu_0 \mu_r \frac{NI}{l}$$

$$\phi = \frac{NIA\mu_0 \mu_r}{l}$$

$$\phi = \frac{NI}{l/(A\mu_0 \mu_r)}$$

The denominator $l/(A\mu_0\mu_r) = S$, where "S" is the reluctance of the circuit. Therefore,

$$\phi = \frac{NI}{S}$$

Since this resembles the Ohm's law of electricity, namely,

$$I = \frac{V}{R}$$

it is called *Ohm's law of magnetism.*

3.2.2 Composite Magnetic Circuit of Iron Ring with Air Gap

An iron ring with a small air gap forms a composite magnetic circuit (Fig. 3). The reluctance offered by the iron ring is different from that offered by the air gap. The total reluctance of the magnetic circuit will be the sum of these two reluctances.

Let A be area of cross-section of the ring in m^2, μ_{r_1} be the relative permeability of iron ring, μ_{r_2} be the relative permeability of air gap which is 1, N be the number of turns of the coil, I be the current carried by the coil in amperes, l_1 be the length of iron ring in meters, l_2 be the length of air gap in meters, S_1 be the reluctance of iron ring in AT/Wb, S_2 be the reluctance of air gap in AT/Wb and S be the total reluctance in AT/Wb. Then,

$$S_1 = \frac{l_1}{\mu_0\mu_{r_1} A}$$

$$S_2 = \frac{l_2}{\mu_0\mu_{r_2} A}$$

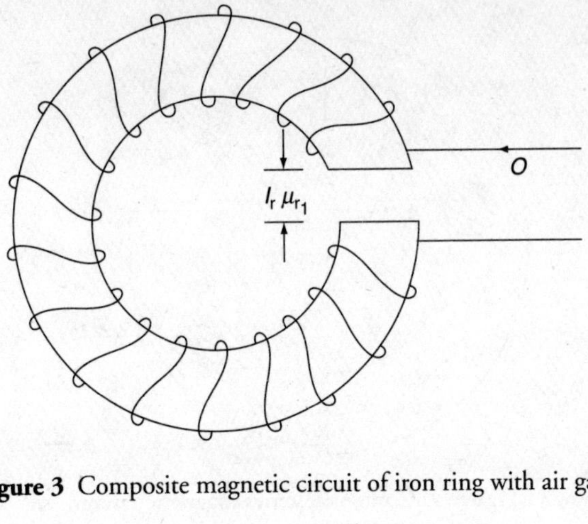

Figure 3 Composite magnetic circuit of iron ring with air gap.

Therefore S, the reluctance of the circuit is given by

$$S = S_1 + S_2 = \frac{l_1}{\mu_0 \mu_{r_1} A} + \frac{l_2}{\mu_0 \mu_{r_2} A}$$

We know that

$$\text{Flux} = \frac{\text{m.m.f.}}{\text{Reluctance}}$$

Therefore

$$\text{Flux} = \frac{NI}{[l_1/(\mu_0 \mu_{r_1} A)] + [l_2/(\mu_0 \mu_{r_2} A)]} \text{ Weber}$$

3.2.3 Composite Series Magnetic Circuit

It consists of different types of iron with different dimensions and relative permeabilities. Since they are in series, the total reluctance "S" can be given by the sum of individual reluctances (Fig. 4). The flux is given by

$$\text{Flux} = \frac{\text{m.m.f.}}{\text{Total reluctance}}$$

where the total reluctance S is equal to

$$S = \frac{l_1}{\mu_0 \mu_{r_1} A_1} + \frac{l_2}{\mu_0 \mu_{r_2} A_2} + \frac{l_3}{\mu_0 \mu_{r_3} A_3}$$

Hence flux is given by

$$\phi = \frac{\text{m.m.f.}}{[l_1/(\mu_0 \mu_{r_1} A_1)] + [l_2/(\mu_0 \mu_{r_2} A_2)] + [l_3/(\mu_0 \mu_{r_3} A_3)]}$$

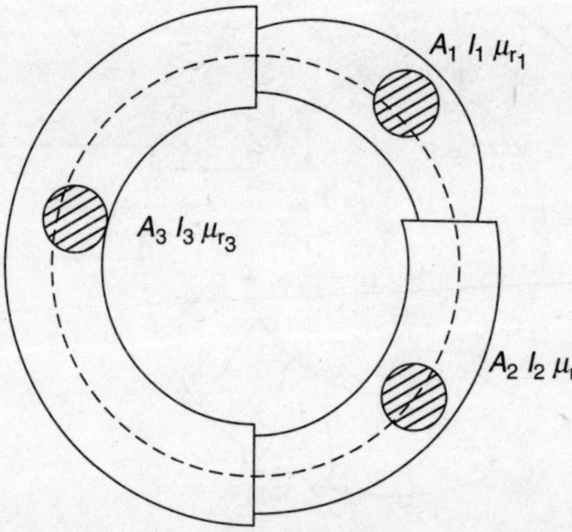

Figure 4 Composite series magnetic circuit.

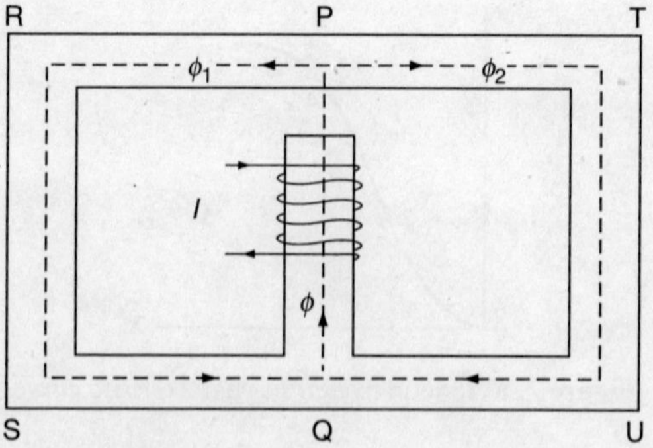

Figure 5 Parallel magnetic circuit.

3.2.4 Parallel Magnetic Circuit

In series circuits, all parts of the magnetic circuit carry same flux and total ampere turns required to create a given flux is the arithmetic sum of the ampere turns required for individual parts of the circuit. However, if the various paths of the magnetic circuit are in parallel (Fig. 5) the ampere turns required for the combination is equal to the ampere turns required to create the given flux in one path.

The parallel magnetic circuit has more than one path for magnetic flux. Flux is established by the ampere turns provided in the central limb (PQ) and divides it into two parts toward PRSQ and PTUQ as ϕ_1 and ϕ_2, respectively, thus we have $\phi = \phi_1 + \phi_2$ (Fig. 5).

3.3 Magnetic Leakage

The flux that follows an undesired path is called the leakage flux (Fig. 6). To utilize the magnetic flux established by the magnetic material we provide an air gap. The flux in the air gap is called useful flux. Let ϕ_t be the total flux established in the iron path, ϕ_l be the leakage flux and ϕ_u be the useful flux. The ratio ϕ_t / ϕ_u is called the coefficient of magnetic leakage "λ" whose value varies from 1.15 to 1.25. The spreading of magnetic flux in the air gap is called *fringing*.

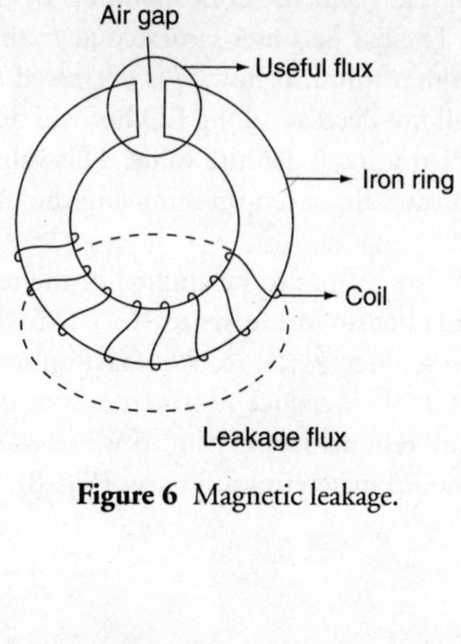

Figure 6 Magnetic leakage.

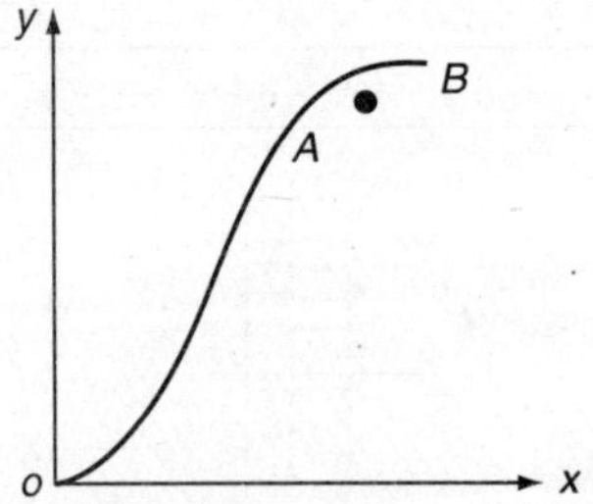

Figure 7 Magnetic hysteresis characteristic curve.

3.4 Retentivity

The retaining capacity of magnetism of a magnetic material after the removal of inducing agency is called retentivity. An example of such a material is hard steel.

3.5 Magnetic Hysteresis

Hysteresis means lagging. The lagging of flux density (B) behind the magnetizing force (H) is called magnetic hysteresis. We know that the substances are generally classified as diamagnetic ($M < 1$) and paramagnetic ($M > 1$) ($M >> 1$ for iron, cobalt and nickel). The magnetization characteristic curve (Fig. 7) magnetic substance is nonlinear. Initially (portion OA) the curve rises rapidly, indicating rapid change in "B" with relatively small change in "H." At portion AB, we can notice gradual decrease in the slope of the characteristic. At saturation the slope of the curve becomes constant.

3.6 Magnetization Cycle

Consider an unmagnetized iron bar PQ (Fig. 8). It can be magnetized by placing it in the magnetizing force "H" produced by the solenoid. This H can be varied from zero to maximum value by varying the current with the help of a rheostat, and the corresponding B values can be noted. The relation between B and H can be plotted. The bar becomes saturated at E; that is, when it becomes H_{max}, the corresponding flux destiny will be maximum. If now H is decreased to zero by decreasing the current with the help of the rheostat, B will not decrease along EO but will decrease along EC. Also when H is zero B does not become zero; instead it has a definite value. This value of B is the measure of residual magnetism or remnence. This indicates that even on removing the magnetizing force H, the bar does not get completely demagnetized.

To remove this residual magnetism, H has to be applied in the reverse direction. When this value of H becomes equal to OG, the flux density decreases to "G". This value of H is called coercive force. If H is further increased in the reverse direction, it reaches maximum value at I. Plotting corresponding B and H values we get the curve GI. If we reduce H to zero again, the B–H plot will be given by the curve IT. When $H = 0$, $B =$ OJ, the residual magnetism. If we reverse H and gradually increase to the same maximum we will get the smooth magnetization curve (Fig. 8).

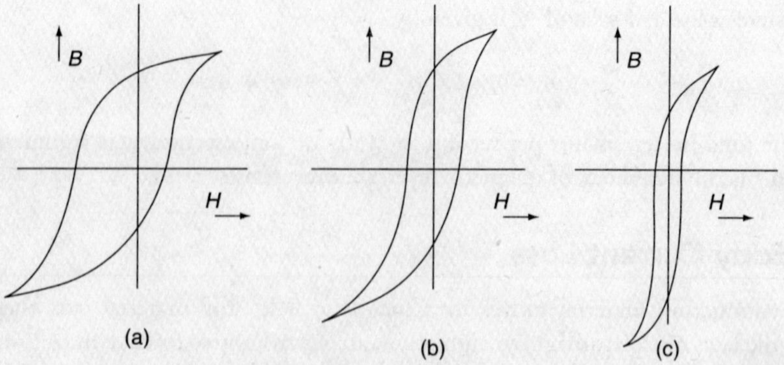

Figure 8 Unmagnetized curve.

Thus it is seen that B always lags behind H. This lagging of flux density B with respect to magnetizing force H is called *hysteresis* and the complete loop is called *hysteresis loop*.

3.7 Hysteresis Loss

The energy expended in taking an iron bar through one cycle of magnetization is proportional to the area of the hysteresis loop. The shape of the hysteresis loop depends on the nature of the magnetic material.

For AC-excited devices the hysteresis loop is repeated every cycle of alternating current. Since hysteresis loop measures the energy dissipated in the form of heat due to hysteresis, hard steel is not suitable for applications where rapid reversal of magnetism takes place. Due to its high retentivity and coercivity, hard steel is used for making permanent magnets.

Cast steel has high permeability and fairly good coercivity, and hence is used for cores of electromagnets. Alloy steel has high permeability and low hysteresis loss and hence is used for making transformer and armature cores.

3.8 Steinmetz Law

Steinmetz experimentally found that hysteresis loss can be approximately expressed by

$$\text{Hysteresis loss} = \eta B_{max} K \ \text{J/m}^3 \ \text{per cycle}$$

where η is a constant called Steinmetz's hysteresis coefficient which depends upon the material and

$$K = 1.6 \text{ for } 0.1 < B_{max} < 1.2 \text{ Wb/m}^2$$

$$K > 1.6 \text{ for } B_{max} < 0.1 \text{ Wb/m}^2 \text{ and } B_{max} > 1.2 \text{ Wb/m}^2$$

Also, total hysteresis loss per second ω_r is given by

$$\omega_r = \eta \times 1.6\, B_{max} \times f \times v \quad \text{Watts}$$

where ω_r is the total hysteresis loss per second in Watts or Joule/second, v is volume of the magnetic flux in m^3 and f is the frequency of magnetic reversal in reversals/second.

3.9 Eddy Current Loss

Whenever a conducting material moves in a magnetic field, the material cuts the magnetic flux. According to the laws of electromagnetic induction, an electromotive force (e.m.f.) is introduced in the conductor. This e.m.f. sets up large currents through the solid mass of the material. There currents are called eddy currents.

The heat liberated by eddy currents is considered as a loss as it serves no purpose. In a solid core, a direct path is offered for the flow of eddy currents whereas a broken path is offered by a laminated core. As the laminated core has much less cross-section, it offers higher resistance for the flow of eddy current. Thus, eddy current loss can be minimized by using a laminated core.

Summary

In this chapter basics of magnetic circuits and their types were presented in detail.

1. The closed path followed by the lines of magnetic flux is called a magnetic circuit.
2. The m.m.f. is the source of producing flux in a magnetic circuit.
3. Definition about various magnetic quantities and simple analysis of magnetic circuit are provided. There are two types of magnetic circuits: simple and composite magnetic circuits.
4. Each magnetic circuit is explained in detail in terms of derivations.
5. Comparison between the series and parallel magnetic circuits has been discussed briefly in the different types of magnet circuits.

Solved Problems

Problem 1

A square cross-section of a magnet of side 4 cm has a strength of 0.5 mWb. Determine the magnetic flux density.

Solution

Flux = 0.5 mWb = 0.5×10^{-3} Wb

Cross-sectional area = 4 cm × 4 cm = 4×10^{-4} m^2

Flux density $B = \phi/A = (0.5 \times 10^{-3})/(4 \times 10^{-4}) = 1.25$ Wb/m^2

Problem 2

A solenoid is wound with a coil of 100 turns (see Fig. 9). The coil is of length 50 cm and is carrying a current of 2 A. Determine the magnetic field strength at the line of the solenoid.

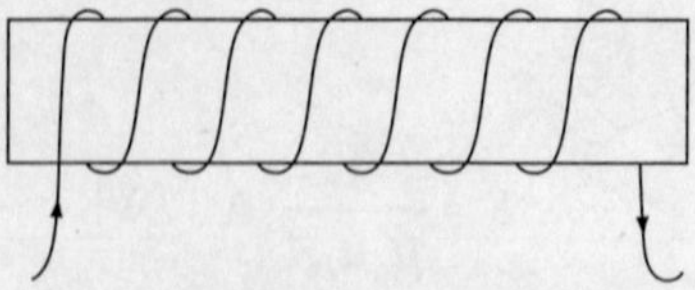

Figure 9 Problem 2.

Solution

We have

$$I = 2\,\text{A}$$
$$l = 50\,\text{cm} = 50 \times 10^{-2}\,\text{N}$$
$$N = 1000\,\text{turns}$$

Therefore, magnetic field strength is

$$H = \frac{NI}{l}\ \text{AT/m}$$
$$= \frac{100 \times 2}{50 \times 10^{-2}}\ \text{AT/m}$$
$$= 400\ \ \text{AT/m}$$

Problem 3

A mild steel ring of circumference 50 cm and cross-sectional area of 5 cm² has a coil of 250 turns wound around it. Calculate

(i) The reluctance.

(ii) The current required to produce a flux of 700 m^3.

The relative permeability $\mu_r = 380$.

Solution

Given,

$$\text{Cross-sectional area of ring } (A) = 5\,\text{cm}^2$$
$$= 5 \times 10^{-4}\,\text{m}^2$$
$$\text{No. of turns of coil } (N) = 250$$
$$\text{Length of the magnetic path} = \text{Circumference of the ring}$$
$$= 50\,\text{cm}$$
$$= 50 \times 10^{-2}\,\text{m}$$

Flux established,

$$\phi = 700 \text{ mWb}$$
$$= 700 \times 10^{-6} \text{ Wb}$$

Relative permeability $\mu_r = 380$.
(i) Reluctance is

$$S = \frac{1}{\mu_0 \mu_r A} \text{ AT/Wb}$$
$$= \frac{50 \times 10^{-2}}{4\pi \times 10^{-7} \times 380 \times 5 \times 10^{-4}}$$
$$= 2{,}094{,}144 \text{ AT/Wb}$$

(ii) We know that the flux is given by

$$\text{Flux} = \frac{\text{m.m.f.}}{\text{Reluctance}}$$
$$\phi = \frac{NI}{s}$$

Therefore the current required is

$$I = \frac{\phi \times s}{N}$$
$$= 5.86 \text{ A}$$

Problem 4

An iron ring of diameter 15 cm is uniformly wound with 450 turns of wire. A current of 2 A when passed through the coil establishes a flux density of 1.2 Wb/m² in the iron ring. Determine the value of relative permeability of iron.

Solution
Given that

$$\text{Diameter, } D = 15 \text{ cm}$$
$$\text{Length } (l) = \text{Mean circumference} = \pi D$$
$$= \pi \times 15 \text{ cm} = \pi \times 15 \times 10^{-2} \text{ m}$$
$$N = 450$$
$$\text{Current passing through the coil } (I) = 2 \text{ A}$$
$$\text{Flux density } (B) = 1.2 \text{ Wb/m}^2$$
$$\mu_0 = 4\pi \times 10^{-7} \text{ H/m}$$

We have to find the relative permeability. We know that $B = \mu_0 \mu_r H$. Therefore,

$$\frac{B}{\mu_0 H} = \mu_r$$

Now

$$H = \frac{NI}{l} \text{ AT/m}$$

$$= \frac{450 \times 2}{\pi \times 15 \times 10^{-2}} \text{ AT/m}$$

Therefore

$$\mu_r = \frac{1.2(\pi \times 15 \times 10^{-2})}{4\pi \times 10^{-7} \times 450 \times 2}$$

$$= 500$$

Problem 5

A flux density of 1.2 Wb/m² is required in the firm 1 mm air gap of an electromagnet having an iron path of length 1.5 m. Calculate the m.m.f. required. Given, relative permeability of iron is 1600. Neglect leakage.

Solution

Given that

$$l = 1.5 \text{ m}$$

$$\mu_r = 1600$$

$$B = 1.2 \text{ Wb/m}^2$$

$$\mu_0 = 4\pi \times 10^{-7} \text{ H/m}$$

Now

$$H = \frac{NI}{l} \Rightarrow NI = lH$$

Again

$$H = \frac{B}{\mu_0 \mu_r} = \frac{1.2}{4\pi \times 10^{-7} \times 1600}$$

Therefore, amperes turn on iron path is equal to

$$NI = \frac{1.5 \times 1.2}{4\pi \times 10^{-7} \times 1600} \text{ AT} = 895.24 \text{ AT}$$

Now

$$\text{Length of air gap} = 1 \text{ mm} = 1\times10^{-3} \text{ m}$$
$$\mu_r \text{ of air} = 1$$

Therefore

$$\text{Ampere turns (required)} = \frac{\text{Length of the air gap}\times B}{\mu_0(\mu_r \text{ of air})}$$
$$= \frac{1\times10^{-3}\times1.2}{4\pi\times10^{-7}\times1}$$
$$= 954.93 \text{ AT}$$

Hence, the total ampere turns required is

$$895.24 + 954.93 = 1850.17 \text{ AT}$$

Problem 6

A magnetic circuit has a uniform cross-sectional area of 5 cm² and a length of 25 cm. A coil of 100 turns is wound uniformly over the magnetic circuit. When the current in the coil is 2 A, the total flux is 0.3 mWb and when the current in the coil is 5 A, the total flux is 0.58 mWb. For each value of current calculate

(i) T the magnetizing force.
(ii) Relative permeability.
(iii) Magnetic flux density.

Solution

Case 1: When the current in the coil is 2 A and the total flux is 0.3 mWb

Given that

$$\text{Cross-sectional area, } A = 5\,\text{cm}^2 = 5\times10^{-4}\text{ m}^2$$
$$l = 25\,\text{cm} = 25\times10^{-2}\text{ m}$$
$$\text{No. of turns, } N = 100$$
$$\text{Current in the coil, } I = 2 \text{ A}$$

Total flux established is given by

$$\phi = 0.3\,\text{mWb}$$
$$= 0.3\times10^{-3}\text{ Wb}$$

(i) Magnetic force,

$$H = \frac{NI}{l}\text{ AT/m}$$
$$= \frac{100\times2}{25\times10^{-2}}$$
$$= 800\text{ AT/m}$$

(ii) Now

$$\phi = \frac{NI}{l/\mu_0\mu_r A}$$

$$\phi = \frac{NI\mu_0\mu_r A}{l}$$

Total relative permeability $\mu_r = ?$

$$\mu_r = \frac{\phi \times l}{NI\ \mu_0 A}$$

$$= \frac{0.3\times10^{-3}\times25\times10^{-2}}{100\times2\times4\pi\times10^{-7}\times5\times10^{-4}}$$

$$= 596.83 \approx 597$$

(iii) Magnetic flux density is given by

$$B = \mu_0\mu_r H$$

$$= 4\pi\times10^{-7}\times597\times800\ \text{Wb/m}^2$$

$$= 0.6\ \text{Wb/m}^2$$

Case 2: When the current in the coil is 5 A and the total flux is 0.58 mWb.
Given that

$$A = 5\times10^{-4}\ \text{m}^2$$

$$l = 25\times10^{-2}\ \text{m}$$

$$N = 100$$

$$I = 5\ \text{A}$$

Total flux established is given by

$$\phi = 0.58\ \text{mWB} = 0.58\times10^{-3}\ \text{Wb}$$

(i) Magnetic force,

$$H = \frac{NI}{l}$$

$$= \frac{100\times5}{25\times10^{-2}}$$

$$= 2000\ \text{AT/m}$$

(ii) Total relative permeability μ_r

$$\mu_r = \frac{\phi\times l}{NI}$$

$$= \frac{0.58\times10^{-3}\times25\times10^{-2}}{100\times5\times4\pi\times10^{-7}\times5\times10^{-4}}$$

$$\mu_r = 461.55 \approx 462$$

(iii) Magnetic flux density is given by

$$B = \mu_0\mu_r H\ \text{Wb/m}^2$$

$$= 4\pi\times10^{-7}\times462\times2000$$

$$= 1.16\ \text{Wb/m}^2$$

Problem 7

A magnetic ring has a mean circumference of 150 cm; cross-section of 0.01 m² and is wound with 200 turns. A saw cut of 2 mm is made in the ring. Calculate the magnetizing current required to produce a flux of 800 mWb in the air gap. Assume relative permeability of iron is 600 and a leakage factor of 1–2.

Solution

Given that

$$A = 0.01 \text{ m}^2$$

$$l = 2 \text{ mm} = 2 \times 10^{-3} \text{ m}$$

$$\mu_r \text{ (for air gap)} = 1$$

$$\mu_0 = 4\pi \times 10^{-7} \text{ H/m}$$

$$\phi = 800 \text{ mWb} = 800 \times 10^{-6} \text{ Wb}$$

Flux density in the air,

$$B = \frac{\phi}{A} = \frac{800 \times 10^{-6}}{0.01}$$

$$= 0.08 \text{ Wb/m}^2$$

Magnetizing force in the air gap is

$$H = \frac{B}{\mu_0 \mu_r} = \frac{0.08}{4\pi \times 10^{-7} \times 1} \text{AT/m}$$

$$= 63661.9 \text{ Tm}$$

Now

$$\text{Ampere turns for air gap} = H \times I$$

$$= 63661.9 \times 2 \times 10^{-3}$$

$$= 127.32 \text{ AT}$$

Now for the iron path

$$\phi_1 = (800 \times 10^{-6}) \times 1.2 = 9.6 \times 10^{-4} \text{ Wb}$$

$$\mu_{r_1} = 600$$

$$I_1 = 150 \times 10^{-2} \text{ N}$$

Therefore,

$$\text{Ampere turns for iron path} = \frac{\phi_1 \times I_1}{A \mu_0 \mu_{r_1}}$$

$$= \frac{(800 \times 10^{-6}) \times (1.2) \times (150 \times 10^{-2})}{0.01 \times 4\pi \times 10^{-7} \times 600}$$

$$= 190.98$$

$$= 191 \text{ AT}$$

So

$$\text{Total ampere turns} = \text{Ampere turns for air} + \text{Ampere turns for iron path}$$
$$= 127.32 + 191 = 318.32$$

No. of turns = 200. Therefore,

$$\text{Magnetizing current} = \frac{\text{Total ampere turns}}{200}$$
$$= \frac{318.32}{200}$$
$$= 1.59 \text{ A}$$

Problem 8

An iron ring with a mean circumference of 200 cm and cross-section of 14 cm^2 *is wound with 200 turns of wire when the exciting current is 4 A and the flux is found to be 1.4 mWb. What is the relative permeability of the iron?*

Solution
We have,

$$B = \mu_0 \mu_r H$$
$$\frac{1.4 \times 10^{-3}}{14 \times 10^{-4}} = 4\pi \times 10^{-7} \mu_r \times \frac{200 \times 2}{2}$$
$$1 = \frac{4\pi}{2} \times 10^{-4} \mu_r$$
$$\mu_r = 1591$$

Problem 9

A glass tube 2.5 cm diameter and 90 cm long is completely and uniformly wound with 900 turns of copper wire 0.9 mm diameter insulated to 1 mm diameter. Find the strength of the magnetic field at the center of the solenoid thus formed when a potential difference of 4 V is applied to its terminals at temperature 20°C.

Solution
Axial distance occupied by each turn is 1 mm = 0.1 cm. Therefore,

$$\text{Number of turns per layer} = \frac{90}{0.1} = 900 \text{ turns}$$

That is only one layer of winding as the number of turns is given to be 900. Diameter to the center of each turn is

$$2.5 + 0.1 = 2.6 \text{ cm}$$

Therefore,

$$\text{Length of each turn } = \pi \times 2.6$$

Therefore,

$$\text{Total length of wire} = \pi \times 2.6 \times 900 \times 10^{-2} = 73.5 \text{ m}$$

$$\text{Cross-section of wire} = \frac{\pi \times 0.9^2}{4} = 0.636 \text{ mm}^2$$

At 20°C copper wire 1 m long and 1 mm^2 in cross-section has a resistance of 1/58 Ω. Therefore, the resistance of winding,

$$R = \frac{1}{58}\frac{l(\text{in m})}{A(\text{in mm}^2)} = \frac{1}{58} \times \frac{73.5}{0.636} \; C = 2\Omega \text{ (nearly)}$$

Therefore,

$$\text{Current} = \frac{V}{R} = \frac{4}{2} = 2 \text{ A}$$

Hence, the magnetizing force is

$$H = \frac{NI}{l} = \frac{900 \times 2}{0.9} = 2008 \text{ AT/m}$$

Problem 10

Calculate the flux density (i) at the surface and (ii) at 10 cm away from the center of a long straight insulated conductor 0.5 cm in diameter carrying a direct of 100 A.

Solution

Given that, $\mu_0 = 4\pi \times 10^{-7}$ and $I = 100$ A we get

$$B = \frac{\mu_0 I}{2\pi r} \text{ Wb/m}^2$$

$$= \frac{200 \times 10^{-7}}{r} \text{ Wb/m}^2$$

$$= \frac{200 \times 10^{-7} \times 10^3}{r} \text{ mWb/m}^2$$

$$= \frac{0.02}{r} \text{ mWb/m}^2$$

(i) At the surface:

$$r = 0.25 \text{ cm} = 0.25 \times 10^{-2} \text{ m}$$

Therefore

$$B = \frac{0.02}{0.25 \times 10^{-2}} = 8 \text{ mWb/m}^2$$

(ii) At 10 cm away:

$$r = 0.1 \text{ m}$$

Therefore

$$B = \frac{0.02}{0.1} = 0.2 \text{ mWb/m}^2$$

Problem 11

A long straight wire of diameter 1 cm carries a direct current of 1000 A. A draw of curve showing the magnetic field strength both inside and outside the wire on a radial line extending 10 cm from the axis of the wire. Points the Inside conductor. Hx = field strength at radius x.

Solution

$$2\pi Hx = \text{Current enclosed}$$

$$= \frac{Ix^2}{R^2}$$

Now $I = 1000$ A, $R = 0.5$ cm. Therefore

$$Hx = \frac{Ix^2}{2\pi R^2}$$

$$= \frac{1000x^2}{2\pi \times 0.5^2}$$

$$= \frac{2000x}{x} = 737\ x \text{ A/cm}$$

This will be a straight line inside the conductor ($x < 0.5$ cm). Outside the conductor,

$$2\pi x\ Hx = I$$

$$Hx = \frac{1000}{2\pi x} = \frac{159}{x} \text{ A/m} \quad (x > 0.5)$$

Outside the conductor this will be a hyperbola.

Problem 12

A wire is bent into a plane surface of side 30 cm and a current of 250 A is circulated round it. What is the magnetic flux density at the center of the square (Fig. 10)?

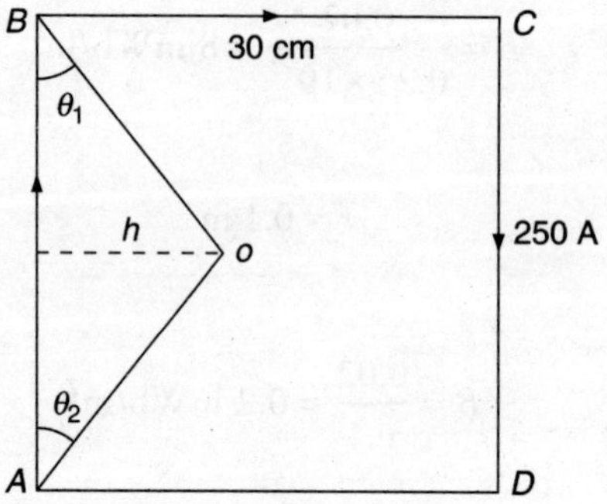

Figure 10 Problem 12.

Solution

The field in intensity at θ due to current inside AB is

$$H = \frac{I}{4\pi h}(\cos\theta_1 + \cos\theta_2)$$

Since $I = 250$ A, so

$$h = \frac{30}{2}\text{ cm} = 0.15\text{ m}$$

$$\cos\theta_1 = \cos\theta_2 = \frac{1}{\sqrt{2}}$$

Therefore

$$H = \frac{200}{4\pi \times 0.15}\left(\frac{1}{\sqrt{2}} + \frac{1}{\sqrt{2}}\right) = \frac{250 \times \sqrt{2}}{0.6\pi}\text{ A/m}$$

That is the intensity due to all the four conductors at O will be additive. Hence the total intensity is

$$4H = \frac{1000\sqrt{2}}{0.6\pi}\text{ A/m}$$

Therefore, the magnetic flux density is given by

$$B = \mu_0 H$$

$$= 4\pi \times 10^{-7} \times \frac{1000\sqrt{2}}{0.6\pi}\text{ Wb/m}^2$$

$$= \frac{0.4\sqrt{2}}{0.6} = 0.943\text{ mWb/m}^2$$

Problem 13

A cast steel electromagnet has an air gap of length 2 mm and an iron path of length 30 cm. Find the number of ampere turns necessary to produce a flux density of 0.8 Wb/m² in the gap. Neglect leakage and fringing.

Solution

For air gap it is given that

$$l = 2 \text{ mm} = 2\times10^{-3} \text{ m}$$

$$B = 0.8 \text{ Wb/m}^2$$

$$\mu_0 = 4\pi\times10^{-7}$$

The required ampere turns for air gap is

$$\frac{B}{\mu_0}l = \frac{0.8}{4\pi\times10^{-7}}\times2\times10^{-3} = 1272.72 \approx 1273$$

For 0.8 Wb/m^2 cast steel requires 750 AT/m (750 is equivalent value for 0.8 Wb/m^2 cast steel). Therefore, ampere turns required for iron path (l = 30 cm = 0.3 m) is

$$750l = 750\times0.3 = 225 \text{ AT}$$

The total ampere turns required is

$$1273 + 225 = 1498 = 1500 \text{ (nearly)}$$

Problem 14

An iron ring of mean length 50 cm has an air gap of 1 mm and a winding of 200 turns. If the permeability of the iron is 300, then find the flux density when a current of 1 A flows through the coil.

Solution

The reluctance is given by

$$R = \frac{1}{A\mu}$$

Therefore, the total reluctance is

$$R_1 + R_2 = \frac{1\times10^{-3}}{A\mu_0} + \frac{0.5}{300A\mu_0}$$

$$= \frac{8\times10^{-3}}{3A\mu_0} \text{ AT/m}$$

Hence,

$$\text{Total flux} = \frac{\text{Total ampere turns}}{\text{Reluctance}}$$

$$= \frac{200 \times 1 \times 3A\mu_0}{8 \times 10^{-3}}$$

Therefore,

$$\text{Flux density} = \frac{600 \times \mu_0}{8 \times 10^{-3}} \text{ Wb/m}^2$$

$$= \frac{600 \times 4\pi \times 10^{-7} \times 10^3}{8 \times 10^{-3}} \text{ mWb/m}^2$$

$$= 94.2 \text{ mWb/m}^2$$

Problem 15

A wire of 1 m length is bend to form a square. The plane of the square is right angled to a uniform field having a flux density of 1 mWb/mm². Determine the work done if the wire carrying a current of 10 A throught it changes to circular shape.

Solution

$$\text{Side of the square} = \frac{1}{4} = 0.25 \text{ m}$$

$$A_1 = \text{Area of square}$$

$$= 0.25^2 = 0.0625 \text{ m}^2$$

Radius of circle,

$$r = \frac{1}{2\pi}$$

Therefore, the area of circle is

$$A_2 = \pi r^2 = \pi \times \frac{1}{4\pi^2} = \frac{1}{4\pi} = 0.0796 \text{ m}^2$$

Now,

$$\text{Work done} = \text{Difference between the two torques}$$

$$= BIA_2 - BIA_1$$

$$= 1 \times 10^{-3} \times 10(0.0796 - 0.0625)$$

$$= 0.0171 \times 10^{-2}$$

$$= 0.0171 \times 10^{-3} \text{ J}$$

Problem 16

Calculate the force necessary to separate two surfaces with 100 cm^2 of contact area when the flux density normal to the surface is 1 Wb/m^2.

Solution

$$\text{Force} = \frac{B^2}{\mu_0} \times \text{Area in m}^2$$

$$= \frac{1}{4\pi \times 10^{-7}} \times 100 \times 10^{-4}$$

$$= \frac{10^5}{8\pi} = 3980 \text{ N}$$

Problem 17

A wire of length 50 cm moves at right angles to its length at 40 m per second in a uniform magnetic field of 1 Wb/m^2. Calculate the e.m.f. induced in the conductor when the direction of motion is

(i) Perpendicular to the field.

(ii) Inclined at 30° to the direction of the field.

Solution

(i) E.m.f. induced in the conductor when the direction of motion is perpendicular to the field ($\theta = 90°$):

$$\text{E.m.f. induced} = Blv\sin\theta$$

$$= 1 + 0.5 \times 40 \times 1 = 20 \text{ V}$$

(ii) E.m.f. induced in the conductor when the direction of motion is inclined at 30° to the direction of the field ($\theta = 30°$):

$$\text{E.m.f.} = 1 \times 0.5 \times 40 \times \sin 30° = 10 \text{ V}$$

Problem 18

A circular coil of 100 turns with a mean diameter of 30 cm is rotated about a vertical axis in the Earth's field. Calculate the instantaneous value of e.m.f. in volts induced in the coil when the plane is

(i) Perpendicular.

(ii) Inclined at 30°.

(iii) Coincident with the magnetic median at 30°.

Take H = 14.3 AT/m.

Solution

The flux density,

$$B = \mu_0 H = 4\pi \times 10^{-7} \times 14.3 = 1.8 \times 10^{-5} \text{ Wb/m}^2$$

The maximum flux linkage,

$$\psi_m = BAT = 1.8 \times 10^{-5} \times \frac{\pi(0.3)^2}{4} \times 100$$

$$= 0.127 \times 10^{-3} \text{ Wb/turns}$$

The flux linkage at any time t is

$$\psi = \psi \cos \omega t$$

$$\omega = 2\pi \times \text{RPS} = 32\pi = 0.127 \times 10^{-3} \cos 32\pi t$$

Therefore, the induced voltage is

$$e = \frac{-d\psi}{dt} = 0.127 \times 10^{-3} \times 32\pi \sin \pi t$$

That is

$$e = 12.8 \times 10^{-3} \sin \theta$$

Now $\theta = 32\pi t$ (at $t = 0$, $\theta = 0$ and the coil right angle to the magnetic meridian).

(i) $\theta = 0$ implies $e = 0$.

(ii) $\theta = 90° - 30° = 60°$. So

$$e = 12.8 \times 10^{-3} \sin 60°$$

$$= 12.8 \times \frac{\sqrt{3}}{2} = 0.0111 \text{ V}$$

(iii) $\theta = 90°$. So

$$e = 12.8 \times 10^{-3} \sin 90° = 0.0128 \text{ V}$$

Problem 19

A copper disc 40 cm in diameter is rotated at 3000 RPM on a horizontal axis perpendicular to and through the center of the disc. The axis lies in the magnetic meridian. Two brushes make contact with the disc, one at the edge and the other at the center. If the horizontal component of the Earth's field is 0.02 Wb/m², then calculate the e.m.f. induced between the brushes.

Solution

The angular speed is

$$\omega = 2\pi \times \text{RPS} = 2\pi \times \frac{3000}{60} 0.100\pi \text{ rad/s}$$

At any radius r the linear velocity is

$$v = \omega r = 100\,\pi r$$

Therefore, the element of voltage induced is

$$\begin{aligned} d\theta &= Blv \\ &= 0.02 \times 10^{-3} \times dr \times 100\pi r \text{ Volts} \\ &= 2\pi r \; dr \text{ mV} \end{aligned}$$

Hence the voltage between the brushes ($d = 40$ cm $= 0.4$ m) is

$$\begin{aligned} e &- \int_0^{r=0.2} 2\pi r \; dr \text{ mV} \\ &= 2\pi \left. \frac{r^2}{2} \right|_0^{0.2} \\ &= \pi \times 0.04 = 0.12 \text{ mV} \end{aligned}$$

Problem 20

A circuit has 1000 turns enclosing a magnetic circuit 20 m² in cross-section. With a current of 4 A, the flux density is 1 Wb/m² and with 9 A, it is 1.4 Wb/m². Find the mean value of inductance between these current limits and the induced electromotive force if the current fell uniformly from 9 A to 4 A in 0.05 s.

Solution

$$\text{Inductance } L = N\frac{A\phi}{AI} = NA\frac{(AB)}{AI} = NA\frac{AB}{AI}$$

$$\begin{aligned} \phi &= AB = \text{Area} \times \text{Flux density} \\ &= 1000 \times 20 \times 10^{-4} \times \frac{1.4 - 1}{9 - 4} \\ &= \frac{2 \times 0.4}{5} = 0.16 \text{ H} \end{aligned}$$

$$e = -L\frac{di}{dt} = -0.16\left(\frac{4-9}{0.05}\right)16 \text{ V}$$

where e is induced e.m.f. and ϕ is flux.

Problem 21

A bar magnet with a length of 6 m and width of 2 m has a magnetic flux density of 0.12 tesla. Calculate the total flux of the magnet.

Solution

The value of area is $A = 6 \times 2 = 12\ \text{m}^2$. The value of magnetic flux is

$$\phi = BA = 0.12 \times 0.12 = 1.44\ \text{Wb}$$

Problem 22

The relative permeability of the rectangular core shown in Fig. 11 is 625. Calculate the magnetic field intensity, permeability and magnetic flux density. The core is wound by 25 turns and carries a current of 2 A.

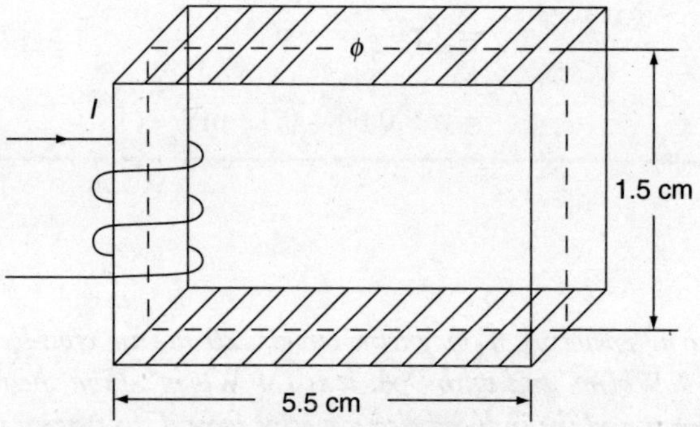

Figure 11 Problem 22.

Solution

The mean length of the core is

$$l = 2(5.5 + 1.5) = 14\ \text{cm}$$

The magnetic field intensity is

$$H = \frac{NI}{l} = \frac{25 \times 2}{0.14} = 357\ \text{Wb/m}$$

The permeability is

$$\mu = \mu_0 \mu_r = 4\pi \times 10^{-7} \times 625 = 7.85 \times 10^{-4}$$

The magnetic flux density is

$$B = \mu H = 7.85 \times 10^{-4} \times 357 = 0.28\ \text{Wb/m}^2$$

Problem 23

A wire consists of six turns and carries a current of 3 A. Calculate the magnetomotive force and the reluctance of the circuit if the flux is 0.089 Wb.

Solution

The magnetomotive force is

$$F = NI = 6 \times 3 = 18 \text{ AT}$$

The value of reluctance is

$$S = \frac{F}{\phi} = \frac{18}{0.089} = 202.2472 \text{ AT/Wb}$$

Problem 24

A coil carries a current of 3 A which results in a magnetomotive force of 12 AT. Calculate the number of turns of the coil and reluctance of the circuit if the flux is 2.2 Wb.

Solution

Given,

$$I = 3 \text{ A} \qquad F = 12 \text{ AT}$$

Therefore number of turns equals

$$N = \frac{F}{I} = \frac{12}{3} = 4 \text{ turns}$$

The value of reluctance is

$$S = \frac{F}{\phi} = \frac{12}{2.2} = 5.45 \text{ AT/Wb}$$

Problem 25

A long straight conductor carries a current of 15 A. The conductor is surrounded by air and is wound by 10 turns. Find the magnetic flux density at a distance of 2.5 cm from the conductor?

Solution

The magnetic flux density can be determined as

$$B = \mu H = \frac{\mu_0 \mu_r NI}{2\pi r}$$

$$B = \frac{4\pi \times 10^{-7} \times 1 \times 10 \times 15}{2\pi \times 0.015} = 2 \times 10^{-3} \text{ T}$$

Multiple Choice Questions

1. The SI unit of reluctance is
   (a) AT/Wb
   (b) AT/m
   (c) AT
   (d) N/Wb

2. A magnetic circuit has an m.m.f. of 400 AT and a reluctance of 2×10^5 AT/Wb. The magnetic flux in the magnetic circuit is
   (a) 3×10^{-5} Wb
   (b) 2×10^{-3} Wb
   (c) 1.5×10^{-2} Wb
   (d) 2.5×10^{-4} Wb

3. A 2 cm long coil has 10 turns and carries a current of 750 mA. The magnetizing force of the coil is
   (a) 225 AT/m
   (b) 675 AT/m
   (c) 450 AT/m
   (d) 375 AT/m

4. A magnetic device has a core with cross-section of 1 inch^2. If the flux in the core is 1 mWb, then flux density (1 inch = 2.54 cm) is
   (a) 2.5 T
   (b) 1.3 T
   (c) 1.55 T
   (d) 0.25 T

5. Permeance of a magnetic circuit is ______ area of cross-section of the circuit.
   (a) inversely proportional to
   (b) directly proportional to
   (c) independent of
   (d) none of the above

6. The reluctance of a magnetic circuit varies as
   (a) Length × area
   (b) Length ÷ area
   (c) Area ÷ length
   (d) (Length)^2 + area

7. The m.m.f. in a magnetic circuit corresponds to ______ in an electric circuit.
   (a) voltage drop
   (b) potential difference
   (c) electric intensity
   (d) e.m.f.

8. The *B–H* curve for ______ will be a straight line passing through the origin.
   (a) air
   (b) soft iron
   (c) hardened steel
   (d) silicon steel

9. A magnetizing force of 800AT/m will produce a flux density of ______ in air.
   (a) 1 mWb/m^2
   (b) 1 Wb/m^2
   (c) 10 mWb/m^2
   (d) 0.5 Wb/m^2

10. Out of the following materials, which one has the least area of hysteresis loop?
    (a) Wrought iron
    (b) Silicon steel
    (c) Hard steel
    (d) Soft iron

Review Questions

1. Define magnetic circuit.
2. Define m.m.f.
3. Write the relation between m.m.f. and flux.
4. Mention the advantages of magnetic flux.
5. State Steinmetz law.
6. Define magnetic field.

7. What is leakage factor?
8. Compare the magnetic and electric circuits.
9. What are magnetic lines of force?
10. Derive an expression of analysis of magnetic circuit in series network.
11. Explain the types of analysis of magnetic circuit in detail.
12. Define permeability.
13. Define relative permeability.
14. Define magnetic field strength.
15. Define retentivity.

Problems

1. A magnetic pole face has a rectangular section having dimensions 200 mm × 100 mm. If the total flux emerging from the pole is 150 μWb, calculate the flux density.
2. The maximum working flux density of a lifting electromagnet is 1.8 T and the effective area of the pole face is circular in cross-section. If the total magnetic flux produced is 353 mWb, determine the radius of the pole face.
3. A magnetizing force of 8000 A/m is applied to a circular magnetic circuit of mean diameter 30 cm by passing a current through a coil wound on the circuit. If the coil is uniformly wound around the circuit and has 750 turns, find the current in the coil.
4. A flux density of 1.2 T is produced in a piece of cast steel by a magnetizing force of 1250 A/m. Find the relative permeability of the steel under these conditions.
5. Determine the magnetic field strength and m.m.f. required to produce a flux density of 0.25 T in an air gap of length 12 mm.
6. A coil of 300 turns is wound uniformly on a ring of nonmagnetic material. The ring has a mean circumference of 40 cm and a uniform cross-sectional area of 4 cm^2. If the current in the coil is 5 A, calculate
   (a) The magnetic field strength.
   (b) The flux density.
   (c) The total magnetic flux in the ring.
7. Determine the reluctance of a piece of mumetal l of length 150 mm and cross-sectional area 1800 mm^2 when the relative permeability is 4000. Find also the absolute permeability of the material.
8. A flux density of 1.2 Wb/m^2 is required in an electromagnet having an iron path. Calculate the magnetic field intensity required assuming a relative permeability of iron as 1500. Neglect leakage.
9. A steel ring of 30 cm mean diameter and of circular cross-section 2 cm in diameter has an air gap 1 mm long. It is wound uniformly with 600 turns of wire carrying a current of 2.5 A. Find
   (a) Total m.m.f.
   (b) Total reluctance.
   (c) Flux.

   Neglect magnetic leakage. The iron path takes 40% of total m.m f.
10. A magnetic circuit is made of silicon steel and has a volume of 2×10^{-3} m^3. The area of hysteresis loop of silicon steel is found to be 7.25 cm^2, the seals being 1 cm = 10 AT/m and 1 cm = 4 Wb/m^2. Calculate the hysteresis

power loss when the flux is alternating with a frequency of 50 Hz.

11. Calculate the loss of energy caused by hysteresis in 1 h in 50 kg of iron if the peak lux density reached is 1.3 T (Wb/m^2) and the frequency is 25 Hz. Assume Steinmetz coefficient as 628 J/m^3 and density of iron as 7.8×10^3 kg/m^3. What will be the area of B–H curve of this specimen if 1 cm = 12.5 AT/m and 1cm = 0.1 T (Wb/m^2)?

12. A power transformer has a core material for which hysteresis coefficient is 180 J/m^3, its volume is 8000 cm^3 and the maximum flux density is 1.25 Wb/m^2. What is the hysteresis loss in watts if the frequency of the alternating current is 50 Hz?

13. An iron magnetic circuit has a uniform cross-sectional area of 5 cm^2 and a length of 25 cm. A coil of 120 turns is wound uniformly over the magnetic circuit when the current in the coil is 5 A and total flux is 9.6 mWb. Calculate
    (a) The magnetic field strength.
    (b) Relative permeability.

14. A circular iron ring of relative permeability 650 and diameter 20 cm is wound with a coil of 300 turns. The area of cross-section of the iron is 3.6 cm^2. Calculate the current required in the coil to send a flux of 0.5 mWb through the air. Neglect fringing.

15. The hysteresis loss of a transformer at a flux density of 1.2 Wb/m^2 is 150 W at 50 Hz. What will be the loss when flux density is reduced to 1 Wb/m^2 at 45 Hz?

16. A toroidal air core coil with 1000 turns has a mean effective radius of 10 cm, diameter of each turn being 3 cm. If current in the coil is 15 A, find **(a)** m.m.f. and **(b)** flux.

17. A toroidal air core coil with 1000 turns has a mean radius of 10 cm, diameter of each turn 2 cm. If current is 20 A, find flux density when flux is 2.119×10^{-5} Wb.

18. A toroidal air core coil with 2000 turns has mean reluctance of 3.47×10^{-8} Wb and flux of 4.48×10^{-5}. Find current in the coil.

19. Find **(a)** m.m.f and **(b)** diameter of a coil having flux 4.48×10^{-5}, flux density 1.6×10^{-2} T, number of turns is 1500 and current flowing through coil is 10 A.

20. A coil of 500 turns and resistance of 10π is wound uniformly over a steel ring of mean circumference 30 cm and cross-sectional area 9 cm^2. If it is connected to supply of 20 V and relative permeability is 1600, find **(a)** reluctance and **(b)** magnetic field intensity.

21. Consider a coil of 600 turn, current 1 A and reluctance 1.657×10^5. Find the **(a)** m.m.f. and **(b)** flux of coil.

22. Find the number of turns of coil when the length is 30 cm and current is 1 A. Then find **(a)** magnetic field intensity when m.m.f. is 600 AT and reluctance is 1.65×10^5 and **(b)** flux.

23. A mild steel ring having a cross-sectional area of 200 mm^2 and a mean circumference of 200 mm has a coil of 100 turns wound uniformly around it [given $u_r = 300$]. Find **(a)** reluctance "S" of ring and **(b)** m.m.f. required to produce flux of 800 μWb in ring.

24. A mild steel ring has a cross-sectional area of 4500 mm^2, a mean circumference of 400 mm, 200 turns, magnetizing current 10.6 A and reluctance 2.65×10^6 A/Wb. Find **(a)** m.m.f. and **(b)** flux.

25. A magnetic circuit has a coil wound around it with 75 turns. The cross-sectional area and length of magnetic circuit are $r = 2.5\times10^{-4}$ m^2 and 12.5×10^{-2} m, respectively. Determine **(a)** H, **(b)** μ_r and **(c)** flux density if current is 2 A, length is 25 cm and flux is 0.3×10^{-3}.

26. An air core coil has 100 turns. The mean length of magnetic flux path is 10 cm and the area of cross section is 1×10^{-4} m^2. If existing current is 2.5 A, determine **(a)** H, **(b)** flux density and **(c)** flux?

27. An air core coil of 1000 turns having mean length 25 cm and flux density 6.283×10^{-3} T. Calculate **(a)** H and **(b)** I.

28. An iron ring of 10 cm mean circumference is made from round iron cross-section of 2×10^{-3} m^2 and its relative permeability is 18500. If it is wound with 100 turns, what current will be required to produce a flux of 2×10^{-3} Wb?

29. An air gap 1.1 mm long and having cross-sectional area of 30 cm^2 exists in a magnetic circuit. Determine reluctance S of the air gap.

30. An air gap of 2 mm long and cross-sectional area of 20 cm^2 exists in magnetic circuit. Determine m.m.f. to create a flux of 10×10^{-4} Wb.

31. A current of 10 A when flowing through a coil of 1000 turns establishes a flux of 0.5 mWb. Calculate inductance of the coil.

Answers

Multiple Choice Questions

1. (a)
2. (b)
3. (d)
4. (c)
5. (b)
6. (b)
7. (d)
8. (a)
9. (a)
10. (b)

Problems

1. 7.5 mT

2. 250 mm

3. 10.05 A

4. 764

5. 198943.6 A/m, 2387A

6. **(a)** 3750 A/m, **(b)** 4.712 mT, **(c)** 1.885 μWb

7. 16580 H, 5.027×10^{-3} H/m

8. 637 AT/m

9. **(a)** 1500 AT, **(b)** 2.283×10^{6} AT, **(c)** 0.657 mWb

10. 290 T/m^3/cycle, 29 W

11. 764 cm^2

12. 102.89×10^{6} W

13. **(a)** 2400 AT/m, **(b)** 6366.19

14. 3.56 A

15. 45 W

16. **(a)** 15000 A, **(b)** 2.119×10^{-5} Wb

17. 0.03

18. 10 A

19. **(a)** 15000A, **(b)** 3×10^{-2} m

20. **(a)** 1.657×10^5 AT/Wb, **(b)** 3333.33

21. **(a)** 600 A, **(b)** 3.62 mWb

22. **(a)** 2000 AT, **(b)** 3.62 mWb

23. **(a)** 2.65×10^6 A/Wb, **(b)** 2.122×10^3

24. **(a)** 2.122×10^3, **(b)** 800×10^{-6}

25. **(a)** 600 A/m, **(b)** 1592.35, **(c)** 1.2 T

26. **(a)** 2500 A/m, **(b)** 31415.92×10^{-7} T, **(c)** 31415.92×10^{-11} Wb

27. **(a)** 0.499×10^4 A/m, **(b)** 1.25 A

28. 16.66 A

29. 2.65×10^5 A/m

30. 795 A

31. 0.05 H

4 Principles of Electrical Machines

LEARNING OBJECTIVES

After reading this chapter, you will be able to:

- Understand the functions of electrical machines.
- Understand the types of magnetic materials and their behavior.
- Apply the right-hand grip rule and screw rule to determine the direction of magnetic field.
- Understand Fleming's left-hand rule to predetermine direction of force in current-carrying conductor.
- Understand Lenz's law.
- Understand Faraday's laws of electromagnetic reduction.
- Understand Fleming's right-hand rule for relative direction.
- Understand mutual inductor.
- Calculate induced e.m.f. given number of turns, time, inductance change of flux or change of current.
- Understand the factors affecting the inductance of an inductor.

4.1 Functions of Electrical Machines

The direct current (DC) machine, the synchronous machine and the induction machine are the major electromechanical conversion devices in industry. The merits of the squirrel cage induction machine are: lightness, simplicity, ruggedness, robustness, less initial cost, higher torque–inertia ratio, capability of much higher speeds, ease of maintenance, etc. The most important feature which declares it as a tough competitor to DC machines in the driving field is that its cost per Kilo Volt Ampere (KVA) is approximately 1/50 of its counterpart and it possesses higher suitability in hostile environment.

Unfortunately, induction machines suffer from the drawback that, in contrast to DC machines, their speed cannot be easily and effectively adjusted continuously over a wide range of operating conditions. Conversely, the synchronous machine has the merit of being operated under a wide range of power factors, both lagging and leading, and are much better suited for bulk power generation. In the induction motor, alternating current (AC) is applied to the stator and ACs are induced in the rotor by transformer action.

In the synchronous machine, DC is supplied to the rotor and AC flows in the stator. Conversely, a DC machine is a machine that is excited from DC sources only or that itself acts as a source of DC. It is a common practice in the industry to employ AC motors whenever they are inherently suitable or can be given appropriate characteristics by means of power electronic devices. Yet, the increasing complexity of industrial processes demands greater flexibility from electrical machines in terms of special characteristics and speed control. It is in this field that the DC machines, fed from the AC supply through rectifiers, are making their mark.

In this chapter, the concepts of applications of magnetic field in electrical machines are discussed.

4.2 Permanent Magnet

A permanent magnet is an object made from a material that is magnetized and creates its own persistent magnetic field. A day-to-day example is a refrigerator magnet used to hold on a refrigerator door. Materials that can be magnetized, which are also the ones that are strongly attracted to a magnet, are called ferromagnetic (or ferrimagnetic). These include iron, nickel, cobalt, some alloys of rare Earth metals and some naturally occurring minerals such as lodestone. Although ferromagnetic (and ferrimagnetic) materials are the only ones attracted to a magnet strongly enough to be commonly considered magnetic, all other substances respond weakly to a magnetic field.

4.2.1 Magnetic Materials

A permanent magnet is made up of cobalt, steel or tungsten steel. It is used in moving coil instruments, energy meters (as a brake magnet), loud speakers, microphones, etc.

4.2.2 Magnetic Field and Its Behavior

The imaginary magnetic lines shown in Fig. 1 that travel from north pole to south pole outside the magnet, and from south pole to north pole inside the magnet, are called magnetic lines of force. These magnetic lines do not intersect with each other, are parallel and appear in the same direction; however, they repel one another. The advantages of permanent magnets are as follows:

1. They are smaller in size.
2. They do not require an extra source (AC or DC source).

The disadvantage of the permanent magnet is that once a magnet is fabricated to a particular size, the magnitude and direction of the magnetic flux is fixed. Hence, control over the flux produced is lost.

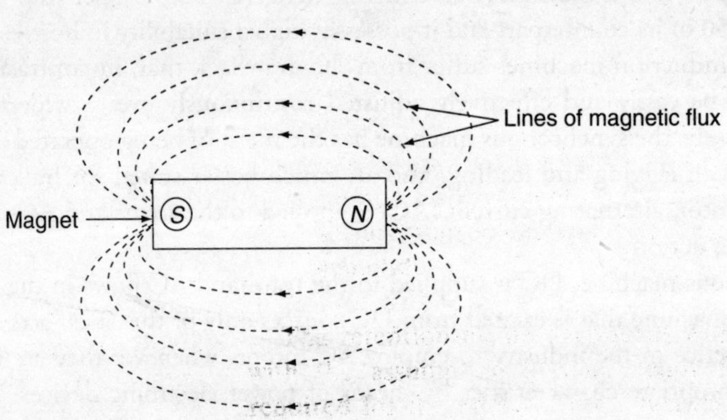

Figure 1 Magnetic field lines.

4.3 Electromagnet

An electromagnet is a type of magnet whose magnetic field is produced by the flow of electric current. The magnetic field disappears when the current ceases.

4.3.1 Magnetic Effect of Electric Current

Whenever a conductor carries a current, a magnetic field is set up as an integral part of the electrical phenomenon of current flow. The magnetic flux lines are considered to be circular in shape and are present in a plane perpendicular to the flow of current. The magnitude and direction of the magnetic field depends on the magnitude and direction of the current flow.

In Fig. 2 the solid circle denotes the cross-sectional view of a round conductor; when current is passed through this conductor, a circular-shaped magnetic field, shown by dotted lines, is set up in a plane perpendicular to the current flow. A circular field can have any one of the two circular directions, namely, clockwise or anticlockwise. In Fig. 2(a) the conductor carries a current in an inward direction which is shown by a cross (×) mark inside the conductor. Corresponding to this direction of current flow, the magnetic field takes a clockwise direction. The direction of the magnetic field can be changed by changing the direction of the current flow through the conductor. For an outward flow of current direction which is shown by a dot (·) mark inside the conductor in Fig. 2(b), the magnetic field takes an anticlockwise direction.

To remember the direction of magnetic field in accordance with the direction of current, any one of the rules explained in the following subsections can be used.

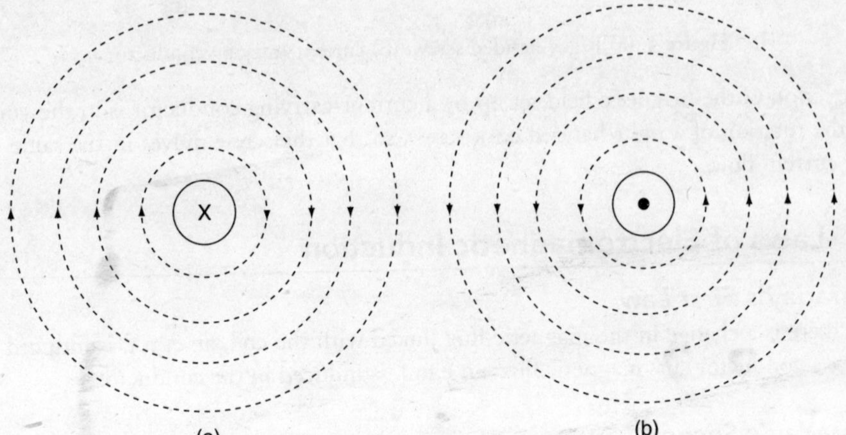

Figure 2 Magnetic field around current-carrying conductor: (a) current inward; (b) current outward.

4.3.2 Right-Hand Grip Rule

The right-hand rule, Fig. 3, simply shows how a current-carrying wire generates a magnetic field. If you point your thumb in the direction of the current, as shown, and let your fingers assume a curved position, the magnetic field circling around those wires flows in the direction in which your four fingers point.

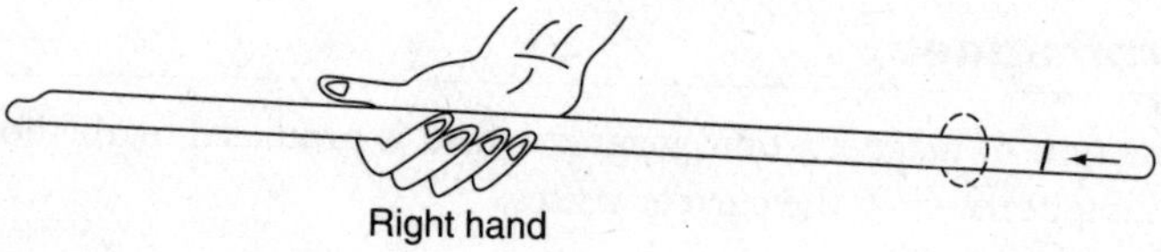

Figure 3 Right-hand rule.

In Fig. 3, the current flow is from right to left. The conductor is gripped by the right hand so that the thumb points toward left. The lines of magnetic field are delivered clockwise when viewed from the right end of the conductor as given by the direction of the four fingers.

4.3.3 Right-Handed Cork Screw Rule

In the conductor shown in Fig. 4, the current direction is shown from left to right. As seen from the left edge of the conductor, the current direction is inward and it sets up a magnetic field in the clockwise direction.

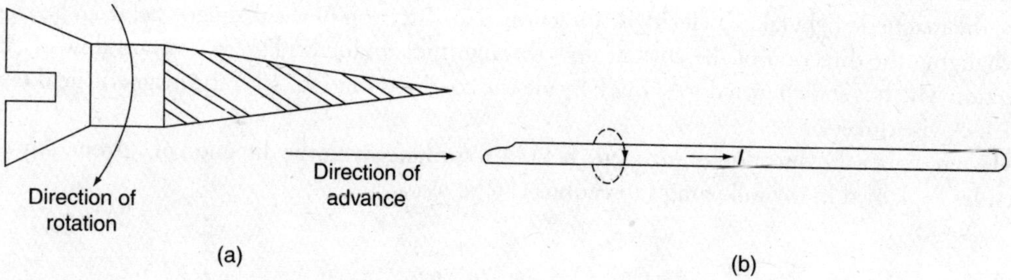

Figure 4 (a) Right-handed screw; (b) current-carrying conductor.

The direction of the magnetic field set up by a current-carrying conductor is in the same direction as that of the rotation of a right-handed cork screw, so that the screw moves in the same direction as that of the current flow.

4.4 Laws of Electromagnetic Induction

4.4.1 Faraday's First Law

Whenever there is a change in the magnetic flux linked with the coil, an e.m.f. is induced in the coil, or whenever a conductor cuts magnetic flux, an e.m.f. is induced in the conductor.

4.4.2 Faraday's Second Law

The magnitude of the induced e.m.f. is equal to the product of number of turns of the coil and the rate of change of flux. Consider a coil of "N" turns. Let initial value of flux be "ϕ_1" (in Wb) and final value of flux be "ϕ_2" (in Wb). Let "e" be the e.m.f. induced in volts and let "t" be the time during which the change in flux takes place. According to Faraday's laws of electromagnet induction,

$$e = N\frac{\phi_2 - \phi_1}{t}\ \text{Volt}$$

In differential form,

$$e = N\frac{d\phi}{dt} \text{ Volt}$$

4.4.3 Lenz's Law

Lenz's law gives the direction of induced e.m.f. in a conducting coil. The statement is: *The direction of the induced e.m.f. due to electromagnetic induction (EMI) is such that the current set up by it tends to oppose the change which causes the induced e.m.f.* Mathematically,

$$e = -N\frac{d\phi}{dt} \text{ Volt}$$

4.4.4 Statically Induced E.M.F.

According to Faraday's laws of electromagnetic induction, to have an e.m.f. induced, the flux linked with a conductive coil should change.

E.m.f. induced this way, that is without involving any mechanical movement, is called a statically induced e.m.f. This is also called transformer e.m.f. and forms the basic principle of transformer action. The statically induced e.m.f. can be classified into two categories, namely, self-induced e.m.f. and mutually induced e.m.f.

4.4.4.1 Self-Induced E.M.F.

The e.m.f. induced in a coil due to change in the value of its own flux is called self-induced e.m.f. It acts in such a way so as to oppose the very cause of its production, that is, to oppose the change in the value of e.m.f. It is denoted by "e_L" and is measured in volts. Mathematically

$$e_L = -N\frac{d\phi}{dt}$$

or

$$e_L = -L\frac{dI}{dt} \text{ Volt}$$

where "L" is self-inductance in H, "I" is the current in ampere and "ϕ" is the flux in webers.

4.4.4.2 Mutually Induced E.M.F.

The e.m.f. induced in a coil (first coil) due to change in the value of flux linked in a nearby coil (second coil) is called mutually induced e.m.f. Mathematically it is given by

$$e_{m1} = -M\frac{dI_2}{dt} = -N_1\frac{d\phi_2}{dt} \text{ Volt}$$

$$e_{m2} = -M\frac{dI_1}{dt} = -N_2\frac{d\phi_1}{dt} \text{ Volt}$$

where e_{m1} is the mutual e.m.f. induced in the first coil in volts; e_{m2} is the mutual e.m.f. induced in the second coil in volts; M is the mutual inductance in henry; I_1, ϕ_1 are the current and flux in the first coil in ampere and weber, respectively and I_2, ϕ_2 are the current and flux in the second coil in ampere and weber, respectively.

4.4.5 Self-Inductance (L)

Consider a coil of "N" turns carrying a current of "I" amperes. This current establishes a flux of "ϕ" weber as shown in Fig. 5. When the current in the coil is changed (increased or decreased), the flux linked with the coil also changes.

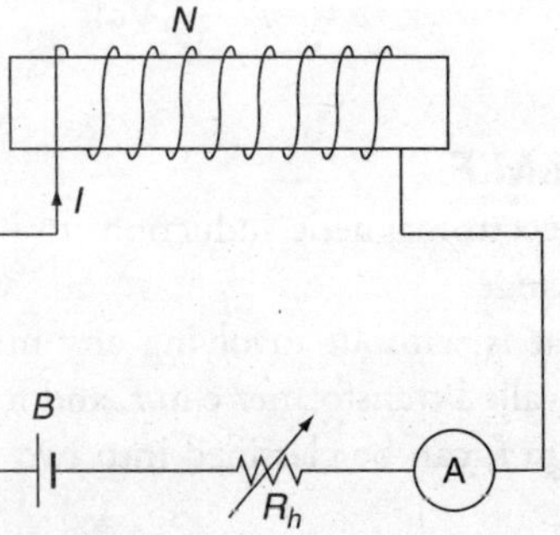

Figure 5 Schematic diagram of self-inductance.

It is also observed that when an effort is made to increase the value of current (flux) through the coil, it is opposed by the instantaneous induction of opposing e.m.f. (induced in the coil). Similarly, when an effort is made to decrease the value of current (flux) through the coil, it is also opposed by the instantaneous induction of opposing e.m.f.

This property of the coil by which it opposes change in value of current flux through it due to the production of self-induced e.m.f. is called self-inductance and is measured in the terms of coefficient of self-inductance (L). Its unit is henry (H). This property is analogous to electrical inertia.

Definition of "L": A coil is said to have 1 H of self-inductance if a current of 1 A flowing through it produces 1 Wb turn flux linkage. Thus,

$$L = \frac{N\phi}{I}\ \text{H}$$

We know that

$$\phi = \frac{NI}{S}$$

and so

$$L = \frac{N}{I}\frac{NI}{S} = \frac{N^2}{S}$$

Therefore

$$L = \frac{N^2}{S}$$

where S is the reluctance given by

$$S = \frac{l}{\mu_0 \mu_r A} \text{ AT/Wb}$$

We now find out the expression for self-induced e.m.f. (e_L). We have,

$$L = \frac{N\phi}{I}$$

$$\Rightarrow LI = N\phi$$

$$-LI = -N\phi$$

Differentiating the above equation on both the sides with respect to "t," we get

$$-L\frac{dI}{dt} = -N\frac{d\phi}{dt}$$

But according to Faraday's law,

$$\text{Induced e.m.f.} = -N\frac{d\phi}{dt}$$

Therefore, the self-induced e.m.f. is

$$e_L = -L\frac{dI}{dt} \text{ Volt}$$

$$L = \frac{-e_L}{dI/dt} \text{ H}$$

Hence we have the following:

If $e_L = 1$ Volt and $dI/dt = 1$ A/s, then $L = 1$ H.

Thus a coil is said to have a self-inductance of 1 H if current through it changes at the rate of 1 A/s inducing an e.m.f. of 1 V in it.

4.4.6 Mutual Inductance

Consider two coils "A" and "B" having N_1 and N_2 number of coils, respectively, placed close to each other as shown in Fig. 6. Let I_1 be the current in amperes carried by a coil establishing a flux of "ϕ_1" webers in it.

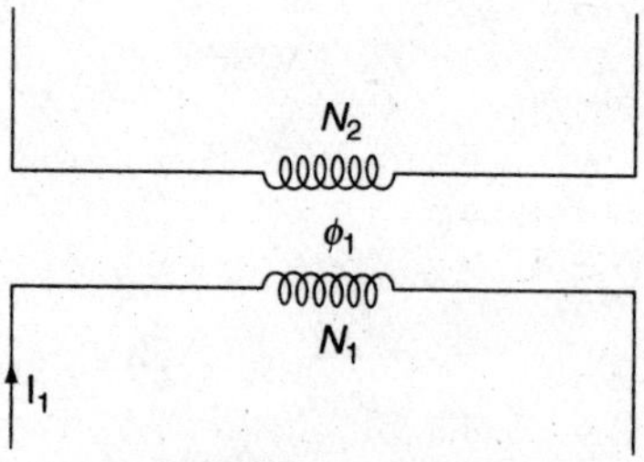

Figure 6 Schematic diagram of mutual inductance.

Now since coil "B" is closer to coil "A" some of the flux established by coil "A" will be linked with coil "B." Thus, a change in current or flux in coil "A" will cause a change in the flux linked with the coil "B." This change in the value of flux current in coil "A" will not only produce an e.m.f. in coil "A" itself but also produce an e.m.f. in coil "B." This e.m.f. induced in coil "B" is called mutually induced e.m.f.

Thus, *mutual inductance* (M) is the ability of the first coil to produce an e.m.f. in the nearby coil (second coil) due to a change of current flux in the first coil. The reciprocal action also holds good. That is, whenever the current of flux in the second coil changes, an e.m.f. is induced in the first coil (nearby coil). This ability is measured in terms of coefficient of mutual inductance (M). Its unit is henry.

The coefficient of mutual inductance between two coils is defined as the weber turns in one coil due to a current of one ampere in the other coil:

$$M = \frac{N_2\phi_1}{I_1} \tag{4.1}$$

If $N_2\phi_1 = 1$ Wb turn and $I_1 = 1$ A, then $M = 1$ H. Hence two coils are said to have a mutual inductance of 1 H, if 1 A current flowing in the first coil produces 1 Wb turn flux linkage in the second coil.

We know that

$$\phi = \frac{NI}{S} \Rightarrow \phi_1 = \frac{N_1 I_1}{S}$$

Substituting in Eq. (4.1) we get

$$M = \frac{N_2}{I_1}\,\frac{N_1 I_1}{S}$$
$$= \frac{N_1 N_2}{S}\ \text{H}$$

where

$$S = \text{Reluctance} = \frac{l}{\mu_0 \mu_r A}\ \text{AT/Wb}$$

Expression for mutually induced e.m.f. (e_m):
We have

$$M = \frac{N_2\phi_1}{I_1}$$
$$MI_1 = N_2\phi_1$$
$$-MI_1 = -N_2\phi_1$$

Differentiating the above equation on both sides with respect to "t," we get

$$-M\frac{dI_1}{dt} = -N_2\frac{d\phi_1}{dt}$$

But

$$-N_2\frac{d\phi_1}{dt} = \text{Mutually induced e.m.f. in the second coil } (e_{m2})$$

Therefore,

$$e_{m2} = -M\frac{dI_1}{dt}$$

Hence we have:

$$\text{If } \frac{dI_1}{dt} = 1 \text{ A/s and } e_{m2} = 1 \text{ V, then } M = 1 \text{ H.}$$

Therefore, two coils are said to have a mutual inductance of 1 H, if current through one coil changes at the rate of 1 A/s inducing an e.m.f. of 1 V in the other.

Again from Eq. (4.1) we have

$$M = \frac{N_2\phi_1}{I_1}$$

Here the assumption is that the entire flux "ϕ_1" produced by coil "A" is linked with coil "B." If it is not the case, then the above equation can be modified as

$$M = \frac{N_2K_1\phi_1}{I_1}$$

where "K_1" is percentage of "ϕ_1," linked with coil "B." Similarly we can write

$$M = \frac{N_1\phi_2}{I_2} \quad \text{or} \quad M = \frac{N_1K_2\phi_2}{I_2} \text{ H}$$

$$e_{m1} = -M\frac{dI_2}{dt} \text{ Volt}$$

where "I_2" is the current carried by coil "B" establishing a flux of "ϕ_2" Wb in it and "e_{m1}" is the mutually induced e.m.f. in coil "A" due to the change of current flux in coil "B."

4.4.7 Coefficient of Coupling

Consider two coils wound around a core as shown in Fig. 7. Let ϕ_1 be the flux set up by the coil due to current I_1 in it. Then the self-inductance of coil 1 is

$$L_1 = N_1 \frac{\phi_1}{I_1} = \frac{N_1^2}{S}$$

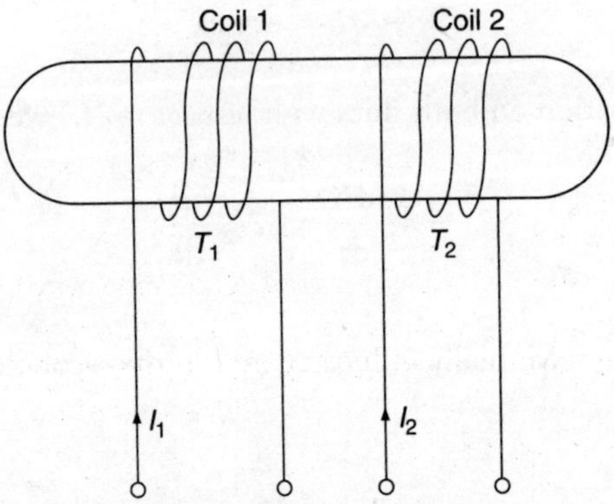

Figure 7 Mutual coupling.

Out of ϕ_1 flux set up, allowing for leakage, let the flux reached and linked by coil 2 be $K_1\phi_1$ where K_1 is less than unity. Then the mutual inductance of coil 2 is

$$M_{21} = N_2 \frac{K_1\phi_1}{I_1} = \frac{N_2 N_1 R_1}{S} \tag{4.2}$$

Let ϕ_2 be the flux set up by coil 2 due to current I_2 in it. Then, self-inductance of coil 2 is

$$L_2 = N_2 \frac{\phi_2}{I_2} = \frac{N_2^2}{S}$$

in the above equation ϕ_2 flux set up, allowing for leakage, let the flux reached and linked by coil 1 be $K_2\phi_2$, where K_2 is less than unity. Then the mutual inductance of coil 1 is

$$M_{12} = N_1 \frac{K_2\phi_2}{I_2} = \frac{N_1 N_2 K_2}{S} \tag{4.3}$$

For a bilateral magnetic circuit,

$$M_{21} = M_{12} = M \tag{4.4}$$

Multiplying Eqs. (4.2) and (4.3) and using Eq. (4.4) we have

$$M_{21}\, M_{12} = M^2 = K_1 K_2 \frac{N_1^2 N_2^2}{S^2} \tag{4.5}$$

Substituting the value of L_1 and L_2in the above Eq. (4.7), we have

$$K_1 = K_2 = K$$

$$\Rightarrow M^2 = K^2 L_1 L_2$$

Thus, coefficient of coupling is

$$K = \frac{M}{\sqrt{L_1 L_2}}$$

For a tightly coupled circuit where there is no leakage flux, which is only theoretically possible, K can take its maximum value of $K = 1$. Therefore, theoretically maximum possible mutual inductance is

$$M = \sqrt{L_1 L_2}$$

The ratio of the actual value of mutual inductance to the theoretical maximum value is called *coefficient of mutual coupling.*

4.4.8 Energy Stored in the Magnetic Field

Consider a coil of inductance "L" henry. Let the current in the coil be raised from 0 to I ampere. This rise in the value of current establishes a self-induced e.m.f. in the coil which opposes the very cause of production of it, namely, the rise in value of current. Any increase in the value of current in the circuit can be achieved only by doing certain work against the self-induced e.m.f. This work done provided through source is stored in the magnetic field.

The self-induced e.m.f. in the coil is given by

$$e = L\frac{\mathrm{d}i}{\mathrm{d}t}$$

We know that

$$\text{Power} = e \times i$$

Also,

$$\text{Energy} = \text{Power} \times \text{Time}$$

Therefore, energy supplied during time dt is

$$ei\mathrm{d}t = Li\frac{\mathrm{d}i}{\mathrm{d}t}\mathrm{d}t$$

$$= Li\,\mathrm{d}i$$

The total energy supplied to the system to establish a current of "I" ampere from "0" value is given by

$$\int_0^I Li\mathrm{d}i = L\left[\frac{i^2}{2}\right]_0^I$$

Therefore, energy stored in a magnetic field is

$$\frac{1}{2}LI^2 \text{ Joules}$$

We know that

$$L = \frac{N^2}{S}$$

where "L" is the self-inductance of the coil in henry, N is the number of turns of the coil and S is the reluctance given by

$$S = \frac{l}{\mu_0 \mu_r A} \text{ AT/Wb}$$

Also,

$$H = \frac{NI}{l} \text{ AT/m}$$

where H is the magnetic field strength. Now,

$$E = \frac{1}{2}LI^2$$

$$= \frac{1}{2} \times \frac{N^2}{l / \mu_0 \mu_r A} \times I^2$$

Therefore,

$$E = \frac{1}{2}\frac{N^2 I^2 \mu_0 \mu_r A}{l}$$

$$= \frac{1}{2}H^2 \mu_0 \mu_r A l \text{ Joules}$$

$$= \frac{1}{2}H^2 \mu_0 \mu_r V$$

where "V" is the volume. Hence, stored energy per m^3 is

$$\frac{1}{2}H^2 \mu_0 \mu_r \text{ Joules}$$

Now

$$B = \mu_0 \mu_r H \Rightarrow H = \frac{B}{\mu_0 \mu_r}$$

Therefore

$$\text{Stored energy/m}^3 = \frac{1}{2}\frac{B^2}{\mu_0^2\mu_r^2}\mu_0\mu_r$$

$$= \frac{1}{2}\frac{B^2}{\mu_0\mu_r}$$

For air $\mu_r = 1$ and so

$$\text{Stored energy/m}^3 = \frac{1}{2}\frac{B^2}{\mu_0}$$

4.4.9 Lifting Power of Magnet

Consider the arrangement of two poles as shown in Fig. 8. Let "P" be the force of attraction between two poles in newtons. Let "A" be area of the magnet. Keeping "S" pole fixed, let the "N" pole be pulled upward through a distance "dx" against the force of attraction. The work done is given by

$$P \times \mathrm{d}x \text{ Joules}$$

This work done is stored in the magnetic field:

$$\text{Stored energy/m}^3 = \frac{1}{2}\frac{B^2}{\mu_0}$$

$$\text{Stored energy} = \frac{1}{2}\frac{B^2}{\mu_0}A\mathrm{d}x$$

$$P\,\mathrm{d}x = \frac{B^2}{2\mu_0}A\,\mathrm{d}x$$

$$P = \frac{B^2 A}{2\mu_0}$$

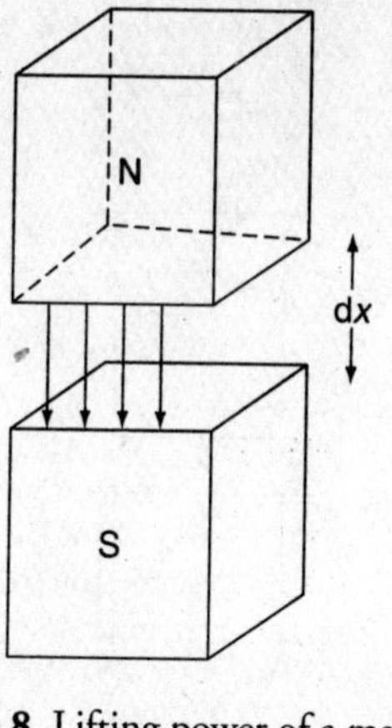

Figure 8 Lifting power of a magnet.

4.4.10 Dynamically Induced E.M.F.

The e.m.f. induced in a coil is said to be a dynamically induced e.m.f. when change in the magnetic flux linked by the coil is caused by the relative mechanical movement between the coil and field system. This forms the principle of all generator actions.

In the experimental setup shown in Fig. 9 the coil was held stationary and mechanical movement was given to the field system. Alternatively, the field system can be held stationary and the coil or conductor can be given a mechanical movement as shown in Fig. 10. Here a stationary uniform magnetic field of average flux density B is provided by two magnetic poles: N – north pole at the top and S – south pole at the bottom. The direction of the magnetic field is from north pole to south pole.

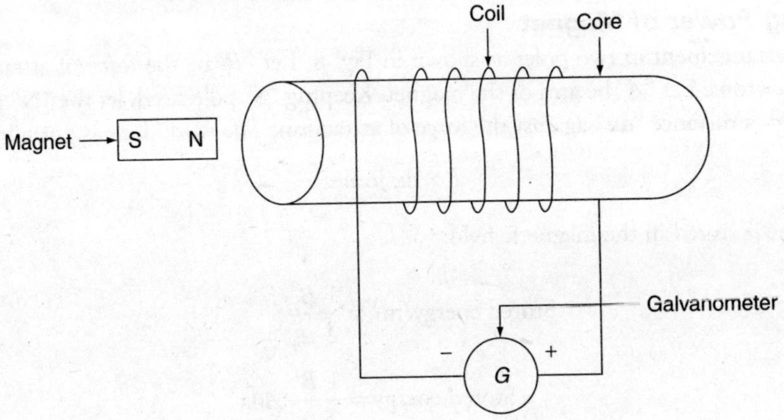

Figure 9 Dynamically induced e.m.f.

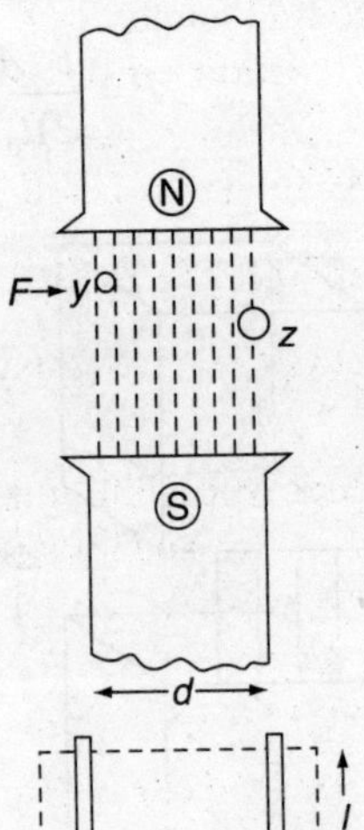

Figure 10 Generator action.

A conductor of active length l meter is initially placed in position Y. The active length mentioned here differs from its physical length as it is the only portion of the conductor which is linked by the magnetic field. Now, a mechanical force F is applied to the conductor such that the conductor is moved from left to right with a velocity of v m/s in a direction perpendicular to the magnetic field. Due to the force, the conductor gets moved from position y to position z, covering a distance of d meter in t seconds.

During the movement, as the flux linked with the conductor changes according to Faraday's law, an e.m.f. is induced in the conductor. Faraday's second law gives the quantitative value of the e.m.f. induced as

$$e = \frac{Td\phi}{dt}$$

Here $T = 1$ for a single conductor. Now

$$d\phi = BA = B \times l \times d$$

$$dt = t$$

$$e = \frac{1 \times Bld}{t}$$

$$e = Blv \text{ Volt} \tag{4.6}$$

Equation (4.6) holds good only if the conductor moves at right angles (90°) to the magnetic field. In case the conductor moves at an angle 0° with respect to the magnetic field, then the equation for the dynamically induced e.m.f. is

$$e = Blv \sin\theta \text{ Volt} \tag{4.7}$$

Lenz's law gives the direction of e.m.f. and hence the current. In the setup shown in Fig. 9, when the conductor is moved from left to right, the current will be in an inward direction. So the current-carrying conductor sets up another magnetic field around the conductor in the clockwise direction. To the right-hand side of the conductor the direction of the field set up by the current is in the same direction as that of the uniform magnetic field, giving a resistive force to the movement.

In this set-up, the input energy is mechanical energy, a force F is given to the conductor causing it to travel a distance d. The output energy is electrical energy available with a voltage and current for a period of t seconds. That is, mechanical energy is converted into electrical energy which forms the basic principle of all types of generator actions.

4.4.11 Fleming's Right-Hand Rule

The thumb, the forefinger and the middle finger of the right hand are held so that the three fingers are mutually perpendicular to each other as shown in Fig. 11.

The forefinger is pointed toward the direction of the main field, the thumb toward the direction of movement of the conductor and the middle finger gives the direction of the current set up by the e.m.f. induced in the conductor.

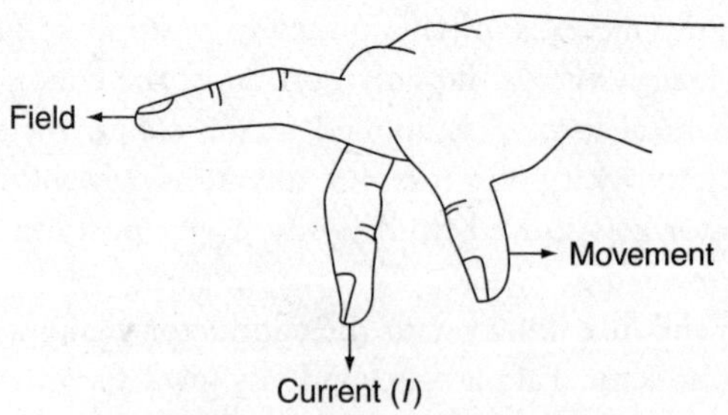

Figure 11 Fleming's right-hand rule.

4.5 Electrochemical Energy Conversion

All DC and AC synchronous motors, which convert electrical energy into mechanical energy, work on the following principle:.

When a current-carrying conductor is placed in a uniform magnetic field, a force is developed in the conductor. Each type of motor differs from other types only by the way in which the current is passed though the conductor and the way in which the magnetic field is set up.

4.5.1 Force on Current-Carrying Conductor

In Fig. 12(a) a uniform magnetic field is provided by two magnetic poles. North pole is at the top and south pole at the bottom. The lines of the magnetic field flow from top to bottom, that is, from north pole to south pole. When a current-carrying conductor is placed in the space between the poles, as indicated in Fig. 2 (b), the magnetic field set up by the inward current is in a clockwise direction.

Figure 12(c) shows the resultant of the two magnetic fields. To the right-hand side of the conductor, as both fields are in the same direction, more lines of field are present that too bent around the conductor, whereas on the left-hand side of the conductor, as both fields are in opposite directions,

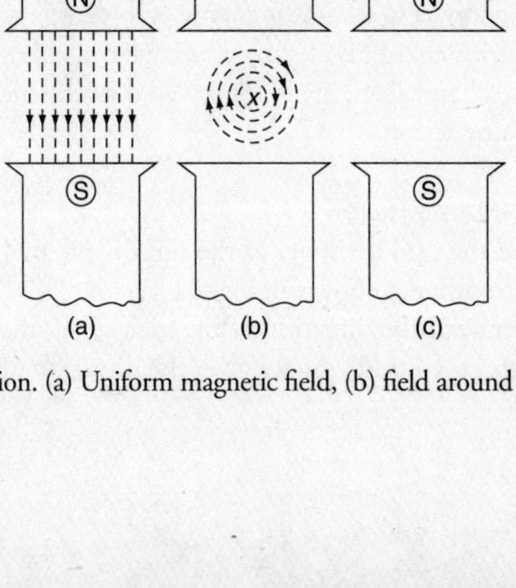

Figure 12 Principle of motor action. (a) Uniform magnetic field, (b) field around conductor and (c) resultant field.

the field strength is less. Hence the resultant field exerts a force on the conductor from right to left. It the conductor is allowed to move due to the force, it will move from right to left.

The amount of force developed by a current-carrying conductor is given by

$$F = Blv \tag{4.8}$$

where F is the force in newton, B is the average flux density in tesla, I is the current through the conductor in ampere and l is the active length of the conductor in meter. The force is developed in a direction perpendicular to both the current and the main field.

Here, the input energy is electrical energy, a current I is given to the conductor by a source voltage E over a specified time t. The input electrical energy over a specified time t is equal to output mechanical energy over a distance d. Neglecting losses, we have

Input electrical energy over a specified time t = Output mechanical energy over a distance d

$$EIt = Fd$$

$$(Blv)It = (BIl)d$$

$$BIld = BIld$$

A single rotating machine is capable of operating as a generator at one time and as motor at another time.

4.5.2 Fleming's Left-Hand Rule

The direction of force developed on a current-carrying conductor can more easily be obtained by this rule. This rule is similar to Fleming's right-hand rule explained in Section 4.4.11 except that instead of the right hand, the left hand is used as shown in Fig. 13.

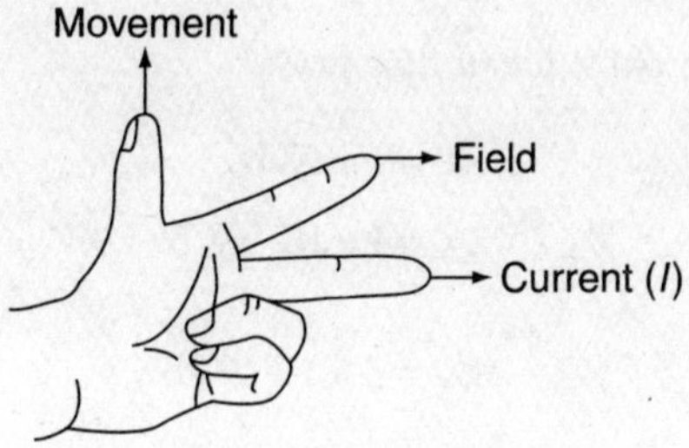

Figure 13 Fleming's left-hand rule.

The thumb, the forefinger and the middle finger of the left hand are held at right angles to each other. If the forefinger is pointed along the direction of the main field and the middle finger along the direction of the current in the conductor, then the thumb gives the direction of force or movement, if allowed, experienced by the conductor.

Summary

In this chapter we have discussed about principles of electric machines related to permanent magnetic field. Some of the definitions are explained below:

1. The total magnetic flux linkage λ in a linear inductor made of a coil is proportional to the current passing through it.
2. A coil containing N turns with magnetic flux ϕ linking each turn has total magnetic flux linkage $\lambda = N\phi$.
3. The sign on a voltage of mutual inductance can be determined if the winding sense is shown in the circuit.
4. A transformer is a device for introducing mutual coupling between two or more electric circuits.
5. An ideal transformer is a hypothetical transformer in which there are no losses and no leakage flux.
6. An auto transformer is an electrically continuous linking, with one or more taps, on a magnetic core.
7. A load Z_2 connected to secondary part of a transformer contributes to its input impedance. This contribution is called reflected impedance.

Solved Problems

Problem 1

Calculate the e.m.f. induced in a coil of 200 turns, when the flux linked with it changes from 1 mWb to 3 mWb in 0.1 s.

Solution

Give no. of turns $N = 200$. Initial value of flux,

$$\phi_1 = 1 \text{ mWb}$$
$$= 1\times 10^{-3} \text{ Wb}$$

Final value of flux,

$$\phi_2 = 3\times 10^{-3} \text{ Wb}$$

Therefore,

$$d\phi = \phi_2 - \phi_1 = 2\times 10^{-3} \text{ Wb}$$

Time of the change = 0.1 s. Therefore,

$$|e| = \left|\frac{-Nd\phi}{dt}\right| = \left|\frac{-200\times 2\times 10^{-3}}{0.1}\right| = 4 \text{ V}$$

Problem 2

Calculate the e.m.f. induced in a 2 m long straight conductor when it is moved at a velocity of 4 m/s at right angles to a uniform magnetic field of flux density of 1.2 Wb/m².

Solution
Given that

$$\text{Flux density, } B = 1.2 \text{ Wb/m}^2$$

$$\text{Velocity of the conductor, } v = 4 \text{ m/s}$$

$$\text{Length of the conductor, } l = 2 \text{ m}$$

Angle between conductor and lines of force, $\phi = 90°$. Therefore, induced e.m.f. is

$$\begin{aligned} e &= Blv \sin\theta \\ &= 4 \times 2 \times 1.2 \sin 90° \\ &= 9.6 \text{ V} \end{aligned}$$

Problem 3

A coil of 1500 turns carries a current of 10 A, establishing a flux of 0.5 mWb. Find the inductance of the coil.

Solution
Given that

$$N = 1500$$

$$I = 10 \text{ A}$$

$$\phi = 0.5 \times 10^{-3} \text{ Wb}$$

Now the inductance is given by

$$\begin{aligned} L &= \frac{N\phi}{I} \\ &= \frac{1500 \times 0.5 \times 10^{-3}}{10} \\ &= 0.075 \text{ H} \end{aligned}$$

Problem 4

An air-cored solenoid has 400 turns, its length is 30 cm and it has a cross-sectional area of 5 cm². Calculate its self-inductance.

Solution

For air-cored solenoid $\mu_r = 1$. Also it is given that

$$N = 400$$
$$l = 30\,\text{cm} = 0.3\,\text{m}$$
$$A = 5\,\text{cm}^2 = 5\times10^{-4}\,\text{m}^2$$

Self-inductance is

$$L = \frac{N^2}{S}$$

where

$$S = \frac{l}{\mu_0\mu_r A} = \frac{30\times10^{-2}}{4\pi\times10^{-7}\times5\times10^{-4}}\ \text{AT/m} = 4.77\times10^{8}\ \text{AT/m}$$

Therefore

$$L = \frac{(400)^2}{4.77\times10^{8}} = 3.35\times10^{-4}\,\text{H}$$

Problem 5

An air-cored toroidal ring with mean diameter of 25 cm and a circular cross-sectional diameter of 4 cm is wound uniformly with 500 turns of copper wire of 1 mm diameter. Calculate the inductance of the coil. Also calculate the e.m.f. induced in it when the current of 5 A is reversed in 50 ms.

Solution

For air-cored toroidal ring, $\mu_r = 1$. Given that

$$\text{Mean diameter, } s = 25\times10^{-2}\,\text{m}$$

$$\text{Length, } l = \pi s = \pi\times25\times10^{-2} = 0.785\,\text{m}$$

$$\text{No. of turns, } N = 500$$

$$\text{Cross-sectional diameter, } d = 4\times10^{-2}$$

The cross-sectional area is

$$A = \pi(d^2/4)$$
$$= \frac{\pi \times 4 \times 10^{-2} \times 4 \times 10^{-2}}{4}$$
$$= 1.256 \times 10^{-3}\,\text{m}^2$$

Self-inductance is given by

$$L = \frac{N^2}{S}$$

where

$$S = \frac{l}{\mu_0 \mu_r A}$$
$$= \frac{0.785}{4\pi \times 10^{-7} \times 1 \times 1.256 \times 10^{-3}}$$
$$= 49.71 \times 10^7\ \text{AT/Wb}$$

Therefore,

$$L = \frac{(500)^2}{49.71 \times 10^7}\ \text{H}$$
$$= 5.029 \times 10^{-4}\ \text{H}$$

Now

$$\text{d}i = 5 - (-5) = 10\,\text{A}$$

$$\text{d}t = 50 \times 10^{-3}\,\text{s}$$

Hence

$$\frac{\text{d}i}{\text{d}t} = \frac{10}{50 \times 10^{-3}}$$
$$= 200\,\text{A/s}$$

Therefore, the induced e.m.f. is

$$e_\text{L} = \left| -L \frac{\text{d}i}{\text{d}t} \right|$$
$$= 5.29 \times 10^{-4} \times 200$$
$$= 0.1005\ \text{V}$$

Problem 6

Repeat Problem 5 when cross-section is square of side 2 cm.

Solution

We have

$$A = 2\times 2\times 10^{-4} = 4\times 10^{-4}\ \text{m}^2$$

Now

$$S = \frac{0.785}{4\pi\times 10^{-7}\times 1\times 4\times 10^{-4}}$$
$$= 1.56\times 10^{9}\ \text{AT/m}$$

Therefore

$$L = \frac{(500)^2}{1.56\times 10^{9}} = 1.6\times 10^{-4}$$

The induces e.m.f. is given by

$$e_L = 1.6\times^{-4}\times 200 = 0.032\ \text{V}$$

Problem 7

Two coils having 50 and 400 turns respectively are wound side by side on a closed iron circuit of cross-sectional area 150 cm² and mean length 200 cm.

(i) Calculate the mutual inductance between the two coils. Take relative permeability of iron as 2500.
(ii) Determine the e.m.f induced in the second coil, if the first coil current changes from 12 to –12 A in 0.03 s.

Solution

Given that

$$N_1 = 50$$
$$N_2 = 400$$
$$A = 150\times 10^{-4}\ \text{m}^2$$
$$l = 200\times 10^{-2}\ \text{m}$$

(i) Relative permeability of iron, $\mu_r = 2500$. Therefore mutual inductance between the two coils is given by

$$M = \frac{N_1 N_2}{S}$$

where

$$S = \frac{l}{\mu_0 \mu_r A}$$

$$= \frac{200 \times 10^{-2}}{4\pi^{-7} \times 10^{-7} \times 2500 \times 150 \times 10^{-4}} = 42441.3$$

Therefore,

$$M = \frac{50 \times 400}{42441.32} = 0.4712\,\text{H}$$

(ii) The e.m.f. induced in the second coil:

$$\text{d}i_1 = 12 - (-12) = 24\ \text{A}$$
$$\text{d}t = 0.03\ \text{s}$$

So

$$\frac{\text{d}i_1}{\text{d}t} = \frac{24}{0.03} = 800\ \text{A/s}$$

Therefore, the mutual induced e.m.f. in the second coil is

$$e_{\text{m}2} = \frac{M\text{d}i_1}{\text{d}t}$$
$$= 0.4712 \times 800$$
$$= 376.96\ \text{V}$$

Problem 8

The inductance of a coil carrying a current of 2 A is 0.5 H. Find the energy stored in it.

Solution

Given that

Inductance of the coil, $L = 0.5$ H

Current, $I = 2$ A

Therefore,

$$\text{Energy stored} = \frac{1}{2} L I^2$$
$$= \frac{1}{2} \times 0.5 \times 2 \times 2$$
$$= 1\ \text{J}$$

Problem 9

A horseshoe magnet has two poles each of area 1.5 cm². Determine the pull between the poles and keeper when the flux density of the contact surface is 1.2 Wb/m².

Solution

Given that

$$\text{Area of each pole, } A = 1.5\ \text{cm}^2 = 15 \times 10^{-4}\ \text{m}^2$$

$$\text{Magnetic flux density, } B = 1.2\ \text{Wb/m}^2$$

$$\text{Relative permeability, } \mu_r = 1$$

Therefore

$$\text{Force of attraction on each pole} = \frac{B^2 A}{2\mu_0}$$

$$\text{Total force on both the poles} = \frac{B^2 A}{2\mu_0}$$

$$= \frac{(1.2)^2 \times 15 \times 10^{-4}}{4\pi \times 10^{-7}}$$

$$= 1718.8\ \text{N}$$

Problem 10

A coil of 500 turns is uniformly wound on a ring of non-magnetic material of mean diameter 25 cm. The cross-sectional area of the coil is 5 cm². A current of 3 A is flowing through the coil. Find the inductance of the coil.

Solution

Given that $N = 500$, $D = 25$ cm $= 0.25$ m, $A = 5 \times 10^{-4}$ m², $I = 3$ A. The length l is given by

$$l = \pi D$$

$$= \pi \times 0.25$$

$$= 0.785\ \text{m}$$

The reluctance S is given by

$$S = \frac{l}{\mu_0 \mu_r A}$$

$$= \frac{0.785}{4\pi \times 10^{-7} \times 1 \times 5 \times 10^{-4}}$$

$$= 12.48 \times 10^8\ \text{A T/Wb}$$

Therefore,

$$L = \frac{N^2}{S}$$

$$= \frac{(500)^2}{12.48 \times 10^8}$$

$$= 2 \times 10^4 \text{ H}$$

Problem 11

An air cored solenoid 50 cm long and 5 cm in radius is wound with 100 turns. Determine (i) the inductance and (ii) the energy stored in the magnetic field when a current of 5 A flows through it.

Solution

We have $\mu_r = 1$, $N = 1000$, $l = 50$ cm $= 0.5$ m, $r = 0.05$ m. Now area is given by

$$A = \pi r^2 = \frac{22}{7} \times (0.05)^2 = 7.86 \times 10^{-3} \text{ m}^2$$

Reluctance equals

$$S = \frac{l}{\mu_0 \mu_r A} = \frac{0.5}{4\pi \times 10^{-7} \times 7.86 \times 10^{-3}}$$

$$= 5.06 \times 10^7 \text{ AT/m}$$

(i) Inductance is

$$L = \frac{N^2}{S} = \frac{(1000)^2}{5.06 \times 10^7}$$

$$= 19.7 \times 10^{-3} \text{ H}$$

(ii) The energy stored is

$$E = \frac{1}{2} L I^2$$

$$= \frac{1}{2} \times 19.7 \times 10^{-3} \times 5^2$$

$$= 0.2467 \text{ J}$$

Problem 12

Two coils A and B have 600 and 1000 turns respectively. If a current of 4 A in coil A establishes a flux of 5 mWb in coil A and 2.5 mWb in coil B, determine (i) self-inductance of coil A and (ii) mutual inductance.

Solution

Given that $N_1 = 600$, $N_2 = 1000$, $I_1 = 4$ A, $\phi_1 = 5$ mWb $= 5\times10^{-3}$ Wb, $\phi_2 = 2.5$ mWb $= 2.5\times10^{-3}$.

(i) Self-inductance of coil A

$$L_1 = \frac{N_1\phi_1}{I_1} = \frac{600\times5\times10^{-3}}{4} = 0.75\,\text{H}$$

(ii) Mutual inductance is given by

$$M = \frac{N_2\phi_1}{I_1} = \frac{1000\times5\times10^{-3}}{4} = 1.25\ \text{H}$$

Problem 13

Two coils have mutual inductance of 0.4 H. If the current in one coil is reduced from 5 A to 1 A in 0.01 s, then find the e.m.f. induced in the second coil. Assume that the whole flux set up by the first coil is linked with the second coil.

Solution

We are given that $M = 0.4$, initial value of current = 5 A and final value of current = 1 A. Hence,

$$di = 4\ \text{A}$$

$$dt = 0.01$$

So,

$$\frac{di}{dt} = \frac{4}{0.01} = 400\ \text{A/s}$$

Therefore, induced e.m.f. is

$$e_{m1} = M\frac{di}{dt} = 0.4\times400 = 160\,\text{V}$$

Problem 14

The length of an air cored coil is 25 cm and the cross-sectional area is 4 cm². The coil is wound by 65 turns. Determine the inductance of the coil.

Solution

The value of the inductance can be calculated as

$$L = \frac{N^2 \mu A}{l}$$

$$= \frac{65^2 \times 1 \times 4\pi \times 10^{-7} \times 4 \times 10^{-4}}{0.25}$$

$$= 8.5\ \mu\text{H}$$

Problem 15

The active length of each conductor placed in a DC generator armature is 0.4 m. The armature is rotated at a speed of 1800 rpm in a magnetic field of average flux density 1.2 T. The diameter of the armature is 0.5 m. Find the e.m.f. induced in the conductor when it cuts the magnetic field at right angles.

Given data: $l = 0.4$ m; $N = 1000$ rpm; $B = 1.2$ T; $D = 0.5$ m; $\phi = 90°$.

Solution

We have

$$e = Blv \sin\phi = Blv$$

where

$$v = \frac{\pi D N}{60}$$

$$= \frac{\pi \times 0.5 \times 1800}{60} = 47.12\ \text{m/s}$$

Therefore

$$E = 1.2 \times 0.4 \times 47.12$$

$$= 22.62\ \text{V}$$

Problem 16

The inductance of a coil carrying a current of 2 A is 0.5 H. Find the energy stored in it.

Solution

Inductance of the coil $L = 0.5$ H. Current $I = 2$ A.

Energy stored is given by

$$\frac{1}{2}LI^2 = \frac{1}{2} \times 0.5 \times 2 \times 2 = 1\ \text{J}$$

Problem 17

A circular deflecting coil having 947 turns has a mean radius of 145 mm and carries a current of 15 mA (Fig. 14). Find the flux density

(i) At the center of the coil.

(ii) At a point 19 mm away, along the axis of the coil.

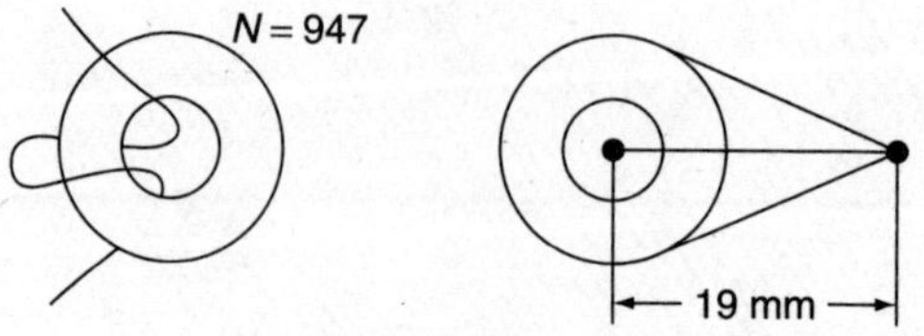

Figure 14 Problem 17.

Solution

Given that $R = 14.5\,\text{mm} = 0.0145\,\text{m}$, $I = 15\,\text{mA} = 0.015\,\text{A}$, $N = 947$.

(i) Flux density at the center of the coil:

$$B = \frac{\mu_0 NI}{2R} = \frac{4\pi\times10^{-7}\times947\times0.015\times(0.0145)^2}{[2(0.0145)^2+(0.019)^2]^{3/2}} = 0.13743\ \text{mWb/m}^2$$

(ii) Flux density at a point 19 mm away along the axis:

$$B = \frac{\mu NIR^2}{2(R^2+D^2)^{3/2}} = \frac{4\times\pi\times10^{-7}\times947\times0.015\times0.145^2}{2(0.9145^2+0.019^2)^{3/2}} = 1.876\ \text{mWb/m}^2$$

Problem 18

Find the force per meter length between two conductors separated by a distance of 17.5 cm (Fig. 15), carrying currents of

(i) $I_1 = I_2 = 100$ A, both in the same directions.

(ii) $I_1 = I_2 = 100$ A and $-I_2$ in the opposite direction of I_1.

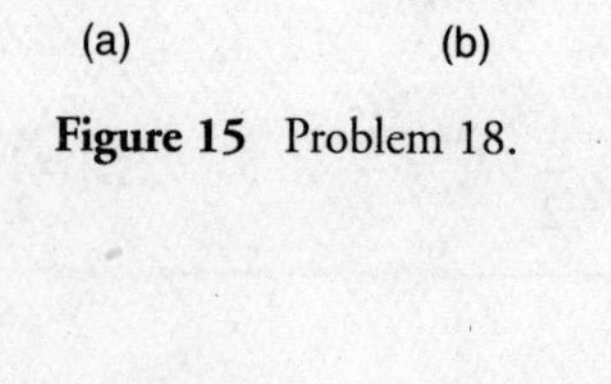

Figure 15 Problem 18.

Solution

Force between parallel conductors:

$$F = \frac{\mu_0 I_1 I_2}{2\pi D} N$$

$$= \frac{4\pi \times 10^{-7} \times 100 \times 100}{2\pi \times 0.175}$$

$$= 13.13 \times 10^{-3} = 1.235 \times 10^{-3}\,\text{kg/mm}$$

(i) The force is attracting 0.01313 N/m (or) 0.001235 kgm/m.

(ii) The force is repulsing 0.01212 N/m (or) 0.001235 kgm/m.

Problem 19

Determine the torque on a 700 turn coil, placed horizontally between the poles of a permanent magnet with a flux density of 0.7 Wb/m^2. The current in the coil is 15 mA. The coil has an average dimension of 17.5 cm × 8.7 cm.

Solution

Given that $N = 700$, $I = 15 \times 10^{-3}$, $B = 0.7$ Wb/m^2, $A = 17.5 \times 8.7 \times 10^{-4}$ m^2.

Torque is

$$\tau = NIBA$$

$$= 700 \times 15 \times 10^{-3} \times 0.7 \times 17.5 \times 8.7 \times 10^{-4}$$

$$= 0.1119\,\text{N m (or) } 0.0114\,\text{kg m}$$

Problem 20

In a test in a laboratory, it is found that 1 kg of silicon steel is equivalent to an area to energy loss of 150 J/m^3. Find the total loss for 7.5 kg of materials in 5 h, when it is subjected to a field strength of μ = 400 AT/m at a frequency of 50 Hz.

Assume density "g" of material = 6.9 g/cm^2.

Solution

Given, energy lost = 150 J/m^3, frequency = 50 Hz, mass of steel = 7.5×1000 = 7500 g.

$$\text{Volume of steel} = \frac{\text{Mass in g}}{\text{Density in g/cm}^2}$$

$$= \frac{7500}{6.9}$$

$$= 1.08695 \times 10^{-3}\,\text{m}^3$$

Now

$$\text{No. of cycles per second} = 50 \text{ numbers of cycles/h}$$
$$= 50 \times 3600 = 180000$$

Therefore

$$\text{Total loss in 5 h} = [\text{Volume} \times \text{Energy lost/m}^3] \times f \times t$$
$$= 1.08695 \times 10^{-3} \times 150 \times 180000 \times 5$$
$$= 1467385$$

Problem 21

A hysteresis loop in drawn to a scale of x-axis 1 cm = 500 AT/m, y-axis 1 cm = 0.1 Wb/m². The area of loop is measured by mid-ordinate method is 15.786 cm².

(i) Determine hysteresis loss/m³/cycle.

(ii) Determine the Steinmetz coefficient for a B_{max} = 1.10 Wb/m² at a frequency of 50 Hz.

(iii) Find the hysteresis loss/m³ for a B_{max} = 1.25 Wb/m² at a frequency of 25 Hz.

Assume the loss to be proportional to $(B_{max})^{1.7}$.

Solution

(i) Hystersis loss/m³/cycle = Area of *B–H* loop

1 cm² area of loop (*xy*) will represent a loss of

$$0.1 \times 500 = 50 \text{ J/m}^3\text{/cycle}$$

$$15.786 \text{ cm}^2 \text{of loop area} = 50 \times 15.786 = 789.3 \text{ J/m}^3\text{/cycle}$$

(ii) Hysteresis loss for a frequency of 50 Hz equals

$$789.3 \times 50 \text{ kJ} = 39465 \text{ J/m}^3\text{/cycle}$$

that is

$$39645 = \eta (B_{\text{max}})^{1.7} \times 50 \Rightarrow \eta = 670.6$$

(iii) Hysteresis loss when $B_{\text{max}} = 1.25 \text{ Wb/m}^2$, $\eta = 670.6$ and $f = 25 \text{Hz}$ is

$$670.6 \times (1.25)^{1.7} \times 25 = 24510.43 \text{ J/m}^3\text{/cycle}$$

Problem 22

A transformer's primary winding has 1250 turns and the secondary winding has 740 turns. The length of magnetic path can be taken as 27.5 cm and the cross-sectional area of the core is 5.5 cm². Assuming $\mu_r = 750$, find the self-inductance of the primary and secondary windings separately and the mutual inductance between them.

Solution

Given that $N_1 = 1250$, $N_2 = 740$, $I = 27.5\times10^{-2}$ m, $A = 5\times10^{-4}$, $\mu_r = 750$.

Reluctance of the path,

$$S = \frac{l}{\mu_0\mu_r A} = \frac{27.5\times10^{-2}}{5\times10^{-4}\times4\pi\times10^{-7}\times750} = 0.5836\times10^6 \text{ H}$$

Self-inductance of the primary winding is

$$\frac{N_1^2}{S} = \frac{1250^2}{0.5836\times10^6} = 2.6773 \text{ H}$$

Self-inductance of the secondary winding is

$$\frac{N_2^2}{S} = \frac{740^2}{0.5836\times10^6} = 0.9383 \text{ H}$$

Mutual inductance is

$$M = \frac{N_1N_2}{S} = \frac{1250\times740}{0.5836\times10^6} = 1.5849 \text{ H}$$

Multiple Choice Questions

1. The basic requirement for inducing e.m.f. in a coil is that
- **(a)** flux should link the coil
- **(b)** there should be charge in flux linking the coil
- **(c)** coil should form a closed loop
- **(d)** none of the above

2. The e.m.f. induced in a coil of N turns is given by
- **(a)** $d\theta/dt$
- **(b)** $Nd\theta/dt$
- **(c)** $-Nd\theta/dt$
- **(d)** $Ndt/d\theta$

3. The direction of induced e.m.f. in a conductor (or coil) can be determined by
- **(a)** Work law
- **(b)** Ampere's law
- **(c)** Fleming's right-hand rule
- **(d)** Fleming's left-hand rule

4. The e.m.f. induced in a ________ is statically induced e.m.f.
- **(a)** DC generator
- **(b)** transformer
- **(c)** DC motor
- **(d)** none of the above

5. The e.m.f. induced in a ________ is dynamically induced e.m.f.
- **(a)** alternator
- **(b)** transformer
- **(c)** DC generator
- **(d)** none of the above

6. Inductance opposes ________ in current in a circuit.
- **(a)** only increase
- **(b)** only decrease

(c) change
(d) none of the above

7. If the number of turns of a coil is increased, its inductance
(a) remains the same
(b) is increased
(c) is decreased
(d) none of above

8. An iron-cored coil of 10 turns has reluctance of 100 A T/Wb. The inductance of the coil is
(a) 1 H
(b) 10 H
(c) 0.1 H
(d) 5 H

9. The mutual inductance between two coils is ________ reluctance of magnetic path.
(a) directly proportional to
(b) inversely proportional to
(c) independent of
(d) none of the above

10. The coefficient of coupling between the primary and the secondary of a practical/ iron-cored transformer is about
(a) 0.9
(b) 0.1
(c) 1.5
(d) 0.2

Review Questions

1. State Faraday's laws of electromagnetic induction.
2. State Fleming's left-hand rule.
3. State Fleming's right-hand cork screw rule.
4. Define statically induced e.m.f.
5. Define dynamically induced e.m.f.
6. Define self-inductance.
7. Define mutual inductance.
8. Derive the expression for mutually induced e.m.f.
9. Explain the concept of energy stored in a magnetic field.
10. Derive the expression of lifting power of a magnet.

Problems

1. A coil of 100 turns is linked by a flux of 20 mWb. If this flux is reversed in a time of 1 ms, calculate the average e.m.f. induced in the coil.

2. A coil of 200 turns of wire is wound on a magnetic circuit of reluctance 2000 AT/mWb. If a current of 1 A flowing in the coil is reversed in 10 ms, find the average e.m.f. induced in the coil.

3. Determine the e.m.f induced in a coil of 200 turns when there is a change of flux of 25 mWb linking with it in 50ms.

4. Calculate the coil inductance when a current of 3A in a coil of 1200 turns produces a flux of 6 mWb linking with the coil.

5. A coil of 300 turns, wound on a core of non-magnetic material, has an inductance of 10 mH. Calculate **(a)** the flux produced by a current of 5 A and **(b)** the average value of emf induced when a current of 5 A is reversed in 8 ms.

6. Define coefficient of coupling and calculate it for two coils having self-inductance of 60 mH and 80 mH. The mutual inductance between them is 40 mH.

7. Two magnetically coupled coils have self-inductances $L_1 = 100$ mH and $L_2 = 400$ mH. If the coefficient of coupling is 0.8, find the value of mutual inductance between the coils.

8. A conductor carries a current of 20 A and is at right angles to a magnetic field having a flux density of 0.9 T. If the length of the conductor in the field is 30 cm, calculate the force acting on the conductor. Determine also the value of force if the conductor is inclined at an angle of 30° to the direction of the field.

9. Two coils having 100 and 50 turns respectively are wound on a core with $\mu = 4000\ \mu_0$. Effective core length = 60 cm and core area = 9 cm^2. Find the mutual inductance between the coils.

10. Find the inductance of a coil in which a current of 0.2 A increasing at the rate of 0.4 A/s represents a power flow of 0.4 W.

11. When a current of 1.3 A flows in a coil the flux linking with the coil is 90 μWb. If the coil inductance if 0.30 H. Calculate the number of turns of the coil.

12. Calculate the e.m.f induced in a coil of inductance 12 H by a current changing at the rate of 4 A/s.

13. A flux of 15 mWb links with a 1200 turn coil when a current of 4A passes through the coil. Calculate **(a)** the inductance of the coil, **(b)** the energy stored in the magnetic field, and **(c)** the average e.m.f. induced if the current falls to zero in 110 ms.

14. Two coils have mutual inductance of 0.4 H. If the current in one coil is reduced from 5 A to 1 A in 0.01 s, find the e.m.f. induced in the second coil. Assume that the whole flux set up by the first coil is linked with the second coil.

15. A meter driving motor consists of a horizontal disc of aluminium 20 cm in diameter, pivoted on a verticle spindle, and lying in a permanent magnetic field of density 0.3 Wb/m^2. The current flow is radical from the spindle to the circumference of the disc. The circuit resistance is 0.225 Ω and potential difference of 230 V is required to pass a current of 10 A through the motor, calculate **(a)** the rotational speed of the disc and **(b)** the power lost in friction.

16. A flat coil of 500 turns of wire and mean area 600cm revolves at a speed of 1600rpm in a uniform magnetic field having a flux density of 80mwb/m^2. Calculate the **(a)** peak, **(b)** average over a half cycle and **(c)** RMS values of the emf induced in the coil.

17. Calculate the **(a)** inductance of a toroid, having 25 cm mean diameter and 6.25 cm^2 circular cross-section, wound uniformly with 1000 turns of wire, and **(b)** also emf induced when a current increasing at the rate of 200 A/s flows in the winding. Develop any formula used.

18. A circuit has 1000 turns enclosing a magnetic circuit 20 m^2 in section. With 4 A the flux density is 1 Wb/m^2 and with 9 A it is 1.4 Wb/m^2. Find the **(a)** mean value of the inductance between these current limits and **(b)** the induced electronic force if the current fell uniformly from 9 A to 6 A in 0.05 s.

19. The mean diameter of a steel ring is 50 cm and a flux density of 1 Wb/m is produced by 40 AT/cm. If the cross-section of the ring is 200 and the number of turns is 500, find

(a) The inductance in energy.

(b) The exciting current and inductance when a gap 1 cm long is cut in the ring, the flux density being maintained at 2 Wb/m^2.

Ignore leakage and ringing.

20. Two identical 1000 turn coils X and Y lie in the parallel planes such that 60% of the magnetic flux produced by one coil links the other. A current of 5 A in X produces in it a flux of 0.05 mWb. If the current in changes from +6 A to –6 A in 0.01 s, what will be the magnitude of the e.m.f. induced in Y? Calculate **(a)** the self-inductance of each coil and **(b)** the mutual inductance.

21. A solenoid consists of 1500 turns of wire wound in a length of 60 cm. A search coil of 500 turns enclosing in mean area of 20 cm^2 is placed centrally in the solenoid.

(a) Find the mutual inductance of the arrangement.

(b) Determine the e.m.f induced in the search coil when the current in the solenoid is changing uniformly at the rate of 250 A/s?

22. A search coil having 50 turns enclosing a cross-sectional area of 2 cm^2 is placed at the middle of a solenoid having 505 turns of wire uniformly wound on a length of 12 cm and carrying an alternating current of 0.2 A at a frequency of 7 kHz. Calculate **(a)** the mutual inductance between the coils and **(b)** the e.m.f. induced in the search coil.

23. Two coils, X of 14,000 turns and Y of 17,000 turns, lie in parallel planes so that 45% of the flux produced by coil X links with coils Y. A current 5 A in X produces 0.05 mWb while the same current in Y produces 0.075 mWb. Calculate **(a)** the mutual conductance and **(b)** the coupling coefficient.

24. Two coils of 30 and 600 turns, respectively, are wound side by side on a closed iron circuit of section 100 cm^2 and mean length 150 cm.

(a) Estimate the mutual inductance between the coils if the relative permeability of the iron is 2000.

(b) A current in the first coil grows steadily from 0 to 5 A in 0.04 s. Find the e.m.f. induced in the other coil.

25. The combined inductance of two coils connected in series is 0.06 H or 0.1 H, depending upon the relative directions of the currents in the coils. If one of the coils when isolated has a self inductance of 0.2 H, calculate **(a)** the mutual inductance and **(b)** coupling coefficient.

26. A solenoid A of 500 turns, winding length 100 cm and diameter 2 cm is placed centrally within a solenoid B of 1000 turns, winding length 100 cm and diameter 6 m. Calculate the total inductance of the circuits when the coils are joined in series with **(a)** cumulative m.m.f., **(b)** opposing m.m.f. **(c)** Also calculate the mutual inductance of the circuit.

27. A solenoid 3 m in length and 10 cm in diameter has 5000 turns. Calculate **(a)** the approximate inductance and **(b)** the energy stored in the magnetic field when a current of 2 A flows in solenoid.

28. The hysteresis loop of sample of sheet steel is carried to a density of 1.375 Wb/m^2. The co-ordinates are such that 1 cm = 10 AT/cm and 1 cm = 1 W/dm^2, while the area of the loop is found to be 0.613 cm^2. Find the loss in watts due to hysteresis if 100 cm^2 of this iron were subjected to an alternative density of 375 Wb/m^2 at 25 Hz.

29. Calculate the hourly loss of energy in a specimen of iron the hysteresis loop of which is equivalent in area to 260 J/m^3. Frequency 50 Hz, density of iron 7500 kg/m^3, weight 10 kg.

30. Calculate the losses per kg in a specimen of alloy steel for a maximum density of 1 Wb/m^2 and a frequency of 50 Hz, using 0.5 mm plates. Resistivity 25 mΩ-cm, density 7.8 g/cm^3, hysteresis 350 J/m^3/cycle.

Answers

Multiple Choice Questions

1. (b) **2.** (c) **3.** (d) **4.** (b) **5.** (c) **6.** (c) **7.** (b) **8.** (a) **9.** (b) **10.** (a)

Problems

1. 2000 V

2. 2 V

3. −100V

4. 2.4 H

5. **(a)** 0.1667 mWb, **(b)** 12.5 V

6. 0.577

7. 0.16 H

8. 5.4 N, 2.7 N

9. 37.7 mH

10. 5 H

11. 4333 turns

12. – 48 V

13. **(a)** 4.5 H, **(b)** 36 J, **(c)** −454.54 V

14. −160 V

15. **(a)** 319 RPM, **(b)** 0.5 W

16. **(a)** 402 V, **(b)** 256 V, **(c)** 284 V

17. **(a)** 1 mH, **(b)** −0.2 V

18. **(a)** 0.16 H, **(b)** 16 V

19. **(a)** 0.15H; **(b)** 28.4 A, 0.0704 H

20. **(a)** 7.2 V, **(b)** each coil 10 mH, 6 mH

21. **(a)** 3.14 mH, **(b)** 0.786 V
22. **(a)** 52.4 μH, **(b)** 460.93×10^{11} V
23. **(a)** 67.5 mH, **(b)** 3.3963
24. **(a)** 0.302 H, **(b)** 37.75 V
25. **(a)** 0.125 H, **(b)** 0.72
26. **(a)** 4.04 mH, **(b)** 3.25 mH, **(c)** 0.197 mH
27. **(a)** 0.0820 H, **(b)** 0.164 J
28. 15.325 W
29. 0.6 μJ
30. 2.76 W/kg

5 AC Circuits

LEARNING OBJECTIVES

After reading this chapter, you will be able to:

- Understand the characteristics of AC circuit.
- Understand the generation of alternating e.m.f. and also equation of alternating voltage.
- Understand cycle, period or periodic time T and frequency f of a waveform.
- Perform calculation involving $T = 1/f$.
- Understand instantaneous, peak, mean and RMS values as well as form factor and peak factor for a sine wave.
- Calculate mean and RMS values and form and peak factors for a given waveform.
- Understand in detail about the phasor representation in R, L and C.
- Understand and analyze the single-phase AC circuit.

5.1 Introduction

Throughout the world, the generation, transmission and distribution of electrical power are in the form of alternating current (AC). AC generator is used in the power station to generate voltage. In the power station, water, natural gas, coal and nuclear fusion are used as a fuel. In this chapter, basics of AC circuits and its time-varying analysis with voltage and current sources will be discussed.

5.2 Generation of Alternating E.M.F.

Alternating e.m.f. can be generated by rotating a coil in a magnetic field (Fig. 1) or by rotating a magnetic field in a stationary coil as shown in Fig. 2.

The magnitude of e.m.f. generated depends on the position of the coil. In Fig. 3, the coil is stationary. The magnitude of voltage induced is 0 V. In Fig. 4, the coil rotates through an angle of 90° generating maximum voltage corresponding to position B in the waveform. In Fig. 5, the coil has rotated through 180° generating 0 V corresponding to position C in the waveform.

Similarly, points D and E correspond to 270° (Fig. 6) and 360° (Fig. 7) rotation of the coil, respectively, generating maximum value and zero value of voltage (Fig. 8).

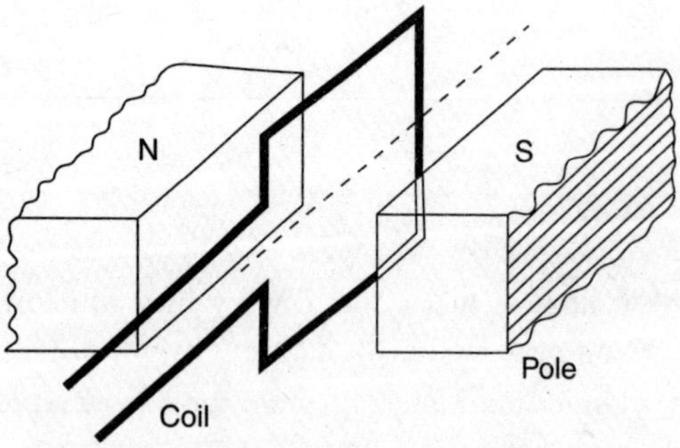

Figure 1 Generation of alternating e.m.f.

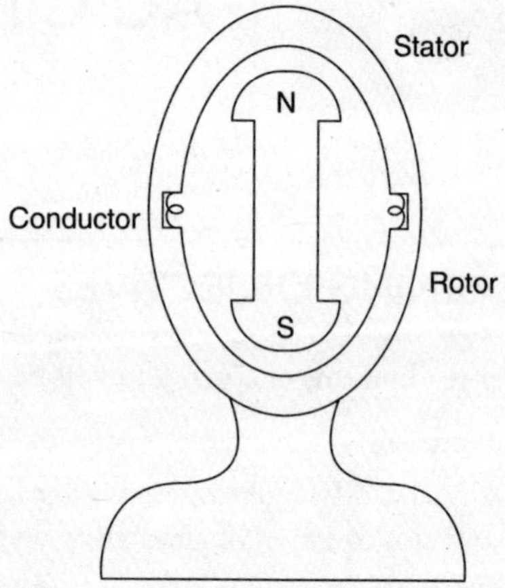

Figure 2 Inner structure of stationary coil.

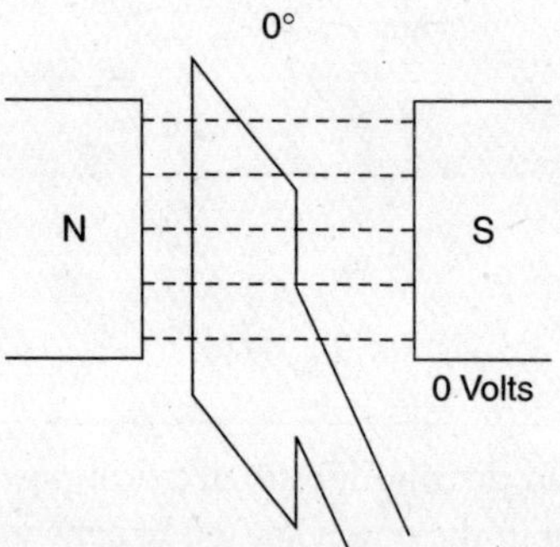

Figure 3 Position of stationary coil at 0°.

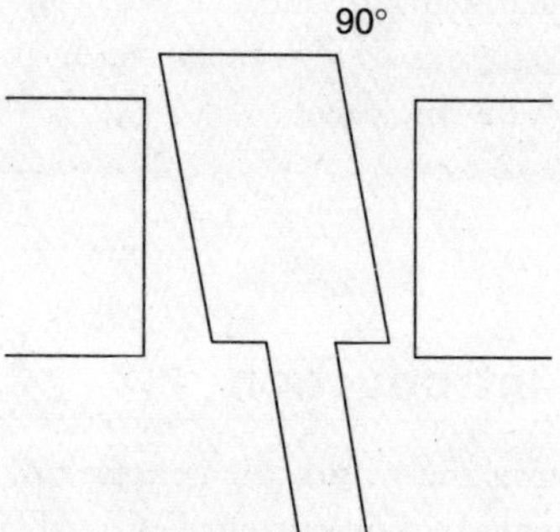

Figure 4 Stationary coil rotation at 90°.

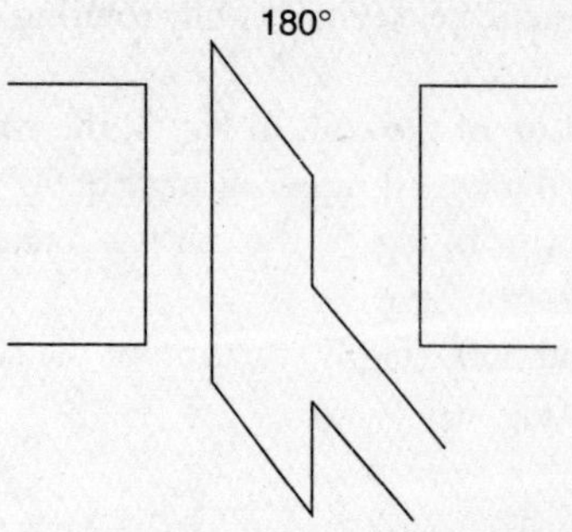

Figure 5 Stationary coil rotation at 180°.

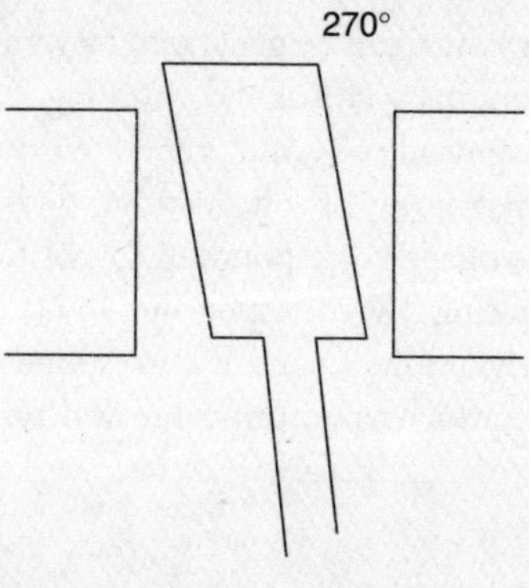

Figure 6 Stationary coil rotation at 270°.

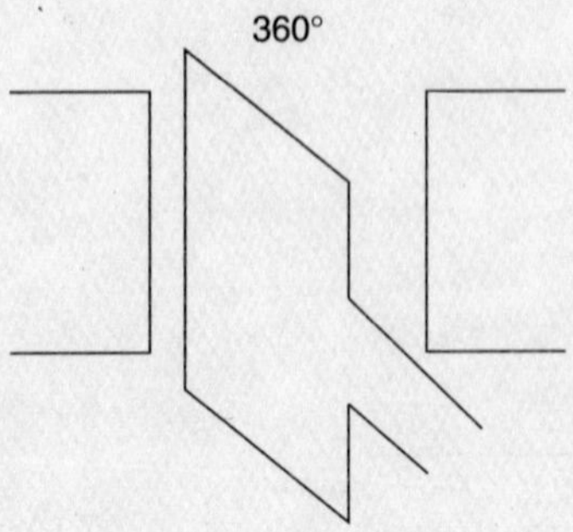

Figure 7 Stationary coil rotation at 360°.

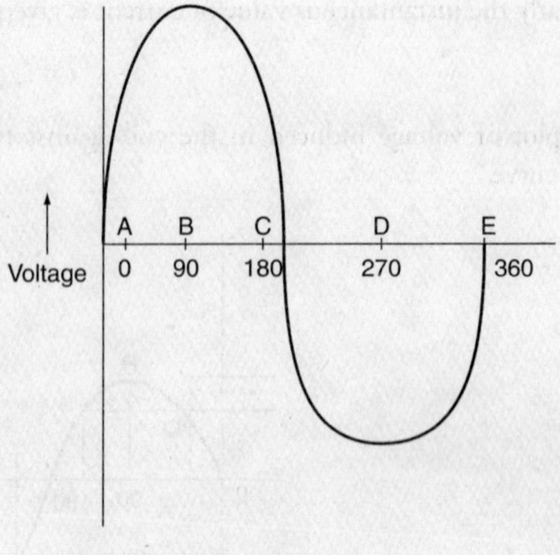

Figure 8 Alternating current waveform.

5.3 Equation of Alternating Voltage

Consider a rectangular coil (of coil side OP) rotating in a magnetic field of "ω" as shown in the Fig. 9. The instantaneous value of e.m.f. is given by projection of vector (showing maximum value of voltage generator) on the *Y*-axis. As OP rotates through an angle θ in counterclockwise direction, it takes the position OQ. The line "OM" gives the projection of "OQ" on *Y*-axis. Now "OMQ" is a right-angled triangle. Thus,

$$\frac{\text{OM}}{\text{OQ}} = \sin\theta \quad \Rightarrow \quad \text{OM} = \text{OQ}\sin\theta$$

where OQ is the maximum value of voltage induced in the coil, $V_{\max}$. OM is the instantaneous value of voltage (V). Therefore,

$$V = V_{\max}\sin\theta \tag{5.1}$$

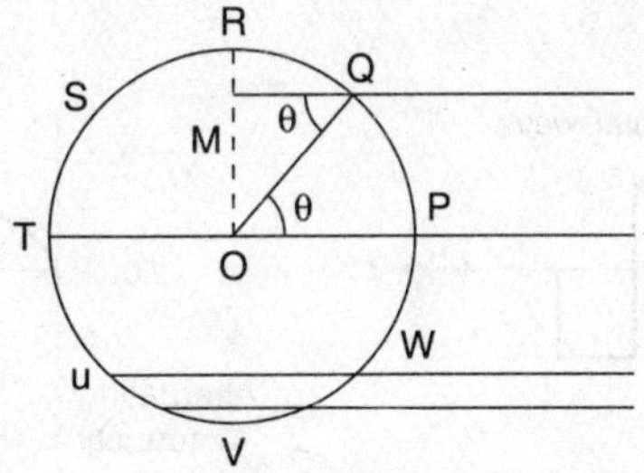

Figure 9 Representation of rectangular coil.

Similarly the instantaneous value of current is given by

$$i = i_{max} \sin\theta \tag{5.2}$$

The plot of voltage induced in the coil against time is shown in Fig. 10. This curve is known as "sine curve."

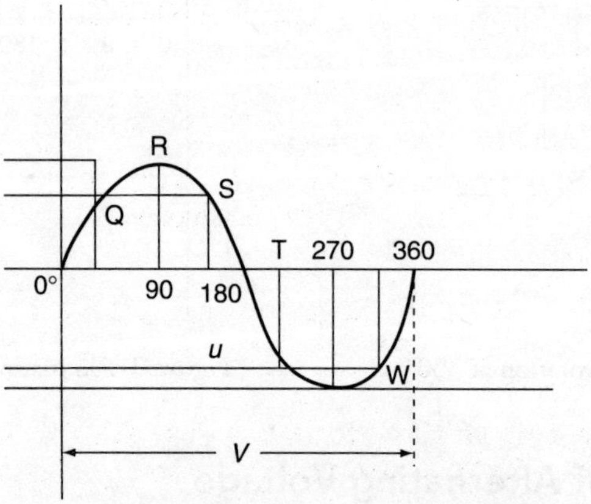

Figure 10 Sine curve.

5.4 Alternating Quantity (Current or Voltage)

It is one in which the magnitude and direction change periodically at regular intervals, as shown in Fig. 11.

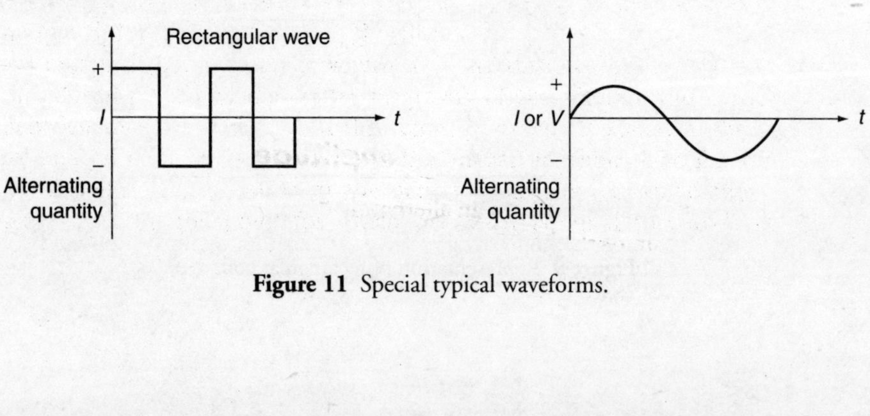

Figure 11 Special typical waveforms.

5.5 Cycle

One complete set of positive and negative values of an alternating quantity constitutes one cycle.

5.6 Time Period (*T*)

It is the time taken by an alternating quantity to complete one cycle. For example, a 50 Hz alternating quantity has a time period of 1/50 s = 20 ms.

5.7 Frequency (*F*)

The number of cycles that occur in 1 s is called frequency. It is measured in cycles/s or hertz (Hz) to honor the scientist Heinrich Hertz (1857–1894):

$$f = \frac{1}{T}\,\text{Hz}$$

$$T = 1/f \tag{5.3}$$

5.8 Angular Frequency

Angular frequency (ω) is given by

$$\omega \cdot 2\pi f = \frac{2\pi}{f}\ \text{rad/s}$$

Example

Supply frequency in our country = 50 Hz
Therefore AC, V and I supplied by Electricity Board takes

$$T = \frac{1}{50} = 0.02\ \text{s}$$

to complete one cycle of waveform. Therefore

$$\omega \cdot 2\pi f = 2 \times \pi \times 50 = 314$$

5.9 Maximum or Peak Value or Amplitude

The maximum value (positive or negative) of an alternating quantity is called amplitude. Amplitude or peak value (I_P) is the maximum or the highest value of the alternating quantity attained by it in a cycle. The peak value is otherwise called as crest value or maximum value.

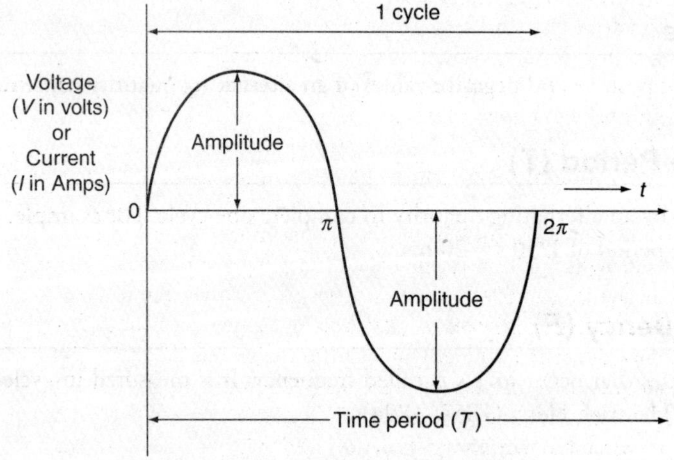

Figure 12 Alternating waveform for one cycle.

The above definition can be understood with the help of Fig. 12. From Eq. (5.1), we have

$$V = V_{max} \sin\theta$$

where V_{max} is the amplitude. The same equation can be rewritten as

$$V = V_{max} \sin \omega t, \quad \text{where } \theta = \omega t$$

$$= V_{max} \sin 2\pi ft, \quad \text{where } \omega = 2\pi f$$

$$= V_{max} \sin\left(\frac{2\pi}{T}\right)t, \quad \text{where } f = 1/T$$

It can be seen that at any instant, the values of voltage and current waveform are either equal to V_m and I_m or a fraction of V_m and I_m. The coefficient of cosine or sine function of the equation represents the maximum or the peak value of the quantity.

Examples

1. If $v = 311 \cos \omega t$ then $v_m = 311$ V.
2. If $i = 10 \sin \omega t$ then $I_m = 10$ A.

5.10 Average Value

It is the mean value of an alternating quantity over one half of a cycle. For a periodic waveform, the average value of an alternating quantity over one or more complete cycles is zero. The average value over half a cycle is, however, not zero and can easily be found knowing the wave shape. For example, consider the sinusoidal current of the form

$$i = I \sin \omega t = I \sin \theta$$

Its area can be obtained as follows:

$$\text{Area of half cycle voltage wave} = \int_0^{\pi} v \, d\theta$$
$$= \int_0^{\pi} v_m \sin\theta \, dt$$
$$= v_m \left[-\cos\theta\right]_0^{\pi}$$
$$= v_m \left[1-(-1)\right]$$
$$= 2v_m$$

Therefore, average value is given by

$$V_{av} = \text{Height of the equivalent rectangle having the same base}$$
$$= \frac{\text{Area}}{\text{Base}}$$
$$V_{av} = \frac{2v_m}{\pi}$$
$$V_{av} = \frac{2}{\pi} v_m = 0.637\, v_m$$

Similarly, for current we have

$$I_{av} = \frac{2}{\pi} I_m = 0.637\, I_m$$

5.11 Root Mean Square Value

The root mean square (RMS) value, also called virtual value, of an alternating current is given by that value of direct current which when flowing through a given circuit for a given time produces the same heating effect as that produced by an alternating current when flowing through the same circuit for the same time.

For an instantaneous current i, the instantaneous real power or heating effect is i^2R. Then the average heating effect of the periodic function is

$$I^2 = \frac{1}{\pi}\int_0^{\pi} i^2 d\theta$$

For sinusoidal current,

$$I^2 = \frac{1}{\pi}\int_0^{\pi} I_m^2 \sin^2\theta d\theta$$

$$= \frac{I_m^2}{\pi}\int_0^{\pi}\left[\frac{1-\cos 2\theta}{\theta}\right]d\theta$$

$$= \frac{I_m^2}{\pi}\left[\theta - \frac{\sin 2\theta}{2}\right]_0^{\pi}$$

$$= \frac{I_m^2}{\pi}\left[\pi - 0 - \frac{\sin 2\pi}{2} + 0\right]$$

This gives

$$I^2 = \frac{I_m^2}{2}$$

$$I = \frac{I_m^2}{\sqrt{2}} = 0.707\, I_m$$

Similarly, we can find the expression for sinusoidal voltage:

$$v = \frac{v_m}{\sqrt{2}} = 0.707\, v_m$$

5.12 Form Factor

The ratio of RMS value to the average value of an alternating waveform is known as form factor and it is denoted by K_f:

$$K_f = \frac{\text{RMS value}}{\text{average value}}$$

For sinusoidal wave,

$$K_f = \frac{\text{RMS value}}{\text{Average value}}$$

$$= \frac{(\text{Maximum value})/\sqrt{2}}{(2/\pi)\text{Maximum value}}$$

$$= \frac{\pi}{2\sqrt{2}} = 1.11$$

5.13 Peak Factor

The ratio of maximum or peak value to the RMS value of an alternating waveform is known as peak value. It is also called the crest factor or the amplitude factor. It is denoted by K_p. Mathematically

$$K_p = \frac{\text{Peak or maximum value}}{\text{RMS value}}$$

For a sinusoidal wave:

$$K_p = \frac{\text{Maximum value}}{(\text{Maximum value}/\sqrt{2})}$$

$$K_p = \sqrt{2} = 1.414$$

5.14 Expression for RMS Value, Average Value of Sinusoidal Alternating Quantity

The sinusoidal AC is given by Fig. 13. The instantaneous value of the current is given by

$$i = I_m \sin\theta$$

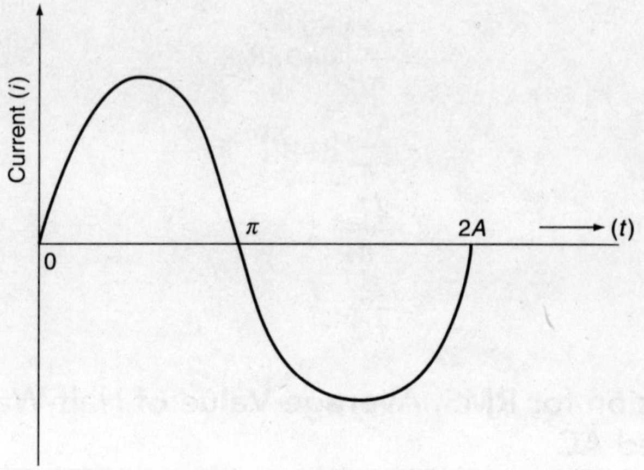

Figure 13 Sinusoidal AC.

5.14.1 RMS Value

$$I_{RMS} = \sqrt{\int_0^{2\pi} \frac{i^2}{2\pi} d\theta}$$

$$= \sqrt{\int_0^{2\pi} \frac{I_m^2 \sin^2\theta}{2\pi} d\theta}$$

$$= \sqrt{\frac{I_m^2}{2\pi}\int_0^{2\pi} \sin^2\theta \, d\theta}$$

$$= \sqrt{\frac{I_m^2}{2\pi}\int_0^{2\pi}\left(\frac{1-\cos 2\theta}{2}\right) d\theta}$$

$$= \sqrt{\frac{I_m^2}{4\pi}\int_0^{2\pi} d\theta - \frac{I_m^2}{4\pi}\int_0^{2\pi} \cos 2\theta d\theta}$$

$$= \sqrt{\frac{I_m^2}{4\pi}[2\pi - 0] - \frac{I_m^2}{8\pi}[\sin 4\pi - \sin\theta]}$$

$$= \sqrt{\frac{I_m^2}{4\pi}\times 2\pi - 0} = \sqrt{\frac{I_m^2}{2}} = \frac{I_m}{\sqrt{2}}$$

5.14.2 Average Value

Since this alternating sine waveform is a symmetrical wave we have to consider only a half cycle. Therefore,

$$I_{av} = \int_0^{\pi} \frac{i d\theta}{\pi}$$

$$= \frac{1}{\pi}\int_0^{\pi} I_m \sin\theta d\theta$$

$$= \frac{I_m}{\pi}\int_0^{\pi} \sin\theta d\theta$$

$$= \frac{-I_m}{\pi}[\cos\theta]_0^{\pi}$$

$$\frac{2I_m}{\pi}$$

$$= \frac{I_m}{\pi/2}$$

5.15 Expression for RMS, Average Value of Half-Wave Rectified AC

The half-wave rectified AC is shown in Fig. 14.

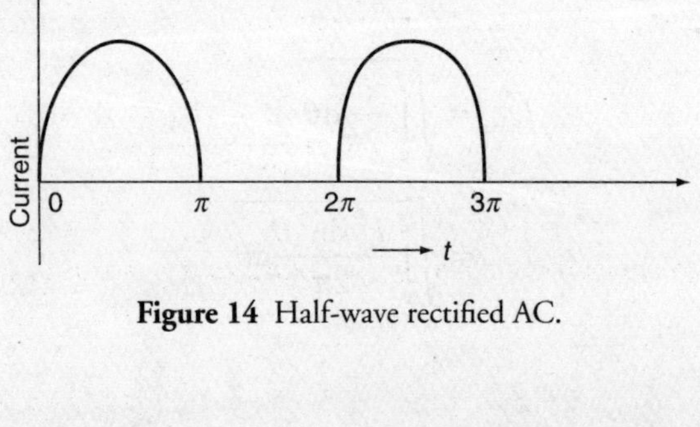

Figure 14 Half-wave rectified AC.

5.15.1 RMS Value

In this case we have

$$i = \begin{cases} I_m \sin\theta, & \theta < \theta \le \pi \\ 0, & \pi < \theta \le 2\pi \end{cases}$$

So

$$I_{RMS} = \sqrt{\int_0^{2\pi} \frac{i^2 d\theta}{2\pi}}$$

$$= \sqrt{\int_0^{\pi} \frac{i^2 d\theta}{2\pi} + \int_{\pi}^{2\pi} \frac{i^2 d\theta}{2\pi}}$$

$$= \sqrt{\int_0^{\pi} \frac{I_m^2 \sin^2\theta d\theta + 0}{2\pi}}$$

$$= \sqrt{\frac{I_m^2}{2}\int_0^{\pi} \sin^2\theta d\theta}$$

$$= \sqrt{\frac{I_m^2}{2\pi}\int_0^{\pi} \frac{(1-\cos 2\theta)d\theta}{2}}$$

$$= \sqrt{\frac{I_m^2}{4\pi}\int_0^{\pi} d\theta - \frac{I_m^2}{4\pi}\int_0^{\pi} \cos 2\theta d\theta}$$

$$= \sqrt{\frac{I_m^2}{4\pi}[\theta]_0^{\pi} - \frac{I_m^2}{4\pi}\left[\frac{\sin 2\theta}{2}\right]_0^{\pi}}$$

$$= \sqrt{\frac{I_m^2}{4\pi}[\pi - 0] - 0}$$

$$= \sqrt{\frac{I_m^2}{4}}$$

Therefore,

$$I_{RMS} = \frac{I_m}{2}$$

5.15.2 Average Value

Since the half-wave rectified current is unsymmetrical, we have to consider full cycle for calculating the average value. Therefore,

$$I_{av} = \int_0^{2\pi} \frac{i d\theta}{2\pi}$$

$$= \int_0^{\pi} \frac{i\,d\theta}{2\pi} + \int_{\pi}^{2\pi} \frac{i\,d\theta}{2\pi}$$

$$= \int_0^{\pi} \frac{I_m}{2\pi} \sin\theta\, d\theta + 0$$

$$= \frac{I_m}{2\pi} \int_0^{\pi} \sin\theta\, d\theta$$

$$= \frac{-I_m}{2\pi} [\cos\theta]_0^{\pi}$$

$$= \frac{-I_m}{2\pi} [-1-1] = \frac{I_m}{\pi}$$

5.16 Phase Representation of Alternating Quantity

The schematic of sinusoidal voltage and current waveform is shown in Fig. 15. The sinusoidal voltage waveform can be mathematically expressed as

$$u = u_m \cos\omega t = u_m \cos\theta$$

The sinusoidal current waveform can be expressed as

$$i = I_m \sin\omega t = I_m \sin\theta$$

So, the alternating current or voltage can be represented by a vector rotating in counterclockwise direction with the same frequency as that of the alternating quantity. This vector representation of alternating quantity simplifies complexity of problems in AC circuit.

5.16.1 Phase

The phase of an alternating quantity in Fig. 16 at point "*P*" is $3\pi/2$. This implies that the phase is nothing but a fraction of time period that has elapsed from reference position (i.e., zero position).

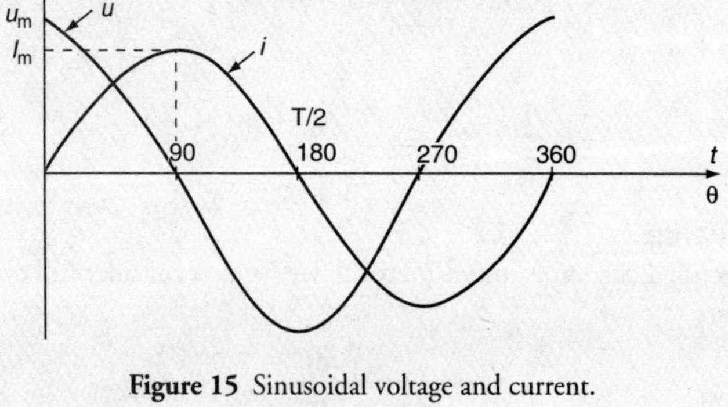

Figure 15 Sinusoidal voltage and current.

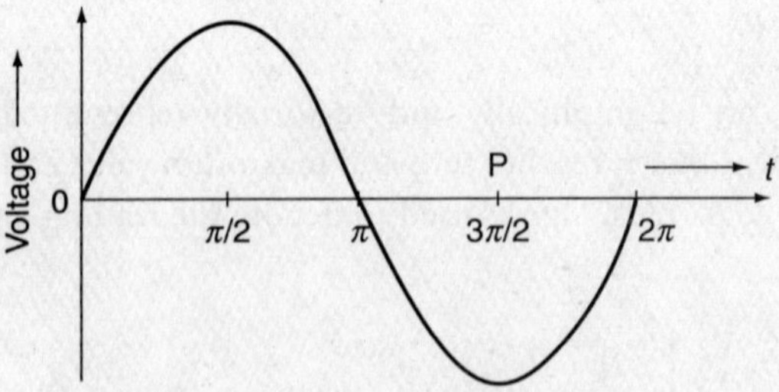

Figure 16 Phase of alternating waveform.

Two alternating quantities are said to be in phase, if they reach their zero value and maximum value at the same time. For example, the two alternating quantities

$$i_1 = I_{m1} \sin \theta$$

and

$$i_2 = I_{m2} \sin \theta$$

can be graphically represented as shown in Fig. 17 and vectorially as shown in Fig. 18. From Fig. 17, it is clear both i_1 and i_2 reach their zero and maximum values at the same time and hence are said to be in-phase. Note that phase difference is 0°.

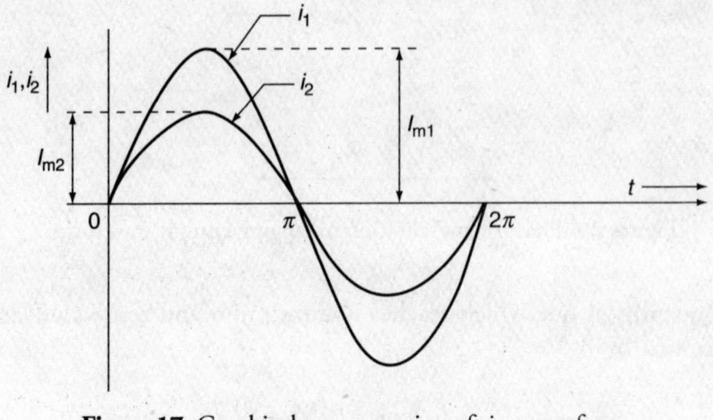

Figure 17 Graphical representation of sine waveform.

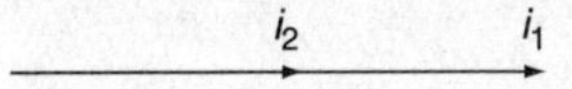

Figure 18 Current direction of sine waveform.

5.16.2 Out-of-Phase or Phase Difference

When the two alternating quantities do not reach their zero and maximum value at the same time, then they are said to be out of phase. This phase difference is measured in electrical degrees or in radians and is given in terms of "lead" or "lag." Leading alternating quantity is one which reaches its maximum or zero value earlier than the other alternating quantity:

$$i_1 = I_{m1} \sin(\omega t + \phi)$$

$$i_2 = I_{m2} \sin(\omega t)$$

Then these two equations can be graphically and vectorially represented as in Figs. 19 and 5.20, respectively. From 19, it is clear that i_1 reaches zero and maximum value earlier than i_2. Hence i_1 is said to be leading i_2 by an angle ϕ. A "plus" sign is used to denote the leading angle.

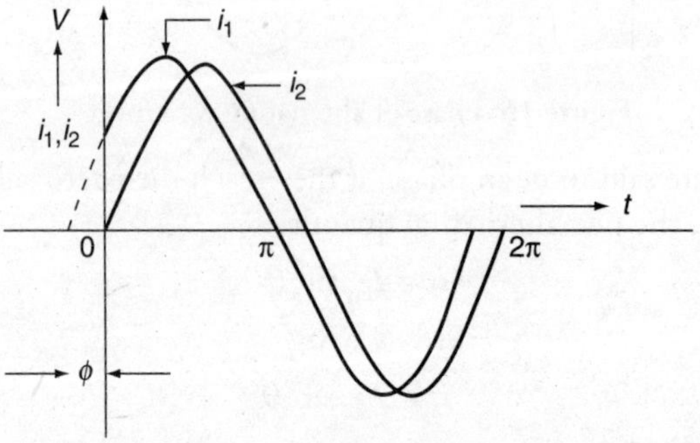

Figure 19 Graphical representation of lagging current waveform.

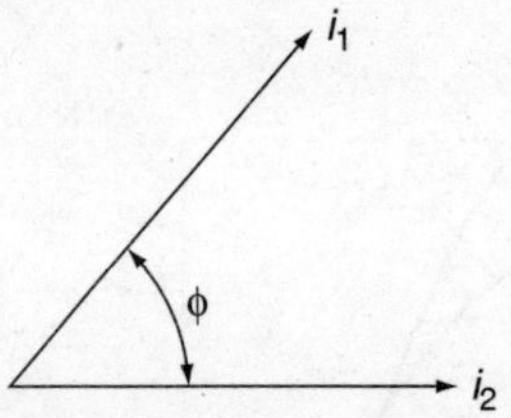

Figure 20 Phasor representation of lagging current waveform.

Lagging alternating quantity is one which reaches its maximum and zero value later than that of the other alternating quantity. If

$$i_1 = I_{m1} \sin(\omega t - \phi)$$

and

$$i_2 = I_{m2} \sin(\omega t)$$

then these two equations can be graphically and vectorially represented as in Figs. 21 and 22, respectively.

From Fig. 21, it is clear that i_1 is reaching zero and maximum value later than i_2. Hence i_1 is said to be lagging i_2 by an angle "ϕ." A "minus" sign is usually given to denote lagging angle.

Note:

1. Two vectors are said to be in antiphase if the phase difference between them is 180°.
2. Two vectors are said to be in quadrature, if the phase difference between them is 90°.

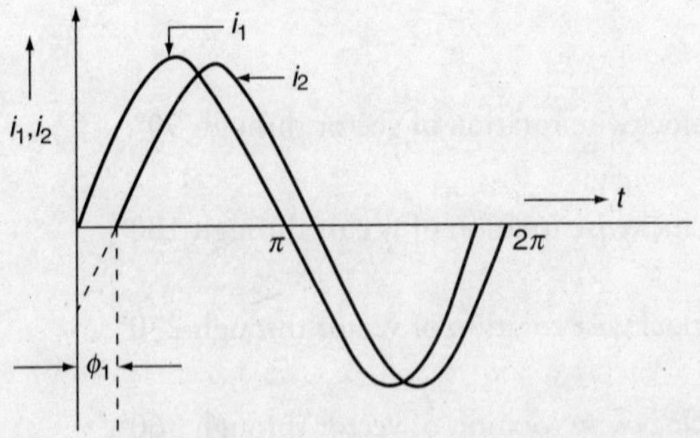

Figure 21 Graphical representation of leading current waveform.

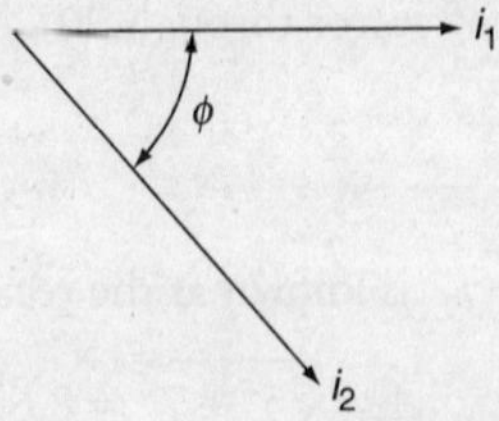

Figure 22 Phasor representation of lagging current waveform.

5.16.3 Complex Algebra

A vector quantity has both magnitude and direction whereas a scalar quantity has only magnitude but no direction. A vector can be written in terms of its horizontal and vertical components. In Fig. 23 "OM" represents a vector "*A*." "*a*" represents the horizontal component and "*b*" represents the vertical component of the vector "*A*." Thus $A = a + jb$ where "*j*" is an operator indicating that "*b*" is in quadrature with "*a*." This form of representing a vector is known as the complex form. Here, "*a*" is the in-phase or active component and "*b*" is the quadrature or reactive component.

The magnitude of the vector is $A = \sqrt{a^2 + b^2}$. The angle made by vector "*A*" with *x*-axis is

$$\phi = \tan^{-1}\left(\frac{b}{a}\right)$$

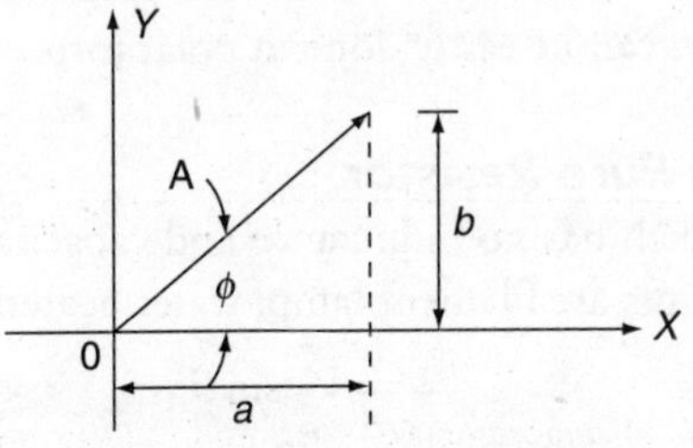

Figure 23 Phasor diagram.

Note:

1. $j = \sqrt{-1}$
   It indicates counterclockwise rotation of vector through 90°.
2. $j^2 = j \cdot j = (-1)$
   It indicates counterclockwise rotation of vector through 180°.
3. $j^3 = j \cdot j \cdot j = -j$
   It indicates counterclockwise rotation of vector through 270°.
4. $j^4 = j \cdot j \cdot j \cdot j = 1$
   It indicates counterclockwise rotation of vector through 360°.
5. $\frac{1}{j} = \frac{1 \cdot j}{j \cdot j} = \frac{j}{j^2} = \frac{j}{-1} = -j$
6. "$-j$" indicates clockwise rotation of vector through 90°.

Example

A vector "A" written in the form $5 + j7$ is known as the rectangular form. The magnitude of A is

$$|A| = \sqrt{5^2 + 7^2} = 8.6$$

The angle α is given by

$$\alpha = \tan^{-1}(7/5) = 54.46°$$

The same vector can be rewritten as

$$A = |A| \angle \alpha$$

that is,

$$A = 8.6 \angle 54.46°$$

This form of representation of a vector is called the polar form.

Polar form: The phasor V may be alternatively expressed as $V < \theta$, where V is the magnitude of the phasor and θ is the angle measured in counter-clockwise direction from reference axis. Such representation is called polar form.

Note:

1. Addition and subtraction can be easily done in rectangular form.
2. Multiplication and division can be easily done in polar form.

5.16.4 Phasor Relation in Pure Resistor

A pure resistive circuit is one which has no inductance and capacitance but only resistance as shown in Fig. 24 (e.g., pure resistance circuits are filament lamp, water heaters, etc.). When an AC voltage of

$$V = V_m \sin\theta$$

or

$$V = V_m \sin \omega t \tag{5.4}$$

Figure 24 Pure resistor circuit.

is applied across such a circuit a current will flow through it. The value of this current is given by Ohm's law. According to that

$$i = \frac{V}{R}$$

Hence we have

$$i = \frac{V_m}{R} \sin \omega t$$

$$\Rightarrow i = I_m \sin \omega t \tag{5.5}$$

where $I_m = V_m / R$ because "i" is maximum at $\sin \omega t = 1$. Comparing Eqs. (5.4) and (5.5), we find that V and i are in-phase. The graphical and vectorial representations of these two quantities are shown in Figs. 25 and 26.

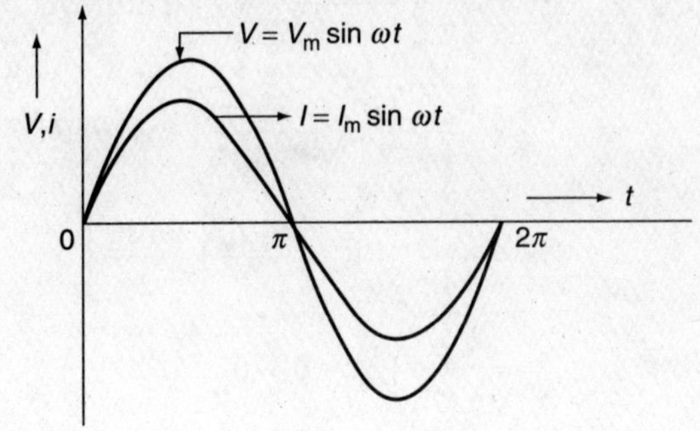

Figure 25 Current and voltage waveforms of pure resistor.

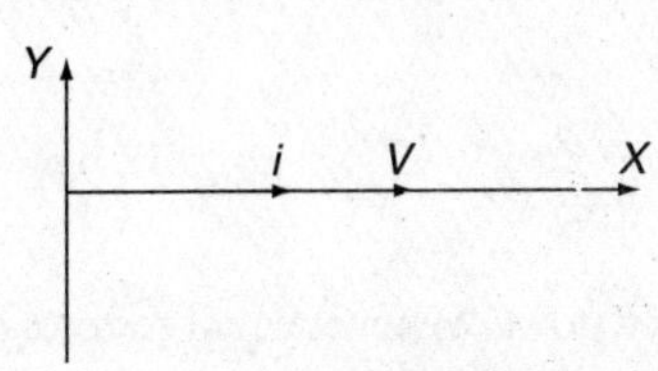

Figure 26 Vector representation of pure resistor.

The instantaneous power is given by

$$\begin{aligned}
P &= v \times i \\
&= (V_m \sin \omega t)(I_m \sin \omega t) \\
&= V_m I_m \sin^2 \omega t \\
&= \frac{V_m I_m}{2}(1 - \cos 2\omega t)
\end{aligned}$$

$$P = \frac{V_m I_m}{2} - \frac{V_m I_m}{2} \cos 2\omega t \tag{5.6}$$

Thus, the power consists of two parts: Constant part

$$\frac{V_m I_m}{2}$$

and fluctuation part

$$\frac{V_m I_m}{2} \cos 2\,\omega t$$

whose average value will be 0 for one complete cycle. Therefore, the average power for one complete cycle is

$$\begin{aligned}
P_{av} &= \frac{1}{2\pi} \int_0^{2\pi} V_{max} I_{max} \sin^2 \theta d\theta \\
&= \frac{1}{2\pi} \int_0^{2\pi} \frac{V_m I_m (1 - \cos 2\theta)}{2} d\theta \\
&= \frac{V_m I_m}{4\pi} \int_0^{2\pi} d\theta - \frac{V_m I_m}{4\pi} \int_0^{2\pi} \cos 2\theta d\theta \\
&= \frac{V_m I_m}{4\pi} [\theta]_0^{2\pi} - \frac{V_m I_m}{4\pi} \left[\frac{\sin 2\theta}{2}\right]_0^{2\pi} \\
&= \frac{V_m I_m}{4\pi} [2\pi - 0] - 0 \\
&= \frac{V_m I_m}{2}
\end{aligned}$$

Hence

$$P_{av} = \frac{V_m I_m}{2} \tag{5.7}$$

Thus in pure resistive circuit, average power is never equal to zero.Therefore the combination of voltage, current and power waveform is shown in Fig. 27.

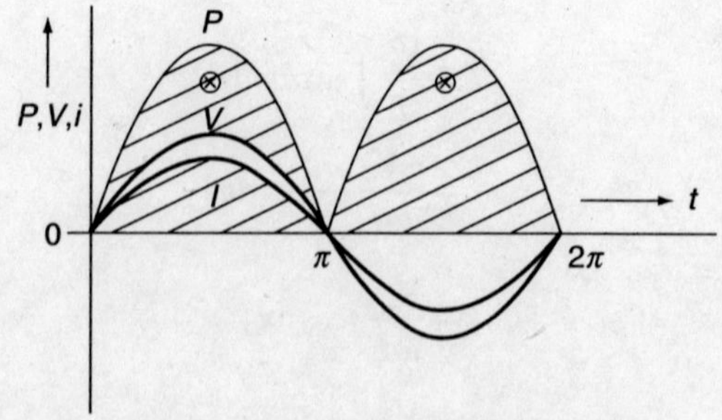

Figure 27 Combination of voltage, current and power waveforms in pure resistor.

5.16.5 Phasor Relation in Pure Inductor

Consider a pure inductive circuit is one which has no resistance and capacitance but only inductance as shown in Fig. 28. When an AC voltage is applied to a pure inductive circuit a current will flow through it. We have

$$V = V_{\mathrm{m}} \sin \omega t \tag{5.8}$$

$$V = L\frac{\mathrm{d}i}{\mathrm{d}t}$$

$$\Rightarrow \frac{V}{L}\,\mathrm{d}t = \mathrm{d}i$$

$$\Rightarrow \mathrm{d}i = \frac{V_{\mathrm{m}}}{L} \sin \omega t \mathrm{d}t$$

$$X_{\mathrm{L}} = \omega L = 2\pi f L$$

$$\Rightarrow I_{\mathrm{L}} = \frac{V_L}{j \times L} = -j\frac{V}{X_{\mathrm{L}}}$$

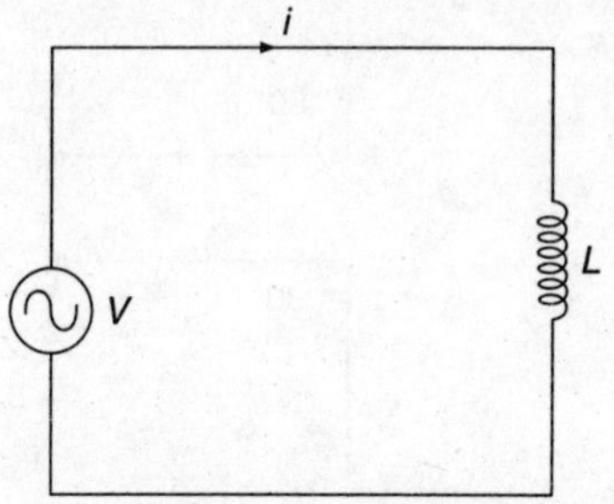

Figure 28 Pure inductor circuit.

Integrating both sides we get

$$I = \frac{V_m}{L}\int \sin \omega t \, dt$$

$$= \frac{-V_m}{\omega L}[-\cos \omega t]$$

$$= \frac{V_m}{\omega L}[\cos \omega t]$$

$$= \frac{V_m}{\omega L}[\sin(\omega t - \pi/2)]$$

$$i = I_m \sin(\omega t - \pi/2) \tag{5.9}$$

where

$$I_m = \frac{V_m}{\omega L} \tag{5.10}$$

because "I" is maximum when $\sin(\omega t - \pi/2) = 1$.

From Eqs. (5.18) and (5.19), it is clear that current lags behind the applied voltage by 90°. This is shown in Figs. 29 and 30.

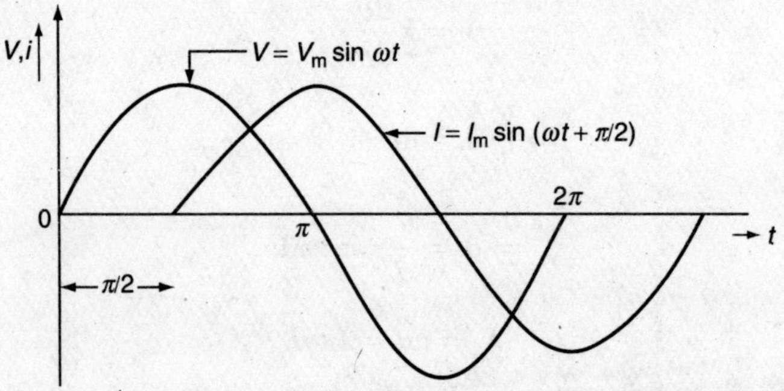

Figure 29 Voltage and current waveforms of pure inductor.

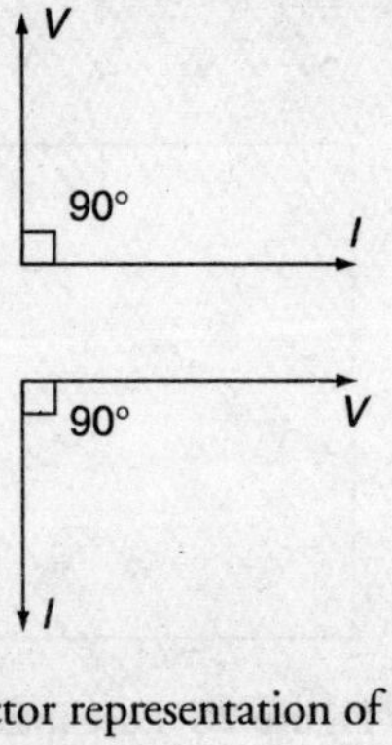

Figure 30 Vector representation of pure inductor.

In Eq. (5.10) "ωL" is called inductive reactance, denoted by "X_L." This acts like resistance. Therefore,

$$I_m = \left(\frac{V_m}{X_L}\right) \tag{5.11}$$

Instantaneous power is given by

$$\begin{aligned} P &= V \times i \\ &= V_m \sin \omega t \times I_m \sin(\omega t - \pi/2) \\ &= -V_m I_m \sin \omega t \cos \omega t \\ &= \frac{-V_m I_m}{2} \sin 2\omega t \end{aligned} \tag{5.12}$$

The combination of current, voltage and power waveforms is shown in Fig. 31. Therefore, the average power over one complete cycle is

$$\begin{aligned} P_{av} &= \frac{-V_m I_m}{2} \int_0^{2\pi} \sin 2\theta d\theta \\ &= \frac{-V_m I_m}{2} \left[\frac{\cos 2\theta}{2}\right]_0^{2\pi} \\ &= \frac{V_m I_m}{4} [\cos 4\pi - \cos 0] \\ &= \frac{V_m I_m}{4} [1 - 1] = 0 \end{aligned}$$

Therefore, in a pure inductive circuit, average power over one complete cycle is zero.

5.16.6 Phasor Relation in Pure Capacitance

Consider a pure capacitive circuit is one which has no resistance and inductance but only capacitance as shown in Fig. 32.

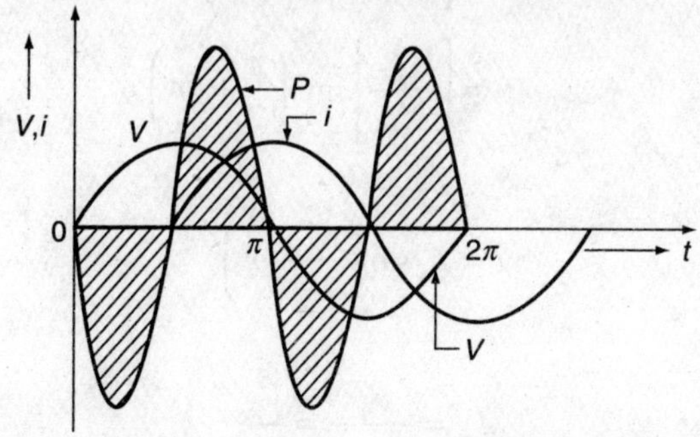

Figure 31 Combination of voltage, current and power waveform of pure inductor.

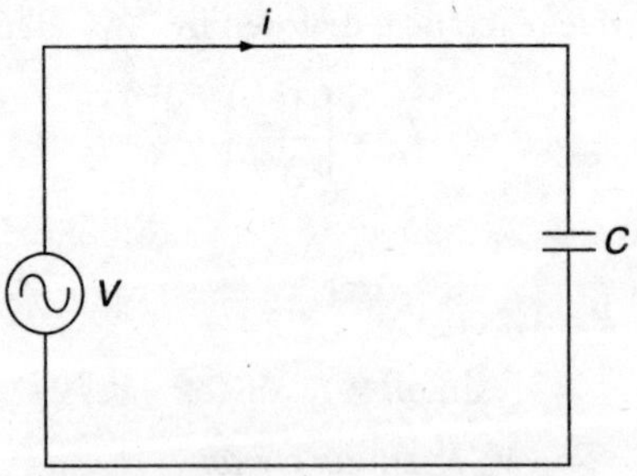

Figure 32 Pure capacitance circuit.

$$V = V_m \sin \omega t \tag{5.13}$$

Also

$$q = CV$$

where V is the voltage between the plates at any instant of time and q is the charge on the plates at that instant of time. Hence

$$q = CV_{max} \sin \omega t$$

We know that

$$i = \frac{dq}{dt}$$

Therefore,

$$\begin{aligned} i &= \frac{d}{dt}(CV_{max} \sin \omega t) \\ &= CV_{max}\, \omega \cos \omega t \\ &= \omega CV_{max} \cos \omega t \\ &= \left[\frac{V_{max}}{1/\omega C}\right] \sin\left(\frac{\pi}{2} + \omega t\right) \\ &= \left[\frac{V_{max}}{X_C}\right] \sin\left(\frac{\pi}{2} + \omega t\right) \end{aligned}$$

Now

$$I = I_m \sin\left(\frac{\pi}{2} + \omega t\right) \tag{5.14}$$

where

$$I_m = \frac{V_m}{1/\omega C} = \frac{V_m}{X_C} \tag{5.15}$$

because "I" becomes I_{max} when $\sin(\pi/2 + \omega t) = 1$. The value

$$X_C = \frac{1}{\omega C}$$

is capacitive reactance which also behaves like resistance and hence its unit is ohm. From Eqs. (5.13) and (5.14), it is clear that current leads voltage by 90°(Fig. 33) and also vector representation is mentioned in Fig. 34.

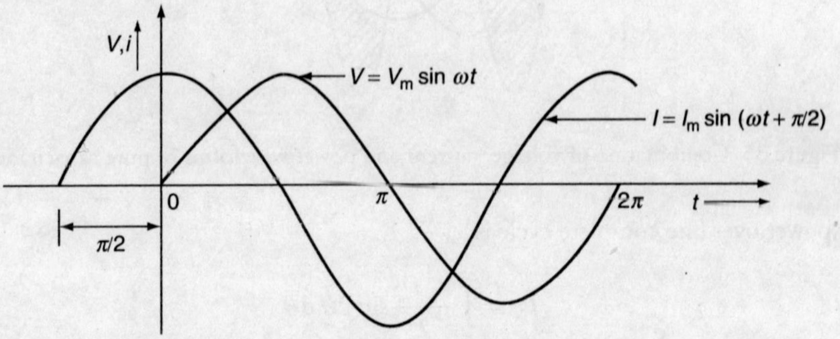

Figure 33 Voltage and current waveforms of pure capacitance.

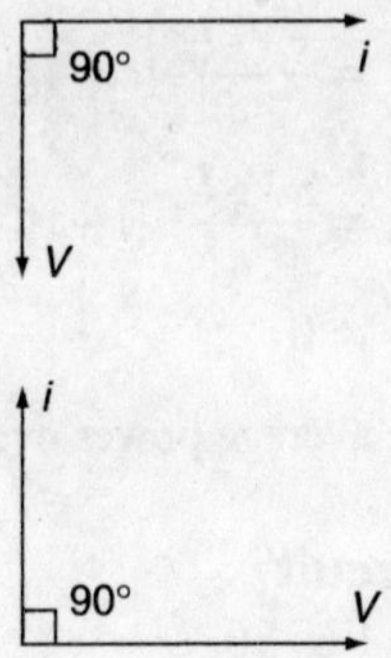

Figure 34 Vector representation of pure capacitance.

Instantaneous power is given by

$$\begin{aligned} P &= V \times i \\ &= V_m \sin \omega t \, I_m \sin(\omega t + 90°) \\ &= V_m I_m \sin \omega t \cos \omega t \end{aligned}$$

$$P = \frac{V_m I_m}{2} \sin 2\omega t \tag{5.16}$$

The combination of voltage, current and power waveform is shown in Fig. 35.

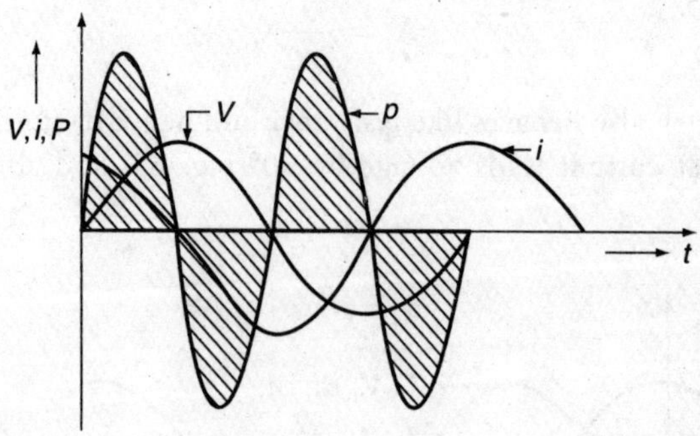

Figure 35 Combination of voltage, current and power waveforms of pure capacitance.

The average power over one complete cycle is

$$P_{\text{av}} = \int_{0}^{2\pi} \frac{V_{\text{m}} I_{\text{m}}}{2} \sin 2\theta \, d\theta$$

$$= \frac{-V_{\text{m}} I_{\text{m}}}{4} \left[\frac{\cos 2\theta}{2} \right]_{0}^{2\pi}$$

$$= \frac{-V_{\text{m}} I_{\text{m}}}{4} [\cos 4\pi - \cos 0]$$

$$= \frac{-V_{\text{m}} I_{\text{m}}}{4} [1 - 1]$$

$$= 0$$

Therefore, in a pure capacitive circuit, the average power over one complete cycle is zero.

5.16.7 Phasor Relation of R–L Circuit

Consider a simple *R–L* circuit is shown in Fig. 36.

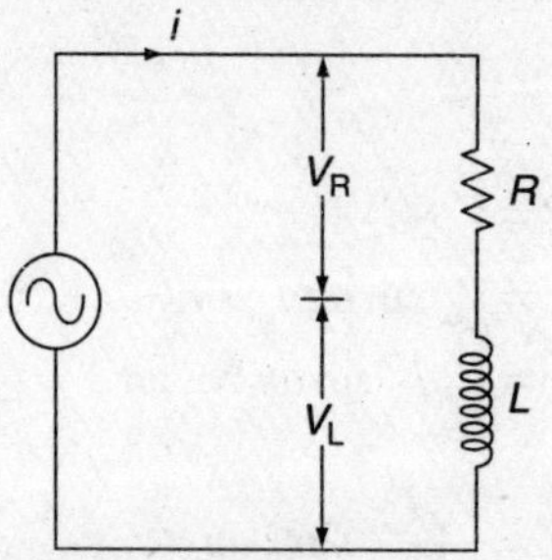

Figure 36 *R–L* circuit.

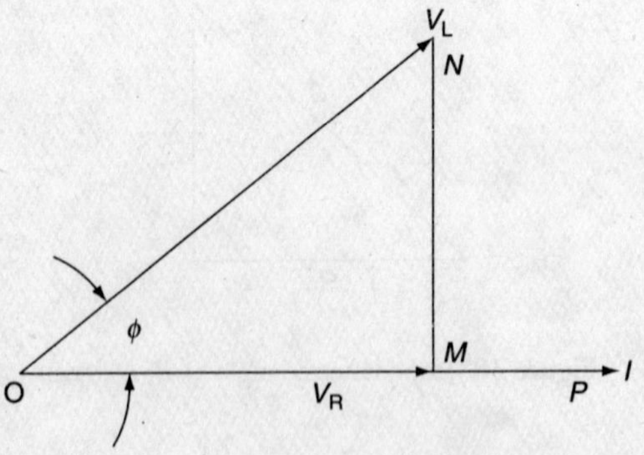

Figure 37 Vector representation of *R–L* circuit given in Fig. 36.

Let

$$V = V_m \sin \omega t \tag{5.17}$$

be the voltage applied across series combination. Let "V" be the RMS value of the applied voltage, "I" be the RMS value of current across "R" and V_L be the RMS value of voltage across "L." Using vector phasor representation of pure resistor and inductor shown in Figs. 26 and 30, we can draw the vector diagram as shown in Fig. 37. Here "OP" depicts the current vector; "OM" depicts V_R, the voltage drop across resistor; "MN" depicts V_L, the voltage drop across inductance; "ON" depicts applied voltage V. Now

$$\text{ON} = \sqrt{(\text{OM})^2 + (\text{MN})^2}$$

$$V = \sqrt{(V_R)^2 + (V_L)^2}$$

But

$$V_R = I \times R$$

$$V_L = I \times X_L$$

Therefore

$$V = \sqrt{(IR)^2 + (IX_L)^2}$$

$$= I\sqrt{(R^2 + X_L^2)}$$

$$\frac{V}{\sqrt{R^2 + X_L^2}} = I$$

The quantity $\sqrt{R^2 + X_L^2}$ is known as impedance and is denoted by "Z." Its unit is ohm. Therefore,

$$I = \frac{V}{Z}$$

From Fig. 38, we have

$$Z^2 = R^2 + X_L^2$$

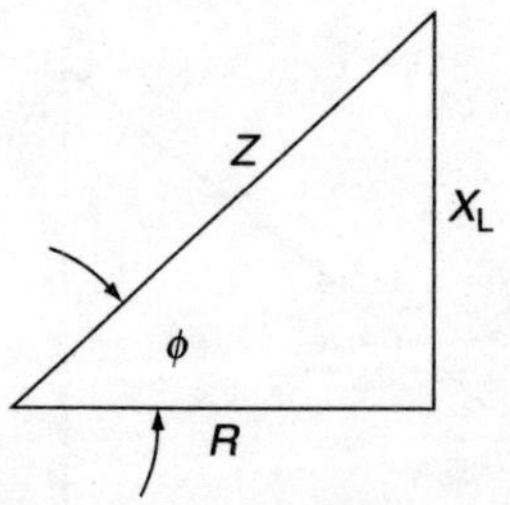

Figure 38 Impedance triangle of *R–L* circuit.

The triangle in Fig. 38 is called impedance triangle. Now

$$\tan\phi = \frac{X_L}{R} \Rightarrow \phi = \tan^{-1}\left(\frac{X_L}{R}\right)$$

$$\cos\phi = \left(\frac{R}{Z}\right) \Rightarrow \phi = \cos^{-1}\left(\frac{R}{Z}\right)$$

$$\sin\phi = \left(\frac{X_L}{Z}\right) \Rightarrow \phi = \sin^{-1}\left(\frac{X_L}{Z}\right)$$

Thus in an *R–L* circuit, if $V = V_m \sin \omega t$ is the applied voltage, then

$$I = I_m \sin(\omega t - \phi)$$

where $0 < \phi < \pi/2$ and

$$I_m = \frac{V_m}{Z}$$

Therefore, "I" lags "V" by an angle "ϕ" (Figs. 39 and 40).

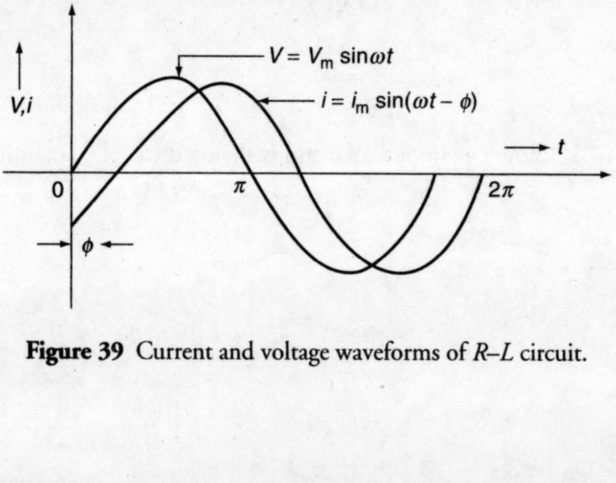

Figure 39 Current and voltage waveforms of *R–L* circuit.

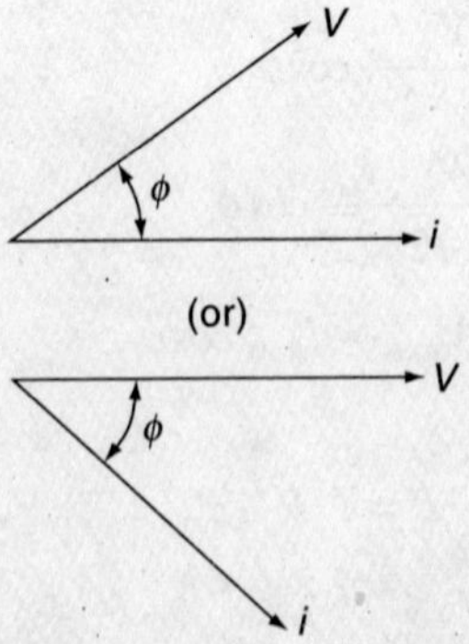

Figure 40 Vector representation of *R–L* circuit for lagging and leading.

Instantaneous power is given by

$$
\begin{aligned}
P &= V \times i \\
&= V_m \sin \omega t \times I_m \sin(\omega t - \phi) \\
&= V_m I_m \sin \omega t \sin(\omega t - \phi) \\
&= \frac{V_m I_m}{2}[\cos \phi - \cos(2\omega t - \phi)] \\
&= \frac{V_m I_m}{2}\cos \phi - \frac{V_m I_m}{2}\cos(2\omega t - \phi) \qquad (5.18)
\end{aligned}
$$

The combination of voltage, current and power waveforms shown in Fig. 41. Average power consumed is given by

$$P_{av} = \int_0^{2\pi} V_m I_m \sin \omega t \sin(\omega t - \phi)\,dt$$

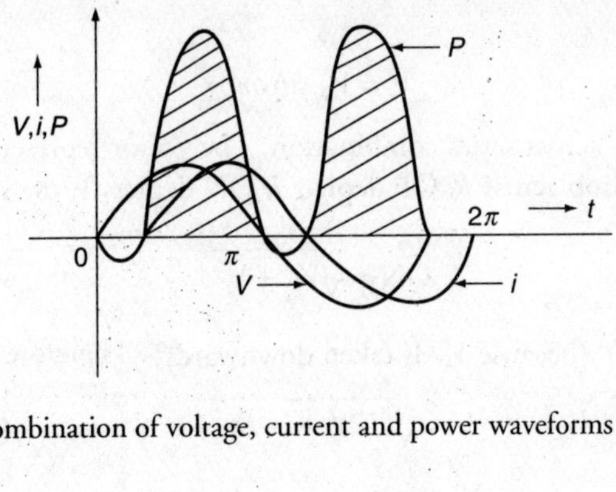

Figure 41 Combination of voltage, current and power waveforms of *R–L* circuit.

$$P_{av} = \frac{V_m I_m}{2} \cos\phi$$

$$= \frac{V_m}{\sqrt{2}} \frac{I_m}{\sqrt{2}} \cos\phi$$

$$P_{av} = V_{RMS} \times I_{RMS} \cos\phi \tag{5.19}$$

In general,

$$P = I^2 R$$

$$= I \times I \times R$$

$$= I \times \frac{V}{Z} \times R$$

$$= I \times V \times \left(\frac{R}{Z}\right)$$

$$= I \times V \times \cos\phi$$

$$= VI \cos\phi$$

5.16.8 Phasor Relation of R–C Circuit

Consider an *R*–*C* circuit as shown in Fig. 42.

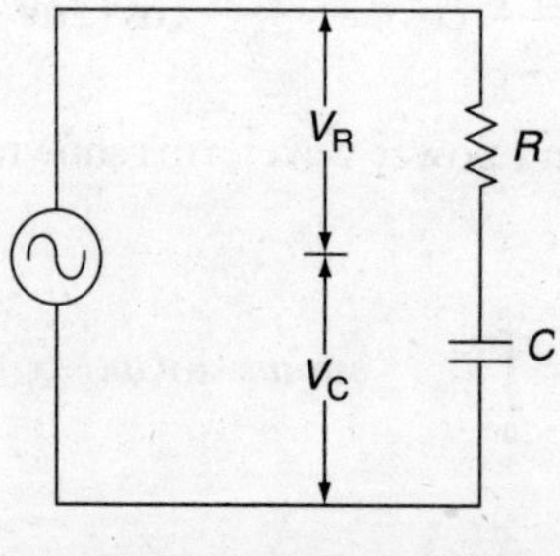

Figure 42 *R*–*C* circuit.

Let

$$V = V_m \sin \omega t \tag{5.20}$$

We have applied the voltage across series combination. The vector representation is shown in Fig. 43 where OM depicts V_R, the drop across *R;* OP depicts *I;* ON depicts *V,* the supply voltage; MN depicts V_C, the drop across *C.* Now

$$\text{ON} = \sqrt{V_R^2 + V_C^2}$$

But $V_R = IR$ and $V_C = -IX_C$ (because V_C is taken downwards) . Therefore

$$V = \sqrt{(IR)^2 + (-IX_C)^2}$$

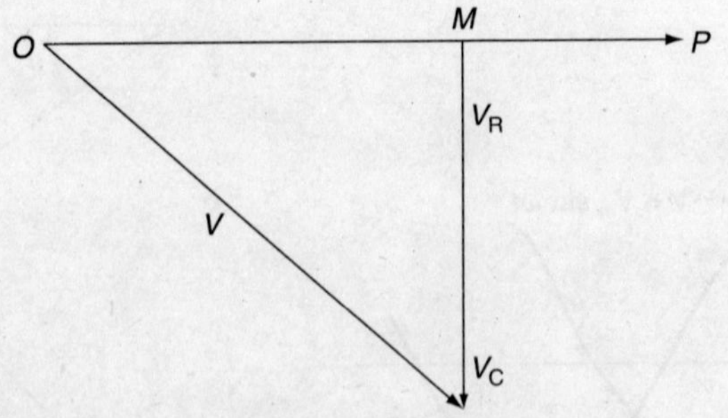

Figure 43 Vector representation of *R–C* circuit given in Fig. 42.

$$V = I\sqrt{R^2 + (-X_C)^2}$$

$$\frac{V}{\sqrt{R^2 + (-X_C)^2}} = I = \frac{V}{Z}$$

where $Z = \sqrt{R^2 + (-X_C)^2}$ is called impedance of the circuit (Fig. 44). Now

$$\cos\phi = \frac{R}{Z} \Rightarrow \phi = \cos^{-1}\left(\frac{R}{Z}\right) \tag{5.21}$$

$$\tan\phi = \left(\frac{-X_C}{R}\right) \Rightarrow \phi = \tan^{-1}\left(\frac{X_C}{R}\right) \tag{5.22}$$

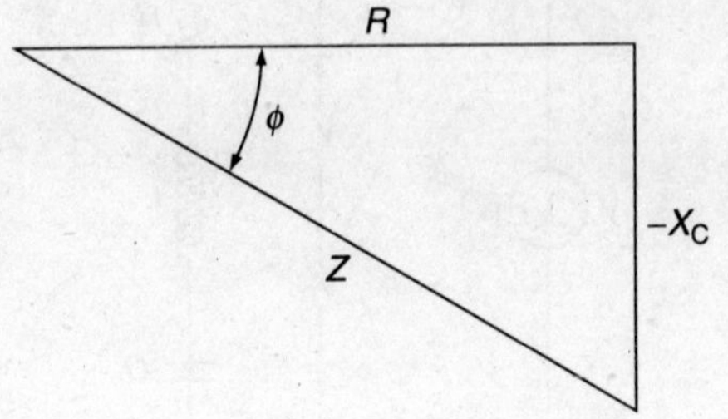

Figure 44 Impedance triangle of *R–C* circuit.

Thus in an *R–C* series circuit, if $V = V_m \sin \omega t$ is the applied voltage, then current

$$I = I_m \sin(\omega t + \phi) \tag{5.23}$$

where

$$0 < \phi < \pi/2$$

"*I*" leads "*V*" by angle ϕ (Fig. 45).

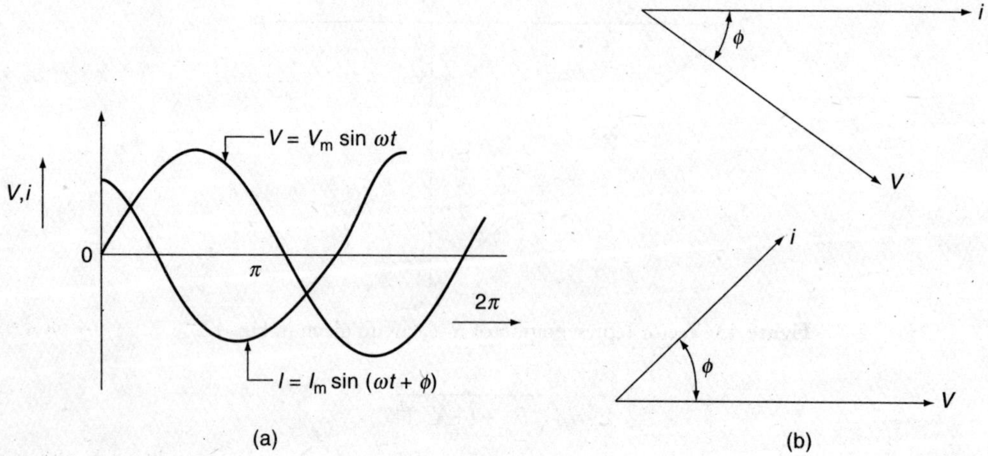

Figure 45 (a) Voltage and current waveforms of *R–C* circuit, (b) vector representation of *R–C* circuit for leading and lagging.

5.16.9 Phasor Relation of R–L–C Circuit

Consider an *R–L–C* series circuit. Let

$$V = V_m \sin \omega t \tag{5.24}$$

be the voltage applied across the series combination as shown in Fig. 46.

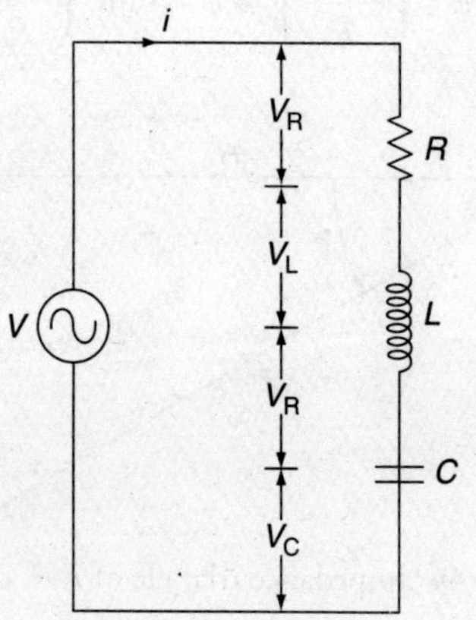

Figure 46 *R–L–C* series circuit.

In vector representation if $V_L > V_C$ and $X_L > X_C$ the circuit can be represented as shown in Fig. 47. On the other hand, if $V_C > V_L$ and $X_C > X_L$ the same can be redrawn as in Fig. 48. From Fig. 47,

$$(\text{OA})^2 = (\text{OB})^2 + (\text{AB})^2$$

$$\text{OA} = \sqrt{(\text{OB})^2 + (\text{AB})^2}$$

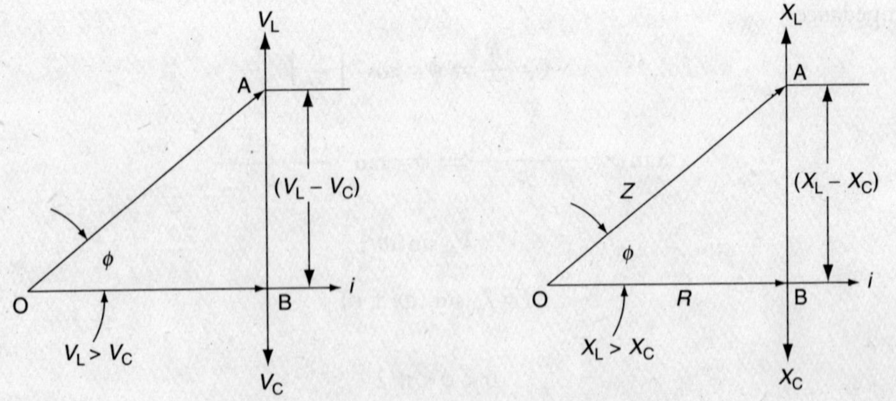

Figure 47 Vector representation of *R–L–C* circuit.

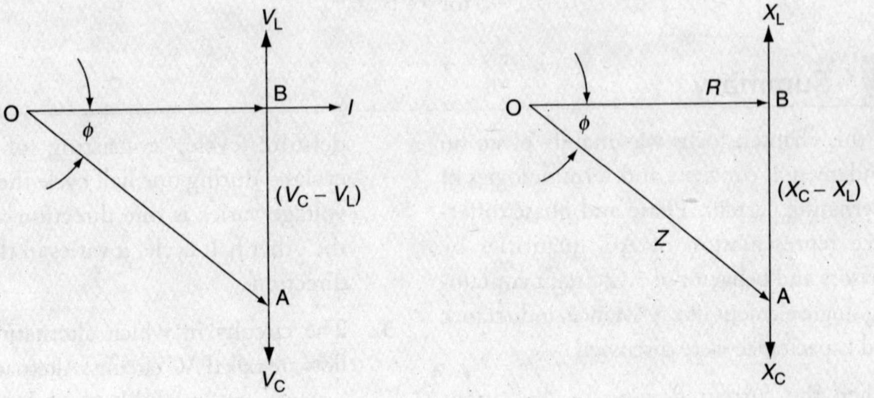

Figure 48 Impedance triangle of *R–L–C* circuit.

where OA depicts V (resultant voltage); OB depicts V_R (voltage across resistance) = IR; AB depicts $(V_L - V_C)$ (net reactive drop) = $I\,(X_L - X_C)$. Hence

$$\text{OA} = \sqrt{(IR)^2 + [I(X_L - X_C)]^2}$$

$$V = I\sqrt{R^2 + (X_L - X_C)^2}$$

$$\frac{V}{\sqrt{R^2 + (X_L - X_C)^2}} = I$$

The quantity $\sqrt{R^2 + (X_L - X_C)^2}$ is called impedance of the circuit.

$$I = \frac{V}{Z}$$

where

$$Z = \sqrt{R^2 + (X_L - X_C)^2} \tag{5.25}$$

From impedance,

$$\cos\phi = \frac{R}{Z} \Rightarrow \phi = \cos^{-1}\left(\frac{R}{Z}\right) \tag{5.26}$$

$$\tan\phi = \frac{(X_L - X_C)}{R} \Rightarrow \phi = \tan^{-1}\frac{(X_L - X_C)}{R} \tag{5.27}$$

$$V = V_m \sin(\omega t) \tag{5.28}$$

$$I = I_m \sin(\omega t \pm \phi)$$

where

$$0 < \phi < \pi/2$$

$$\phi \text{ for } X_C > X_L$$

$$-\phi \text{ for } X_L > X_C$$

Summary

1. In this chapter, focus was mainly given on fundamental concepts and terminologies of alternating current. Phase and phase difference representation of AC quantities by phasors and behavior of AC circuit containing single element like resistance, inductance and capacitance were discussed.
2. When the current flowing in the circuit varies in magnitude as well as direction periodically, it is called alternating current. Thus, an alternating current or voltage is one which periodically passes through a definite cycle, consisting of two half cycles – during one half cycle the current or voltage varies is one direction and during the other half cycle, it varies in the opposite direction.
3. The circuits in which alternating currents flow are called AC circuits. Alternating current systems are invariably used for large-scale power generator, transmission and distribution. Alternating voltage can easily be stepped up and stepped down efficiently with the use of transformers.

Solved Problems

Problem 1

The period of sine wave is 20 ms. What is the frequency?

Solution

The frequency is given by

$$f = \frac{1}{T} = \frac{1}{20}\ \text{ms} = 50\ \text{Hz}$$

Problem 2

The frequency of a sine wave is 30 Hz. What is the period?

Solution

The period is given by

$$T = \frac{1}{f} = \frac{1}{30} = 33.33 \text{ ms}$$

Problem 3

A sinusoidal voltage of 20 V is applied in series with resistance 2 kΩ. Determine I_{RMS}, I_{av}, I_p, I_{pp}, where I_p is the peak current and I_{pp} is the peak to peak current.

Solution

The function given to the circuit is

$$V(t) = V_{\text{p}} \sin \omega t = 20 \sin \omega t$$

where ω is the angular frequency. The current passing through the resistor is

$$I_{\text{p}}(t) = \frac{V_{\text{p}}(t)}{R} = \frac{20}{2\times 10^3} \sin \omega t = 10\times 10^{-3} \sin \omega t$$

Therefore,

$$I_{\text{p}} = 10 \times 10^{-3} \text{ A}$$

The peak current $I_{\text{p}} = 10$ mA.
Peak-to-peak current $I_{\text{pp}} = 20$ mA.
RMS value

$$I_{\text{RMS}} = 0.707\ I_{\text{p}} = 0.707 \times 10 \text{ mA} = 7.07 \text{ mA}$$

Average value

$$I_{\text{av}} = 0.637\ I_{\text{p}} = 0.637 \times 10 \text{ mA} = 6.37 \text{ mA}$$

Problem 4

A sinusoidal voltage is applied in series to a capacitor of 0.01 μF. The frequency of the sine wave is 2 kHz. Determine the capacitive reactance.

Solution

We have

$$X_{\text{C}} = \frac{1}{2\pi f C} = \frac{1}{2\pi \times 2\times 10^3 \times 0.01\times 10^{-6}} = 7.96 \text{ k}\Omega$$

Problem 5

A sinusoidal voltage is applied in series with inductance 2 mH. The frequency is 3 kHz. Determine the inductive reactance.

Solution
We have

$$X_L = 2\pi f\, L = 2\pi \times 3 \times 10^3 \times 2 \times 10^{-3} = 37.69\ \Omega$$

Problem 6

Add the following vectors and hence find the magnitude and scope of resultant vector.

$$A = 3 + j5$$
$$B = 8 - j6$$

Solution
Adding the given vectors we get

$$A + B = (3 + j5) + (8 - j6)$$
$$= 11 - j1$$

$$\text{Magnitude} = |A + B| = \sqrt{(11)^2 + (-1)^2} = 11.045$$

$$\text{Scope of } (A + B) = \tan^{-1}\left(\frac{-1}{11}\right) = -5.19^\circ$$

Problem 7

Subtract the vectors given in Problem 6 and hence find the magnitude and scope of resultant vector.

Solution
Given

$$A = 3 + j5$$
$$B = 8 - j6$$

Subtracting we get

$$A - B = (3 + j5) - (8 - j6)$$
$$= 3 + j5 - 8 + j6 = -5 + j11$$

Therefore

$$\text{Magnitude} = |A - B| = \sqrt{(-5)^2 + (11)^2} = 12.08$$

$$\text{Slope of } (A - B) = \tan^{-1}\left(\frac{11}{-5}\right) = -65.55^\circ$$

Problem 8

Consider the following two vectors: $A = 60\angle 20°, B = 30\angle 15°$. *Determine*

(i) $\dfrac{A}{B}$

(ii) $A \times B$

Solution

Given that

$$A = 60\angle 20°$$

$$B = 30\angle 15°$$

(i) We have

$$\frac{A}{B} = \frac{60\angle 20°}{30\angle 15°} = 2\angle(20° - 15°) = 2\angle 5°$$

(ii) Also,

$$A \times B = 60\angle 20° \times 30\angle 15°$$

$$= 1800\angle(20° + 15°)$$

$$= 1800\angle 35°$$

Problem 9

For the following data

$$A = 12\angle -10°, B = 16 + j8, C = 0 - j20, D = 2 + j0$$

perform the following operations:

(i) $A \times B$

(ii) $B \times C$

(iii) $C \times D$

(iv) A/B

(v) B/C

(vi) C/D

(vii) $ABCD$

(viii) AB/CD

(ix) CD/AB

(x) BC/AD

(xi) $A + B$

(xii) $C + D$

(xiii) $A - B$

Solution

Given $A = 12\angle -10°$. Therefore, rectangular form of A is

$$A = 11.81 - j2$$

Now, $B = 16 + j8$. The polar form of B is

$$B = 17.88\angle 26.56°$$

Given, $C = 0 - j20$. The polar form of C is

$$C = 20\angle -90°$$

Finally $D = 2 + j0$. The polar form of D is

$$D = 2\angle 0°$$

Note: Addition and subtraction can be easily done in rectangular form and multiplication and division can be easily done in polar form.

(i) $A \cdot B = 12\angle -10° \times 17.88\angle 26.56°$
$= 12 \times 17.88\angle(-10° + 26.56°) = 214.56\angle 16.5°$

(ii) $B \cdot C = 17.88\angle 26.56° \times 20\angle -90°$
$= 17.88 \times 20\angle(26.56° - 90°)$
$= 357.6\angle -63.44°$

(iii) $C \cdot D = 20\angle -90° \times 2\angle 0° = 20 \times 2\angle(-90° + 0°) = 40\angle -90°$

(iv) $\dfrac{A}{B} = \dfrac{12\angle -10°}{17.88\angle 26.56°} = \dfrac{12}{17.88}\angle(-10° - 26.56°)$
$= 0.671\angle -36.56°$

(v) $\dfrac{B}{C} = \dfrac{17.88\angle 26.56°}{20\angle -90°} = \dfrac{17.88}{20}\angle(26.56° + 90°)$
$= 0.894\angle 116.5°$

(vi) $\dfrac{C}{D} = \dfrac{20\angle -90°}{2\angle 0°} = \dfrac{20}{2} = \angle(-90° - 0°) = 10\angle -90°$

(vii) $ABCD = 12\angle -10° \times 17.88\angle 26.56° \times 20\angle -90° \times 2\angle 0°$
$= 12 \times 17.88 \times 20 \times 2\angle(-10° + 26.56° - 90° + 0°)$
$= 858.24\angle -73.44°$

(viii) $\dfrac{AB}{CD} = \dfrac{214.56\angle 16.5°}{40\angle -90°} = 5.364\angle(16.5° + 90)$
$= 5.364\angle 106.5°$

(ix) $\frac{CD}{AB} = \frac{40\angle -90^\circ}{214.56\angle 16.5^\circ} = 0.1864\angle(-90^\circ + 16.5^\circ)$

$= 0.1864\angle -106.5^\circ$

(x) $\frac{BC}{AD} = \frac{357.6\angle -63.44^\circ}{12\angle -10^\circ \times 2\angle 0^\circ} = 14.9\angle -53.44^\circ$

(xi) $A - B = (11.81 - j2) - (16 + j8)$

$= 11.81 - j2 - 16 + j8$

$= 27.81 - j6$

(xii) $C - D = (0 - j20) - (2 - j0)$

$= 2 - j20$

(xiii) $A - B = (11.81 - j2) - (16 - j8)$

$= 11.81 - j2 - 16 - j8$

$= -4.14 - j10$

Problem 10

Find the magnitude and phase of Z where

$$Z = \frac{10 + j3}{6 + j12}$$

Solution

Given

$$Z = \frac{10 + j3}{6 + j12}$$

$10 + j3$ in polar form is $16.44\angle 16.7^\circ$

$6 + j12$ in polar form is $13.41\angle 63.4^\circ$

Therefore,

$$Z = \frac{16.44\angle 16.7^\circ}{13.41\angle 63.4^\circ} = 0.7785\angle -46.7^\circ$$

Hence, magnitude of Z is $|Z| = 0.7785$.

Phase angle of Z is -46.7°.

Problem 11

Determine the resultant of two vectors

$$i_1 = 5\sin\omega t$$
$$i_2 = 10\sin\omega t + 60°$$

Solution

Draw

$$\text{OA} = 5\text{ A along } X\text{-axis}$$
$$\text{OD} = 10\text{ A at } 60° \text{ to } X\text{-axis}$$

with suitable scale. Draw

$$\text{AC} = 10\text{ A parallel to "OD" at "A"}$$

Join "O" and "C." OC gives the resultant (i_R) of i_1 and i_2. Therefore,

$$i_R = \text{vector sum of } i_1 \text{ and } i_2 = \text{OC}$$

Referring to the vector diagram (Fig. 49), magnitude of OC is

$$\text{OC} = \sqrt{\text{OB}^2 + \text{BC}^2}$$
$$\alpha = \tan^{-1}\left(\frac{\text{BC}}{\text{OB}}\right)$$

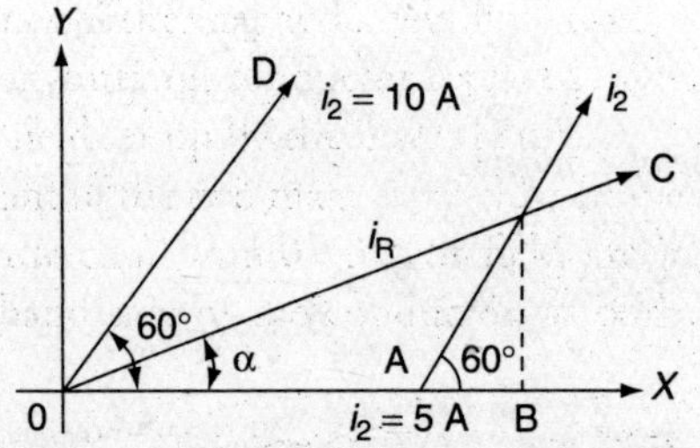

Figure 49 Solution to Problem 11.

From ΔABC,

$$\sin 60° = \frac{\text{BC}}{\text{AC}}$$
$$\Rightarrow \text{BC} = \text{AC}\sin 60°$$
$$\Rightarrow \text{BC} = 10\sin 60°$$
$$= 8.66\text{ A}$$

Consider ΔABC,

$$\cos 60° = \frac{\text{AB}}{\text{AC}}$$

Therefore,

$$AB = (AC)\cos 60° = 10 \times 0.5 = 5$$

$$OB = OA + AB = 5 + 5 = 10$$

$$OC = \sqrt{(10)^2 + (8.66)^2} = 13.228$$

$$\alpha = \tan^{-1}\left(\frac{8.66}{10}\right) = 40.89°$$

Resultant current is

$$i_R = |OC|\sin(\omega t + \alpha)$$

$$i_R = 13.22\,8 \sin(\omega t + 40.89°)$$

Problem 12

An alternating current of $3\sin(4t + \pi/4)$ passes through

(i) An inductance of 0.1 H.

(ii) A capacitance of 100 μF.

Find the voltage applied in each case.

Solution

(i) Given current is

$$I = 3\sin(4t + \pi/4)$$

The voltage magnitude is given by $V/I = X_L$ and it leads the current by $\pi/2$. Hence,

$$V = IX_L$$

Now

$$X_L = \omega L = 4 \times 0.1 = 0.4$$

Therefore $V = 0.4 \times 3\sin(4t + \pi/4)$ in magnitude or

$$V = 1.2\sin\left(4t + \frac{\pi}{4} + \frac{\pi}{2}\right) = 1.2\sin\left(4t + \frac{3\pi}{4}\right)$$

The addition of π/2 is because voltage leads current by π/2 in an inductance.

(ii) $V/I = X_C$ and $X_C = 1/\omega C$. Therefore

$$X_C = \frac{1}{\omega C} = \frac{1}{4 \times 100 \times 10^{-6}} = 2500$$

V in magnitude $= IX_C = 2500 \times 3$. Therefore

$$V = 7500\sin\left(4t + \frac{\pi}{4} - \frac{\pi}{2}\right)$$

The subtraction of $\pi/2$ is because voltage lags current by $\pi/2$ in capacitance.

Problem 13

A choke coil has a resistance of 9 Ω is connected in series with an inductance of 0.04 H. An alternating voltage of 150 sin 300t is applied. Find the current.

Solution

Given that $V = 150 \sin 300t$, $\omega = 300$. Therefore

$$X_L = \omega L = 300 \times 0.04 = 12$$

$$Z = \sqrt{R^2 + X_L^2} < \tan^{-1}(X_L/R)$$

$$= [9^2 + 12^2]^{1/2} < \tan^{-1} 12/9 = 15 < 0.93 \text{ rad}$$

Now $V = ZI$ and so

$$I = V/Z$$
$$= \frac{150}{15}\sin(\omega t - \phi)$$
$$= 10\sin(300t - 0.93)$$

The – sign with ϕ is because the current lags behind the voltage by the single angle.
Alternately,

$$I = \frac{V}{Z}$$

$$= \frac{V}{R + j\omega L}$$

$$= \frac{150}{9 + j12}$$

$$= \frac{150(9 - j12)}{9^2 + 12^2}$$

$$= \frac{150}{15^2}(9 - j12)$$

The current magnitude is

$$\frac{10}{15}(9^2 + 12^2)^{1/2} = \frac{10 \times 15}{15} = 10$$

The phase angle $\phi = \tan^{-1}(-12/9) = -0.93$ rad and $I = 10 \sin (300t - 0.93)$ as before.

Problem 14

Determine the inductance, reactance, impedance and power factor of a circuit which takes 10 A and dissipates 1390 W when connected to a 250 V 50 Hz main supply.

Solution

$$\text{Power} = VI\cos\phi = 1390$$

$$\text{Power factor} = \cos\phi = \frac{1390}{VI} = \frac{1390}{V(250\times 10)} = 0.555$$

Impedance,

$$Z = \frac{V}{I} = \frac{250}{10} = 25\ \Omega$$

Inductance (L): We know that

$$R = Z\cos\phi,\ X = Z\sin\phi$$

where

$$\phi = \cos^{-1} 0.555 = 56°18'$$

Hence

$$X = Z\sin\phi$$
$$= 25\sin\phi$$
$$= 25\sin 56°18'$$
$$= 25\times 0.832$$
$$= 20.85\ \Omega$$

Therefore

$$L = \frac{X}{\omega} = \frac{20.85}{2\pi f} = \frac{20.85}{2\pi\times 50} = 0.0663\ \text{H}$$

Inductive reactance = 20.85 Ω, impedance = 25 Ω.

Problem 15

A capacitance of 20 μF and a resistance of 100 Ω are connected is series across a 120 V 60 cycles/s mains. Determine the average power in the circuit.

Solution

Reactance of a capacitor is given by

$$X_C = \frac{1}{\omega C}$$

$$= \frac{1}{2\pi fC}$$

$$= \frac{1}{2\pi(60)(20\times10^{-6})}$$

$$= \frac{10^4}{24\pi} = 132.8$$

Impedance is

$$Z = (R^2 + X^2)^{1/2} = (100^2 + 132.8^2)^{1/2} = 166.5\ \Omega$$

Current is

$$I = V/Z = 120/166.5 = 0.721\ \text{A}$$

Hence

$$\text{Circuit power} = \text{Power consumed by resistance}$$
$$= (0.721)^2 \times 100 = 52\ \text{W}$$

Problem 16

A 100 V 80 W lamp is to be operated on 300 V 60 Hz AC supply. Calculate the inductance of the choke required to be connected in series with the lamp for this operation. The lamp can be taken as equivalent to a non-inductive resistance.

Solution

We know that

$$V^2/R = \omega \Rightarrow R = V^2/\omega = 100^2/80 = 125\ \Omega$$

$$Z = \sqrt{(R^2 + L^2\omega^2)}$$

$$I = V/Z = 300/Z$$

Now

$$\text{Power of lamp} = 100\ \text{W}$$
$$= \text{Volt across lamp} \times \text{Current } I$$
$$= 100 \times I$$

This gives $I = 1$ and so

$$Z = V/I = 300\ \Omega$$

Hence substituting the values in $Z = \sqrt{(R^2 + L^2\omega^2)}$ we get

$$300^2 = 125^2 + (\omega L)^2$$

$$\omega L = \sqrt{300^2 - 125^2} = 273$$

$$L = 273/\omega = 273/(2\pi \times 60) = 0.724\ \text{H}$$

Problem 17

A coil of resistance 20 Ω and an inductance of 60 mH is connected in series with a 130 μF capacitor across a 230 V 5 Hz supply. Calculate

(i) The impedance.
(ii) The power factor of the circuit.
(iii) The current flowing in it.
(iv) The voltage across the coil.
(v) The voltage across the capacitor.

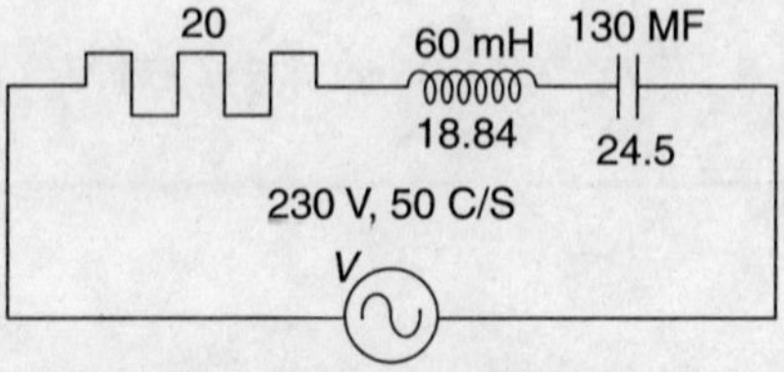

Figure 50 Solution to Problem 17.

Solution

See Fig. 50. Reactance of the inductance,

$$X_L = 2\pi fL$$
$$= 2\pi \times 50 \times 60 \times 10^{-3}$$
$$= 18.84\ \Omega$$
$$X_C = 1/(2\pi fC)$$
$$= 1/(2\pi \times 50 \times 130 \times 10^{-6})$$
$$= 24.5\ \Omega$$

(i) Impedance is given by

$$Z = \sqrt{R^2 + (X_L - X_C)^2}$$
$$= \sqrt{20^2 + (24.5 - 18.84)^2}$$
$$= 20.8\ \Omega$$

(ii) The power factor of the circuit

$$\cos\phi = \frac{R}{Z} = \frac{20}{20.8} = 0.96$$

$$\phi = \cos^{-1} 0.96 = 16.26$$

(iii) The current flowing in the circuit:

$$\frac{V}{Z} = \frac{230}{20.8} = 11.05\,\text{A}$$

(iv) Voltage across the coil:

$$IX_{\text{C}} = 11.05 \times 18.84 = 208.5\ \text{V}$$

(v) Voltage across the capacitor:

$$IX_{\text{L}} = 11.05 \times 24.5 = 271\ \text{V}$$

Problem 18

An alternating voltage $80 + j60$ is applied to a circuit and the current flowing is $-4 + j10$ A. Find

(i) The circuit impedance.
(ii) Power consumed.
(iii) The phase angle.

Solution

(i) Impedance:

$$Z = \frac{V}{I} = \frac{80 + j60}{-4 + j10} = \frac{(80 + j60)(-4 - j10)}{(-4)^2 + 10^2}$$

$$= \frac{[(-320 + 600) - 240j - 800j]}{116} = 2.41 - j8.95$$

$$Z \text{ in magnitude } = [(2.41)^2 + (8.95)^2]^{1/2} = 9.27\ \Omega$$

(ii) Power consumed:

$$\text{Power} = VI = (80 - j60)(-4 + j10)$$
$$= 280 + 1040j$$

(iii) Phase angle:

$$\cos\phi = R/Z = 2.41/9.27 = 0.26 \text{ leading}$$

Problem 19

The equation for a voltage wave is $V = 0.02\sin(4000t + 30°)$. Find the frequency, the instantaneous voltage when $t = 300$ μs. What is the time represented by 20° phase difference?

Solution

We know that

$$V = V_{\mathrm{m}} \sin(2\pi ft + \theta)$$

We calculate the frequency:

$$2\pi f = 4000$$
$$f = \frac{4000}{2\pi} = 637 \text{ Hz}$$

Now $t = 300$ μs, hence i instantaneous voltage is given by

$$\begin{aligned} V &= 0.02\sin\left(4000 \times 300 \times 10^{-6} \times \frac{180}{\pi} + 20°\right) \\ &= 0.02\sin(68.7 + 20) \\ &= 0.02\sin(88.7) \\ &= 0.0199 \text{ V} \end{aligned}$$

Now since $f = 637$ Hz, therefore

$$\frac{1}{f} = \frac{1}{637}$$

and the time represented by 20° phase difference is

$$\frac{20}{360} \times \frac{1}{637} \text{ s} = 0.863 \text{ ms}$$

Problem 20

An alternating current takes 3.375 ms to reach 15 A for the first time after becoming instantaneously zero. The frequency of the current is 40 Hz. Find the maximum value of the alternating current.

Solution

We have $i = 15$ A, $t = 3.375$ ns, $f = 40$ Hz. Now

$$i = I_m \sin 2\pi ft$$
$$15 = I_m \sin(2\pi \times 40 \times 3.375 \times 10^{-3})$$
$$15 = I_m \times 0.75$$
$$I_m = 20\text{ A}$$

Problem 21

An alternating current with the frequency 50 cycles/s has a maximum value of 100 A.

(i) Calculate its value 1/600 s after the instant the current is zero.
(ii) In how many seconds after the zero value will the current attain the value of 86.6 A?

Solution

(i) We have $f = 50$ cycles/s, $I_m = 100$ A. Then

$$I = I_m \sin 2\pi ft$$
$$= 100 \sin\left(2\pi \times 50 \times \frac{1}{600}\right)$$
$$= 100 \sin(30°) = 50\text{ A}$$

(ii) Now

$$I = I_m \sin 2\pi ft$$
$$86.6 = 100 \sin(2\pi \times 50 \times t)$$
$$\sin(100\pi t) = 0.866$$
$$100\pi t = 60°$$
$$t = \frac{60}{100 \times 180} = \frac{1}{300}\text{ s}$$

Problem 22

Find the following parameters of a voltage V = 200 sin 310 t:
(i) Frequency.
(ii) Form factor.

Solution

We have

$$V = 200 \sin 310t$$

Now

$$V = V_m \sin 2\pi ft$$

(i) Comparing both equations we get

$$2\pi f = 310$$

$$f = \frac{310}{2\pi} = 49 \text{ Hz}$$

(ii) **Form factor:** Now

$$V_{av} = \frac{2V_m}{\pi}$$

$$V_{RMS} = \frac{V_m}{\sqrt{2}}$$

$$\text{Form factor} = \frac{V_{RMS}}{V_{av}} = \frac{V_m/\sqrt{2}}{2V_m/\pi} = 1.11$$

Problem 23

Find the following parameters of voltages V = 200 sin 300 t:

(i) Frequency.

(ii) Crest factor.

Solution

We are given that

$$V = 200 \sin 300t$$

and we know that

$$V = V_m \sin 2\pi ft$$

(i) Comparing the two equations we get

$$2\pi f = 300$$

$$f = \frac{300}{2\pi} = 47 \text{ Hz}$$

(ii) Again

$$V_{av} = \frac{2V_m}{\pi}$$

$$V_{RMS} = \frac{V_m}{\sqrt{2}}$$

Therefore

$$\text{Crest factor} = \frac{V_m}{V_{RMS}} = \frac{V_m}{V_m/\sqrt{2}} = 1.414$$

Problem 24

A non-sinusoidal voltage is having a form factor of 1.2 and peak factor of 1.5. If the average value of the voltage is 10 V, calculate

(i) RMS value.
(ii) Maximum value.

Solution

We have $k_f = 1.2;\ k_p = 1.5,\ V_{av} = 10.$

(i) The RMS value

$$k_f = \frac{V_{RMS}}{V_{av}}$$

$$1.2 = \frac{V_{RMS}}{10}$$

$$V_{RMS} = 12\,\text{V}$$

(ii) The maximum value

$$k_p = \frac{V_m}{V_{RMS}}$$

$$1.5 = \frac{V_m}{12}$$

$$V_m = 18\,\text{V}$$

Problem 25

Find the RMS and average values of the waveform shown in Fig. 51.

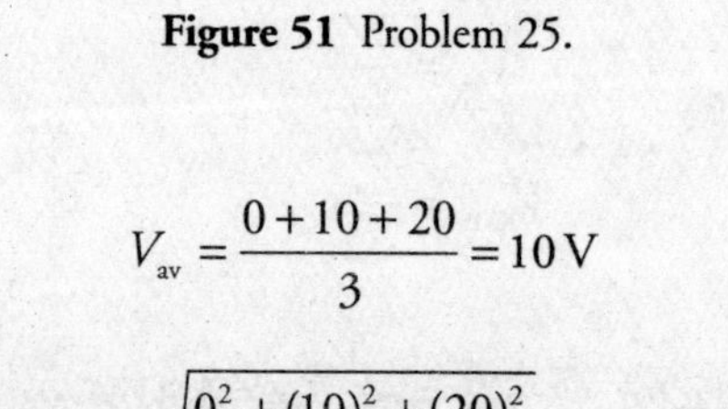

Figure 51 Problem 25.

Solution

We have

$$V_{av} = \frac{0 + 10 + 20}{3} = 10\,\text{V}$$

$$V_{RMS} = \sqrt{\frac{0^2 + (10)^2 + (20)^2}{3}} = 12.9\,\text{V}$$

Problem 26

Find the relative heating affects of two current waves of equal peak value, one sinusoidal and the other rectangular in shape (Fig. 52).

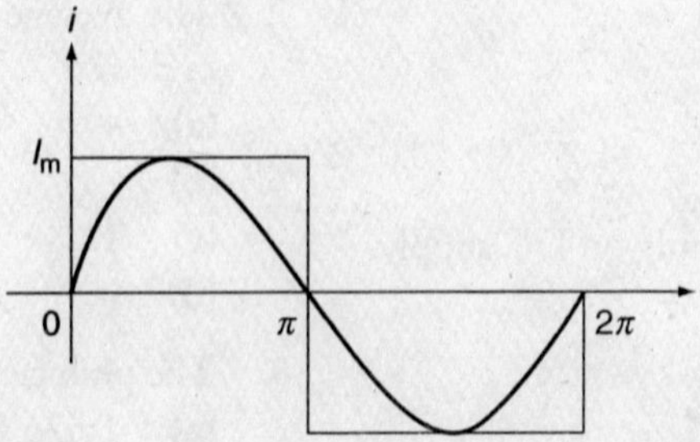

Figure 52 Problem 26.

Solution

RMS value of the rectangular wave is I_m. RMS value of sinusoidal current wave is $I_m/\sqrt{2}$.

Heating effect due to rectangular current wave is $(I_m)^2 RT$. Heating effect due to sinusoidal current wave is

$$\left(\frac{I_m}{\sqrt{2}}\right)^2 RT = \frac{(I_m)^2}{2} RT$$

So the relative heating affects of two current waves of equal peak value is

$$\frac{(I_m)^2}{2} RT : (I_m)^2 RT$$

$$= \frac{1}{2} : 1$$

$$= 1 : 2$$

Problem 27

An alternating current takes 3.375 ms to reach 10 A for the first time after becoming instantaneously zero. The frequency is 30 Hz. Find the maximum current.

Solution

We have $i = 10$ A, $t = 3.375$ ms, $f = 30$ Hz. Now

$$i = I_m \sin 2\pi ft$$

$$10 = I_m \sin(2\pi \times 30 \times 3.375 \times 10^{-3})$$

$$10 = I_m (0.0111)$$

$$I_m = 10/0.0111 = 900 \text{ A}$$

Multiple Choice Questions

1. In A.C. circuit, the power is consumed only in
   (a) inductance
   (b) capacitance
   (c) resistance
   (d) susceptance

2. __________ will work only on DC supply.
   (a) electric lamp
   (b) refrigerator
   (c) heater
   (d) electroplating

3. A sine wave has a maximum value of 20V. Its value at 135° is
   (a) 10 V
   (b) 14.14 V
   (c) 15 V
   (d) 5 V

4. A sinusoidal current has a magnitude of 3 A at 120°. Its maximum value will be
   (a) $\sqrt{3}$ A
   (b) $\frac{\sqrt{3}}{2}$ A
   (c) $2\sqrt{3}$ A
   (d) 6 A

5. An electric bulb rated 220 V is connected 200 V, 5 Hz source. Then, the bulb _____
   (a) does not glow
   (b) glows continously
   (c) glows intermittently
   (d) fuses

6. The RMS value of sinusoidally varying current is ____ that of its average value.
   (a) more than
   (b) less than
   (c) same as
   (d) none of the above

7. The average value of $\sin^2\theta$ over a complete cycle is
   (a) +1
   (b) −1
   (c) 1/2
   (d) 0

8. The peak factor of a half-wave rectified AC is
   (a) 1.57
   (b) 2
   (c) 1.11
   (d) 1.4142

9. The form factor of a half-wave rectified AC is
   (a) 2
   (b) 1.11
   (c) 1.414
   (d) 1.57

10. The average value of a sinusoidal current is 100 A. Its RMS value is
    (a) 63.7 A
    (b) 70.7 A
    (c) 141.4 A
    (d) 111 A

11. Which of the following wave has the highest peak value?
    (a) sinusoidal
    (b) square
    (c) rectangular
    (d) triangular

12. A 100 V peak AC is as effective as __________ DC.
    (a) 100 V
    (b) 50 V
    (c) 70.7 V
    (d) None of the above

13. The form factor __________ of a wave is 1.
    (a) sinusoidal
    (b) square
    (c) triangular
    (d) tooth

14. The peak factor of a sine wave form is
    (a) 1.11
    (b) 1.414
    (c) 2
    (d) 1.5

Review Questions

1. Explain the concept of generation of alternating e.m.f.
2. Define alternating voltage and current.
3. Define frequency and time period.
4. Define peak value.
5. Derive an expression of RMS value, average value, form factor and peak factor of sinusoidal alternating quantity.
6. Define form factor and peak factor.
7. Explain the concept of root mean square value.
8. Write an expression of phasor representation in series *R–L* circuit.
9. Write an expression of phasor representation in series *R–C* circuit.
10. Define phase.
11. Explain in detail about the complex algebra in electrical engineering.
12. Write an expression of behavior of AC through pure capacitance.
13. Write an expression of phasor representation in series *R–L–C* circuit.

Problems

1. Determine the periodic time for frequencies of **(a)** 50 Hz and **(b)** 20 kHz.
2. Determine the frequencies for periodic time of **(a)** 4 ms and **(b)** 4 μs.
3. An alternating current completes 5 cycles in 8 ms. What is its frequency?
4. Determine the **(a)** peak value and **(b)** mean value for a 240 V mains supply.
5. An alternating current is given by $i = 20 \sin 600t$ amperes. Find the **(a)** frequency, **(b)** peak value of current.
6. An alternating current of frequency 50 Hz has a maximum value of 100 A. **(a)** Write down the equation for its instantaneous value. **(b)** Also find the time taken to reach 70A for the first time.
7. An alternating voltage is given by $v = 141.4 \sin 314t$. Find **(a)** frequency, **(b)** RMS value, **(c)** average value.
8. The equation of an alternating current $i = 42.42 \sin 628t$. Determine **(a)** its maximum value, **(b)** frequency, **(c)** RMS value, **(d)** average value, **(e)** form factor.

9. An AC circuit consists of a pure resistance of 10 Ω and is connected across an AC supply of 230 V, 50 Hz. Calculate **(a)** current, **(b)** power consumed.

10. A 4 Ω resistor is connected to a 10 mH inductor across a 100 V, 50 Hz voltage source. Find **(a)** input current, **(b)** voltage drops across resistor and inductor, **(c)** power factor of the circuit and **(d)** real power consumed in the circuit.

11. A coil has an inductance of 40 mH and negligible resistance. Calculate its inductive reactance and the resulting current if connected to **(a)** a 240V, 50 Hz supply and **(b)** a 100V, 1 kHz supply.

12. A circuit consists of 20 Ω resistance in series with capacitance of 200 μF, connected across 50 Hz supply. The current through the circuit is 10.8 sin 314t amperes. Determine the voltage across each component and across the circuit.

13. For a supply sine wave peak value of 100 V, determine **(a)** the RMS value, **(b)** the average value and **(c)** crest factor.

14. The minimum value of sinusoidal alternating current having frequency 150 Hz is 75 A. Write down the equation for instantaneous value of alternating current. Determine i at 0.0035.

15. The expressions of two voltage waves are given below:

$$V_A = 200\sin(\omega t + 60°)$$
$$V_B = 150\sin(\omega t - 30°)$$

Determine the phase angle between V_A and V_B.

16. Find the root mean square value of resultant current in a wire which simultaneously directs a current of 22.2 A and sinusoidal alternating current of peak value 22.2 A.

17. Determine the effective RMS value of voltage waveform represented by

$$V = 400\sin\omega t + 200\sin 3\omega t + 100\sin 5\omega t$$

18. Find average value of the trapezoidal waveform shown in Fig. 53.

19. Determine the RMS value of half-wave rectified alternating current represented by $i = 10\sin\omega t$.

20. Determine **(a)** the average value and **(b)** form factor of a half-wave rectified current represented by $i = 20\sin\omega t$ where RMS value is given as 10 A.

21. If a delayed full-wave rectifier sinusoidal current has an average value equal to 1/5 times its maximum value, find delay angle θ.

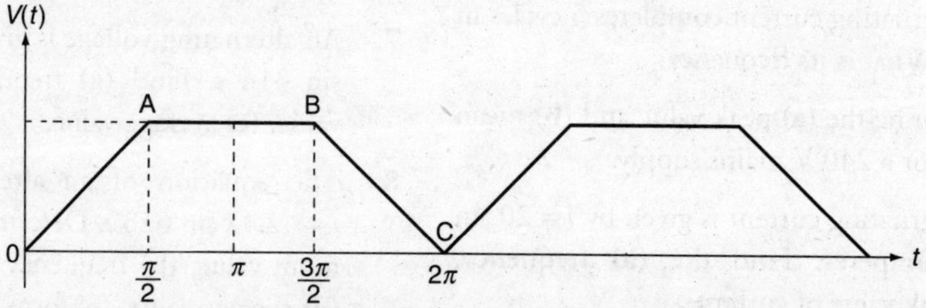

Figure 53 Problem 18.

22. A half-wave rectified sinusoidal current having $I_m = 10$ A is delayed by an angle of $\theta = 60°$. Find **(a)** the average value and **(b)** the RMS value.

23. If a half-wave rectified sinusoidal current having $I_m = 20$ A is delayed by angle $\theta = 45°$, find **(a)** the average current and **(b)** form factor if the RMS value is given as 9.535 A.

24. Determine frequency if V_m is 62.8 V, number of turns is 100 and $\phi_m = 0.002$ Wb.

25. The expressions of two voltage wave are given below:

$$V_A = 75\sin(\omega t + 60°)$$
$$V_B = 125\sin(\omega t - 30°)$$

Determine the phase angle between V_A and V_B.

26. Determine the effective, that is, RMS value of the voltage waveform represented by

$$V = 100\sin\omega t + 50\sin 3\omega t$$

27. Find the average value of trapezoidal waveform shown in Fig. 54.

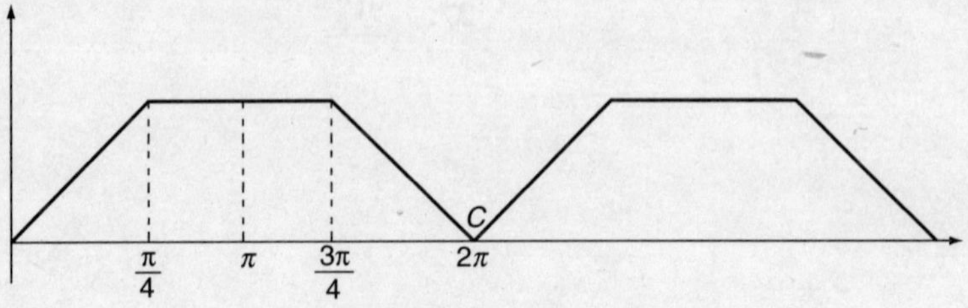

Figure 54 Problem 27.

Answers

Multiple Choice Questions

1. (c)
2. (d)
3. (b)
4. (c)
5. (c)
6. (a)
7. (c)
8. (b)
9. (d)
10. (d)
11. (d)
12. (c)
13. (b)
14. (b)

Problems

1. **(a)** 20 ms, **(b)** 50 μs

2. **(a)** 250 Hz, **(b)** 0.25 MHz

3. 625 Hz

4. **(a)** 339.41 V, **(b)** 216.3 V

5. **(a)** 89.5 Hz, **(b)** 20 A

6. **(a)** 120 sin (314)t, **(b)** 0.1414 sec

7. **(a)** 50 Hz, **(b)** 100 V, **(c)** 90 V
8. **(a)** 42.42 V, **(b)** 100 Hz, **(c)** 30 A, **(d)** 27 A, **(e)** 1.11
9. **(a)** 23 A, **(b)** 5290 W
10. **(a)** 19.66 A, **(b)** 78.64 V, 61.77 V, **(c)** 0.786, **(d)** 1.546 W
11. **(a)** 12.57 Ω, 19.99 A, **(b)** 2513 Ω, 0.398 A
12. 152.74 V, 121.55 V, 195.2 V
13. **(a)** 70.7, **(b)** 63.7, **(c)** 1.415
14. $i = 75\sin 300\pi t$, 3.6977 A
15. 90°
16. 27.18
17. 324.03 V
18. $\dfrac{5V_m}{8}$
19. 5 A
20. **(a)** 6.369 A, **(b)** 1.570
21. 111.7°
22. **(a)** 2.388 A, **(b)** 20.114 A
23. **(a)** 5.436 A, **(b)** 1.754
24. 50 Hz
25. 90°
26. 79.05
27. $\dfrac{9V_m}{4}$

6 Polyphase Systems

LEARNING OBJECTIVES

After reading this chapter, you will be able to:

- Understand the two- and three-phase systems and their advantages.
- Understand the three-phase supply.
- Understand a star connection, and recognize that $I_l = I_{ph}$ and $V_L = \sqrt{3}\, V_{ph}$.
- Draw a complete phase diagram for balanced and unbalanced loads for star- and delta-connected systems.
- Understand how power is measured in a three-phase system by one-, two- and three-wattmeter methods.
- Understand star and delta connections.
- Understand the symmetrical components.

For generation, transmission and distribution of electrical power, a single-phase system is not sufficient. Hence, a polyphase system was developed. There are a number of systems available apart from the single-phase system. The systems available are as follows:

1. Two-phase system.
2. Three-phase system.
3. Six-phase system.

By experience it has been found that a three-phase system is more economical, convenient and advantageous for generation, transmission and distribution of electrical power whereas a six-phase system is more advantageous for static rectifiers, welding transformers, etc. The cost of generation of two-phase power is almost equal to that of the three-phase system. Hence, generation of two-phase systems has been discontinued.

6.1 Two-Phase System

A pair of coils placed perpendicular to each other, if rotated in a constant magnetic field, will generate a two-phase power. The voltage induced in each coil will have 90° phase displacement with each other. If the coils have equal number of turns, the generated voltage in each coil will have equal magnitude, both instantaneous and the average.

In Fig. 1, AA′ is coil A; BB′ is coil B and AA′ and BB′ are joined at *N*. If $V_{BN} = V\angle 0°$, $V_{AN} = V\angle 90°$, then the line voltage is

$$V_{AB} = V_{BN} + V_{AN} = \sqrt{2}\, V\angle 135°$$

(Reversing the vector BB′ and adding vectorially with AA′ yields the above result.)

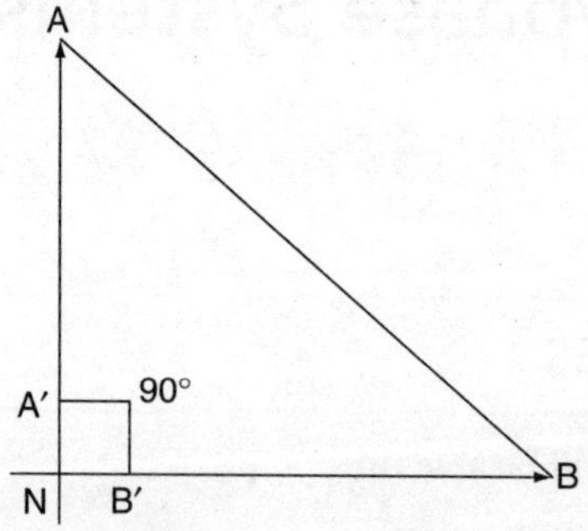

Figure 1 Vectorial representation of two-phase system.

6.2 Three-Phase System

Three coils placed 120° with each other and rotated in an anticlockwise direction at a uniform speed in a constant magnetic field generate three-phase power (Fig. 2). If the coils have equal number of turns, the generated voltage in each coil will have equal magnitude both instantaneous average and RMS values of each phase displaced by 120° (Fig. 3).

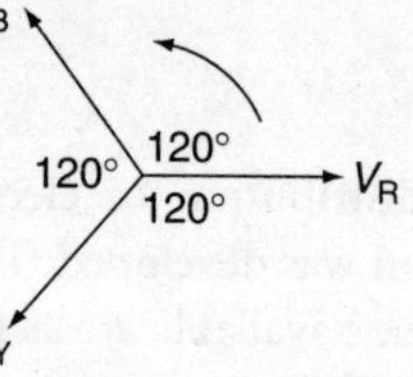

Figure 2 Three-phase positive phase sequence.

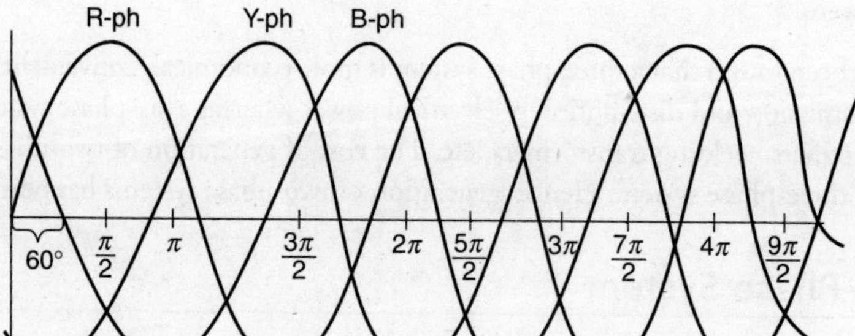

Figure 3 Three-phase waveform for positive phase sequence.

6.2.1 Three-Phase Negative Phase Sequence

Three coils placed 120° with each other and rotated in a clockwise direction at a uniform speed in a constant magnetic field generate three-phase power (Fig. 4). If the coils have equal number of turns, the generated voltage in each coil will not have equal magnitude in both instantaneous average and RMS values of each phase displaced by 120°. This concept is known as negative phase sequence (Fig. 5).

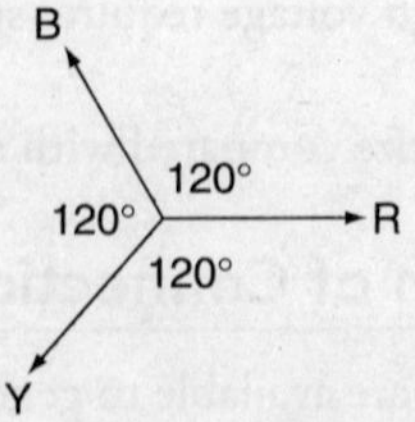

Figure 4 Three-phase negative phase sequence.

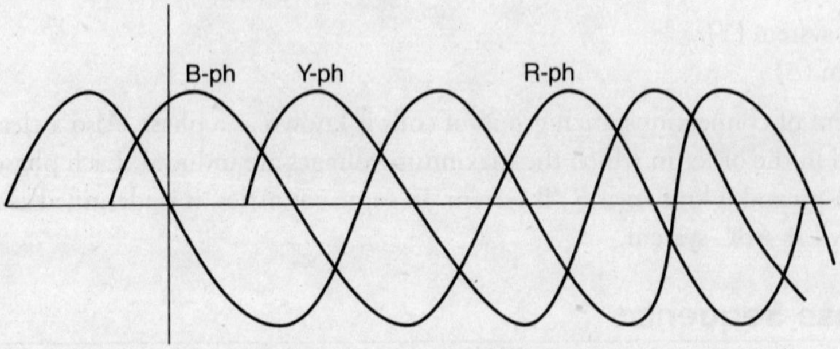

Figure 5 Three-phase waveform for negative phase sequence.

6.2.2 Two-Phase Positive Phase Sequence

Two coils placed 90° with each other and rotated in an anticlockwise direction at a uniform speed in a constant magnetic field generate two-phase power (Fig. 6) so that the generated voltages in the two phases have 90° phase displacement as shown in Fig. 7. This concept is known as two-phase positive phase sequence.

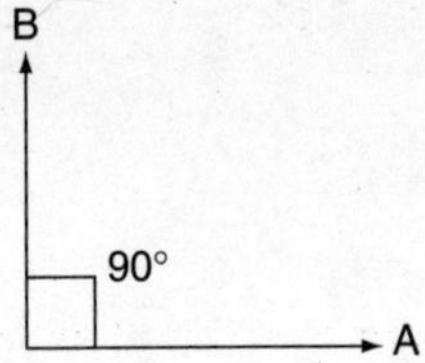

Figure 6 Two-phase positive phase sequence.

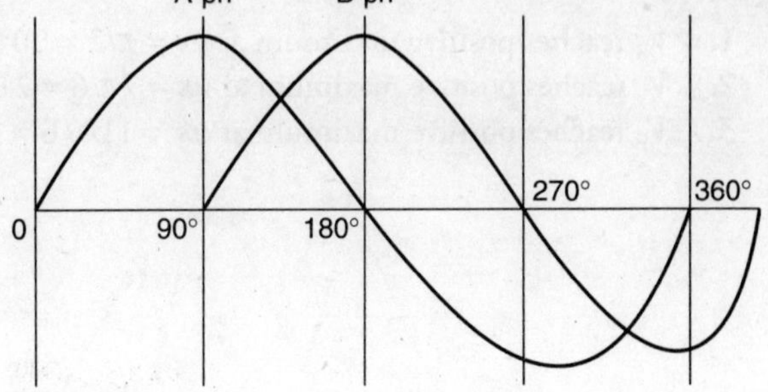

Figure 7 Two-phase waveform for positive phase sequence.

6.2.3 Advantages of Three-Phase System

1. Both lighting and motor loads can be operated from a single three-phase four-wire system.
2. Three-phase motor loads of any hp can be operated from a three-phase three-wire system.
3. Voltage stabilization on fluctuating loads will be maximum.
4. Three-phase voltages as high as 600 kV with a power of 1000 MW can be easily transmitted over long distances with minimum line loss.

5. For given power, a three-phase high voltage requires smaller conductor size for transmission and distribution.
6. A three-phase motor is smaller in size compared with an equivalent single-phase motor.

6.3 Three-Phase System of Connections

In the generator, a large number of coils are available to generate large amount of power. All these coils will be separated into three fixed groups with 120° angular displacement with each group. Ultimately these three groups will be connected in any one of the following ways:

1. Star or wye system (Y).
2. Delta system (Δ).

In the above system of connections, each group of coils is known as a phase. Also a clear universal procedure is followed in the order in which the maximum voltages are induced. Each phase is identified as R-ph, Y-ph and B-ph and is known as RYB system. In some countries, it is identified as A-ph, B-ph and C-ph and is known as ABC system.

6.4 Phase Sequence

Phase sequence or phase order is the order in which the three voltages V_R, V_Y and V_B reach their maximum values or pass through a zero or through any given point. There are two methods of representing this order:

1. Positive phase sequence;
2. negative phase sequence.

6.4.1 Positive Phase Sequence

Figure 8 shows the positive phase system. An anticlockwise rotation of a vector is positive phase sequence. In this system:

1. V_R reaches positive maximum at $\omega t = \pi/2 = 90°$ (ω is angular displacent and t is time).
2. V_Y reaches positive maximum at $\omega t = 7\pi/6 = 210°$.
3. V_B reaches positive maximum at $\omega t = 11\pi/6 = 330°$.

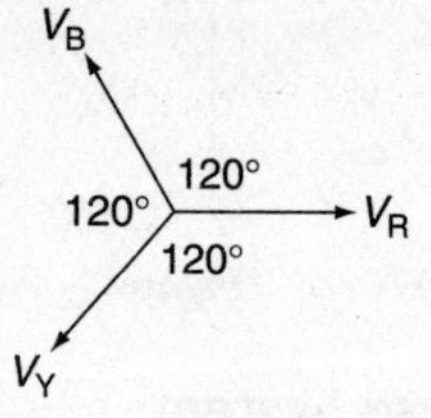

Figure 8 Positive phase sequence.

6.4.2 Negative Phase Sequence

Figure 9 shows the negative phase system. A clockwise rotation of a vector is negative phase sequence. In this system:

1. V_B reaches positive maximum at $\omega t = \pi/6 = 30°$.
2. V_Y reaches positive maximum at $\omega t = 5\pi/6 = 150°$.
3. V_R reaches positive maximum at $\omega t = 3\pi/6 = 270°$.

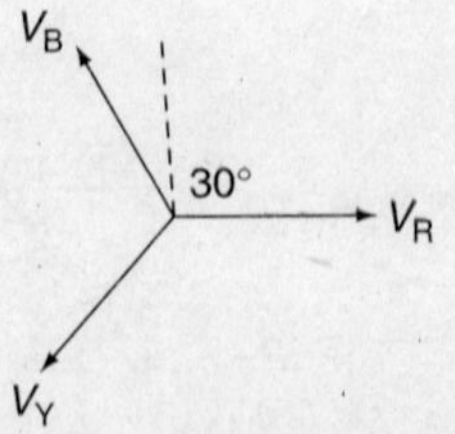

Figure 9 Negative pahse sequence.

Note: Also one should not think that letters RYB can be played arbitrarily on the symmetrically placed vectors and fix the phase sequence. It is totally wrong, because the direction of rotation and the torque developed on a motor depends on the phase sequence of the supply and the end connection of motor winding.

Throughout this book, only positive phase sequence is dealt, unless otherwise specified.

6.5 Balanced or Unbalanced Three-Phase System

In a balanced three-phase system, the following conditions do exist:

1. A single machine of any hp has the same impedance in all the three phases.
2. Lighting/other heater or single-phase loads are equally distributed in all the three phases.

In a unbalanced three-phase system, an important property is that the phasor sum of the three line or phase voltages is not zero, and the phasor sum of the three line or phase currents is not zero. When the three load impedances are not equal to one another, the phasor sums and the neutral current (I_n) are not zero, and the load is, therefore, unbalanced. The imbalance occurs when an open or short circuit appears at the load. If a three-phase system has an unbalanced load and an unbalanced power source, the methods of fixing the system are complex. Therefore, we will only consider an unbalanced load with a balanced power source.

6.6 Symmetrical or Unsymmetrical Systems

In a symmetrical system, the three voltage vectors are having 120°E displacement with each other, but in an unsymmetrical system it is not so.

In symmetrical systems

$$V_R = V\angle 0°$$
$$V_Y = V\angle -120° \quad \text{or} \quad V\angle +240°$$
$$V_B = V\angle -240° \quad \text{or} \quad V\angle +120°$$

In unsymmetrical systems

$$V_R = V_1\angle\theta_1$$
$$V_Y = V_2\angle\theta_2$$
$$V_B = V_3\angle\theta_3$$

6.7 Voltages and Currents

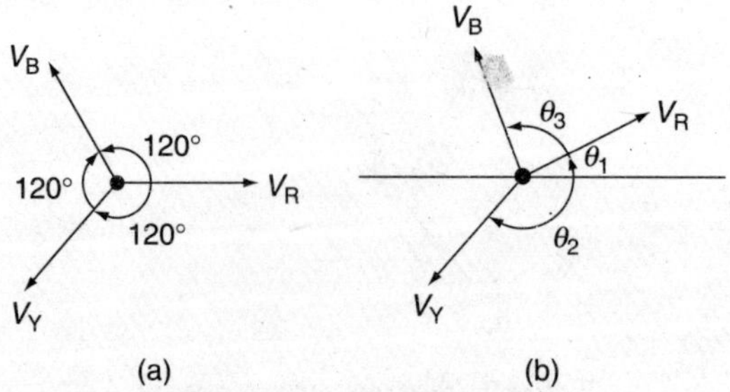

Figure 10 Symmetrical three-phase system. (a) Symmetrical three-phase system at 120° and (b) symmetrical three-phase system phase shift at 30°.

There are two different voltages and currents in any symmetrical three-phase system (Figs. 9–10). They are as follows:

1. **Phase voltages/currents:** The voltage between any line and the neutral point is called the *phase voltage*. The current flowing through the phases are called the *phase current*.
2. **Line voltages/currents:** The voltage between any two lines is called *line voltage*. The current flowing in the lines is called *line currents.*

6.8 Systems of Connections

Consider the three coils shown in Fig. 11. They can be electrically connected to form a star connection [Figs. 12(a) and (c)] or to form a delta [Figs. 12(b) and (d)].

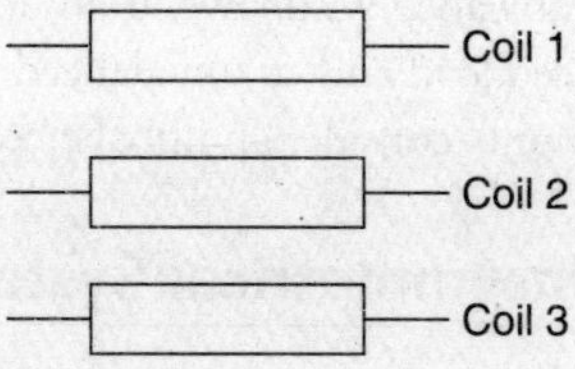

Figure 11 Three coils.

Throughout this book, connections shown in Figs. 12(a) and (b) are followed. Here afterward it is assumed that coil *A* represents R-ph, coil *B* represents Y-ph, coil *C* represents B-ph, and the corresponding vector diagram is shown in Fig. 13. Also

$$V_R = V\angle 0° = V[1 + j0]$$

$$V_Y = V\angle -120° = V[-0.5 - j0.866]$$

$$V_B = V\angle -240° = V[-0.5 + j0.866]$$

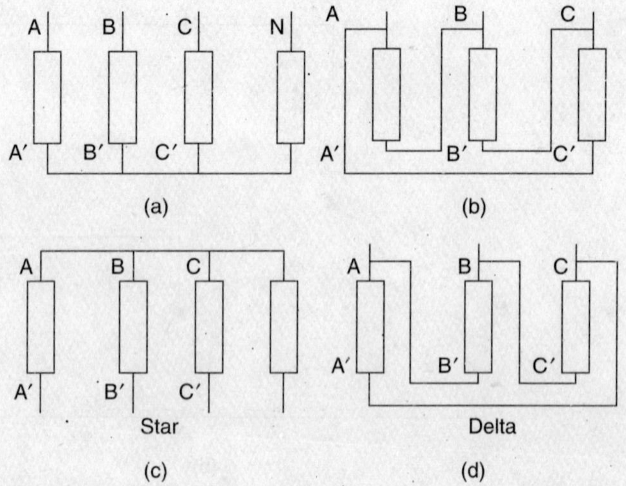

Figure 12 (a), (c) Star connections; (b), (d) delta connections.

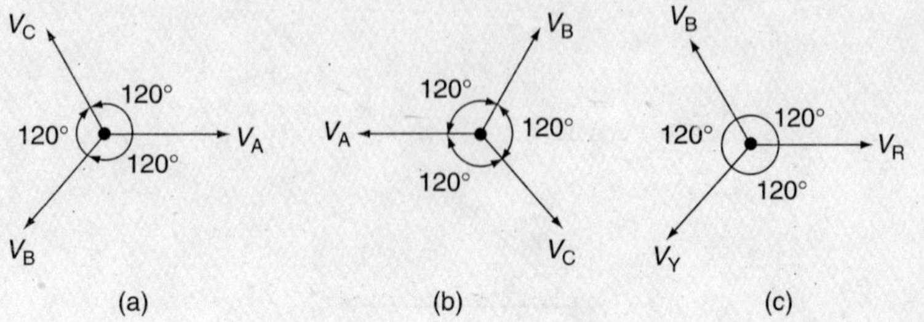

Figure 13 Vector diagram.

Net voltage is

$$V = V_R + V_Y + V_B = 0$$

satisfying Kirchohoff's voltage law (energy can neither be stored nor be destroyed).

6.9 Star- or Wye-Connected System

For a symmetrical 3Φ star-connected system [Fig. 14(b)], the vector diagram is drawn for phase voltages [Fig. 14(a)]. As seen from the star system, the phase voltage is not same as that of line voltage. But the same phase current flows in the line also. To get the line voltage V_{R-Y}, one has to add vertically the phase voltages V_R and V_R. To get the line voltage V_{R-Y}, one has to reverse V_Y and add vectorially with V_R.

Note: Vector algebra informs that whenever two vectors, whose included angle is more than 90°, are to be added, one of the vectors is to be reversed and added with the other (Fig. 15). This can be verified

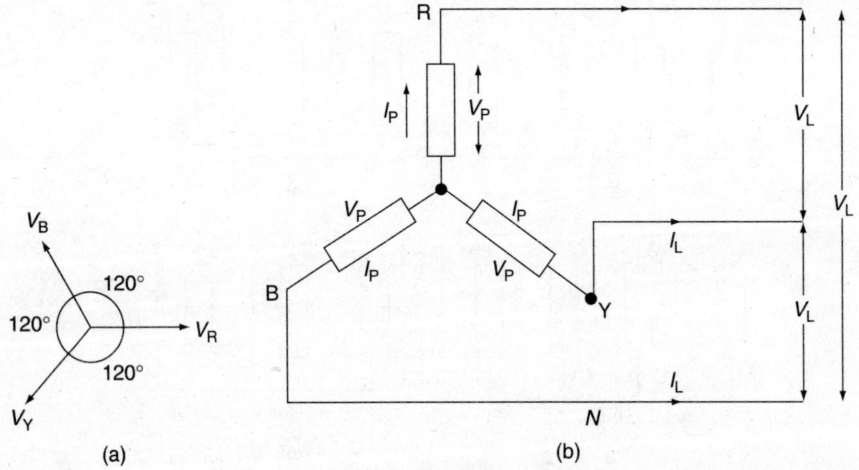

Figure 14 Symmetrical three-phase system (a) vector diagram, (b) star connected three-phase system.

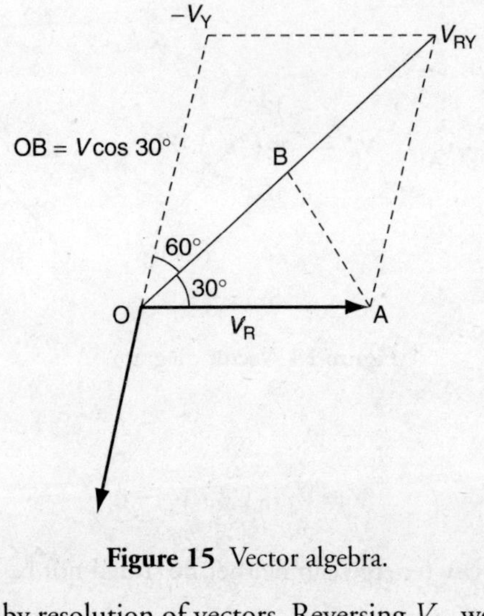

Figure 15 Vector algebra.

either by applying sine rule or by resolution of vectors. Reversing V_R, we get $-V_Y$ and now the included angle between V_R and $-V_Y$ is 60°. The diagonal V_{RY} in the parallelogram is given by

$$OC = 2V_R \cos 30 = 3V_R \frac{\sqrt{3}}{2} = \sqrt{3}V_R = V_L$$

Since

$$V_R = V_Y = V_P$$

so the three-phase power is

$$P = 3 \times V_P \times I_P \cos\theta = 3 \times \frac{V_L}{\sqrt{3}} \times I_2 \cos\theta$$
$$= \sqrt{3} V_L I_L \cos\theta$$

6.10 Delta Connected Systems

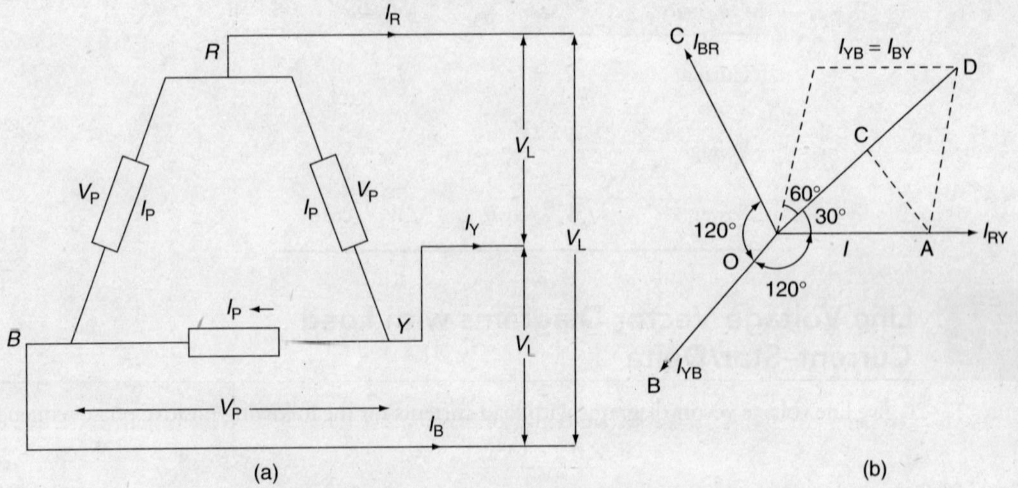

Figure 16 (a) Delta connected system and (b) vector diagram.

As seen from Fig. 16, phase voltage

$$V_{RY} = V_{YB} = V_{BR}$$

and line voltage

$$V_{RY} = V_{YB} = V_{BR}$$

But line currents $I_R = I_Y = I_B$ are different from $I_{RY} = I_{YB} = I_{BR}$. As before, to get I_R, I_{YB} is reversed and added with I_{RY} to get I_R:

$$I_{RY} = -I_{YB} + I_{BR} = I_R$$

$$I_R = I_Y = I_B = I_L$$

$$I_R = 2I_{RY}\cos 30 = 2 \times I_{RY}\frac{\sqrt{3}}{2} = \sqrt{3}I_P = I_L$$

Three-phase power is given by

$$P = 3 \times V_P \times I_P \cos\theta$$

$$= 3 \times V_L \times \frac{I_L}{\sqrt{3}} \cos\theta$$

$$= \sqrt{3}V_L I_L \cos\theta$$

6.11 Comparison Between Star and Delta

Quantity	*Star*	*Delta*
Current	$I_P = I_L$	$I_P = \frac{V_L}{\sqrt{3}}$
Voltage	$V_P = \frac{V_L}{\sqrt{3}}$	$V_P = V_L$
Power	$\sqrt{3}V_L I_L \cos\theta$	$\sqrt{3}V_L I_L \cos\theta$

6.12 Line Voltage Vector Diagrams with Load Current–Star/Delta

Figures 17–20 give line voltage vector diagrams with load currents for the following positive phase sequences:

(a)

$$V_R\angle 0°$$

$$V_Y\angle -120° = V_Y\angle 240°$$

$$V_B\angle 120°$$

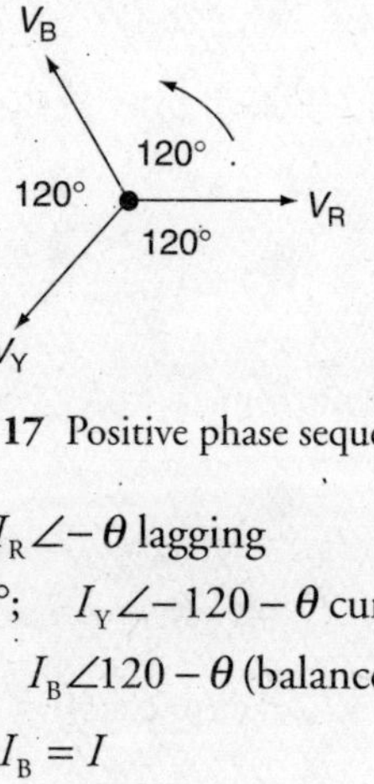

Figure 17 Positive phase sequence.

(b)

$$V_R\angle 0°;\quad I_R\angle -\theta \text{ lagging}$$
$$V_Y\angle -120°;\quad I_Y\angle -120-\theta \text{ current}$$
$$V_B\angle 120°;\quad I_B\angle 120-\theta \text{ (balanced load)}$$
$$I_R = I_Y = I_B = I$$

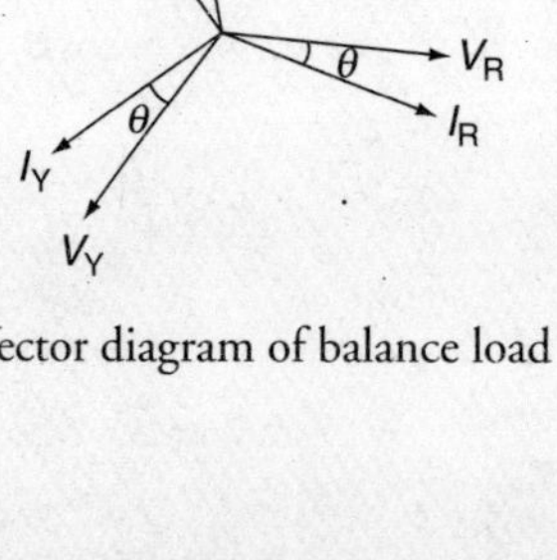

Figure 18 Vector diagram of balance load for lagging.

(c)

$V_R\angle 0°;\quad I_R\angle\theta$, leading

$V_Y\angle -120°;\quad I_Y\angle -120+\theta$, current

$V_B\angle 120°;\quad I_B\angle 120+\theta$ (balaned load)

$I_R \neq I_Y \neq I_B$ (unbalaned load)

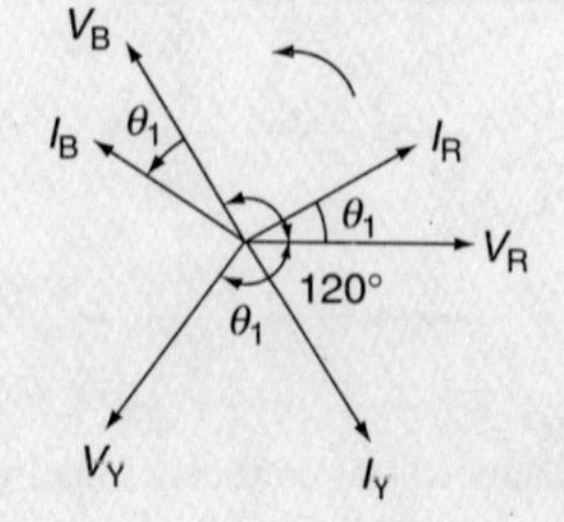

Figure 19 Vector diagram of unbalanced load for leading.

(d)

$I_R \neq I_Y \neq I_B\angle\theta_1 \neq \angle\theta_2 \neq \angle\theta_3$

$V_R\angle 0°;\quad I_R\angle\theta_1$

$V_Y\angle -120°;\quad I_Y\angle -120-\theta_2$

$V_B\angle 120°;\quad I_B\angle 120$

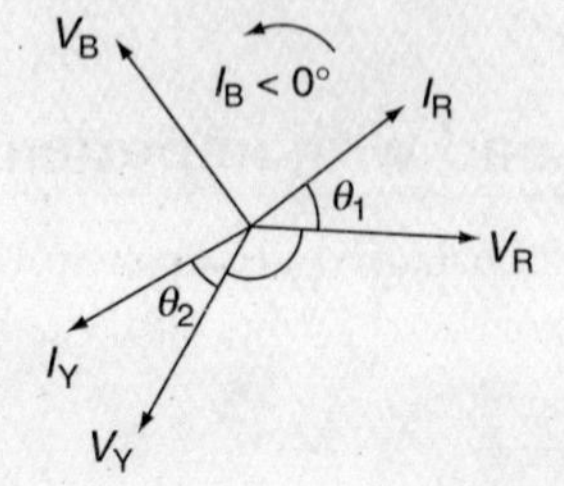

Figure 20 Vector diagram for both lagging and leading.

Note:

1. For star:
   - V_R, V_Y, V_B: each represents phase voltage.
   - I_R, I_Y, I_B: each represents phase/line current.
2. For delta:
   - V_{RY}, V_{YB}, V_{BR}: each represents phase/line voltage.
   - I_R, I_Y, I_B: each represents line current.

6.13 Balanced Star Load with Impedance

Let Z be the impedance per phase of a balanced star system (Fig. 21). It is required to find the phase currents, line currents and power factor. Let V_{RN}, V_{YN} and V_{BN} be the respective phase voltages. Then

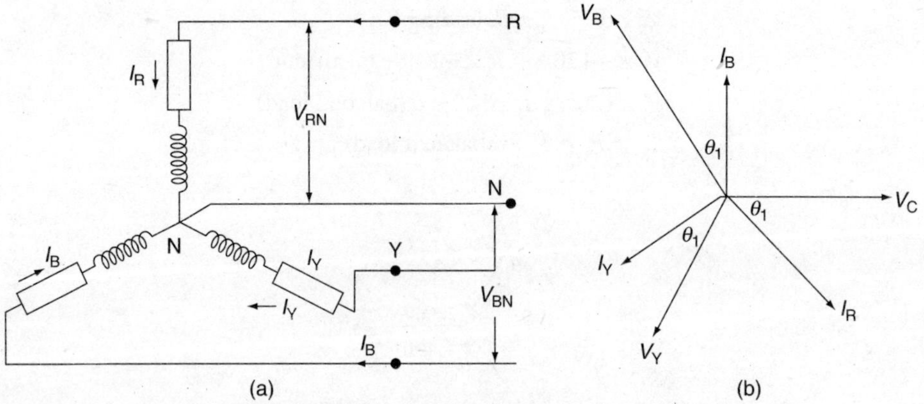

Figure 21 Balanced star connected load with impedance.

$$I_R = \frac{V_{RN}}{Z_{EN}} = \frac{V\angle 0^\circ}{Z\angle\theta_1} = \frac{V}{Z}\angle -\theta_1$$

$$I_Y = \frac{V_{YN}}{Z_{YN}} = \frac{V\angle -120}{Z\angle\theta_1} = \frac{V}{Z}\angle -120-\theta_1 = \frac{V}{Z}\angle 240-\theta_1$$

$$I_B = \frac{V_{BN}}{Z_{BN}} = \frac{V\angle +120}{Z\angle\theta_1} = \frac{V}{Z}\angle 120-\theta_1$$

6.14 Balanced Delta Load with Impedance

Let Z be the impedance per phase of a balanced delta system (Fig. 22). It is required to find the phase currents, line currents and power factor. Let V_{RY}, V_{YB} and V_{BR} be the respective phase voltages.

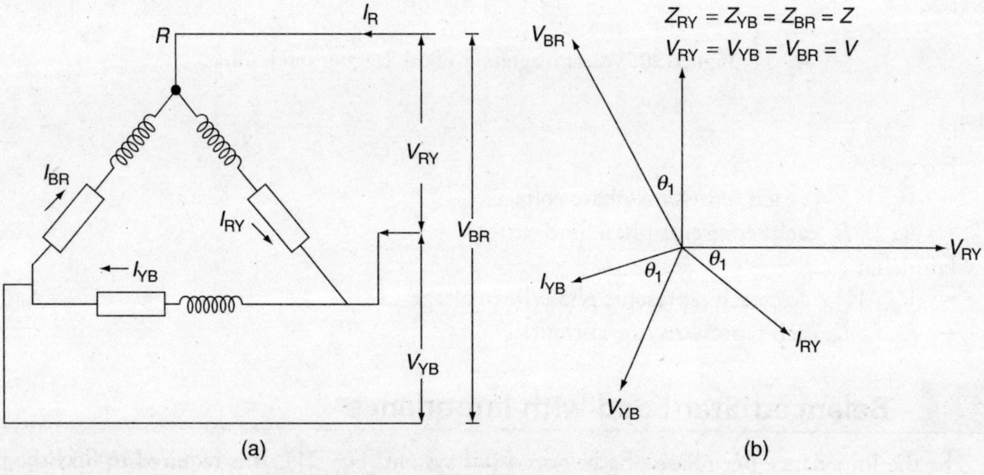

Figure 22 Balance delta load with impedance.

$$I_{RY} = \frac{V_{RY}\angle 0°}{Z\angle\theta_1} = \frac{V\angle -\theta_1}{Z}$$

$$I_{YB} = \frac{V_{YB}\angle 240°}{Z\angle\theta_1} = \frac{V}{Z}\angle 240 - \theta_1$$

$$I_{BR} = \frac{V_{BR}\angle 120°}{Z\angle\theta_1} = \frac{V}{Z}\angle 120 - \theta_1$$

Applying Kirchhoff's current law (KCL) at each node, the line currents I_R, I_Y and I_B are found. To be as follows:

$$I_R + I_{BR} - I_{RY} = 0 \quad \text{or} \quad I_R = I_{RY} - I_{BY} = \sqrt{3}\, I_{RY} \angle -30°$$

$$I_Y + I_{RY} - I_{YB} = 0 \quad \text{or} \quad I_Y = I_{YB} - I_{RY} = \sqrt{3}\, I_{YB} \angle -30°$$

$$I_B + I_{YB} - I_{BR} = 0 \quad \text{or} \quad I_B = I_{BR} - I_{YB} = \sqrt{3}\, I_{BR} \angle -38°$$

6.15 Power Measurement in Single-Phase System

6.15.1 Three-Ammeter Method

An *ammeter* is a measuring instrument used to measure the electric current in a circuit. Consider the two resistance connected in parallel with voltmeter (voltage supply) shown in Fig. 23. In parallel connection, current value if different but voltage value is same, then we need to find the current flowing through these two resistance by using ammeter instrument.

Resistance R is purely a non-inductive one. From vector diagram shown in Fig. 24 we have

$$I_1^2 = I_2^2 + I_3^2 + 2I_2I_3\cos\theta \tag{6.1}$$

But

$$I_2 = \frac{V}{R} \tag{6.2}$$

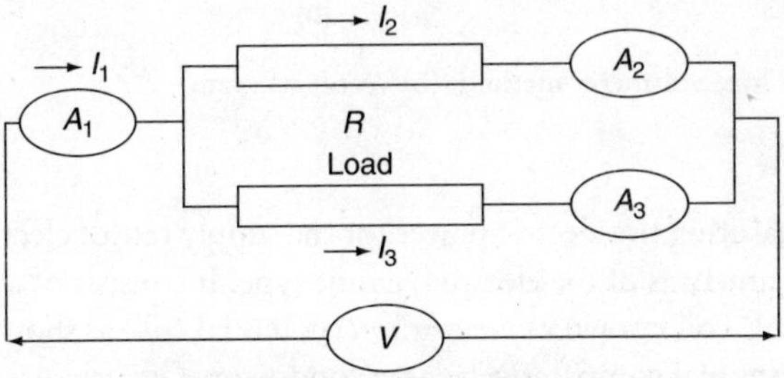

Figure 23 Three ammeter method.

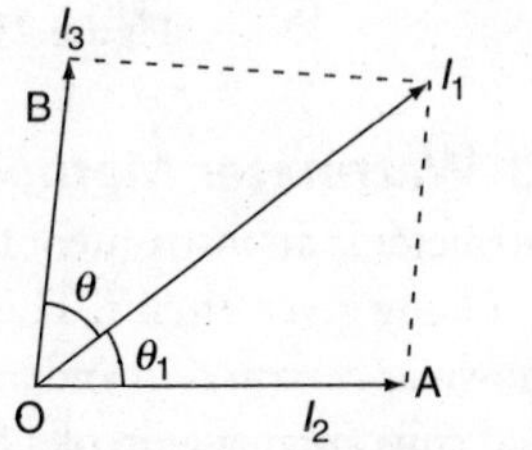

Figure 24 Vector diagram.

Therefore, load power is

$$P = VI_3 \cos\theta \tag{6.3}$$

Substituting the value of I_2 in the cosine term of Eq. (6.1), we get

$$I_1^2 = I_2^2 + I_3^2 + 2\frac{V}{R} I_3 \cos\theta$$

$$I_1^2 = I_2^2 + I_3^2 + 2\frac{P}{R}$$

Therefore,

$$P = \frac{R}{2}(I_1^2 - I_2^2 - I_3^2) \text{ (parallel branch voltage is same)}$$

6.15.2 Three-Voltmeter Method

A voltmeter is a measuring instrument used to measure the voltage in a circuit. Consider the two resistance connected in series with voltmeter. In series connection, current is same but voltage is different, so we need to find the voltage across two resistance by using voltmeter as shown in Fig. 25(a).

From vector diagram shown in Fig. 25(b).

$$V_3^2 = V_1^2 + V_2^2 + 2V_1 V_2 \cos\theta$$

But $V_1 = 1R$. Therefore,

$$V_3^2 = V_1^2 + V_2^2 + 2R_1V_2 \cos\theta = V_1^2 + V_2^2 + 2RP$$

or

$$P = \frac{V_3^2 - V_1^2 - V_2^2}{2R}$$

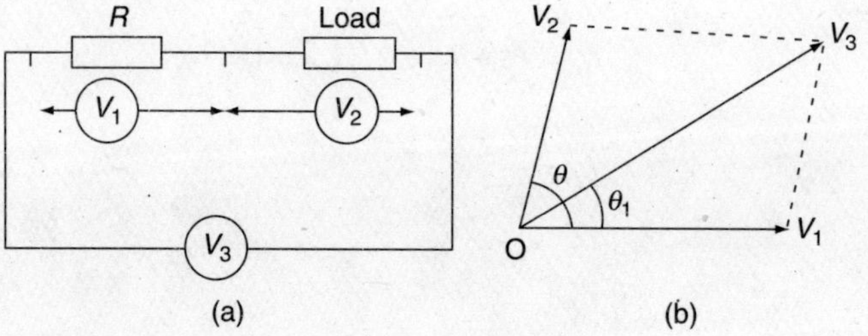

Figure 25 (a) Three voltmeter method, (b) vector diagram.

6.15.3 Wattmeter Method

The wattmeter is an instrument for measuring the electric power (or the supply rate of electrical energy) in watts of any given circuit. This instrument is of the electrodynamic type. It consists of a pair of fixed coils known as *current coils* and a movable coil known as *pressure coil* (potential coil) as shown in Fig. 26. The fixed coils are made up of a few turns of a comparatively large conductor. The pressure coil consists of many turns of fine wire. It is mounted on a shaft, carried in jeweled bearings, so that it may turn inside the stationary coils. The movable coil carries a needle which moves over a suitably marked scale. Spiral coil springs hold the needle to a zero position. The current coil (stationary coil) of the wattmeter is connected in series with the circuit (load), and the potential coil (movable coil) is connected across the line. When line current flows through the current coil of a wattmeter, a field is set up around the coil. The strength of this field is proportional to the line current and in phase with it. The potential coil of the wattmeter generally has a high-resistance resistor connected in series with it. This is for the purpose of making the potential-coil circuit of the meter as purely resistive as possible. As a result, current in the potential circuit is practically in phase with line voltage. Therefore, when voltage is applied to the potential circuit, current is proportional to and in phase with the line voltage.

The actuating force of a wattmeter comes from the field of its current coil and the field of its potential coil. The force acting on the movable coil at any instant (tending to turn it) is proportional to the instantaneous values of line current and voltage. In Fig. 26, ML is the current coil (CC), CV is the pressure coil (PC). Power is given by

$$\text{Power} = \text{Wattmeter reading} \times \text{Multiplying factor}$$

The multiplying factor (MF) is 1 for rated voltage and current terminal connection and higher for high currents and voltages.

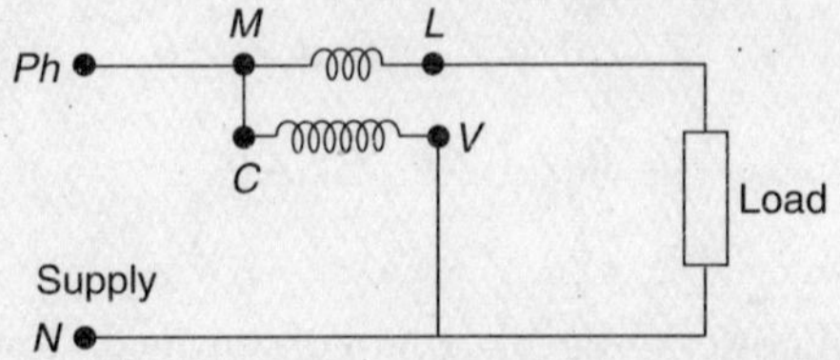

Figure 26 Wattmeter method.

6.16 Power Measurement in Three-Phase System

6.16.1 Single-Wattmeter Method

In a balanced 3-wire, 3-phase load circuit the power in each phase is equal and, therefore, total power of the circuit can be determined by multiplying the power measured in any one phase by three. Figure 27(a)

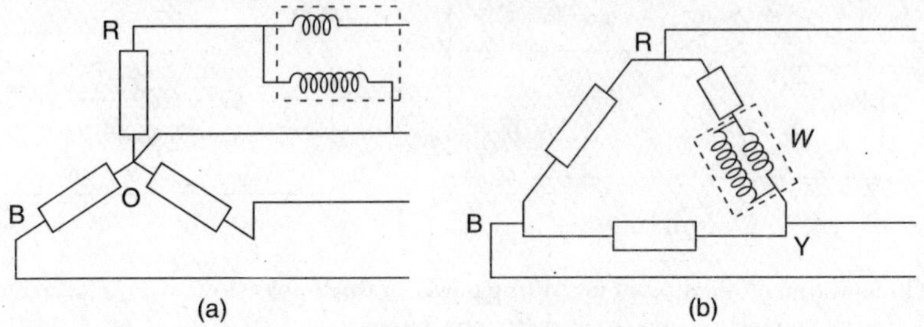

Figure 27 Single wattmeter method of measuring power in three-phase three-wire balanced load circuits: (a) in star connection and (b) in delta connection.

shows connections for measurement of power in one phase of the 3-phase, 3-wire star-connected balanced load circuit. Total power of the load circuit will be three times the reading of wattmeter. Figure 27(b) shows connections for measurement of power in 3-phase, 3-wire delta-connected balanced load circuits by one wattmeter method. In this method of measurement of power, resistance coils of resistance equal to that of pressure coil is connected in each of remaining two phases as shown in Fig. 27(b). The pressure coil (potential coil) and resistamce coils form a balanced star-connection.

It should be noted that the one-wattmeter method has the disadvantage that even a slight degree of unbalance in the loading produce a large error in the measurement. Refer to Fig. 27, here total power is given by

$$\text{Total power} = \text{Wattmeter reading} \times \text{MF} \times 3$$

6.16.2 Three-Wattmeter Method

This method is employed for measurement of power in three-phase, four-wire circuits. The connection is as shown in Fig. 28(a). As the neutral wire is common to the three phases, each wattmeter reads power in its own phase, and the total power of the load circuit is given by the sum of the readings of the three wattmeters. In case of delta-connected circuit [Fig. 28(b)], the difficulty in adopting is that the phase coils are required to be broken for inserting current coils of wattmeter.

Unbalanced load,

$$Z_1 \angle\theta_1 \neq Z_2 \angle\theta_2 \neq Z_3 \angle\theta_3$$

and

$$I_1 \angle\theta_1 \neq I_2 \angle\theta_2 \neq I_3 \angle\theta_3$$

The total power is given by

$$\text{Total power} = \text{Algebraic sum of wattmeter reading}$$
$$= (W_1 + W_2 + W_3) \times \text{MF}$$

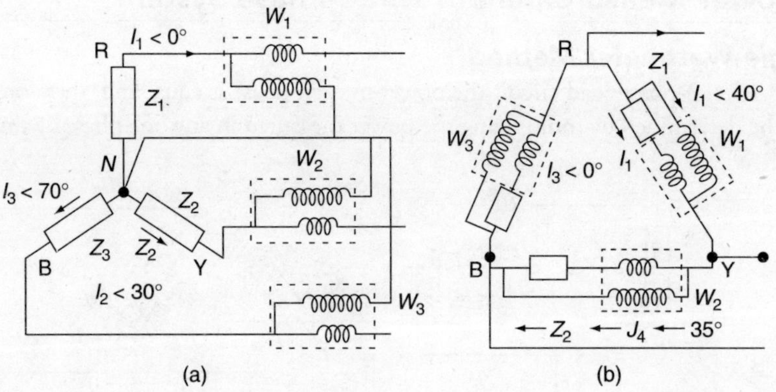

Figure 28 Three-wattmeter method of measuring power in three-phase four-wire balanced load circuits (a) in star connection and (b) in delta connection.

6.16.3 Two-Wattmeter Method

This is generally used method for measurement of power in three-phase, three-wire load circuit in star connection [Fig. 29(a)] and in delta connection (Fig. 30). The current coils of wattmeters are inserted in any two lines and pressure coil is connected from its own current coil to the line without a current coil.

From vector diagram [Fig. 29(b)], V_R, V_T, V_P are phase voltages.

$$\text{Total Power} = (W_1 + W_2) \times \text{MF}$$

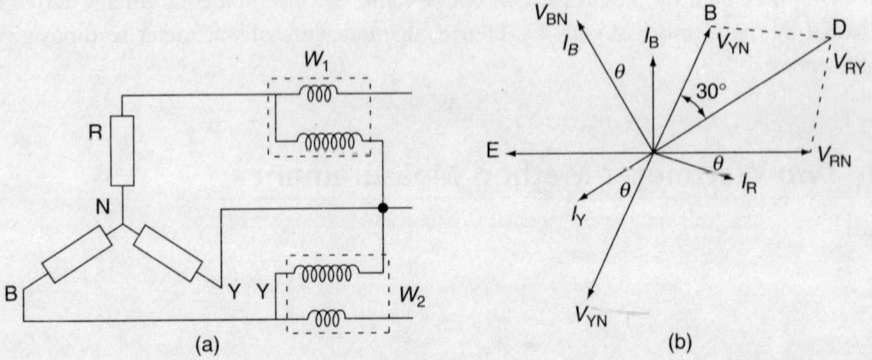

Figure 29 Two wattmeter method of measuring power in three-phase three-wire balanced load circuits: (a) in star connection and (b) in vector diagram.

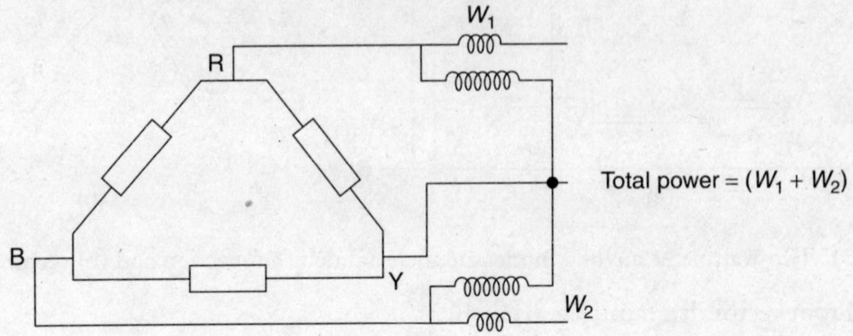

Figure 30 Two wattmeter method of measuring power in three-phase three-wire balanced load circuits in delta connection.

6.17 Two-Wattmeter Method of Power Measurement

Let V_R, V_Y and V_B be the phase voltages across the load and i_1, i_2 and i_3 be the load currents. Instantaneous power is given by

$$p = V_R i_1 + V_Y i_2 + V_B i_3 \tag{6.4}$$

There is no wattmeter connected in the Y-phase. Hence, in Eq. (6.4) eliminate i_2 by applying KCL:

$$i_1 + i_2 + i_3 = 0 \text{ or } i_2 = -(i_1 + i_3) \tag{6.5}$$

Putting Eq. (6.5) in Eq. (6.4) for i_2, we have the instantaneous power as

$$\begin{aligned} P &= V_R i_1 - V_Y(i_1 + i_3) + V_B i_3 \\ &= V_R i_1 - V_Y i_1 - V_Y i_3 + V_B i_3 \\ &= i_1(V_R - V_Y) + i_3(V_B - V_Y) \end{aligned}$$

That is instantaneous power is $i_1 \times$ the instantaneous voltage across R and Y phase + $i_3 \times$ instantaneous voltage across B and Y phase.

From the circuit diagram shown in Fig. 31(a), it is evident that the instantaneous voltage across R and Y is being measured by the pressure of coil W_1 and the instantaneous voltage across B and Y is being measured by the pressure of coil W_2. Hence, algebraic sum of wattmeter readings give the total instantaneous power

$$P = W_1 + W_2$$

6.18 Two-Wattmeter Method Measurement

Delta load:

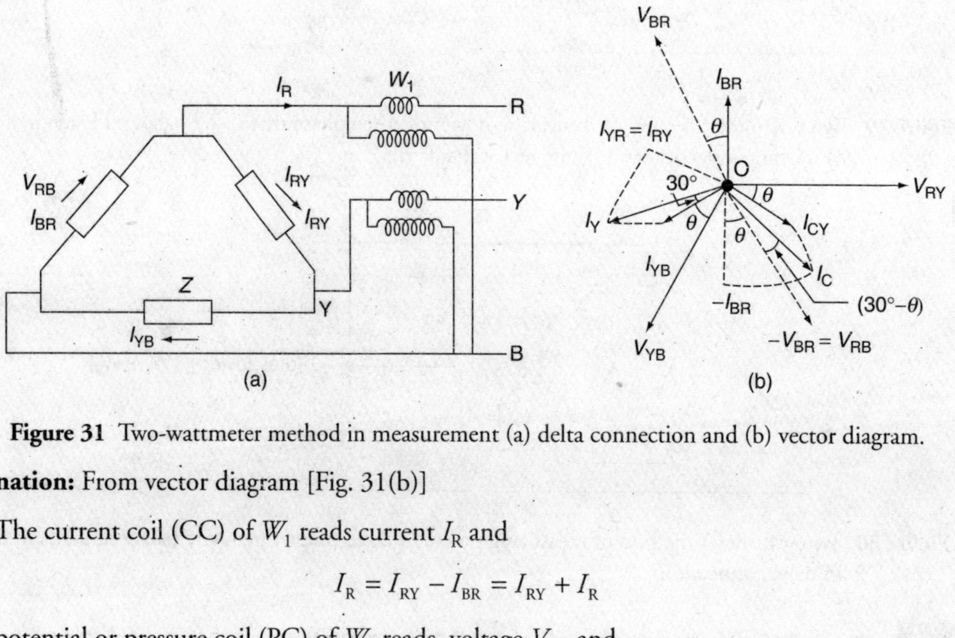

Figure 31 Two-wattmeter method in measurement (a) delta connection and (b) vector diagram.

Explanation: From vector diagram [Fig. 31(b)]

1. The current coil (CC) of W_1 reads current I_R and

$$I_R = I_{RY} - I_{BR} = I_{RY} + I_R$$

potential or pressure coil (PC) of W_1 reads voltage V_{RB} and

$$V_{RB} = -V_{BR}$$

Phase angle between V_{RB} and I_R is (30 – θ).

2. The CC of W_2 reads current I_Y and

$$I_Y = I_{YB} - I_{RY} = I_{YB} + I_{YR}$$

PC of W_2 reads voltage V_{YB}. Phase angle between V_{YB} and I_Y is (30 + θ).

Star load:

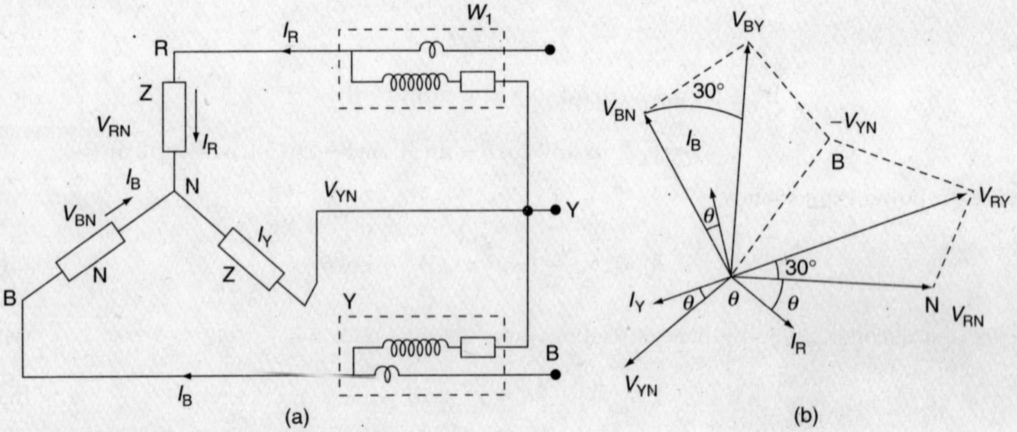

Figure 32 Two-wattmeter method in measurement: (a) star connection and (b) vector diagram.

Explanation:

1. The CC of W_1 reads current I_R and

$$I_R = I_{RN} = I_{LINE}$$

PC of W_1 reads voltage V_{RY} and V_{RY}:

$$V_{RN} + V_{NY} = V_{RN} - V_{YN}$$

Phase angle between V_{YR} and I_R is $(30 + \theta)$.

2. The CC of W_2 reads current and

$$I_B = I_{BN} = I_{LINE}$$

PC of W_2 reads voltage

$$V_{BY} = V_{BN} + V_{YN} = V_{BN} - V_{YN}$$

Phase angle between V_{BY} and I_B is $(30 - \theta)$.

Proof:

From the circuit diagram and the vector diagram [Figs. 32(a) and (b)] it is seen that wattmeter W_2 reads the vector current I_R and voltage V_{RY}:

$$W_1 = V_{RY} I_R \cos(30 - \theta) \tag{6.6}$$

W_1 reads the vector current I_B and voltage V_{BY}. Now

$$W_2 = V_{BY} I_B \cos(30 - \theta) \tag{6.7}$$

Therefore,

$$W_1 + W_2 = V_{RY} I_R \cos(30 - \theta) + V_{BY} I_B \cos(30 + \theta)$$

Now,

$$V_{RY} = V_{BY} = V_2$$

and

$$I_R = I_B = I_L$$

$$W_1 + W_2 = V_L I_L\,[\cos(30+\theta) + \cos(30-\theta)]$$

$$= V_L I_L\,[\cos 30\,\cos\theta - \sin 30\,\sin\theta + \cos 30\cos\theta \sin 30 \sin\theta] \tag{6.8}$$

Hence the power expression is

$$V_L I_L \times 2 \times \frac{\sqrt{3}}{2}\cos\theta = \sqrt{3}V_L I_L\,\cos\theta \tag{6.9}$$

Hence, the algebraic sum of wattmeter readings gives the total power:

$$W_1 - W_2 = VI\,\sin\theta \tag{6.10}$$

Dividing Eq. (6.10) by Eq. (6.9), we get

$$\frac{VI\,\sin\theta}{\sqrt{3}VI\,\cos\theta} = \frac{W_1 - W_2}{W_1 + W_2}$$

or

$$\tan\theta = \sqrt{3}\,\frac{W_1 - W_2}{W_1 + W_2}$$

A higher reading wattmeter will be considered as W_1.

$$\tan\theta = \sqrt{3}\,\frac{W_1 - W_2}{W_1 + W_2}$$

$$\theta = \tan^{-1}\left(\sqrt{3}\,\frac{W_1 - W_2}{W_1 + W_2}\right)$$

6.19 Unbalanced Loads in Star

If an unbalanced load is in delta and connected to a balanced three-phase supply, the phase currents can be easily found out as

$$I_{ph} = \frac{V_L - V_{ph}}{Z_{ph}}$$

But if an unbalanced load is in star and connected to a balanced three-phase supply, the phase currents or line currents cannot be found out so easily, since

$$E_p \neq \frac{E_2}{\sqrt{3}}$$

and also

$$E_{RN} \neq E_{YN} \neq E_{BN}$$

To solve this unbalancing load in star, the following four techniques are available:

1. Graphical method.
2. Applying Kirchhoff's laws.
3. Analytical method.
4. Displacement neutral method.
5. Conversion of the given star load to an equivalent delta load.

6.19.1 Graphical Method

Application of this method is restricted due to the following reasons:

1. One must be well versed in drawing.
2. One must be able to select appropriate scales for different parameter involved, viz. voltage, current, impedances, etc.
3. All the three loads must have the same power factor. The individual load may have different Ohmic values.

The procedure involved is as follows:

Step 1: Construct supply voltage (line) triangle taking a convenient scale.

Step 2: Mark the neutral point N. Since the load is unbalanced, the neutral point will not be at the geometrical center of the triangle.

Step 3: Divide the side KY in the ratio of Z_R:Z_Y, the side YB in the ratio of Z_Y:Z_B and the side RB in the ratio of Z_B:Z_R. Mark these division points as E, F, G.

Step 4: Join RF, YG and BE. Mark the meeting point of these lines as "*O*."

Step 5: Measure the lengths RD, YO and BO. These lengths will give the phase voltages of the respective phases for balanced load.

Step 6: Measure the length NO. This length gives the potential difference between the neutral point of supply and load.

Result:

Once phase voltages are known, further solutions become very simple (Fig. 33).

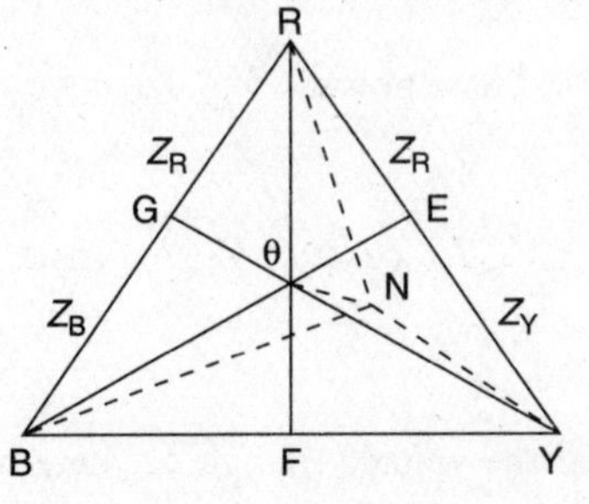

Figure 33 Graphical method.

6.19.2 Applying Kirchhoff's Laws

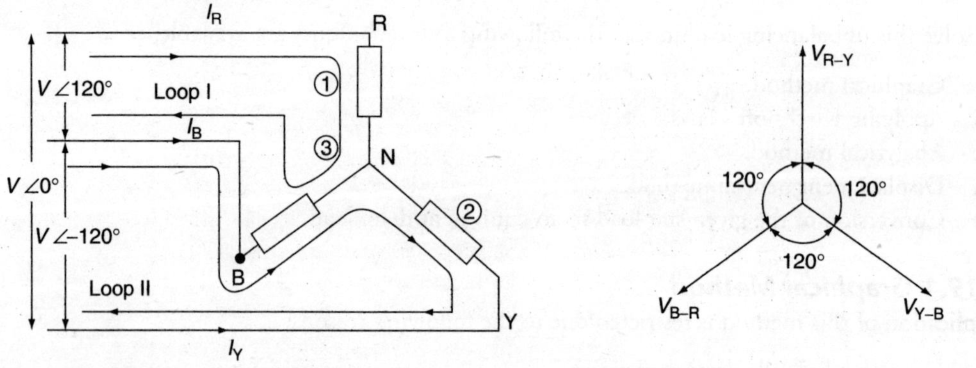

Figure 34 Star connection balanced system. Here ① $Z_1 \angle\theta_1$, ② $Z_2 \angle\theta_2$, ③ $Z_3 \angle\theta_3$.

I_1 and I_2 are the loop currents. Writing Kirchhoff's voltage law equation for loop I and loop II (Fig. 34), we have

$$V\angle -120° = I_1(Z_1\angle\theta_1 + Z_3\angle\theta_3) - I_L Z_3\angle\theta_3 \quad (6.11)$$

$$V\angle -120° = I_L(Z_3\angle\theta_3 + Z_2\angle\theta_2) - I_1 Z_3\angle\theta_3 \quad (6.12)$$

Solving Eqs. (6.11) and (6.12) we get loop currents I_1 and I_L. From loop current vectors I_1 and I_L we have

$$I_R = I_1 - I_L$$
$$I_Y = I_L$$
$$I_B = I_L - I_1$$

Knowing the phase currents, we find the phase voltage as

$$V_{RN} = I_R Z_R$$
$$V_{YN} = I_Y Z_Y$$
$$V_{BN} = I_B Z_B$$

and

$$\text{Phase power} = V_{RN}\ I_{RN} \cos\phi_R$$

Therefore,

$$\text{Total power} = V_{RN} I_{RN} \cos\phi_R + V_{YN} I_{YN} \cos\phi_Y + V_{BN} I_{BN} \cos\phi_B$$

6.19.3 Analytical Method

In this method, the coordinates for the voltage triangle are determined taking the neutral point as the origin, since the neutral point will not be at the geometric center of the triangle for unbalanced loads.

Let the line voltage be 400 V and let the load impedances be Z_R, Z_Y and Z_B. The voltage triangle is constructed as shown in Fig. 35.

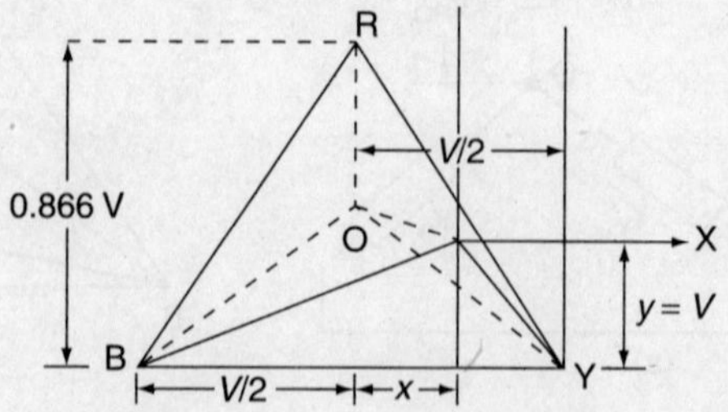

Figure 35 Analytical method.

Let O be the neutral point if the load is a balanced one. Let x be the shift in the X direction and y be the shift in the Y direction, measured from the horizontal BY. Taking N as the origin, the coordinates for points R, Y and B are fixed as follows.

$$\mathrm{R} = -x, (0.866\, V - y)$$
$$\mathrm{Y} = [-(V/2) - x], -y \text{ and } \mathrm{B} = [(V/2) + x], -y$$

where V is line voltage.
For the above coordinates, the voltage vectors are written as follows:

$$V_R = -x + y(0.866v - y)$$
$$V_Y = (v/y - x) - jy$$
$$V_B = (v/y + x) - jy$$

The respective currents will be

$$I_R = \frac{V_R}{Z_R} = -x + j\,\frac{(0.866\ v - y)}{r_1 + jx_1}$$

$$I_Y = \frac{V_Y}{Z_Y} = \frac{(v/2 - x) - jy}{r_2 + jx_2}$$

and

$$I_B = \frac{V_B}{Z_B} = \frac{-(v/2 + x) - jy}{r_3 + jx_3}$$

The numerical values of V, R, X, etc. are substituted and solved for the coordinates x and y.

6.19.4 Displacement Neutral Method

In this method, the line currents are represented in terms of load voltages and load admittances is shown in three-phase balanced star connected load in terms of admittance (Y) [Fig. 36(a)]:

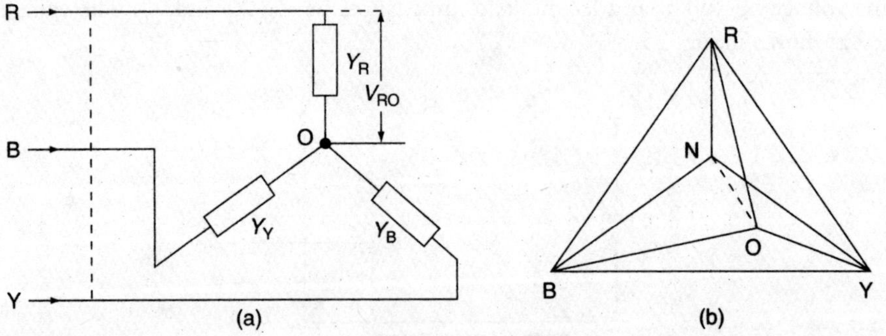

Figure 36 (a) Three-phase balanced star connected load in terms of admittance and (b) vector diagram.

$$I_R = V_{RO}Y_R$$
$$I_Y = V_{YO}Y_Y$$
$$I_B = V_{BO}Y_B$$

Applying KCL to point O, we get

$$I_R + I_Y + I_N = 0$$

or

$$V_{RO}Y_R + V_{YO}Y_Y + V_{BO}Y_B = 0$$

Referring to the vector diagram [Fig. 36(b)], we get

$$V_{RO} = V_{RN} + V_{NO}$$
$$V_{YO} = V_{YN} + V_{NO}$$
$$V_{BO} = V_{BN} + V_{NO}$$

Substituting the above,

$$(V_{RN} + V_{NO}) + (V_{YN} + V_{NO})\,V_Y + (V_{BN} + V_{NO})\,Y_B = 0$$

$$V_{NO}Y_R + Y_{NO} + V_Y + V_{NO}Y_N = -V_{RN}Y_R + V_{YN}Y_B$$

$$V_{NO}Y_R + Y_Y + Y_B = -V_{RN}Y_R + V_{BN}Y_B$$

$$V_{NO} = \frac{-(V_{RN}Y_R + V_{YN}Y_Y + V_{BN}Y_B)}{Y_R + Y_Y + Y_B}$$

where Y is admittance and V is voltage. Absorbing the negative sign, V_{NO} becomes V_{ON}:

$$V_{ON} = \frac{(V_{RN}Y_R + V_{YN}Y_Y + V_{BN}Y_B)}{Y_R + Y_Y + Y_B}$$

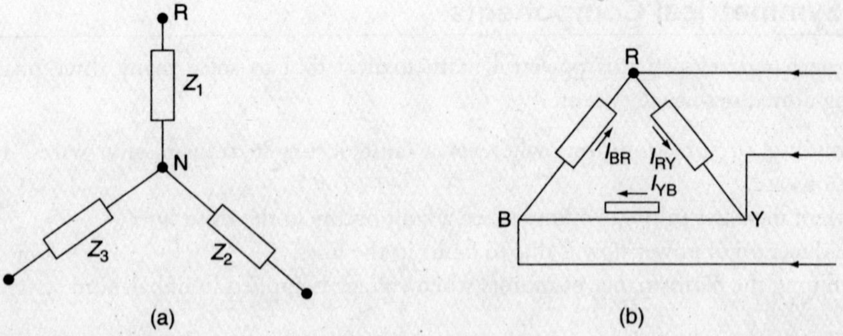

Figure 37 Star connection in terms of impedance and (b) delta connection in terms of impedance.

From the known values V_{RN}, V_R, etc., V_{ON} is computed, then V_{RO}, Y_{NO} and V_{BO} are found out, from which again I_{RO}, I_{YO}, and I_{BO} are found out.

6.19.5 Conversion of Star to Delta

In this section star delta conversion has been dealt with in detail. Here resistances are replaced by impedances [Figs. 37(a) and (b)] and the final expression will be

$$Z_{RY} \text{ in } \Delta = Z_1 + Z_2 + \frac{Z_1 Z_2}{Z_3}$$

$$Z_{YB} \text{ in } \Delta = Z_2 + Z_3 + \frac{Z_2 Z_3}{Z_1}$$

$$Z_{BR} \text{ in } \Delta = Z_3 + Z_1 + \frac{Z_3 Z_1}{Z_2}$$

After conversion to Δ, we get

$$I_R = \frac{V_{RY}}{Z_{RY}}$$

$$I_Y = \frac{V_{YB}}{Z_{YB}}$$

$$I_B = \frac{V_{BR}}{Z_{BR}}$$

and the line currents

$$I_R = I_{RY} - I_{BR}$$
$$I_Y = I_{YB} - I_{RY}$$
$$I_B = I_{BR} - I_{YB}$$

6.20 Symmetrical Components

C.L. Fortesque has developed this powerful mathematical tool to solve many three phase circuits. The following difficulties usually occur:

1. Determining circuit conditions wherever a fault occurs in transmission wires. The circuit conditions are:
   - current increases in the two lines when a fault occurs in the third line;
   - the direction of power flow is due to faults in the lines.
2. Determining the performance of motors when voltage is applied in unbalanced system.

6.20.1 Statement of Theorem

Any unbalanced three-phase systems of vectors (voltage or current or fluxes) can be resolved into three balanced systems of vectors (voltage or currents or fluxes) without any change in magnitude and frequency.

The three resolved vectors are named as positive, negative and zero sequence components and these are called symmetrical components.

6.20.2 "a" Operator

This is similar to j operator. A number multiplied by j will make it to rotate by 90° in an anti-clockwise direction. Here, a number multiplied by "a" will rotate it by 120° in an anticlockwise direction.

$$a = e^{j120} = 1\ \angle 120° = -0.5 + j0.866$$

$$[a^2 e^{j120}]^2 = 1\ \angle 240° = \angle -120° = 0.5 - j0.866$$

that is

$$Ve^{j0} = V\angle 0° = av = v = [-0.5 + j0.866]$$

$$Ve^{j120} = V\angle 120 = av = v[-0.5 + j0.886]$$

$$Ve^{j240°} = V\angle -120 = a^2 v = v[-0.5 - j0.886]$$

Adding, we have

$$V\angle 0 + V\angle 120° + V\angle -120° = V[1 + a + a^2] = 0$$

6.20.3 Positive Phase Sequence Components

This is formed by three vectors A_1, B_1 and C_1, which are equal in magnitude but have a phase displacement of 120° between each other and have the same phase sequence as the original three-phase unbalanced vector (Fig. 38). Example

$$V_1 = V_{A1}\angle 0°$$

$$V_{B1} = a^2 V_{A1} = V_{A1}\ \angle -120°$$

$$V_{C1} = aV_{A1} = V_{A1}\ \angle 120°$$

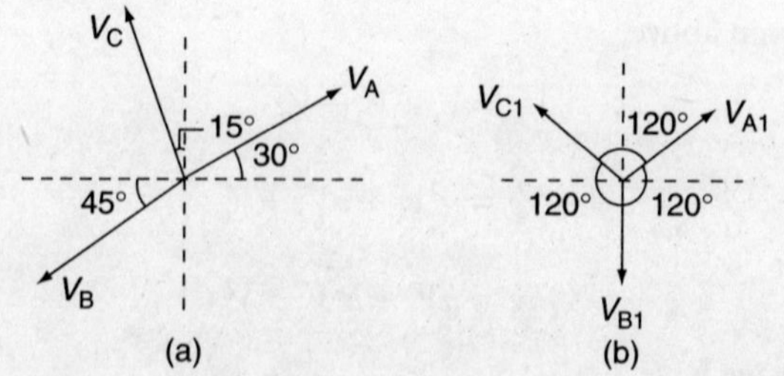

Figure 38 Positive phase sequence components. (a) Given unbalanced voltage system and (b) symmetrical components, positive phase sequence.

6.20.4 Negative Phase Sequence Components

This is formed by three vectors A_2, B_2 and C_2 which are equal in magnitude and have a phase displacement of 120° but have the phase sequence opposite to the original unbalanced vector (Fig. 39). Example

$$V_{A2} = V_{A2}\angle 0^\circ$$
$$V_{B2} = aV_{A1} = V_{A2}\angle 120^\circ$$
$$V_{C2} = a^2V_{A2} = V_{A2}\angle -120^\circ$$

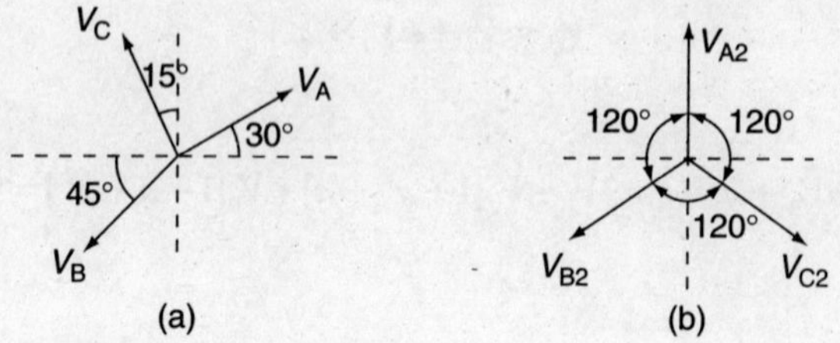

Figure 39 Negative phase sequence components. (a) Given unbalanced voltage system and (b) symmetrical components, negative phase sequence.

6.20.5 Zero-Phase Sequence Components

This is formed by three vectors A_3, B_3 and C_3 which are equal in magnitude and have zero phase displacement, that is they are in phase with each other.

$$\text{Zero phase sequence, } V_{AO} = V_{BO} = V_{CO}$$

Note: For simplicity, hereafter, we write

$$V_{A1} = V_1$$
$$V_{A2} = V_L$$
$$V_{A0} = V_0$$

Also,

$$1 + a^2 + a = 0$$

To find V_1:V_2 and V_0. As given above,

$$V_A = V_1 + V_2 + V_0 \tag{6.13}$$

$$V_B = a^2V_1 + aV_2 + V_0 \tag{6.14}$$

$$V_B = aV_1 + a^2V_2 + V_0 \tag{6.15}$$

Multiplying Eq. (6.14) by a we have

$$a^3V_1 + a^2V_2 + aV_0 = V_1 + a^2V_2 + aV_0 \quad [\because a^3 = 1] \tag{6.16}$$

Multiplying Eq. (6.15) by a^2V_C, we get

$$a^3V_1 + a^4V_2 + a^2V_0 = V_1 + aV_2 + a^2V_0 \quad [\because a^4 = a]$$

Rewriting Eq. (6.13), we get

$$V_A = V_1 + V_2 + V_0$$

Rewriting Eq. (6.14), we get

$$aV_B = V_1 + a^2V_2 + aV_0$$

Rewriting Eq. (6.15), we get

$$a^2V_C = V_1 + aV_2 + a^2V_0$$

Adding all we get

$$V_A + aV_B + a^2V_C = 3V_1 + V_2[1 + a^2 + a] + V_0[1 + a + a^2] = 3V_1$$

Hence

$$V_1 = \frac{1}{3}[V_A + aV_B + a^2V_C]$$

$$= \frac{1}{3}[V_A + V_B\angle 120 + V_C\angle 240]$$

$$= \frac{1}{3}[V_A + V_B(-0.5 + j0.866 + V_C(-0.5 - j0.86t))]$$

To get V_2, multiply Eq. (6.15) by a and Eq. (6.14) by a^2 and simplify:

$$V_2 = \frac{1}{3}[V_A + V_B(-0.5 - j0.866) + V_C(-0.5 - j0.866)]$$

To get V_0 add Eqs. (6.13) and (6.14) and simplify:

$$V_0 = \frac{V_A + V_B + V_C}{3}$$

Conclusion:

$$+\text{ve}\quad V_1 = \frac{1}{3}[V_A + aV_B + a^2V_C]$$

$$-\text{ve}\quad V_2 = \frac{1}{3}[V_A + a^2V_B + aV_C]$$

$$\text{Zero}\quad V_0 = \frac{1}{3}[V_A + V_B + V_C]$$

$$= \frac{1}{3}\begin{bmatrix} 1 & a & a^2 \\ 1 & a^2 & a \\ 1 & 1 & 1 \end{bmatrix}\begin{bmatrix} V_A \\ V_B \\ V_C \end{bmatrix} = \begin{bmatrix} V_1 \\ V_2 \\ V_0 \end{bmatrix}$$

Summary

In this chapter we discussed the polyphase circuits in detail.

1. This chapter mainly discussed three-phase circuits which are the industry standard. However, examples of two-phase circuits were also presented.
2. A balanced two-phase generator has two voltage sources producing the same amplitude and frequency displaced by 90° or 180° out-of-phase.
3. Three-phase generators contain three sinusoidal voltage sources with voltages of same frequency displaced a 120° phase shift with respect to each other.
4. Three identical impedances connected make up a balanced Y-connected load.
5. Three identical impedances connected in a delta configuration, and three impedances connected in a Y configuration.
6. An AC wattmeter has a potential coil and a current coil and responds to product of voltage, effective current cosine of phase angle between them.

Solved Problems

Problem 1

Three similar coils each of resistance 10 Ω and an inductance of 0.05 H are connected (i) in star, (ii) in delta to the three-phase 400 V, 50 Hz symmetrical system. Find the phase current, line currents, total phase power and total line power.

Solution

Given $R = 10\,\Omega$; $L = 0.05\,\text{H}$; $X_L = 15.7\,\Omega$; line voltage, $V = 400\,\text{V}$. Now in star connection

$$\text{Line voltage} = \sqrt{3} \times \text{Phase voltage}$$

Therefore,

$$\text{Phase voltage} = \frac{\text{Line voltage}}{\sqrt{3}} = \frac{400}{\sqrt{3}} = 231\text{ V}$$

Hence

$$\text{Impedence, } Z = \sqrt{R^2 + X_L^2} = \sqrt{10^2 + 15.7^2} = 18.61\angle 57.6°$$

(i) **Star** (Fig. 40):

$$\text{Power/phase} = V_P I_P \cos\theta$$

$$= 231 \times \frac{231}{18.61}\cos 57.6 = 1536.0\text{ W}$$

$$\text{Total three-phase power} = 3 \times V_P I_P \cos\theta$$

$$= 1536 \times 3 = 4608\text{ W}$$

Power drawn from the mains or power in the line is

$$\sqrt{3} \times 400 \times \frac{231}{18.61} \times \cos 57.6 = 4608\text{ W}$$

Phase power factor = Line power factor

Phase power factor angle = Line power factor angle

$$\text{Phase voltage} = \frac{\text{Line voltage}}{\sqrt{3}}$$

Phase current = Line current

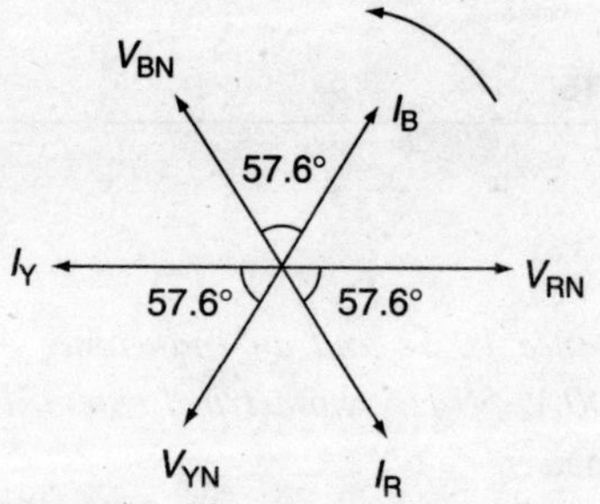

Figure 40 Solution to Problem 1, part (i).

(ii) **Delta** (Fig. 41):

$$\text{Power/phase} = V_P I_P \cos\theta$$
$$= 400 \times \frac{400}{18.61} \cos 57.6$$
$$= 4608\ \text{W}$$

$$\text{Total three-phase power} = 3 \times 4608 = 13824\ \text{W}$$

Power drawn from the mains or power in the line is

$$P = \sqrt{3} \times 400 \times \left(\frac{400}{18.61}\right) \cos 57.6$$
$$= 7970\ \text{W}$$

$$\text{Phase power factor} \neq \text{Line power factor}$$

$$\text{Line power factor angle} = \text{Phase power factor angle}$$

$$\text{Phase voltage} = \text{Line voltage}$$

$$\text{Phase current} = \frac{\text{Phase voltage}}{Z_{\text{phase}}}$$

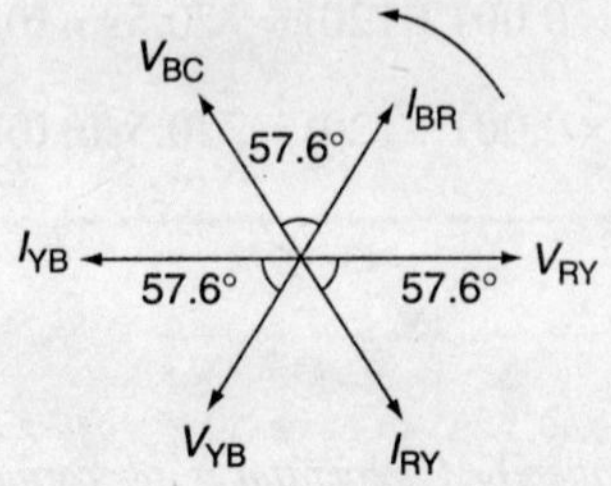

Figure 41 Solution to Problem 1, part (ii).

Problem 2

The currents drawn by a balanced three-phase load at 150 Hz are given by

$$I_R = 720.5 \sin 942.48\, t$$

$$I_Y = 720.5 \sin [942.48 + (-120°)]$$

$$I_B = 720.5 \sin (942.48t + 120°)$$

Find I_R, I_Y and I_B at the following instants of time, after the currents pass through zero and increase in the positive direction:
(i) $t = 0$.
(ii) $t = 0.001$.

Solution
Given,

$$f = 150\,\text{Hz}$$
$$\omega = 942.48 \Rightarrow 2\pi f = 942.48$$
$$V = V_m \sin \omega t$$

(i) At $t = 0$

$$I_R = 720.5 \sin 942.48 \times 0 = 0$$
$$I_Y = 720.5 \sin (0 - 120°) = -623.97\text{ A}$$
$$I_B = 270.5 \sin (0 + 120°) = 623.97\text{ A}$$

(ii) At $t = 0.001$

$$I_R = 720.5 \sin 2\pi \times 150 \times 0.001 = 720 \sin 0.3 \times 180° = 572.89\text{ A}$$
$$I_Y = 720.5 \sin (2\pi \times 150 \times 0.001 - 120) = 720.5 \sin (0.3 \times 180° - 120) = 658.20\text{ A}$$
$$I_B = 720.5 \sin (2\pi \times 150 \times 0.001 + 120) = 720.5 \sin (0.3 \times 180° + 120) = 75.312\text{ A}$$

Problem 3

A 10.6 kV, 100 MVA (mega volt ampere) AC generator is star connected. Find the line current, coil current and voltage at the terminal and voltage per coil. Find the same values, if the generator is delta connected.

Solution
Given, 10.6 kV, 100 MVA, Y connected. Therefore,

$$I_L = \frac{100 \times 18}{\sqrt{3} \times 10.6 \times 10^3} = 5.4\,\text{kA}$$
$$\text{Voltage at the terminal} = \frac{10.6}{\sqrt{3}} = 6.12\text{ kV}$$

For delta connected:

$$\text{Voltage at the coil} \neq \text{Voltage at the terminal}$$
$$\text{Coil current} = \frac{\text{Line current}}{\sqrt{3}} = 3.117\text{ A}$$

Problem 4

Three capacitors each of 50 μF is in star to improve the power factor of a three-phase 1.1 kV, 50 Hz system. Determine the KVARs they take. Find the same, if the capacitors are in delta.

Solution

Given $C = 50\ \mu\text{F}$, $f = 50\,\text{Hz}$.

$$X_C = \frac{1}{2\pi fC} = 63.66\ \Omega$$

$$V_{ph} = \frac{1.1}{\sqrt{3}} = 0.6351\ \text{kV}$$

$$I_{ph} = \frac{V_{ph}}{X_C} = 9.976\ \text{A}$$

Now $I_{ph} = I_L$, since it is star connected and $\theta = 90^\circ$, since $K = 0$ (from Fig. 42). Therefore

$$\begin{aligned}\text{KVAR} &= \sqrt{3}\,V_L I_L \sin\theta \\ &= \sqrt{3}\times 1.1\times 9.976\times \sin 90^\circ \\ &= 19\ \text{KVAR}\end{aligned}$$

If the capacitors are in delta:

$$I_{ph} = 1100\times\frac{1}{63.66} = 17.279\ \text{A}$$

$$I_L = \sqrt{3}\,I_{ph} = 29.92\ \text{A}$$

Therefore KVAR taken is

$$\sqrt{3}\,V_L I_L \sin\theta_{LINE} = 49.379\ \text{KVAR}$$

$$\theta_{LINE} = 90 - 30 = 60^\circ$$

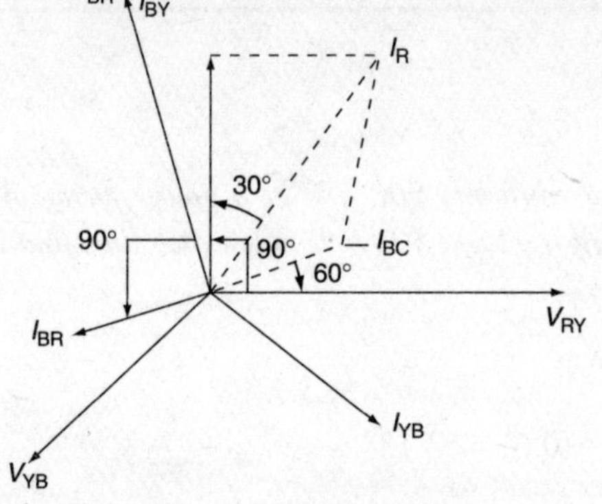

Figure 42 Solution to Problem 4.

Problem 5

The connected loads in a three-phase 6.6 kV, 50 Hz system are:

(i) 500 KVA, 0.94 efficiency 10.65 power factor transformer.
(ii) 100 hp motor, 0.90 efficiency, 0.85 power factor.
(iii) 100 kW heater, 0.95 efficiency.
(iv) 100 KVAR 0.8 power factor lead capacitor bank.

Find the total power in the system in kW, KVA, KVAR, the total current and power factor.

Solution

The solution is given by Table 1.

Table 1

S. No.	*Load*	*Efficiency*	*Power factor*	$\theta = \cos^{-1}()$	*Power line*			*Line current*
					K VA	*KVAR*	*kW*	
1.	500 KVA transformer	0.94	0.65 lag	−49.45°	532.0	−404.106	354.74	46.53
2.	100 hp motor	0.90	0.8 lag	−31.78°	96.143	−50.666	81.722	8.410
3.	100 kW heater	0.95	1.0	0°	105.26	0	105.26	9.28
4.	100 K VAR	1.0	0.8 lead	36.87°	166.66	100.00	133.33	11.66
	Total		0.8258 lag	−34.33°	806.48	−354.842	666.05	64.88

Now

$$I_{\text{Total}} = 46.53\angle -49.45 + 8.410\angle -31.78 + 9.28\angle 0 + 11.66\angle 36.87$$

Expanding and simplifying, we get

$$I_{\text{Total}} = 64.88 \text{ A at } 34.33° \text{ log } (0.8258 \text{ power factor})$$

Problem 6

A three-phase balanced load consumes 500 kW at a power factor of 0.6 load. Construct the power triangle. If the voltage applied is 416 V, 50 Hz between each line, find the active and reactive current.

Solution

Power consumed = 500 kW
Power factor = 0.6 lead
Power factor angle $\theta = \cos^{-1}(0.6) = 53.13°$

From the triangle in Fig. 43, we have

$$\text{KVA} = \frac{\text{Power consumed}}{\cos 53.13^\circ} = \frac{500}{0.6} = 833.33$$

$$\text{KVAR} = \text{kW} \tan 53.13^\circ = 666.664 \text{ KVAR}$$

$$\text{Active component of current} = I\cos\theta = \frac{\text{Power consumed}}{\sqrt{3}\times KV} = \frac{500}{\sqrt{3}\times 0.416}$$
$$= 693.95\,\text{A}$$

where

$$KV = 416\times 10^{-3} = 0.416 \text{ V}$$

$$\text{Reactive component of current} = I\sin\theta = \frac{\text{KVAR}}{\sqrt{3}\times KV} = \frac{666.64}{\sqrt{3}\times 0.416}$$
$$= 925.20\,\text{A}$$

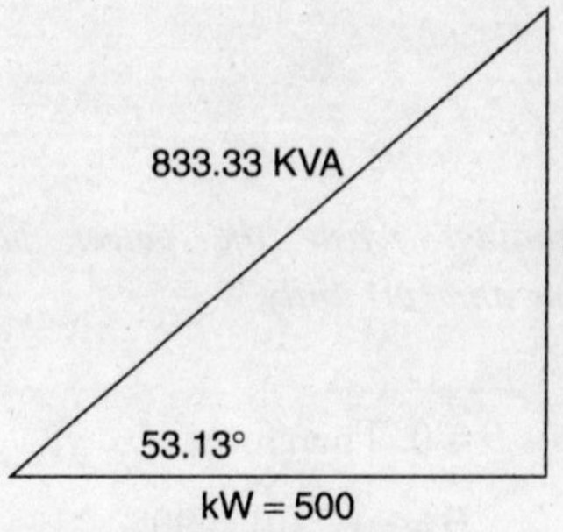

Figure 43 Solution to Problem 6.

Problem 7

A 10 hp, 400 V, three-phase, 50 Hz system has an efficiency of 80% and a power factor of 0.75. Find the reading each of the two wattmeters connected to measure the power when it is delivering half load with the same efficiency and power factor.

Solution

$$\text{Output} = 5 \text{ hp (half-load)} = 5\times 735.5 = 3677.87 \text{ W}$$

$$\text{Input} = \frac{\text{Output}}{\text{Efficiency}} = \frac{3677.5}{0.8} = 4596.87 \text{ W}$$

$$W_1 + W_2 = 4596.87 \text{ W} \tag{6.17}$$

Power factor (given) = 0.75; cos $\theta = 0.75$; $\theta = 41.41$. Therefore,

$$\tan\theta = \sqrt{3}\,\frac{W_1 - W_2}{W_1 + W_2}$$

$$\tan 41.41 = \sqrt{3}\,\frac{W_1 - W_2}{4{,}596.87} = 0.8819$$

Hence

$$W_1 - W_2 = \frac{0.8819 \times 4{,}596.87}{\sqrt{3}} = 2340.68 \qquad (6.18)$$

Solving Eqs. (6.17) and (6.18), we have

$$W_1 = 3468.77 \text{ W}$$

$$W_2 = 1128.09 \text{ W}$$

Problem 8

Find the ratio of wattmeter readings when the power factor is (i) zero, (ii) infinity, (iii) 0.3 lag, (iv) 0.5 lag, (v) 0.707 lag and (vi) unity.

Solution

(i) When the power factor is zero, cos $\theta = 0$. Therefore,

$$\theta = \cos^{-1}(0) = 90°$$

$$\tan 90 = \sqrt{3}\,\frac{W_1 - W_2}{W_1 + W_2}$$

$$\frac{(W_1 + W_2)}{(W_1 - W_2)} = \frac{\sqrt{3}}{\infty} = 0$$

Hence

$$W_1 + W_2 = 0$$

$$W_1 = -W_2$$

$$\frac{W_1}{W_2} = -1$$

Both readings are equal but one of them gives a negative reading.

(ii) When the power factor is infinity, θ is indeterminate. Hence no readings on W_1 and W_2. This condition cannot exist physically.

(iii) When power factor = 0.3 lag; $\theta = \cos^{-1}(0.3) = 72.5$. Now

$$\tan 72.5 = \sqrt{3}\,\frac{W_1 - W_2}{W_1 + W_2} = 3.18$$

or

$$\frac{W_1 - W_2}{W_1 + W_2} = \frac{3.18}{\sqrt{3}} = 1.835$$

This implies

$$W_1 - W_2 = 1.835\,W_1 + 1.835W_2$$

$$-0.835\,W_1 = 2.835\,W_2\ \frac{W_1}{W_2} = \frac{2.835}{-0.835} = -3.395$$

that is, one of the wattmeters gives a negative reading.

(iv) When power factor = 0.5 lag, $\theta = \cos^{-1}(0.5) = 60°$. Then

$$\tan 60° = \sqrt{3}\ \frac{W_1 - W_2}{W_1 + W_2}$$

$$\sqrt{3} = \sqrt{3}\ \frac{W_1 - W_2}{W_1 + W_2}$$

Hence

$$W_1 + W_2 = W_1 - W_2$$

$$\Rightarrow 2W_2 = 0$$

that is one of the meters gives zero reading.

(v) When power factor = 0.707 lag; $\theta = \cos^{-1}(0.707) = 45$. Then

$$\tan 45° = \sqrt{3}\,\frac{W_1 - W_2}{W_1 + W_2}$$

$$\frac{1}{\sqrt{3}} = \frac{W_1 - W_2}{W_1 + W_2}$$

Hence

$$W_1 + W_2 = \sqrt{3}\,W_1 - \sqrt{3}\,W_2$$

$$(\sqrt{3} - 1)\,W_1 = (\sqrt{3} + 1)W_2$$

$$\frac{W_1}{W_2} = \frac{\sqrt{3} + 1}{\sqrt{3} - 1}$$

that is, both meters give unequal reading.

(vi) When power factor = unity; $\theta = \cos^{-1}(1.0) = 8$. Then

$$\tan 0° = \sqrt{3}\,\frac{W_1 - W_2}{W_1 + W_2}$$

$$\left(\frac{W_1 + W_2}{\sqrt{3}}\right) \times 0 = W_1 - W_2$$

$$W_1 = W_2 \text{ or } \frac{W_1}{W_2} = 1$$

that is both meters read equal and positive.

Problem 9

Calculate the (i) positive, (ii) negative and (iii) zero phase sequence voltages for the unbalanced voltage system:

$$V_A = 50\angle 0°;\quad V_B = 100\angle -140°;\quad V_C = 70\angle 90°$$

Solution

(i) Positive phase sequence vectors:

$$V_1 = \frac{1}{3}[V_A + aV_B + a^2V_C]$$

$$= \frac{1}{3}[(50\angle 0°) + (a100\angle -140°) + (a^2 70\angle 90°)]$$

$$= \frac{1}{3}[50\angle 0° + 100\angle -140° \times 1\angle 120° + 70\angle 90° \times 1\angle 240°]$$

$$= \frac{1}{3}[50\angle 0° + 100\angle -20° + 70\angle 330°] = 70\angle -18.7°$$

where $a = \angle 120$; $a^2 = \angle 240 = \angle -120°$. The components are:

$$V_{A1} = V_{A1}\angle 0°;\ V_{B1} = a^2 V_{A1}\angle -120°;\ V_{C1} = aV_{A1} = V_{A1}\angle +120°$$

$$V_{B1} = a^2 = V_{A1} = 72\angle -18.7° \times 1\ \angle -120° = 72\ \angle -138.7°$$

$$V_{C1} = aV_{A1} = V_{A1} = 72\angle -18.7° \times 1\angle 120° = 72\angle 101.3°$$

(ii) Negative phase sequence vectors:

$$V_2 = \frac{1}{3}[V_A + a^2V_B + aV_C]$$

$$= \frac{1}{3}[50\angle 0° + 100\angle -140° \times 1\angle 240° + 70\angle 90° \times 1\angle 120°]$$

$$= 23.12\angle 113.8°$$

The components are:

$$V_{A2} = V_2 = \angle 0° = 23.12\angle 113.8°$$

$$V_{B2} = aV_2 = 23.12\angle 113.8° \times 1\angle 120° = 23.12\angle 233.8°$$

$$V_{C2} = a^2V_2 = 23.12\angle 113.8° \times 1\angle -120° = 23.12\angle -6.2°$$

(iii) Zero sequence vectors:

$$V_0 = \frac{1}{3}[V_A + V_B + V_C]$$

$$= \frac{1}{3}[50\angle 0° + 100\angle -140° + 70\angle 90°]$$

$$= 9.07\angle 167.85°$$

The components are:

$$V_{A0} = V_{B0} = V_{C0} = 9.07\angle 167.85°$$

Problem 10

Three non-reactive resistors of 5, 10 and 15 Ω are star connected to R, Y and B of a 440 V symmetrical system. Determine the current and power in each resistor and the voltage between star point and neutral point (assume RYB).

Solution

Given, $R_1 = 5\ \Omega$, $R_2 = 10\ \Omega$, $R_3 = 15\ \Omega$, $V_L = 440$ V. Now

$$V_L = \sqrt{3} \times V_{ph}$$

Therefore

$$V_{ph} = \frac{440}{\sqrt{3}} = 254$$

Also

$$I_R = \frac{V_{ph}}{R_1} = \frac{254\angle 0°}{5} = 50.8\angle 0°$$

$$I_Y = \frac{V_{ph}}{R_2} = \frac{254\angle -120°}{10} = 25.4\angle -120°$$

$$I_B = \frac{V_{ph}}{R_3} = \frac{254\angle -240°}{5} = 16.936\angle -240°$$

Now

$$P_R = (I_R)^2 \times R_R = (16.936)^2 \times 15 = 4312.5 \text{ W}$$

$$P_Y = (I_Y)^2 \times R_Y$$

$$= (25.4)^2 \times 10$$

$$= 6451.6 \text{ W}$$

Problem 11

A three-phase 3-wire system with a symmetrical 440 V source is supplying power to an unbalanced, delta-connected load in which $Z_{RY} = 20\angle 30°\ \Omega$, $Z_{YB} = 20\angle 0°$ and $Z_{BR} = 20\angle -30°\ \Omega$ if the phase sequence is RYB. Find line currents.

Solution
Given

$$V_L = 440 \text{ V}, Z_{RY} = 20\angle 30°\ \Omega, Z_{YB} = 20\angle 0°\ \Omega, Z_{BR} = 20\angle -30°\ \Omega$$

Now

$$I_R = \frac{V_{RY}}{Z_{RY}} = \frac{440\angle 0°}{20\angle 30°} = 22\angle -30° = 19 - j11$$

$$I_Y = \frac{V_{YB}}{Z_{YB}} = \frac{440\angle -120°}{20\angle 0°} = 22\angle -120° = -11 - j19$$

$$I_B = \frac{V_{BR}}{Z_{BR}} = \frac{440\angle 240°}{20\angle -30°} = 22\angle -270° = 0 + j22$$

Line current,

$$I_1 = I_R - I_B = (19.0 - j11) - (0 + j22)$$

$$= 19.0 - j33$$

$$= 38.07\angle -60.60°$$

$$I_2 = I_Y - I_R = (-11 - j99) - (19.0 - j11)$$

$$= -30 \angle j8$$

$$= 31.04 \angle -165.06$$

$$I_3 = I_B - I_R$$

$$= j22 - (-11 - j19)$$

$$= 11 + j44$$

$$= 45.35 \angle 75.96°$$

Problem 12

Three equal resistances connected in star across a three-phase balanced supply consume 1000 W. If the resistors were connected in delta across the same supply, determine the power consumed.

Solution

In the delta connection the power is three times the star connection; so the power in the delta connection is 3000 W.

Problem 13

A three-phase four-wire symmetrical 440 V RYB system supplies a star-connected load in which $Z_R = 10 \angle 0°\ \Omega$, $Z_Y = 10\angle 26.8°\ \Omega$ $Z_B = 10\angle -26.8°\ \Omega$. Find the I_Ls, the neutral currents and the load power.

Solution

Given

$$V_L = 440 \text{ V}$$

$$Z_R = 10 \angle 0° = 10 + j0$$

$$Z_Y = 10 \angle 26.8°\ \Omega = 8.92 + j4.5$$

$$Z_B = 10 \angle -26.8°\ \Omega = 8.92 - j4.5$$

$$V_{RN} = \frac{440}{\sqrt{3}} = 254.03 \angle 0°$$

$$V_{YN} = 254.03 \angle -120°$$

$$V_{BN} = 254.03 \angle -240°$$

Therefore

$$I_R = \frac{V_{RN}}{Z_R} = \frac{254.03\angle 0°}{10} = 25.4\angle 0° = 25.4 + j0$$

$$I_Y = \frac{V_{YN}}{Z_Y} = \frac{254.03\angle -120°}{10\angle 26.8°} = 25.4\angle -146.8° = -21.25 - j13.9$$

$$I_B = \frac{V_{BN}}{Z_B} = \frac{254.03\angle -240°}{10\angle -26.8°} = 25.4\angle -213.2° = -21.25 + j13.9$$

Now

$$\begin{aligned} I_N &= -(I_R + I_4 + I_B) \\ &= -(25.4 - 42.5) \\ &= 17.1\angle 0° \end{aligned}$$

Also

$$\begin{aligned} P_1 &= I_R^{\ 2} \times R_R \\ &= (645.16) \times 10 = 6451.6 \text{ W} \end{aligned}$$

$$\begin{aligned} P_2 &= I_Y^{\ 2} \times R_Y \\ &= (645.16) \times (8.92) = 5754.8 \text{ W} \end{aligned}$$

$$P_3 = (645.16) \times (8.92) = 5754.8 \text{ W}$$

Therefore, total power is 17961.2 W.

Problem 14

A 440 V, three-phase supply feeds an unbalanced three-wire star-connected load consisting of impedances $Z_R = 7\angle 10°\Omega$, $Z_Y = 8\angle 30°\Omega$, $Z_B = 8\angle 50°\ \Omega$. The phase sequence is RYB. Determine the line currents and total power taken by the load?

Solution

Given that $V_L = 400$ V. Then

$$V_{Ph} = \frac{V_L}{\sqrt{3}} = \frac{400}{\sqrt{3}} = 230.9 \text{ V}$$

$$I_R = \frac{V_{ph}}{Z_R} = \frac{230.9\angle 0^\circ}{7\angle 10^\circ} = 32.99\angle -10^\circ \text{ A}$$

$$I_Y = \frac{V_{ph}}{Z_Y} = \frac{230.9\angle -120^\circ}{8\angle 30^\circ} = 28.86\angle -150^\circ \text{ A}$$

$$I_B = \frac{V_{ph}}{Z_B} = \frac{230.9\angle -240^\circ}{8\angle 50^\circ} = 28.86\angle -290^\circ \text{ A}$$

Now

$$Z_R = 7\angle 10^\circ = 6.89 + j1.2$$

$$Z_Y = 8\angle 30^\circ = 6.92 + j4$$

$$Z_B = 8\angle 50^\circ = 5.14 + j6.1$$

Also

$$\begin{aligned} P_R &= (I_R)^2 \times R_R \\ &= (32.99)^2 \times 6.89 \\ &= 7498.66 \text{ W} \end{aligned}$$

$$\begin{aligned} P_Y &= (I_Y)^2 \times R_Y \\ &= (28.86)^2 \times 6.92 \\ &= 5763.6 \text{ W} \end{aligned}$$

$$\begin{aligned} P_B &= (I_B)^2 \times R_B \\ &= (832.8)^2 \times 5.14 \\ &= 4281.10 \text{ W} \end{aligned}$$

Therefore, total power is 17543.3 W.

Problem 15

Two wattmeters are connected to measure power in a three-phase circuit. The reading of one of the meters is 5 kW when the load power factor is unity. If the power factor of load is changed to 0.707 lagging, without changing the total input power, calculate the reading of both the wattmeters.

Solution

Given,

$$\text{Power factor} = \text{Unity}$$

$$\cos\phi = 1$$

$$\phi = 0°$$

Then we will get $W_R = W_Y$. Given one wattmeter reading is 5 kW. So,

$$W_R + W_Y = 5 + 5 = 10 \text{ kW}$$

Now, the factor of load is changed to 0.707 lagging. Let W_Y be higher reading wattmeter. Then,

$$\cos\phi = 0.707$$

$$\phi = 45°$$

$$\tan\phi = \sqrt{3}\,\frac{W_Y - W_R}{W_R + W_Y}$$

$$1 = \sqrt{3}\,\frac{W_Y - W_R}{10}$$

Hence

$$W_Y - W_R = 5.773 \text{ kW} \quad (6.19)$$

We know that

$$W_Y + W_R = 10 \text{ kW} \quad (6.20)$$

Solving both Eqs. (6.19) and (6.20), we get

$$\begin{aligned} W_Y - W_R &= 5.773 \\ W_Y + W_R &= 10 \\ \hline 2W_Y &= 15.773 \end{aligned}$$

Therefore

$$W_Y = 7.886 \text{ W}$$

$$W_R = 2.113 \text{ kW}$$

Problem 16

Calculate total power input of the readings of two voltmeters connected to measure power in a three-phase balanced load, if the reactive power input is 15 KVAR, and load power factor is 0.8.

Solution

Let W_1 and W_2 be the two wattmeter readings:

$$W_1 \rightarrow \text{Higher}$$

$$W_2 \rightarrow \text{Lower}$$

Given load power factor = 0.8. So,

$$\cos\phi = 0.8$$

$$\phi = 36.86°$$

We know that

$$\tan\phi = \frac{\text{Reactive power}}{\text{Active power}}$$

$$\tan\phi = \sqrt{3}\,\frac{W_1 - W_2}{W_1 + W_2}$$

$$W_1 - W_2 = 8660.5$$

$$\sqrt{3}\,(W_1 - W_2) = 15 \text{ KVAR}$$

$$0.75 = \frac{\sqrt{3}(8660.5)}{W_1 + W_2}$$

Now

$$W_1 + W_2 = 34641.0 \text{ W}$$

$$W_1 - W_2 = 8660.5 \text{ W}$$

$$W_1 = 21650.7 \text{ W}$$

Therefore,

$$W_1 = 21650.7 \text{ W}$$

$$W_2 = 12990.2 \text{ W}$$

Problem 17

A three-phase delta-connected RYB system with an effective voltage of 400 V has a balanced load with impedance 3 + j4. Calculate phase currents and line currents in each phase.

Solution

Given,

$$V_L = V_{ph} = 400 \text{ V}$$
$$Z = 3 + j4 = 5\angle 53.1^\circ$$
$$V_{RY} = 400\angle 0^\circ$$
$$V_{YB} = 400\angle -120^\circ$$
$$V_{BR} = 400\angle -240^\circ$$

Phase currents,

$$I_R = \frac{V_{RY}}{Z} = \frac{400\angle 0^\circ}{5\angle 53.1^\circ} = 80\angle -53.1^\circ = 48.03 - j63.17$$

$$I_Y = \frac{V_{YB}}{Z} = \frac{400\angle -120^\circ}{5\angle 53.1^\circ} = 80\angle -173.1^\circ = -79.42 - j9.61$$

$$I_B = \frac{V_{YB}}{Z} = \frac{400\angle -240^\circ}{5\angle 53.1^\circ} = 80\angle -293.1^\circ = -31.38 - j73.58$$

Line currents,

$$I_1 = I_R - I_B = 138.55\angle -83.09^\circ$$
$$I_2 = I_Y - I_R = 138.55\ \angle 156.9^\circ$$
$$I_3 = I_B - I_Y = 138.55\ \angle 36.9^\circ$$
$$\cos\phi = P/Z = 3/5 = 0.6$$
$$\text{Phase to power} = 3 \times V_{ph} \times I_{ph} \times \cos\phi$$
$$= 3 \times 400 \times 80 \times 0.6$$
$$= 57600 \text{ W}$$

Problem 18

The three impedances $Z_1 = 20\angle 30^\circ\Omega$, $Z_2 = 40\angle 60^\circ\ \Omega$, $Z_3 = 10\angle -90^\circ\ \Omega$ are delta-connected to a 400 V three-phase system. Determine I_{ph}, I_L and total power consumed.

Solution

$$V_{RY} = 400\angle 0^\circ$$

$$V_{YB} = 400\angle -120^\circ$$

$$V_{BR} = 400\angle -240^\circ$$

Now phase currents are given by

$$I_R = \frac{V_{RY}}{ZL\phi_1} = \frac{400\angle 0^\circ}{20\angle 30^\circ} = 20\angle -30^\circ = 17.3 - j10 \text{ A}$$

$$I_Y = \frac{V_{YB}}{ZL\phi_2} = \frac{400\angle -120^\circ}{40\angle 60^\circ} = 10\angle -180^\circ = -10 + j10 \text{ A}$$

$$I_B = \frac{V_{RB}}{ZL\phi_3} = \frac{400\angle -240^\circ}{10\angle -90^\circ} = 40\angle -150^\circ = -34.6 - j20 \text{ A}$$

Line currents are given by

$$\begin{aligned} I_1 &= I_R - I_B \\ &= 51.96 + j10 \\ &= 52.91\angle 10.89 \text{ A} \end{aligned}$$

$$\begin{aligned} I_2 &= I_Y - I_R \\ &= -27.3 + j20 \\ &= 29.09\angle 159.89 \text{ A} \end{aligned}$$

$$\begin{aligned} I_3 &= I_B - I_Y \\ &= -24.6 - j20 \\ &= 31.73\angle -140.93 \end{aligned}$$

$$Z_1 = 20\angle 30^\circ = 17.32 + j10\ \Omega$$

$$Z_2 = 40\angle 60^\circ = 20 + j34.62\ \Omega$$

$$Z_3 = 10\angle -90^\circ = 0 - j10\ \Omega$$

Power consumed is

$$R_{phase} \text{ Power} = (I_R)^2 \times R_R = 6928 \text{ W}$$

$$Y_{phase} \text{ Power} = (I_Y)^2 \times R_Y = 2000 \text{ W}$$

$$B_{phase} \text{ Power} = 0$$

Therefore, total power is 8928 W.

Problem 19

Three impedances of (7 + j4) Ω, (3 + j2) Ω and (9 + j2) Ω are connected between neutral and the R, Y, B phases, respectively, of a three-phase four-wire system; the line voltage is 440 V. Calculate the current in each line.

Solution
In star connection,

$$V_{\text{Ph}} = \frac{440}{\sqrt{3}} = 254.03\text{ V}$$

$$Z_{\text{R}} = 7 + j4 = 8.06\angle 29.7°\text{ V}$$

$$Z_{\text{Y}} = 3 + j2 = 3.6\angle 33.6°\text{ V}$$

$$Z_{\text{B}} = 7 + j2 = 9.2112.5°\angle 12.5°\text{ V}$$

Current in each line is given by

$$I_{\text{R}} = \frac{V}{Z} = \frac{254.03\angle 0°}{8.06\angle 29.7°} = 31.51\angle -29.7\text{A}$$

$$I_{\text{Y}} = \frac{V}{Z} = \frac{254.03\angle -120°}{36\angle 33.6°} = 83.0\angle -153.6\text{ A}$$

$$I_{\text{B}} = \frac{V}{Z} = \frac{254.03\angle -240°}{7.21\angle 12.5°} = 27.5\angle -252.5\text{ A}$$

Problem 20

A delta-connected load has the following impedance in R–Y–Y–B and B–R phase. Find the equivalent star impedance.

Solution
In delta connection (Fig. 44),

$$R - Y = 20 + j30 = 36\angle 53.3°$$

$$Y - B = 30 - j25 = 39.05\angle -39.8°$$

$$B - R = 40 - j40 = 56.566\angle -45°$$

From delta to star:

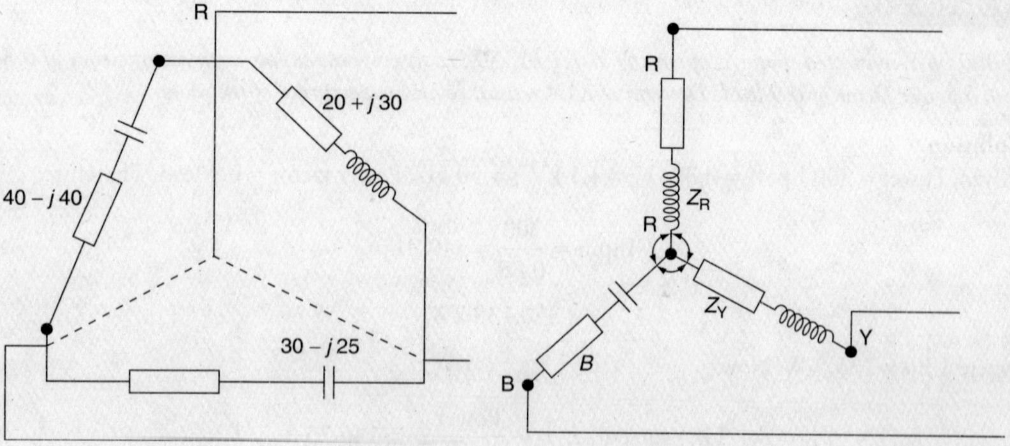

Figure 44 Solution to Problem 20.

$$Z = \frac{Z_{RY} Z_{BR}}{\sum Z_{RY}}$$

$$= \frac{36\angle 53.3 \times 56.56\angle -45^\circ}{96.5666\angle -21.25^\circ}$$

$$= 21.08\angle 32.55^\circ$$

$$= 17.76 + j11.34$$

$$Z_Y = \frac{Z_{YB} Z_{RY}}{\sum Z_{RY}}$$

$$= \frac{39.05^\circ\angle -39.8^\circ \times 36\angle 56.56^\circ}{96.566\angle -21.25^\circ}$$

$$= 14.558\angle 37.75^\circ$$

$$= 11.51 + j8.912$$

$$Z_B = \frac{Z_{BR} Z_{YB}}{\sum Z_{RY}}$$

$$= \frac{56.56\angle -45^\circ \times 39.05\angle -39.8^\circ}{96.566\angle -21.25^\circ}$$

$$= 22.872\angle -63.55^\circ$$

$$= 10.187\angle - j20.477^\circ$$

Problem 21

A 300 hp Y-connected motor is operating at 1.1 kV, 50 Hz three-phase supply with an efficiency of 0.88 and a power factor of 0.9 lead. Determine KVA resistance and capacitance of the motor.

Solution

Given, Power = 300 hp, $\eta = 0.88$, $V_L = 1.1$ kV, $f = 50$ Hz, Power factor = 0.9 lead. Therefore,

$$\text{Input} = \frac{300}{0.88} = 340.9\,\text{hp}$$
$$= 250.738\ \text{kW}$$

where 1 hp = 735.5 W. Now

$$\text{Line current, } I_L = \frac{\text{Power}}{\sqrt{3}\,V_L \cos\theta} = 146.23\,\text{A}$$

$$\text{Impedance per phase of motor } = \frac{V_{ph}}{I_{ph}}$$
$$= \frac{1{,}100}{\sqrt{3}\times 146.23}$$
$$= 4.343\,\Omega$$

$$\cos\theta = \frac{R}{Z} = 0.9$$
$$R = Z\cos\theta$$
$$= 4.343\times 0.9$$
$$= 3.908\ \Omega$$
$$X_C = Z\sin\theta$$
$$= 4.343\times 0.4358$$
$$= 1.893\ \Omega$$

X_C of the motor is 1.893 Ω and

$$C = \frac{10^6}{2\pi\times 50\times 1.893}$$
$$= 16824\ \mu\text{F}$$

Therefore

$$\text{KVA required} = \sqrt{3}\times V_L\times I_L$$
$$= \sqrt{3}\times 1.1\times 146.23$$
$$= 278.59\,\text{KVA}$$

Problem 22

Each phase of a balanced star-connected load consists of a resistance of 80 Ω in parallel with a capacitor 30.4 F capacitance(Fig. 45). If this load is connected to a 400 V, 50 Hz supply, find (i) line current, (ii) system KVA, (iii) system power, (iv) system reactive power and (v) power factor.

Solution

$$X_C = \frac{1}{2\pi fC}$$

$$= \frac{10^6}{2\pi \times 50 \times 30}$$

$$= 106.1\Omega$$

Total admittance/phase,

$$Y_{ph} = \frac{1}{80} + j2\pi fC$$

$$= \frac{1}{80} + j2\pi \times 50 \times 30 \times 10^{-6}$$

$$= 9.125 + j0.0924$$

$$= 0.09506\angle 82.44°$$

Voltage/phase is

$$V_{ph} = \frac{400}{\sqrt{3}} = 231 \text{ V}$$

Phase current,

$$I_{ph} = V_{ph}Y_{ph} = 231\angle 0° \times 0.09506\angle 82.44°$$

$$= 21.959\angle 82.44°$$

Line current = phase current = 21.958 A

Power factor angle $\theta = 82.44°$ and power factor = 0.131510. The system KVA is

$$\text{KVA} = \sqrt{3}\, V_L I_L$$

$$= \sqrt{3} \times 400 \times 21.958$$

$$= 15.212 \text{ KVA}$$

System power is given by

$$\text{KVA} \times \cos\theta = 15.212 \times 0.1315 = 2.00 \text{ kW}$$

System reactive power is given by

$$\text{KVA} \times \sin\theta = 15.212 \times \sin 82.44 = 15.079 \text{ KVAR}$$

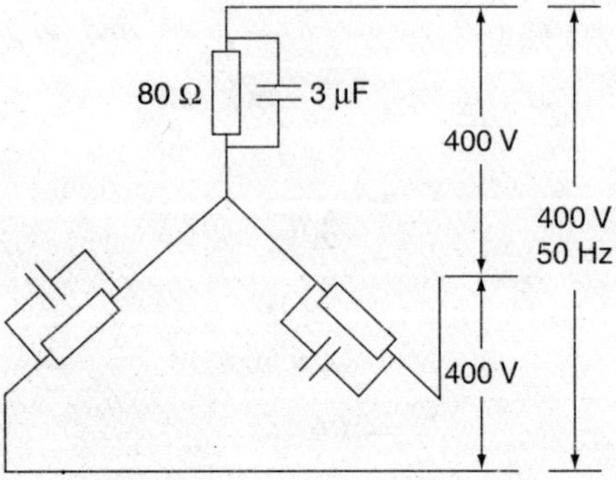

Figure 45 Solution to Problem 22.

Problem 23

A 1,000 hp motor in a crushing industry is operated from a 6.6 kV, three-phase 50 Hz supply with an efficiency of 0.85 and a power factor of 0.9 lag. If the motor is operated at 125% full load, with the above efficiency and power factor, determine the current drawn, input kW, KVA and KVAR.

Solution:

Name plate detail: 1000 hp: 6.6 kV three phase, frequency 50 Hz, efficiency $\eta = 0.85$, power factor = 0.9. Therefore,

$$\text{Output of motor} = 125\% \text{ of } 1000 \text{ hp}$$

$$= 1250 \text{ hp}$$

$$\text{Input of motor} = \frac{1250}{0.85} = 1081.617 \text{ kW}$$

and this is the power drawn. Now

$$\text{Power factor} = 0.9 \quad \text{and} \quad \theta = 25.84°$$

So

$$\text{KVA input} = \frac{\text{kW input}}{\cos\theta}$$

$$= \frac{1081.617}{0.9}$$

$$= 1201.796 \text{ KVA}$$

$$\text{KVAR drawn} = \text{KVA} \sin\theta = 523.814 \text{ KVAR}$$

$$\text{Current drawn} = \frac{\text{kW}}{\sqrt{3} \times V_L \cos\theta}$$

$$= \frac{1081.617}{\sqrt{3} \times 6.6 \times 0.9}$$

$$= 105.133 \text{ A}$$

Problem 24

The two wattmeters connected to measure power on a 400 V three-phase balanced delta load, give the following readings: $W_1 = 1250$ W, $W_2 = 700$ W. Find the load impedance per phase.

Solution

For balanced loads:

$$\tan\theta = \sqrt{3}\frac{W_1 - W_2}{W_1 + W_2}$$

Therefore,

$$\tan\theta = \sqrt{3}\frac{1250 - 700}{1250 + 700}$$

$$= \sqrt{3}\frac{550}{1950} = 0.4885$$

$$\theta = \tan^{-1}(0.4885)$$

$$\theta = 26°$$

Total power is given by

$$\sqrt{3} V_L I_L \cos\theta = W_1 + W_2 = 1950$$

Line current

$$I_L = \frac{1950}{\sqrt{3} \times 400 \times 10526} = 3.132 \text{ A}$$

Now

$$I_{ph} = \frac{I_L}{\sqrt{3}} = 1.807 \text{ A}$$

Therefore phase impedance,

$$Z = \frac{V_{ph}}{I_{ph}}$$

$$= \frac{400}{1.807}$$

$$= 221.34\ \Omega$$

Problem 25

A balanced star-connected load (Fig. 46) has the following elements connected in series per phase R = 25 Ω, L = 0.5 F, C = 30 MF. It is connected to a 500 V three-phase 50 Hz supply. Find the readings of two wattmeters connected to measure the power.

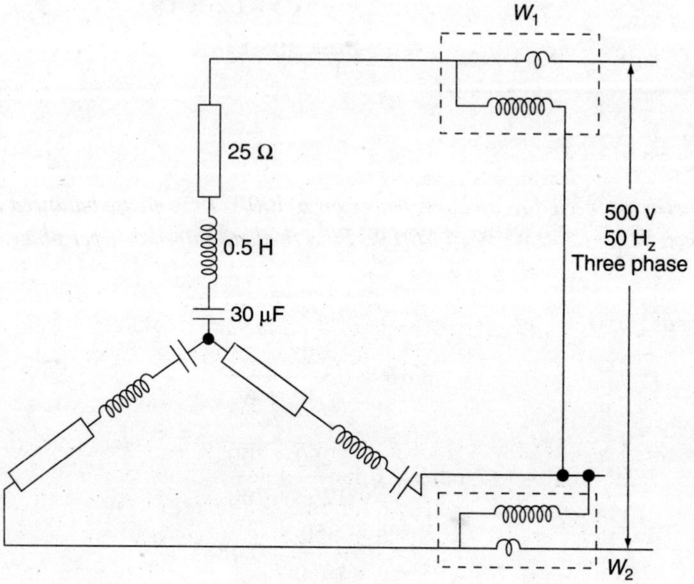

Figure 46 Problem 25.

Solution

Given $R = 25\Omega$. Hence

$$X_L = 2\pi fL = 100\pi \times 0.5 = 157\,\Omega$$

$$X_C = \frac{1}{2\pi fC} = \frac{10^6}{100\pi \times 30} = 106.1\Omega$$

$$\text{Net reactance/phase} = X_L - X_C = 50.9\,\Omega$$

$$\text{Impendance/phase} = \sqrt{R^2 + (X_L - X_C)^2} = 56.7\angle 63.84°$$

$$\text{Voltage phase} = \frac{500}{\sqrt{3}}$$

$$\text{Current/phase} = \frac{V_{ph}}{Z_{ph}}$$

$$= \frac{500\angle 0°}{\sqrt{3} \times 56.7\angle 63.84°}$$

$$= 5.09\angle 63.84°$$

$$\text{Total power} = \sqrt{3}\, V_L I_L \cos\theta$$
$$= \sqrt{3} \times 500 \times 5.09 \cos 63.84$$
$$= 1943.42 \text{ W}$$

that is

$$W_1 + W_2 = 1943.42 \text{ W} \tag{6.21}$$

Now

$$\tan\theta = \sqrt{3}\,\frac{W_1 - W_2}{W_1 + W_2}$$

$$\tan 63.84 = 2.035 = \sqrt{3}\,\frac{W_1 - W_2}{1943.42}$$

Therefore

$$W_1 - W_2 = 2283.41 \text{ W} \tag{6.22}$$

Solving Eqs. (6.21) and (6.22), W reads 2113.4 W and W_2 reads −170.0 W.

Problem 26

A three-phase balanced star load of 10 kW at a power factor of 0.6 lag is in parallel with another balanced star load of 7 kW at power factor of 0.9 lead and the whole combination is connected to a 416 V, three-phase 50 Hz supply. Find the total current, power and power factor of the system.

Solution:

Load 1

Power factor = 0.6 lag; $\phi = 53.1$; kW = 10; $\cos\phi = 0.6$; $\sin\phi = 0.8$. Therefore

$$\text{KVA} = \frac{\text{kW}}{\text{Power factor}} = 16.66$$

$$\text{KVAR} = \text{KVA} \times \sin\phi = 13.328$$

$$16.66 = 10 - j13.328 \tag{6.23}$$

Therefore,

$$\text{KVA} = \sqrt{\text{kW} - \text{KVAR}}$$

Load 2:
Power factor = 0.9 lead. $\phi = 25.8$; kW = 7; $\cos\phi = 0.9$; $\sin\phi = 0.4352$. Therefore

$$\text{KVA} = \frac{\text{kW}}{\text{Power factor}} = 0.77$$

$$\text{KVAR} = \text{KVA} \ \sin\phi = 7.77 \times 0.4352 = 3.384$$

that is

$$7.77 = 7 + j3.384 \tag{6.24}$$

Now

$$\text{Load (1)} + \text{Load (2)} = \text{Total load}$$

So adding Eqs. (6.23) and (6.24) we get

$$24.43 = 17 - j9.944 \tag{6.25}$$

From Eq. (6.25), total power = 17 kW.

$$\text{Overall power factor} = \frac{\text{kW}}{\text{KVA}} = \frac{17}{24.43} = 0.7 \text{ lag}$$

(lagging, since KVAR is –ve). Therefore total current,

$$I_L = \frac{\text{kW} \times 1000}{\sqrt{3} V_L \cos\phi} = \frac{17 \times 1000}{\sqrt{3} \times 416 \times 0.7} = 33.7 \text{ A}$$

Problem 27

A balanced three-phase star-connected load of 43 kW takes a leading current of 75 A with the line voltage of 415 V, 50 Hz. Find the parameter of the load.

Solution

Balanced star load:

Power = 43 kW; current, $I = 75$ A. Therefore,

$$\text{Power per phase} = \frac{43}{3} = 14.333 \text{ kW}$$

$$\text{Voltage per phase} = \frac{416}{\sqrt{3}} = 240 \text{ V}$$

$$\text{Current per phase} = 75 \text{ A}$$

$$\text{Power factor} = \cos\phi$$

$$= \frac{\text{Power}}{V \times I}$$

$$= \frac{14.333 \times 1000}{245 \times 75}$$

$$= 0.796$$

Hence, $\phi = 38.3°$ and $\sin\phi = 0.606$. Now

$$\text{Impedance } Z \text{ per phase} = \frac{V(\text{phase})}{I(\text{phase})} = \frac{249}{75} = 3.2\ \Omega$$

Also

$$\cos\phi = \frac{R}{Z} = \frac{R}{3.2} = 0.796$$

$$\Rightarrow R = 3.2 \times 0.796 = 2.54\ \Omega$$

and

$$\sin\phi = \frac{x}{Z} = \frac{x}{3.2} = 0.606$$

$$\Rightarrow x = 3.2 \times 0.606 = 1.94\ \Omega$$

Again

$$2\pi fL = 1.94$$

$$\Rightarrow L = \frac{1.94}{100\pi} = 6\,\text{mH}$$

The parameters/phase, $R = 2.54\Omega$ and $L = 6\,\text{mH}$.

Multiple Choice Questions

1. In a two-phase generator, the electrical displacement between the two phases or windings is ________ electrical degrees.
   (a) 120
   (b) 90
   (c) 180
   (d) none of the above

2. In a six-phase generator, the electrical displacement between different phases of windings is ________ electrical degrees.
   (a) 60
   (b) 90
   (c) 120
   (d) 45

3. The phase sequence of a three-phase system is RYB. The other possible phase sequence can be
   (a) BRY
   (b) YRB
   (c) RBY
   (d) none of the above

4. The advantage of star-connected supply system is that
   (a) line current is equal to phase current
   (b) two voltages can be used
   (c) phase sequence can be easily changed
   (d) it is a simple arrangement

5. In a balanced star-connected system, line voltages are ________ ahead of their respective phase voltages.
   (a) 30°
   (b) 60°
   (c) 120°
   (d) none of the above

6. A three-phase load is balanced if all the three phases have the same
   (a) impedance
   (b) power factor
   (c) impedance and power factor
   (d) none of the above

7. If one line conductor of a three-phase line is cut, the load is then supplied by ________ voltage.
   (a) single-phase
   (b) two-phase
   (c) three-phase
   (d) none of the above

8. The resistance between any two terminals of a balanced star-connected load is 12 Ω. The resistance of each phase is
   (a) 12 Ω
   (b) 24 Ω
   (c) 6 Ω
   (d) none of the above

9. In a star-connected system, the relation between the line voltage V_L and phase voltage V_{ph} is
   (a) $V_L = V_{ph}$
   (b) $V_L = V_{ph} / \sqrt{3}$
   (c) $V_L = \sqrt{3} V_{ph}$
   (d) none of the above

10. For the same rating, the size of a three-phase motor will be ________ single-phase motor.
    (a) less than that of
    (b) more than that of
    (c) same as that of
    (d) none of the above

Review Questions

1. Describe the various advantages of three-phase systems over single-phase systems.
2. With a neat phasor diagram describe how three-phase voltages are generated.
3. Describe briefly **(a)** phase sequence and **(b)** naming of phases.
4. Derive the phase and line relationship of voltages in a star-connected three-phase system.
5. Derive the phase and line relationship of currents in delta-connected three-phase system.
6. Give a comparative study of star and delta connections.
7. Which are the different methods of power measurement in three-phase circuits?
8. Why is two-wattmeter method popular for measuring three-phase power?
9. With a phasor representation, show that two wattmeters are sufficient to measure the power in a three-phase balance load.
10. Explain the effect of power factor of the load on the wattmeter reading.

Problems

1. Calculate the **(a)** active and **(b)** reactive components of current in each phase of a star-connected 10 000 V, three-phase generator supplying 5.000 kW at a lagging power factor of 0.8. **(c)** Find the new output if the current is maintained at the same value but the power factor is raised to 0.9 lagging.
2. A balanced star-connected load of impedance $(16 + j8)\ \Omega$ per phase is connected to a three-phase, 230 V, 50 Hz, 0.8 power factor supply. Find the **(a)** line current and **(b)** power absorbed by each phase.
3. Three similar coils connected in star take a total power 1.5 kW at a power factor of 0.2 lagging from a three-phase, 400 V, 50 Hz supply. Calculate the resistance of coil.
4. A balanced three-phase, delta-connected load per phase impedance of $(25 + j40)\ \Omega$. If 400 V, three-phase supply is connected to this load, find **(a)** phase current, **(b)** line current, **(c)** power supplied to the load.
5. A balanced delta-connected load takes line current of 18 A from a 400 V, three-phase, 50 Hz supply. Calculate the resistance of each leg of the load.
6. Two-wattmeter method is used to measure the power absorbed by a three-phase induction motor. The wattmeter readings are 12.5 kW and –4.8 kW. Find **(a)** the total active power absorbed by the machine, **(b)** load power factor and **(c)** reactive power taken by the load.
7. A three-phase motor load has a power factor of 0.397 lagging. Two wattmeters connected to measure power show the input as 30 kW. Find the reading on each wattmeter.

8. Each of two wattmeters connected to measure the input to three-phase induction motor reads 10 kW. If the power factor of the motor is charged to 0.866 lagging, determine the readings of the two wattmeters, the total input power remaining unchanged.

9. Find the power and power factor of the balanced circuit in which the wattmeter readings are 5 kW and 0.5 kW, the latter being obtained after the reversal of the current coil terminals of the wattmeter.

10. Three 50 Ω resistors are connected in star across 400 V, three-phase supply. Find the line current.

11. A symmetrical star-connected system. Calculate the three line voltages given $V_{RN} = 230\angle 0°$.

12. Calculate the value of three line currents assuming RYB sequence.

13. A 400 V, three-phase balanced supply in connected to 8 + *j*6 Ω balanced star load with phase sequence R, Y, B. Determine the **(a)** values of the current and **(b)** power consumed by load.

14. Calculate the **(a)** active and **(b)** reactive components in each phase of a star-connected 10000 V, three-phase alternator supplying 5000 kW, at a power factor of 0.8. **(c)** If the total current remains the same when the load power factor is raised to 0.9, find the new output.

15. Voltage across the terminals R and Y is $400\angle 0°$. Calculate the value of three line voltages assuming R, Y, B phase sequence.

16. A balanced delta-connected load of 2 + *j*3 Ω per phase is connected to a balanced three-phase, 440 V supply and the phase current is 10 A. Find **(a)** the total active power, **(b)** the reactive power, **(c)** apparent power.

17. The readings of two wattmeters used to measure power in a capacitive load (leading) are −3000 and 8000 W, respectively. Calculate **(a)** input power and **(b)** power factor at load. Assume RYB sequence.

18. Two wattmeters connected to measure power input to a three-phase circuit indicate 15 and 1.5 kW, respectively, the latter reading being obtained after reversing the current coil connection. Calculate the **(a)** power and **(b)** power factor of the load.

19. A balanced star-connected load is supplied from a symmetrical three-phase, 400 V system. The current in each phase is 30 A and lags 30° behind the phase voltage. Find the **(a)** phase voltage and **(b)** total power.

20. Three 200 Ω non-inductive resistances are connected in star across a 400 V, three-phase supply sequence RYB. Find **(a)** line currents and **(b)** the total power.

21. A balanced three-phase star-connected load of 210 kW takes a leading current of 160 A when connected across a symmetrical 1.1 kV, 50 Hz supply. Find the load circuit passive parameter per phase.

22. The ratio of the readings of two wattmeters connected to measure power in a balanced three-phase, three-wire load in 5:3. The load is known to be inductive with a lagging power factor. Calculate the power factor of the load.

23. In a three-phase system, a balanced star-connected load is connected across a 400 V

supply with the phase sequence RYB. The line current is found to be 9.24 A and the total power consumed by the load is 5.12 kW. Determine the values of the load.

24. A balanced delta-connected load is connected to a three-phase, 230 V supply. The line current is 39.8 A and the total power consumed by the load is 9.52 kW. Determine the elements of the load.

25. In a three-phase balanced circuit the readings of two wattmeters are 5.2 and 1.7 kW. The later reading has been obtained after reversal of the current coil connected to the supply voltage of 400 V. Determine the **(a)** total power factor and **(b)** inductance.

Answers

Multiple Choice Questions

1. (b) **2.** (a) **3.** (c) **4.** (b) **5.** (a) **6.** (c) **7.** (a) **8.** (c) **9.** (c) **10.** (a)

Problems

1. **(a)** 289 A, **(b)** 216 A, **(c)** 5625 W
2. **(a)** 7.47 A, **(b)** 2380 W
3. 7.11 Ω
4. **(a)** 8.48 A, **(b)** 14.68 A, **(c)** 3112.16 W
5. 38.49 Ω
6. **(a)** 7.7 kW, **(b)** 0.2487 lag, **(c)** 29.96 KVAR
7. 24.995 kW, 5.005 kW
8. 3.335 kW, 6.665 kW
9. 4.5 kW, 0.43
10. 8 A
11. $V_{RY} = 398.37\angle 30°, V_{YB} = 398.37\angle -90°,$
    $V_{BR} = 398.37\angle -210°$
12. $I_R = 10\angle 20°$ A, $I_Y = 10\angle -100°$ A,
    $I_B = 10\angle 140°$
13. **(a)** $I_{Rph} = 231\angle -37°, I_{Yph} = 231.1\angle -157°,$
    $I_{Bph} = 231\angle -277°$; **(b)** $P = 12800$ W
14. **(a)** 289 A, **(b)** 216 A, **(c)** 5625 kW
15. $V_{RY} = 400\angle 0°, V_{YB} = 400\angle -120°,$
    $V_{BR} = 400\angle -240°$
16. **(a)** 7259.78 W, **(b)** 10955.67 VAR, **(c)** apparent power = 13199.61 VA
17. **(a)** 5000 W, **(b)** 0.25
18. **(a)** 13.5 kW, **(b)** 0.48
19. **(a)** 231 V, **(b)** 18 kW

20. **(a)** $V_{ph} = 231$ V, $I_{ph} = 1.15$ A, **(b)** 800 W

21. 1105 μF

22. 0.918 (lagging)

23. $Z_{ph}\cos\theta = 200\ \Omega$, $Z_{ph}\sin\theta = 150\ \Omega$

24. $R_{ph} = 6\ \Omega$, $Z_{ph} = 10\ \Omega$, Reactance of load = 8 Ω

25. **(a)** 0.281, **(b)** 16.31 A

7 Electrical Machines

LEARNING OBJECTIVES

After reading this chapter, you will be able to:

- Understand the classification of electrical machines.
- Understand the principle of operation and construction of DC generator and DC motor.
- Understand the concept of transformer and alternates as well as synchronous motor.
- Understand the three-phase single phase induction motors.

7.1 Introduction

An electrical machine converts energy from one form to another. If it converts mechanical energy into electrical energy, it is called a *generator*. On the other hand, if it converts electrical energy into mechanical energy, it is called a *motor*. Since there are two types of currents, direct current (DC) and alternating current (AC), the generators and motors are also of two types, namely DC and AC generators and DC and AC motors. AC motors are again of two types: single-phase and three-phase motors. A motor which requires both DC and AC supply is known as a *synchronous motor*. A motor which can work either on AC or on DC is called a *universal motor*.

A broad classification of electrical machines in given in Fig. 1.

7.2 DC Machines

Depending on the mode of operation, a DC machine can be classified into DC generator or DC motor. A DC generator is one which converts mechanical energy into electrical energy whereas a DC motor is one which converts electrical energy to mechanical energy. The broad classification of DC machines in given in Fig. 2.

7.2.1 DC Generator

7.2.1.1 Construction

A DC generator consists of the following parts:

1. Yoke.
2. Pole core and pole shoe.
3. Pole coil.
4. Armature.

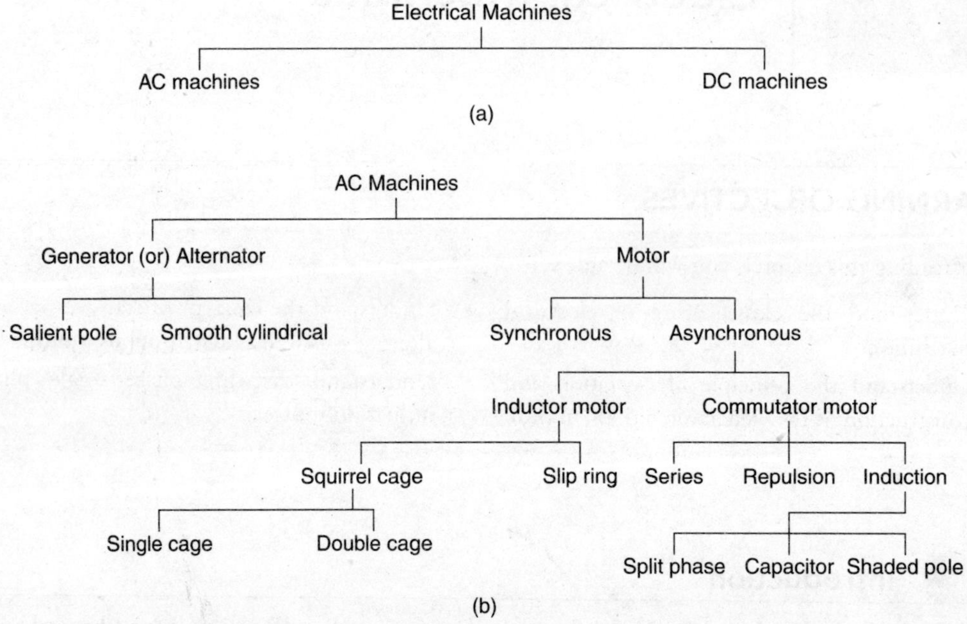

Figure 1 Classification of (a) electrical machines, (b) AC machines.

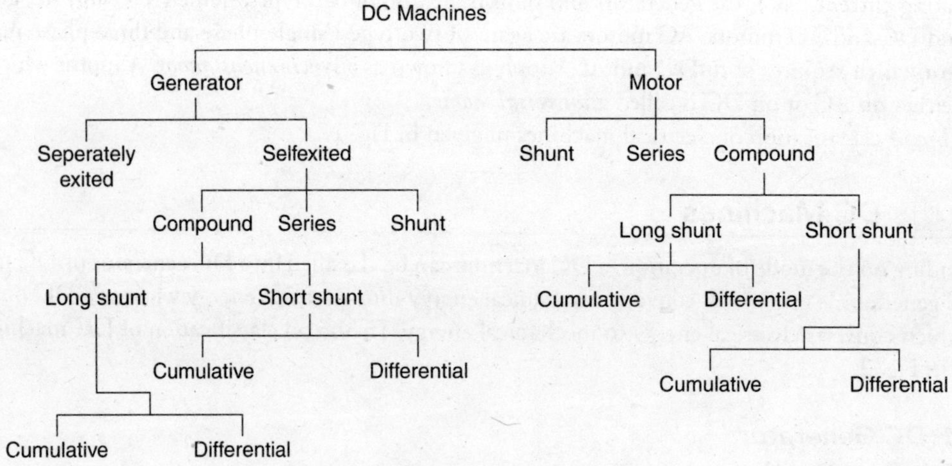

Figure 2 Classification of DC machines.

5. Commutator.
6. Brushes.
7. Bearings.

The DC generator is illustrated in Fig. 3.

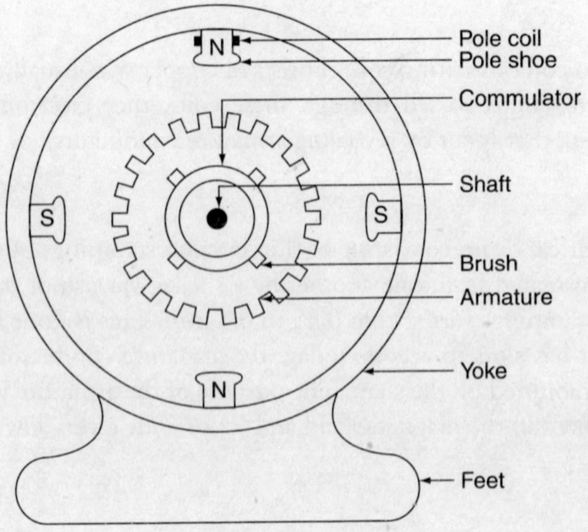

Figure 3 DC generator.

7.2.1.1.1 Yoke

The yoke is made of a magnetic material. In smaller machines, where cost is the main consideration rather than weight, cast iron is used; for larger machines where weight and efficiency are the main considerations, cast steel is used. The feet and terminal boxes are welded to the yoke. The purpose of the yoke is to provide support to the armature and field windings and also to carry the magnetic flux produced by the pole.

7.2.1.1.2 Pole Core and Pole Shoe

Pole core and pole shoe are made up of cast iron or cast steel (Fig. 4). For small machines the poles are cast with the yoke, but for large machines they are screwed with the yoke. The pole shoe is laminated. It not only supports the pole coil but also spreads the flux in the air gap.

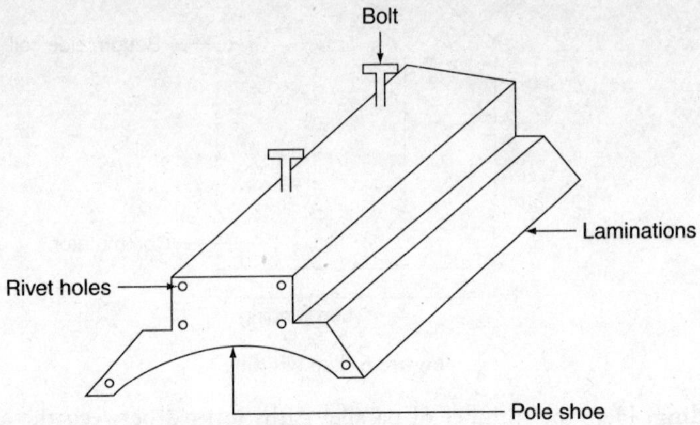

Figure 4 Pole core and pole shoe.

7.2.1.1.3 Pole Coil

Pole coil, also called field coil or exciting coil, consists of copper wire usually former wound and placed over the pole. When current is passed through these coils, they electromagnetize the poles which produce the necessary flux that is cut by revolving armature conductors.

7.2.1.1.4 Armature

The armature is a cylindrical drum consisting of thin circular stampings of silicon steel (Fig. 5). These discs or stampings are insulated from one another by a shellac varnish or paper or oxide film coating. The thickness of these stampings varies from 0.35 to 0.5 mm. This is done to reduce the eddy-current loss. On its periphery it has slots to accommodate the armature conductors. The stampings are held firmly by a key and are mounted on the shaft. The purpose of the armature is to (a) house the armature conductors, (b) rotate them in the magnetic field and (c) provide a very low reluctance path.

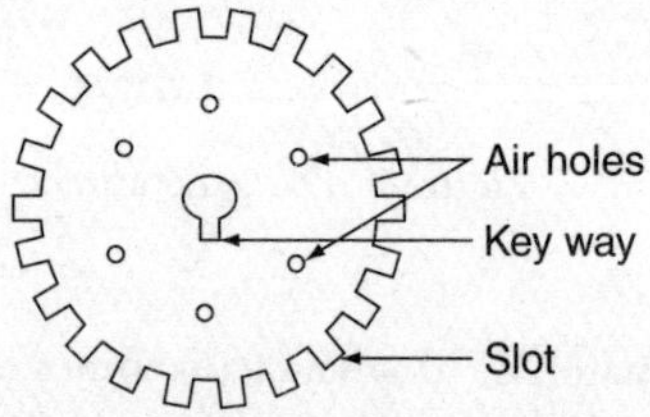

Figure 5 Armature.

There are mainly two types of armature windings:

1. **Lap winding:** The numbers of parallel paths are equal to number of poles and the total current divides equally among them. The numbers of brushes are equal to the numbers of poles. Lap winding is suitable for large current and low voltage requirements. Whenever lap winding is employed, equalizer rings are used to avoid unequal distribution of current at the brushes (Fig. 6).

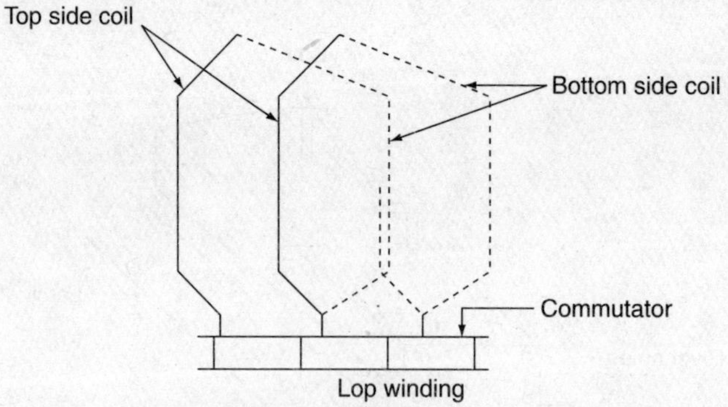

Figure 6 Lap winding.

2. **Wave winding:** Here the number of parallel paths formed between the armature terminals is two irrespective of the number of poles. The number of brushes used to collect current is 2. The advantage of wave winding is that, for a given number of armature conductors and poles,

it gives more e.m.f. than lap winding; hence wave winding is suitable for small generators. The equalizer ring is not used in wave winding. The wave winding is suitable for high-voltage and low-current machines.

7.2.1.1.5 Commutator

The commutator is cylindrical in shape and consists of a number of high-conductivity, hard-drawn copper segment insulated from each other by a thin layer of mica (Fig. 7). The number of segments is equal to the number of armature coils. A copper lug (riser) is used to connect the commutator segment to the armature conductor. The function of the commutator is to convert the alternating current induced in the armature to unidirectional current in the external circuit.

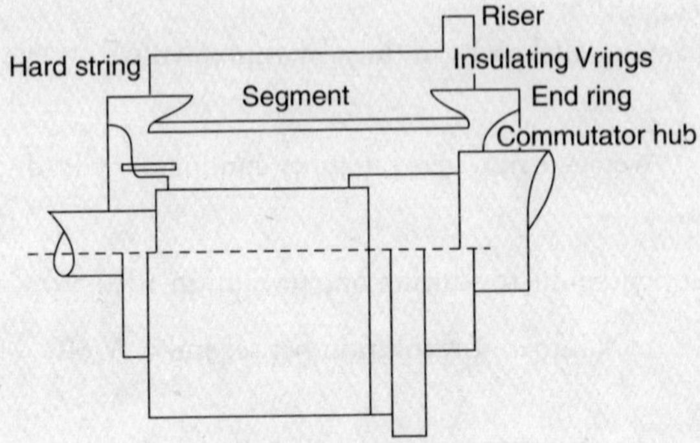

Figure 7 Commutator.

7.2.1.1.6 Brushes

The function of the brushes is to collect current from the commutator. These are made of carbon (for small machines) or copper (for large machines) and are placed in hollow brass bars called brush holders as shown in Fig. 8. The brushes rest on the commutator segments against spring tension. A lever is used to adjust the spring tension. The pigtail made of copper wire connects the brush with the outside circuit.

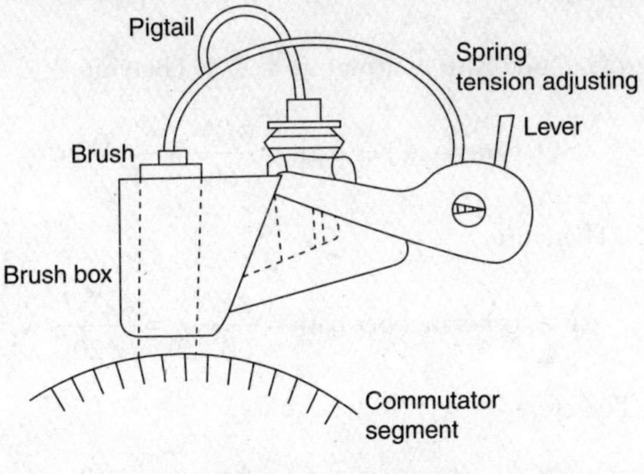

Figure 8 Brushes.

7.2.1.1.7 Bearings

Bearings are fitted inside the cover with the purpose to provide free and smooth rotation to the armature. The different types of bearings used are *brush bearings, ball bearings* and *roller bearings*. Because of their reliability, ball bearings are frequently used. However, for heavy duty rollers, hearings are used.

We now find out the **e.m.f. equation:** Let

P = number of poles
ϕ = flux/pole in weber
N = speed of armature of in r.p.m.
$N/60$ = speed of armature in r.p.s.
Z = total number of armature conductors
A = number of parallel paths
E = e.m.f. induced in any parallel paths in the armature in volts

Then

$$\text{Average e.m.f. generated per conductor} = \mathrm{d}\phi/\mathrm{d}t$$

But

$$\text{Flux cut per conductor during one revolution } (\mathrm{d}\phi) = \phi \times p \text{ weber}$$

$$\text{Number of revolution per second} = N/60$$

Therefore,

$$\text{Time for one revolution } (\mathrm{d}t) = 60/N \text{ s}$$

Hence, according to Faraday's laws of electromagnetic induction,

$$\begin{aligned}\text{e.m.f. generated per conductor} &= \frac{\mathrm{d}\phi}{\mathrm{d}t}\\ &= \frac{\mathrm{d}\phi}{60/N}\\ &= \frac{N\phi P}{60}\text{ Volt}\end{aligned}$$

Now number of conductors in one path of armature = Z/A. Therefore,

$$\text{e.m.f. generated per path} = \frac{\phi PN}{60} \times \frac{Z}{A}\text{ Volt}$$

For wave winding, $A = 2$. Therefore,

$$\text{e.m.f. generated per path} = \frac{\phi ZNP}{60 \times 2} = \frac{\phi ZNP}{120}\text{ Volt}$$

For lap winding, $A = P$. Therefore,

$$\text{e.m.f. generated per path} = \frac{\phi ZNP}{60 \times P} = \frac{\phi ZN}{60}\text{ Volt}$$

7.2.1.2 Principle of DC Generator

A generator is a device that converts mechanical energy into electrical energy. It works on the principle of Faraday's laws of electromagnetic induction. Accordingly, whenever a conductor cuts a magnetic flux, an e.m.f. is induced in the conductor causing a current flow through it, when the circuit is closed. The direction of the e.m.f. induced is given by Fleming's right-hand rule.

The essential parts of a generator are

1. a conductor,
2. a magnetic field.

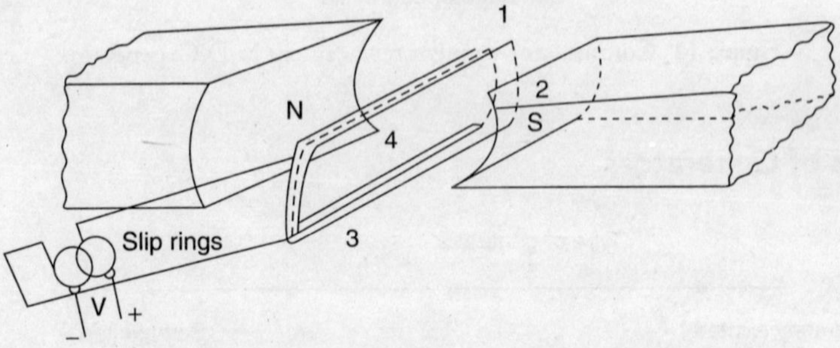

Figure 9 Principle of operation of a DC generator.

The principle of operation can be clearly understood with the help of the arrangement shown in Fig. 9. A single-turn coil is allowed to rotate in a uniform magnetic field at a constant speed. As the coil takes different positions in the magnetic field, the flux linked with the coil changes and hence the magnitude of e.m.f. induced in it changes. When the coil is at position 1, the rate of change of flux linkage is minimum and hence the e.m.f. induced is also minimum. At position 2, the e.m.f. induced is maximum because the rate of change of flux linkages is maximum. Again at position 3, the e.m.f. is minimum due to minimum rate of change of flux linkages. This process repeats in the next half cycle. Thus even in DC generator, the e.m.f. induced in the coil is alternating (Fig. 10).

In order to obtain a unidirectional voltage at the coil terminals, the slip rings are to be replaced by commutator segments (Fig. 11).

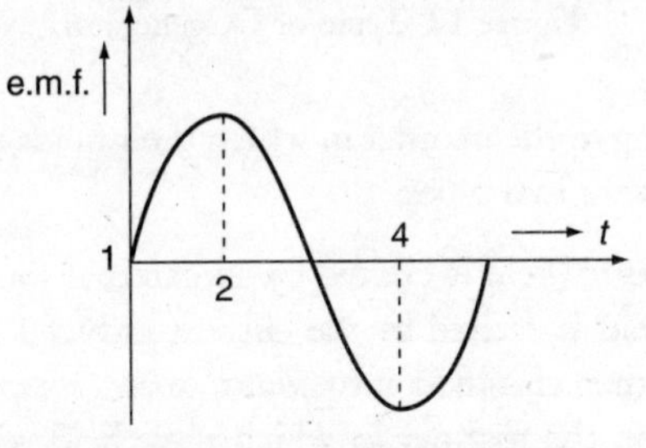

Figure 10 Alternating waveform.

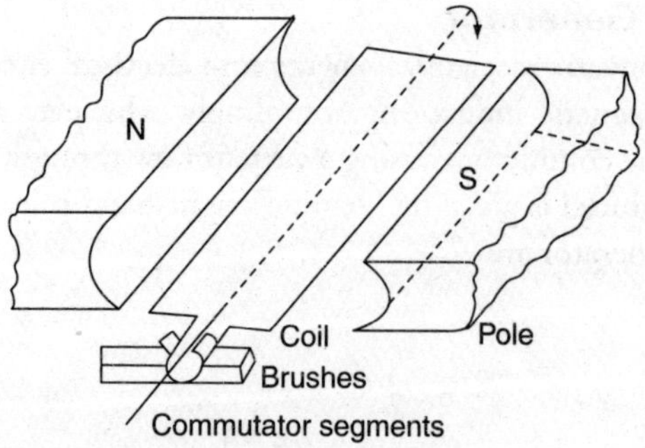

Figure 11 Commutator segment replacement in D.C. generator.

7.2.1.3 Types of Generators

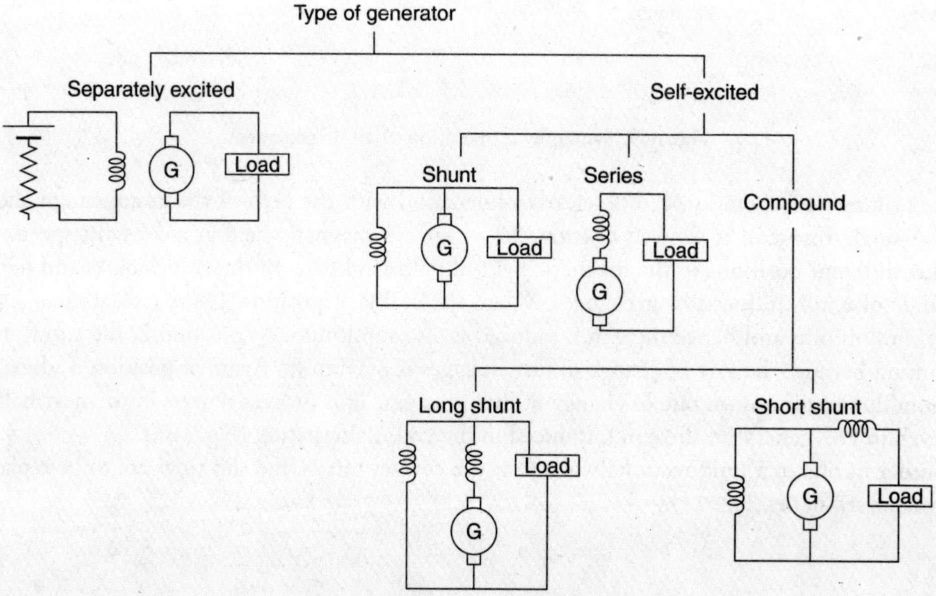

Figure 12 Types of DCgenerator.

Generators are classified according to the manner in which their fields are excited (Fig. 12). Accordingly, they are classified into the following two types:

1. **Separately excited generator** (field is excited by an external source).
2. **Self-excited generator** (field is excited by the current induced in the generator itself). The self-excited generators are further classified into *shunt, series, compound long shunt* and compound *short shunt* depending upon the manner in which their field windings are connected with their armature.

7.2.1.3.1 Shunt Generator

Let I_a, I_{sh}, I_L be the armature, shunt field and load currents. Let R_a, R_{sh}, R_L be armature, shunt field and load resistances.

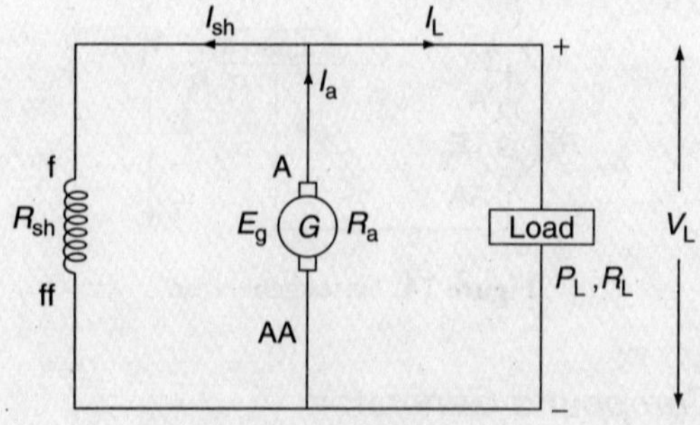

Figure 13 Shunt generator.

In Fig. 13, A–AA are the armature terminals; G is the generator; f–ff are the shunt field terminals; P_L is the load (power) in watt; V_L is the terminal voltage; B.C.D. is the brush contact drop in volt; E_g is the generated e.m.f. in volt and R_L is the load resistance in ohm. We have,
Current equation is

$$I_a = I_L + I_{sh}$$

Also,

$$I_L = \frac{P_L}{V_L} = \frac{V_L}{R_L}$$

and

$$I_{sh} = \frac{V_L}{R_{sh}}$$

Therefore, voltage equation is

$$E_g \text{ (generated e.m.f.)} = V_L + I_a R_a + \text{B.C.D. Volts}$$

7.2.1.3.2 Series Generator

A series generator is as shown in Fig. 14. Let R_a, R_{se} and R_L be the armature, series field and load resistances. Let I_a, I_{se} and I_L be the armature, series field and load currents. Let V_L be the terminal voltage, P_L be the load (power) in Watt, A–AA are the armature terminals and y–yy are the series field terminals. Then we have,
Current equation:

$$I_a = I_{se} = I_L = \frac{P_L}{V_L}$$

Voltage equation:

$$E_g = V_L + I_a R_a + I_{se} R_{se} + \text{B.C.D. Volts}$$

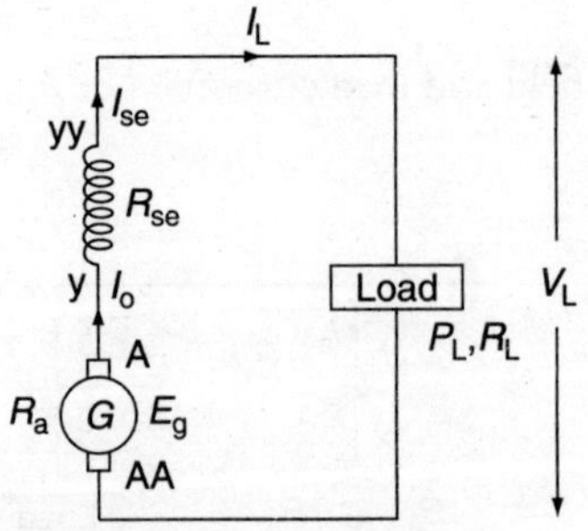

Figure 14 Series generator.

7.2.1.3.3 *Long Shunt Compound Generator*

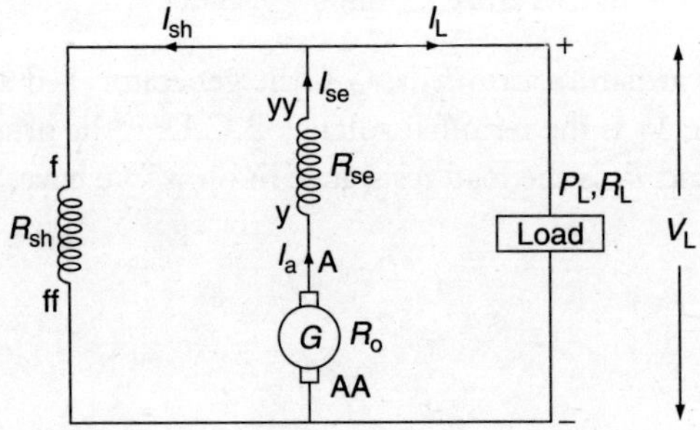

Figure 15 Long shunt compound generator.

A schematic of long shunt compund generator is shown in Fig. 15. The shunt field winding is connected across the series combination of armature and series field windings and across the load. This is characterized by the fact that series field current equals the armature current. That is

$$I_{se} = I_L + I_{sh} = I_{se} + I_a$$

$$I_L = \frac{P_L}{V_L} = \frac{V_L}{R_L}$$

$$I_{sh} = \frac{V_L}{R_{sh}}$$

$$E_g = V_L = R_a I_a = I_{se} R_{se} + \text{B.C.D.}$$

7.2.1.3.4 *Short Shunt Generator*

A schematic of short shunt compund generator is shown in Fig. 16. Shunt field winding is connected across the series combination of load and series field winding, across the armature terminal. This is characterized by the fact that the series field current equals the load current. That is

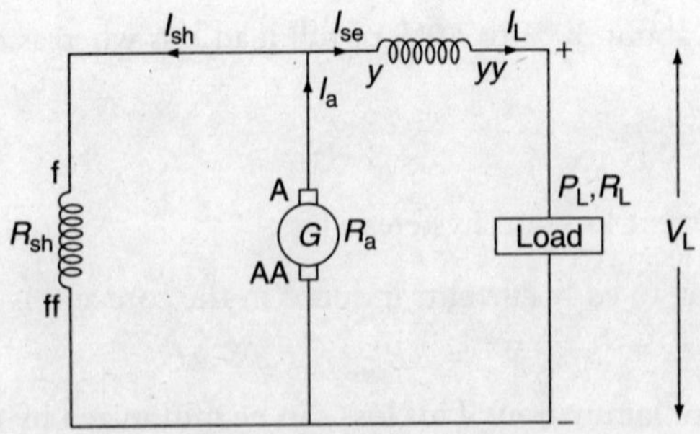

Figure 16 Short shunt compound generator.

$$I_a = I_{sh} + I_{se}$$

$$I_{se} = I_L = \frac{P_L}{V_L} = \frac{V_L}{R_L}$$

$$I_{sh} = V_L + \frac{(I_{se} R_{se})}{R_{sh}}$$

$$E_g = V_L + I_a R_a + I_L I_{se}$$

7.2.1.4 Losses in a Generator

Figure 17 depicts various types of losses in a generator. We will explain them in the following subsections. Note that:

1. Iron loss is also called core loss or magnetic loss.
2. Armature copper loss is also called variable loss.
2. Iron loss and mechanical loss are together called stray loss.
4. The shunt, field and copper loss together is called constant loss.

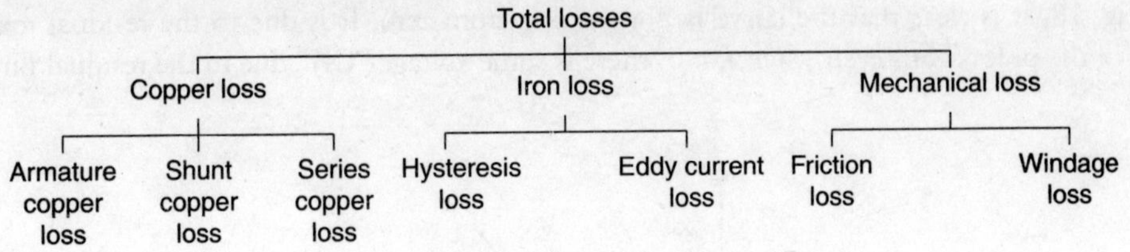

Figure 17 Losses in a generator.

7.2.1.4.1 Copper Loss

It is due to the resistance of armature, shunt and series field windings.

$$\text{Armature copper loss} = I_a^2 R_a \text{ Watt}$$

$$\text{Shunt field copper loss} = I_{ash}^2 R_{sh} \text{ Watt}$$

$$\text{Series field copper loss} = I_{se}^2 R_{se} \text{ Watt}$$

The armature copper loss is about 30% to 40% of full load loss whereas the field copper loss is about 20% to 30% of full load loss.

7.2.1.4.2 Iron Loss

Iron loss consists of eddy-current loss and hysteresis loss.

1. Eddy-current loss is due to eddy currents induced in the core and is given by

$$W_e = KB_{max}^2 f^2 t^2 \text{ Watt}$$

where "*t*" is thickness of lamination. This loss can be minimized by using a laminated core.

2. Hysteresis loss is due to the magnetic reversal and is given by

$$W_h = B_{max}^{1.6} \times f \times V \text{ Watt}$$

This loss can be minimized by selecting a material for the core which has low hysteresis coefficient. For example, in silicon steel, iron loss in about 20% to 30% of full load loss. For shunt and compound generators, these losses practically remain constant.

7.2.1.4.3 Mechanical Loss

This consists of frictional loss at the bearings and the windage loss of the rotating armature. This loss is about 10% to 20% of full load loss.

7.2.1.5 Characteristics of Shunt Generator

There are three characteristics of a generator, namely,

1. open circuit characteristics (O.C.C.) or magnetic characteristic;
2. external characteristics or voltage regulation curve;
3. total or internal characteristics.

7.2.1.5.1 Magnetic Characteristics of DC Shunt Generators

It is a curve drawn by taking no load generator e.m.f. "E_0" on *Y*-axis and field current (I_f) on *X*-axis. From Fig. 18, it is clear that the curve is not starting from zero. It is due to the residual magnetism present in the poles. Thus even when $I_f = 0$, there is some voltage "OA" due to the residual flux.

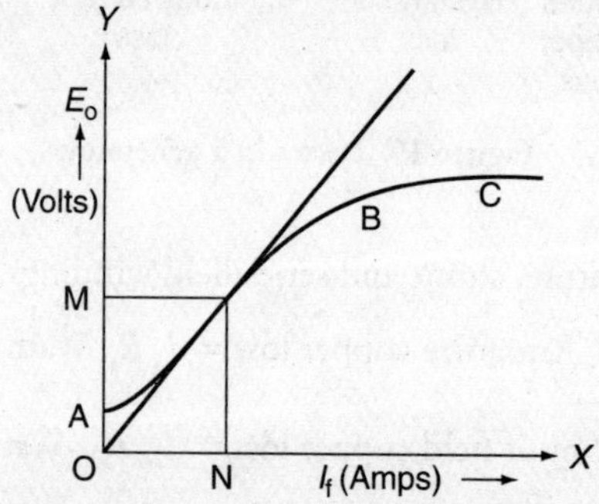

Figure 18 Graph depicting magnetic characteristics of DC shunt generator.

The portion AB is practically a straight line. It is due to the fact that at low flux densities, the reluctance offered by iron path is negligible. Hence the total reluctance is only due to the air gap which is constant. Thus as "I_f" increases "E_0" increases, i.e., $E_0 \propto I_f$ or $E_0 = kI_f$ which is an equation of a straight line, passing through the origin, hence the portion AB. The portion BC refers to saturation, i.e., at higher flux densities, iron path reluctance becomes appreciable and hence E_0 will no more be proportional to I_f.

Critical Resistance (R_C): The critical resistance is the slope of the tangent drawn to the linear portion of the O.C.C. from origin. It is the value of field resistance where the generator will just excite. It is given by

$$R_C = \frac{OM}{ON}$$

The conditions for build-up of a self-excited shunt generator are as follows:

1. There must be some residual magnetism.
2. For a given direction of rotation, shunt field coils should be properly connected to the armature terminals.
3. Shunt field resistance should be less than the critical resistance (when excited on open circuit).

7.2.1.5.2 External Characteristics of DC Shunt Generator

The graph depicting external characteristics is a drawn by taking terminal voltage (V_L) on *Y*-axis and load current (I_L) on *X*-axis (Fig. 19). It is clear from Fig. 19 that as the load current increases, the terminal voltage drops.

1. **Armature resistance drop:** As "I_L" increases, I_a increases because $I_a = I_L + I_{sh}$. Therefore "I_aR_a" drop increases. For shunt generator, $V_L = E_g - I_aR_a$. Hence "V_L" decreases.
2. **Armature reaction:** Due to demagnetizing effect of armature reaction (here armature flux opposes the main field flux), main pole weakens and the induced e.m.f. decreases. When I_L increases, I_a increases. As I_a increases, armature reaction increases. "E_g" decreases and hence "V_L" decreases with increase in I_L.

Due to the above two reasons field current decreases causing decrease in "E_g," therefore "V_L" decreases.

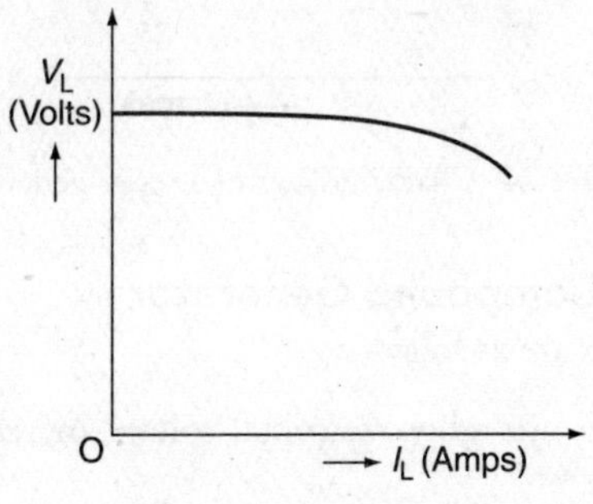

Figure 19 Graph depicting external characteristics of DC shunt generator.

7.2.1.5.3 Internal Characteristics of Shunt Generator

The graph depicting internal characteristics is a curve drawn by taking "E_g" (= $V_L = I_a R_a$) on *Y*-axis and "I_a" (= $I_L = I_{sh}$) on *X*-axis as shown in Fig. 20. Here, E_g is the generated e.m.f. and I_a is the armature current.

The argument made for external characteristics holds good for internal characteristics also.

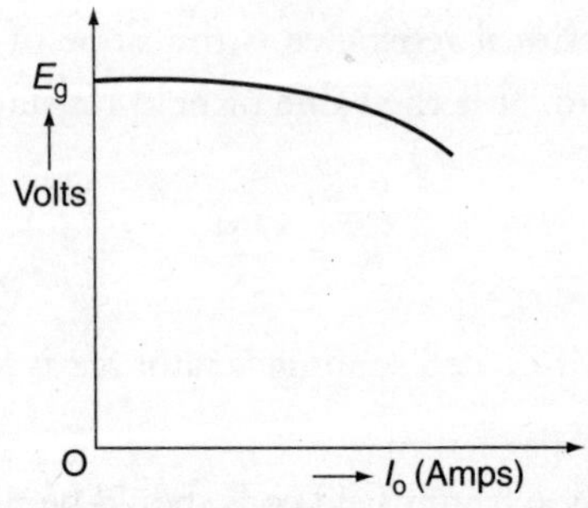

Figure 20 Graph depicting internal characteristics of DC shunt generator.

7.2.1.6 Characteristics of Series Generator

In Fig. 21, curve 1 represents the O.C.C. of series of generator (i.e., E_0/I_f), curve 2 represents the internal characteristics (i.e., E_g/I_a) and curve 3 represents the external characteristics (i.e,, V_L/I_L). It is seen from Fig. 21 that the series generator has a rising characteristics, that is, the voltage increases with increase in load. However, at high loads due to demagnetizing effect to armature reaction the voltage starts decreasing with increase in load.

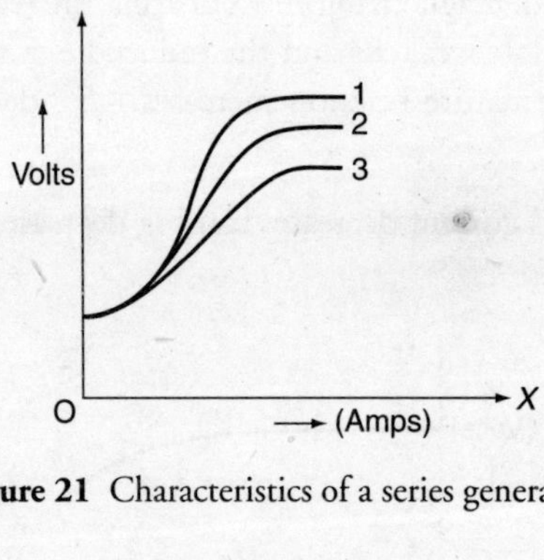

Figure 21 Characteristics of a series generator.

7.2.1.7 Characteristics of Compound Generator

The various compound generators are as follows:

1. An *overcompound generator* is one whose terminal voltage increases with the load or full voltage is greater than no load voltage.
2. A *flat compound generator* is one whose full load voltage is equal to no load voltage.

3. An *undercompound generator* is one whose full load voltage is less than no load voltage.
4. When the series field m.m.f. helps the shunt field m.m.f., it is called *cumulative compound generator.*
5. When the series field m.m.f. opposes the shunt field m.m.f., it is called *differential compound generator.*

The external load voltage characteristics of various compound generators are shown in Fig. 22.

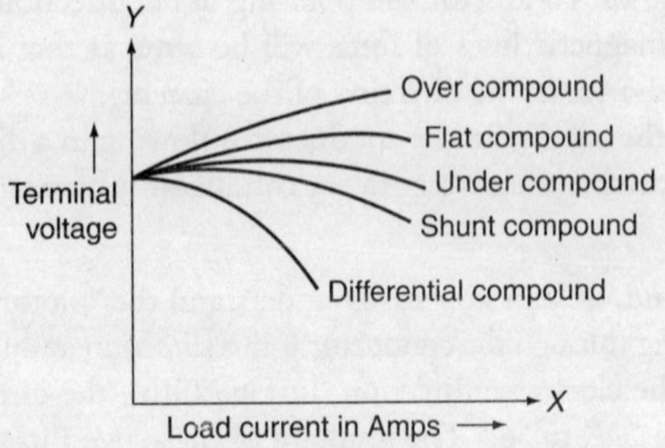

Figure 22 External load voltage characteristics of compound generators.

7.2.1.8 Applications of DC Generator

Applications of various types of DC generators are as follows.

DC shunt generator:

1. Because of its slightly drooping characteristics, the shunt generator is suited for charging batteries and also for parallel operation.
2. It is also used for lighting and power supply purposes.

DC series generator:

1. Because of its raising characteristics, it is widely used as boosters (railway service).
2. It is also used for supplying the field current for regenerative breaking of DC locomotives.

DC compound generator:

1. They are used to supply power to railway circuits, elevator motors, incandescent lamps, etc.
2. Differentially compound generator is used as an arc welding generator.

7.2.2 DC Motor

7.2.2.1 Principle of DC Motor

A motor is one which converts electrical energy into mechanical energy. Whenever a current-carrying conductor is placed in a magnetic field, a force (F) will act on the conductor whose magnitude is given by

$$F = BIl \sin\theta \text{ Newton}$$

where "B" is the flux density in Wb/m^2, "I" is the current in amperes, "l" is the length of conductor in meters and "θ" is the angle between the conductor and the field. The direction of the force is given by Fleming's left-hand rule. Two points worth noting are:

1. We know that whenever a conductor carries a current, it establishes its own magnetic flux whose direction is given by using either right-hand rule (hold the wire or conductor in the right hand with the thumb pointing in the direction of current, the fingers then curl round in the direction of the lines of force) or Maxwell's corkscrew rule (when a right-handed corkscrew is held such that its axis is parallel to the conductor and pointing in the direction in which current is flowing, then the direction of magnetic lines of force will be same as that in which corkscrew must be rotated to cause it to advance in the direction of the current).
2. A "dot" indicates that the current in the conductor is flowing in a direction towards the observer and a cross "⊗" indicates that the current in the conductor is flowing in a direction away from the observer.

With the above things in mind, we can now easily understand the "motoring action." In Fig. 23(a) the direction of the current flowing through the conductor is in a direction away from the observer; hence the flux set up by it will be in the clockwise direction. In Fig. 23(b), the direction of flux lines set up by the pair of poles will be from "N" to "S." The resultant of these two fluxes is given in Fig. 23(c)which is strengthened on the right-hand side and weakened on the left-hand side. Since the flux lines aligned in the same direction repel each other, the conductor will experience a force in the direction shown in Fig. 23(c) (which is just similar to the force exerted by the bent elastic rubber band of catapult on the stone piece). If a number of such conductors are placed, the force exerted on them will cause rotation; this is called "motor action."

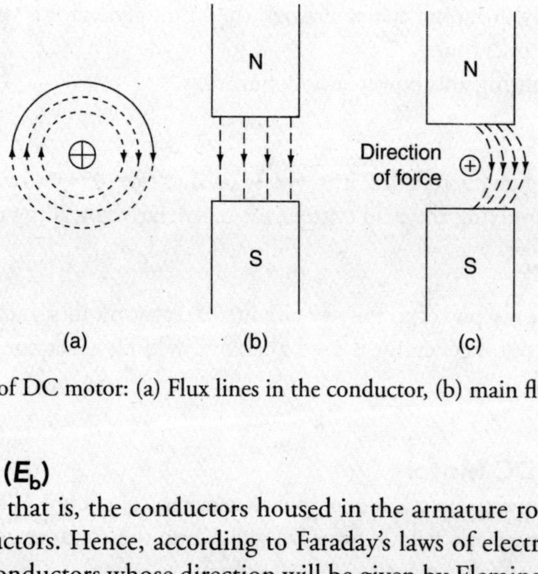

Figure 23 Principle of DC motor: (a) Flux lines in the conductor, (b) main flux, (c) armature flux.

7.2.2.2 Back E.M.F. (E_b)

When the motor rotates, that is, the conductors housed in the armature rotate, the magnetic flux is being cut by these conductors. Hence, according to Faraday's laws of electromagnetic induction, an e.m.f. is induced in the conductors whose direction will be given by Fleming's right-hand rule. When we apply this rule, the direction of the e.m.f. induced in the armature conductor is found to be acting opposite to the applied voltage and hence it is called back e.m.f. (E_b). The value of back e.m.f. is given by

$$E_b = \frac{\phi ZNP}{60\,A} \text{ Volt}$$

This back e.m.f. is always less than the applied voltage. When it becomes equal to the applied voltage "V," the motor action will stop because no current will flow through the armature. Therefore, the current flowing through the armature is

$$I_a = \frac{V - E_b}{R_a}$$

$$E_b = V - I_a R_a$$

$$V = E_b + I_a R_a$$

where R_a is the armature resistance. Therefore,

$$VI_a = E_b I_a + I_a^2 R_a$$

This is called the *power equation* of the motor. Here, VI_a is the electrical input to the armature, $E_b I_a$ is the mechanical power developed in the armature and $I_a^2 R_a$ is the copper loss in the armature.

7.2.2.3 Types of Motors

Similar to generators, motors are also classified depending on the manner in which their field windings are connected to their armature windings. Accordingly the types of DC motors are

1. Shunt motor.
2. Series motor.
3. Compound motor: The compound motor is further subdivided into the following:
   - Cumulative compound motor.
   - Differential compound motor.

7.2.2.3.1 Shunt Motor

See Fig. 24. Let P_L be the input power, V_L the applied voltage, I_L the input current, I_a the armature current, R_a the armature resistance, R_{sh} the shunt field resistance, I_{sh} the shunt field current and E_b the back e.m.f. Then we have

$$I_L = \frac{P_L}{V_L}$$

$$I_{sh} = \frac{V_L}{R_{sh}}$$

$$I_a = I_L - I_{sh}$$

$$E_b = V_L - I_a R_a - \text{B.C.D.}$$

where B.C.D. is the brush contact drop.

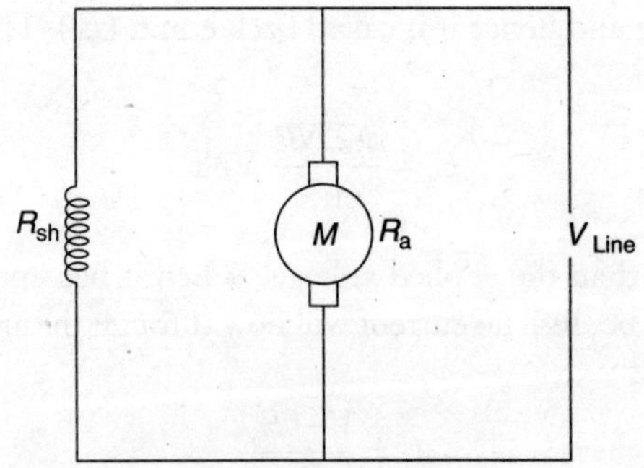

Figure 24 Shunt motor.

7.2.2.3.2 Series Motor

See Fig. 25. We know that

$$I_L = I_{se} = I_a$$

where I_{se} is the series field current. Then

$$I_L = \frac{P_L}{V_L}$$

$$E_b = V_L - I_a R_a - I_{se} R_{se} - \text{B.C.D.}$$

where R_{se} is the series field resistance.

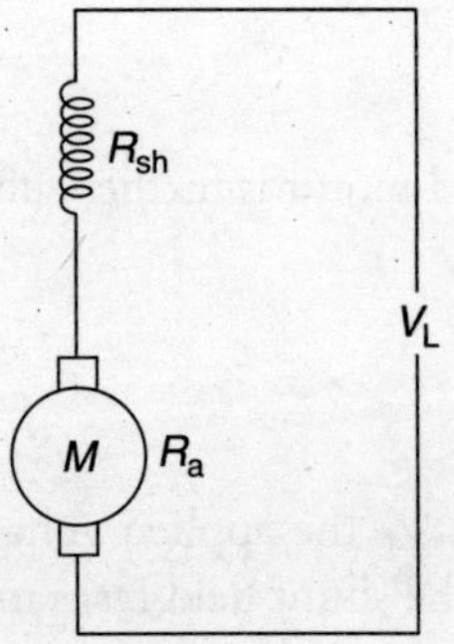

Figure 25 Series motor.

7.2.2.3.3 Long Shunt Motor (Cumulative)

See Fig. 26. Here

$$I_L = \frac{P_L}{V_L}$$

$$I_a = I_L - I_{sh}$$

$$I_{se} = I_a$$

$$I_{sh} = \frac{V_L}{R_{sh}}$$

$$E_b = V_L - I_a R_a - I_{se} R_{se} - \text{B.C.D.}$$

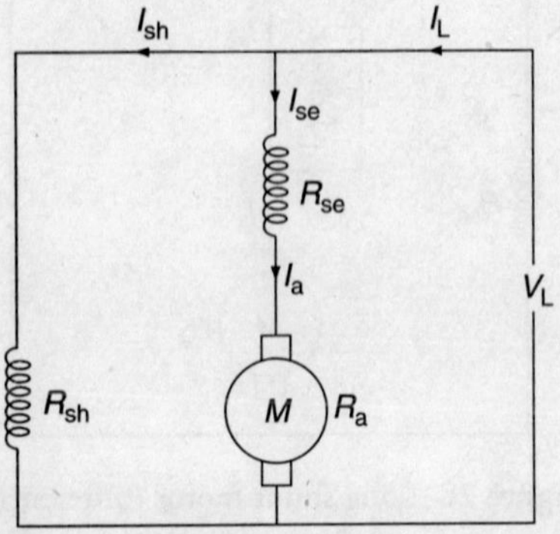

Figure 26 Long shunt motor (cumulative).

7.2.2.3.4 Short Shunt Motor (Cumulative)

See Fig. 27. Here

$$I_L = \frac{P_L}{V_L}$$

$$I_{se} = I_L$$

$$I_a = I_L - I_{sh}$$

$$I_{sh} = \frac{(V_L - I_{se} R_{se})}{R_{sh}}$$

$$E_b = V_L - I_a R_a - I_{se} R_{se} - \text{B.C.D.}$$

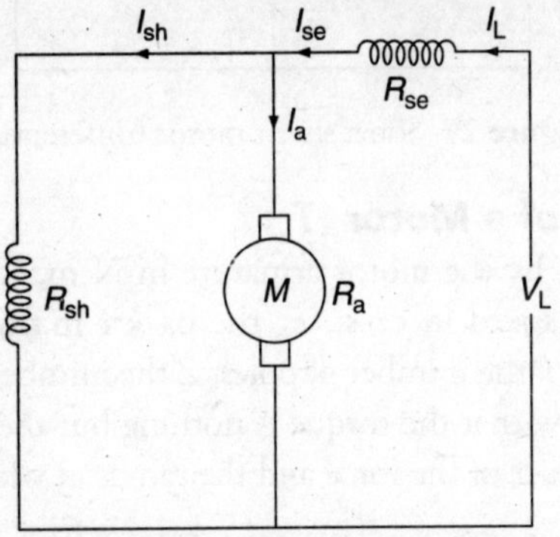

Figure 27 Short shunt motor (cumulative).

7.2.2.3.5 Long Shunt Motor (Differential)

The electrical equivalent of long shunt compound motor is shown in Fig. 28. Depending upon the shunt field winding connection, with respect to the series field winding connection, we can get differential actions.

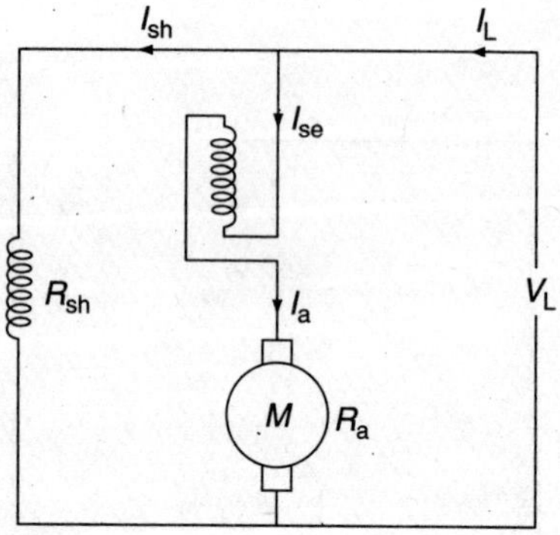

Figure 28 Long shunt motor (differential).

7.2.2.3.6 Short Shunt Motor (Differential)

The electrical equivalent of short shunt compound motor is shown in Fig. 29. Here the effect of V_e and V_{at} is neglected. Depending upon the series field winding connection, with respect to the shunt field windings connection, we can get differential and cumulative compounding actions.

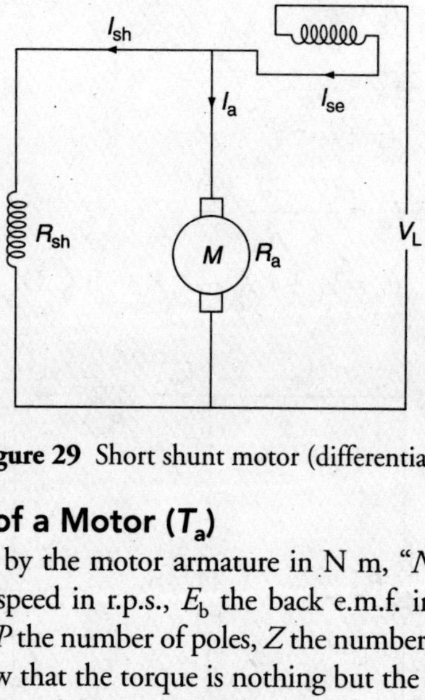

Figure 29 Short shunt motor (differential).

7.2.2.4 Armature Torque of a Motor (T_a)

Let "T_a" be the torque developed by the motor armature in N m, "N" be the speed of rotation of the armature in r.p.m., $N/60$ be the speed in r.p.s., E_b the back e.m.f. in volt, I_a the armature current in amperes, ϕ the flux/pole in weber, P the number of poles, Z the number of armature conductors and A the number of parallel paths. We know that the torque is nothing but the twisting moment of a force about an axis which is given by the product of the force and the radius at which the force acts. Therefore,

$$\text{Torque} = \text{Force} \times \text{Distance}$$
$$= F \times R \ \text{N m}$$

Also,

$$\text{Work done/second} = F \times (2\pi R) \times \frac{N}{60}$$

Therefore, work done/second or power developed equals

$$T_a \times \frac{2\pi N}{60} \text{ T/s or Watt}$$

Also, the electrical equivalent of mechanical power developed in the armature is $E_b I_a$ Watts. Therefore,

$$T_a \times \frac{2\pi N}{60} = E_b I_a$$

where

$$E_b = \frac{\phi ZNP}{60A}$$

Therefore,

$$T_a \times \frac{2\pi N}{60} = \frac{\phi ZNP}{60A} \times I_a$$

$$T_a \times 2\pi = \frac{\phi ZP}{A} \times I_a$$

$$T_a = \frac{1}{2\pi} \frac{\phi ZPI_a}{A}$$

Hence,

$$T_a = \frac{0.159 \phi ZPI_a}{A}$$

$$T_a = \left(\frac{0.159\, ZP}{A} \right) \phi I_a$$

This implies that

$$T_a \propto \phi I_a \Rightarrow T_a \propto I_a$$

For a series motor, before saturation, as the field winding carries armature current, we have

$$\phi \propto I_c$$

Therefore,

$$T_a \propto I_a^2$$

7.2.2.5 Speed of a DC Motor

We know that

$$V = E_b + I_a R_a$$

which gives

$$E_b = V - I_a R_a$$

Now

$$E_b = \frac{\phi ZNP}{60\,A}$$

where ϕ is flux in webers, Z is the total number of armature conductors, P the number of poles, N the speed of rotation of armature in r.p.m., A the number of parallel paths formed by the armature winding between the armature terminals. Therefore

$$\frac{\phi ZNP}{60A} = V - I_a R_a$$

Hence

$$N = \frac{(V - I_a R_a) 60A}{\phi ZP}$$

Since

$$V - I_a R_a = E_b$$

we can write

$$N = \frac{E_b 60A}{\phi ZP} = \frac{60A}{ZP} \frac{E_b}{\phi}$$

or

$$N = \frac{KE_b}{\phi}$$

where

$$K = \frac{60A}{ZP}$$

Hence

$$N \propto \frac{E_b}{\phi}$$

If we consider first and second case as N_1 and N_2 (where N_1 and N_2 are the speed of rotation in r.p.m. in first case and second case), then we get

$$\frac{N_2}{N_1} = \frac{E_{b2}}{E_{b1}} \times \frac{\phi_1}{\phi_2}$$

For a shunt motor, flux is constant. So,

$$\phi_1 = \phi_2 = \phi$$

Therefore,

$$\frac{N_2}{N_1} = \frac{E_{b2}}{E_{b1}}$$

For series motor, since ϕ is proportional to "I_a" for a series motor, before saturation, we have

$$N_1 \propto \frac{E_{b1}}{I_{a1}}$$

$$N_2 \propto \frac{E_{b2}}{I_{a2}}$$

Therefore,

$$\frac{N_2}{N_1} = \frac{E_{b2}}{E_{b1}} \times \frac{I_{a1}}{I_{a2}}$$

7.2.2.6 Characteristics of Motor

A motor has both electrical and mechanical characteristics. There are three important characteristics worth noting:

1. Armature torque versus armature current is T_a/I_a. This is also called electrical characteristic.
2. Speed versus armature current (N/I_a).
3. Speed versus armature torque (N/T_a). This is also called mechanical characteristic.

The following relations are to be recalled while discussing the characteristics of motors:

1. $T_a \propto \phi I_a$
2. $N \propto \dfrac{E_b}{\phi}$
3. "ϕ" is constant for shunt motor.
4. ϕ is proportional to "I_a" for a series motor, before saturation.

7.2.2.6.1 Characteristics of Series Motor

T_a/I_a Characteristics

We know that $T_a \propto \phi I_a$. Before saturation,

$$\phi \propto I_a$$

Therefore

$$T_a \propto I_a^2$$

or

$$T_a = KI_a^2$$

In Fig. 30 "T_a" is taken on Y-axis and "I_a" is on X-axis. Therefore prior to saturation, $y = mx^2$ which is the equation of a parabola. Thus a series motor is used where very high starting torque is required for accelerating heavy masses quickly as in case of cranes, hoists and electric trains.

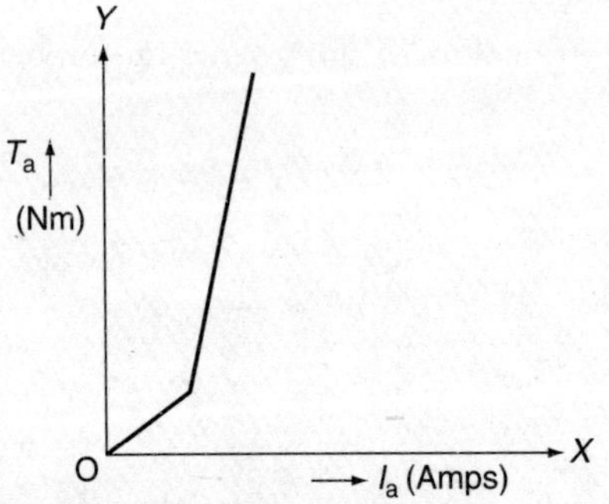

Figure 30 Graph depicting T_a/I_a characteristics of series motor.

After saturation, ϕ is independent of "I." Therefore,

$$T_a \propto I_a$$

only. This implies

$$T_a = K'I_a \tag{7.1}$$

Since "T_a" is on Y-axis and "I_a" on X-axis, Eq. (7.1) resembles $y = mx$, which is the equation of a straight line passing through the origin. Thus the curve becomes a straight line, after saturation.

N/I_a Characteristics

See Fig. 31. We know that

$$N \propto \frac{E_b}{\phi}$$

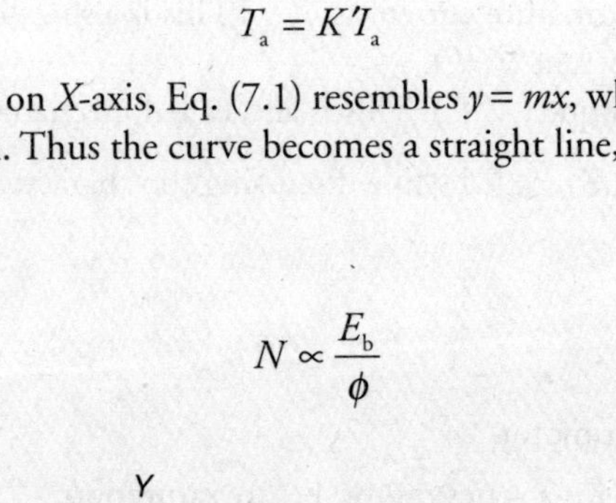

Figure 31 Graph depicting N/I_a characteristics of series motor.

where

$$E_b = V - I_a R_a - I_{se} R_{se}$$

Since R_a, R_{se} are very small, I_a, R_a, $I_{se}R_{se}$ drops do not change much with variation of load. Hence "E_b" is approximately constant. Therefore $N \propto 1/\phi$ also in a series motor since $\phi \propto I_a$; we can write $N \propto 1/I_a$. Thus as load increases, I_a increases and hence speed decreases.

Note: However, when the load is completely removed the armature current falls to a very small value and the speed becomes dangerously high. Therefore, the series motors should never be started on no load. From the characteristics it is clear that a series motor is a variable speed motor which adjusts its speed according to load.

N/T_a Characteristics

The mechanical power developed in the armature is given by

$$\frac{T_a \times 2\pi N}{60}$$

that is,

$$\text{Output of the motor} = \frac{T_a \times 2\pi N}{60}$$

Therefore,

$$\frac{\text{Output of the motor} \times 60}{2\pi N} = T_a$$

For a given load, the term output of the motor $\times 60/(2\pi)$ is a constant. Therefore,

$$T_a \propto \frac{1}{N}$$

or

$$N \propto \frac{1}{T_a}$$

Therefore, for a constant load, if the speed increases, the torque decreases and vice versa. Hence the N/T_a characteristic is as shown in Fig. 32.

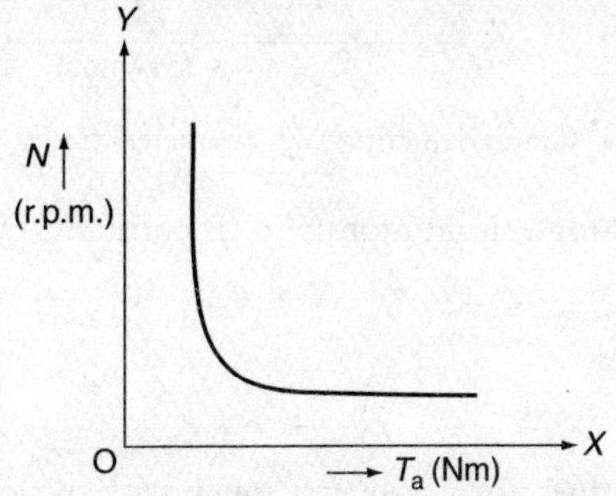

Figure 32 Graph depicting N/T_a characteristics of series motor.

7.2.2.6.2 Characteristics of Shunt Motor

T_a/I_a Characteristics

We know that $T_a \propto \phi I_a$. But, for a shunt motor "ϕ" is constant. Therefore,

$$T_a \propto I_a$$

$$T_a = KI_a \tag{7.2}$$

In Fig. 33, "T_a" is taken on Y-axis and "I_a" on X-axis. Therefore Eq. (7.2) resembles $y = mx$ which is the equation of a straight line passing through the origin. Shunt motors are never started on heavy load because a heavy load needs a heavy starting current which may affect the voltage regulation of the supply and also the electrical installation.

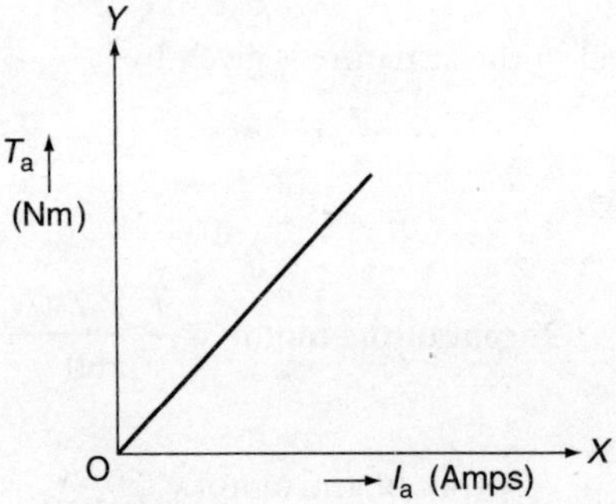

Figure 33 Graph depicting T_a/I_a characteristics of shunt motor.

N/I_a Characteristics

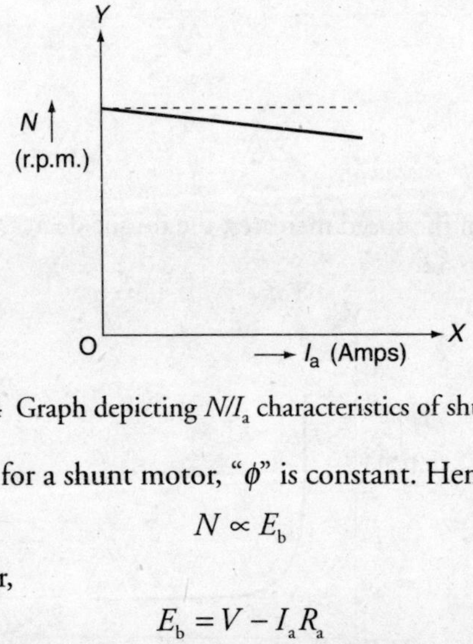

Figure 34 Graph depicting N/I_a characteristics of shunt motor.

We know that $N \propto E_b / \phi$, but for a shunt motor, "ϕ" is constant. Hence,

$$N \propto E_b$$

Also, we have for a shunt motor,

$$E_b = V - I_a R_a$$

Since "R_a" is small, "$I_a R_a$" drop does not clearly vary much with the load and hence "E_b" is also constant. Therefore,

$$N = K$$

In Fig. 34, "N" is taken on Y-axis. Therefore the above equation resembles $y = K$, which is the equation of a straight line parallel to X-axis as shown by the dotted line. Hence shunt motors are used for constant-speed applications (lathes, waters pumps, etc.). However, strictly speaking, both "ϕ" and "E_b" decrease with the load. In fact, decrease in the value of E_b is slightly more than the decrease in the value of "ϕ." Hence the speed of the motor decreases slightly with increase in load (Fig. 34).

N/T_a Characteristics

This characteristic can be easily obtained from the above two characteristics. When the load is increased, motor speeds falls slightly. When "N" decreases slightly "E_b" also does the same. "V" dominates over "E_b" and hence more current flows through the armature (I_a). But

$$T_a \propto I_a$$

Hence "N" decreases (slightly) when T_a increases and vice versa. N/T_a characteristic is as shown in Fig. 35.

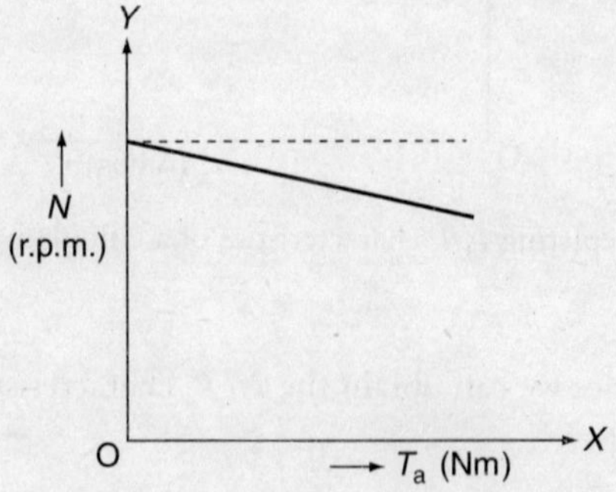

Figure 35 Graph depicting N/T_a characteristic of a shunt motor.

7.2.2.6.3 Characteristics of Cumulative Compound Motor

T_a/I_a Characteristics

We know that in a cumulative compound motor, the series flux helps the shunt field flux. As "I_a" increases, I_{se} increases and the series flux increases. We have $T_a \propto I_a$. Therefore "T_a" also increases. Hence the characteristics are as shown in Fig. 36.

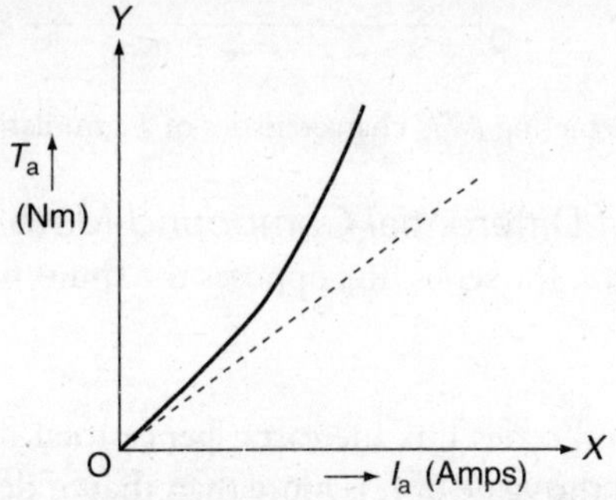

Figure 36 Graph depicting T_a/I_a characteristics of a cumulative compound motor.

N/I_a Characteristics

We have

$$N \propto \frac{E_b}{\phi}$$

As I_a increases, I_{se} increases, the series flux increases, and hence the total flux increases. Thus speed decreases with increase in "I_a." However, the decrease in speed is less rapid than that of series motor and more rapid than that of shunt motor (Fig. 37).

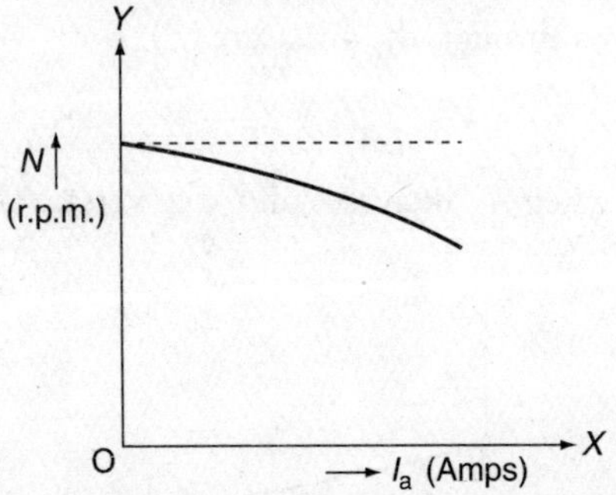

Figure 37 Graph depicting N/I_a characteristics of a cumulative compound motor.

N/T_a Characteristics

From the above two characteristics we can obtain the N/T_a characteristics as shown in Fig. 38.

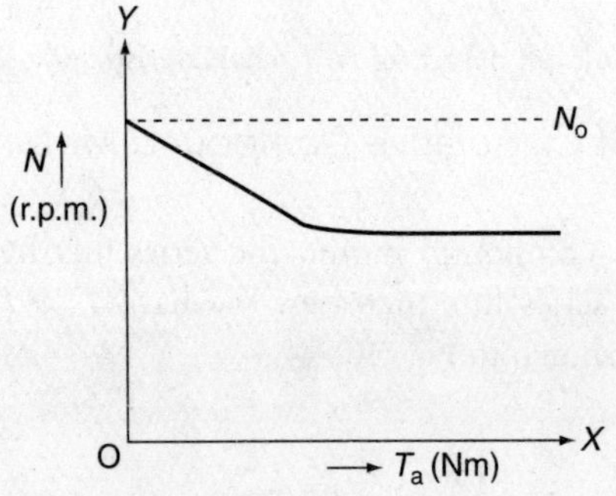

Figure 38 Graph depicting N/T_a characteristics of a cumulative compound motor.

7.2.2.6.4 Characteristics of Differential Compound Motor

In a differential compound motor, the series flux opposes the shunt field flux.

T_a/I_a Characteristics

As "I_a" increases, I_{se} increases and series flux increases; hence total flux decreases to some extent. We have $T_a \propto \phi I_a$. Since increase in the value of I_a is more than that of decrease in the value of total flux, I_a increases less rapidly (than that of series motor). The graph is shown in Fig. 39.

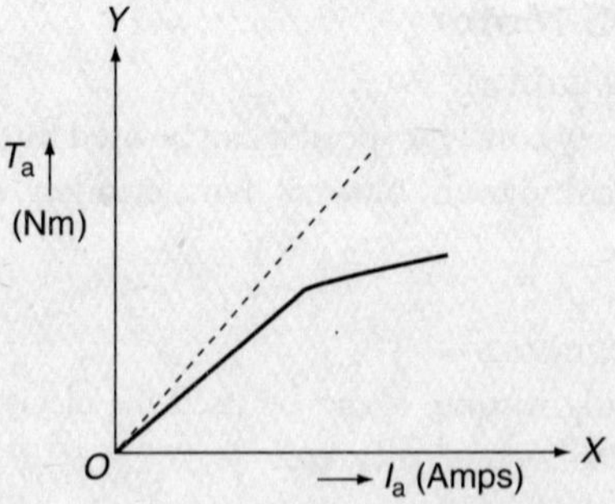

Figure 39 Graph depicting T_a/I_a characteristics of a differential compound motor.

N/I_a Characteristics

We have,

$$N \propto \frac{E_b}{\phi}$$

Just now we have seen that increase in the value of "I_a" causes decrease in the value of total flux. Hence the speed increases slightly. The graphical representation is given in Fig. 40.

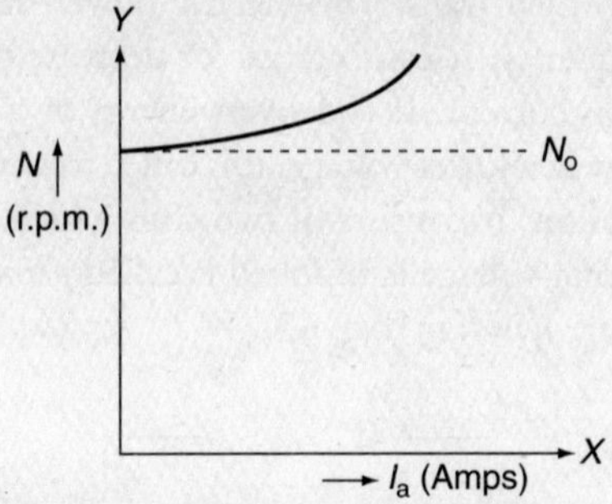

Figure 40 Graph depicting N/I_a characteristics of a differential compound motor.

N/T_a Characteristics

This characteristic can be obtained from the above two characteristics. As the motor speed increases, torque also increases slightly. The graphical representation is given in Fig. 41.

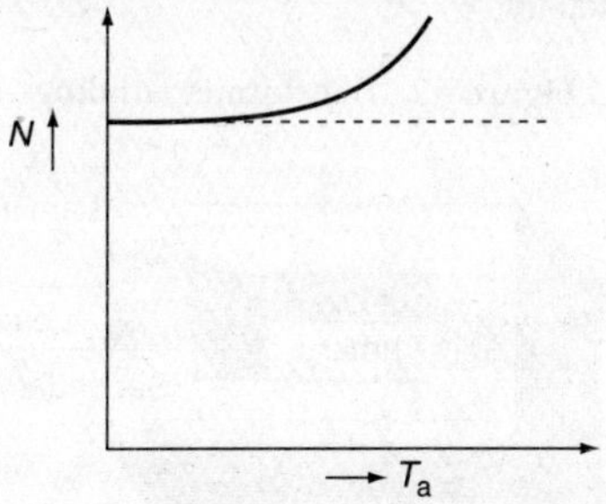

Figure 41 Graph depicting N/T_a characteristics of a differential compound motor.

7.2.2.7 Application of a DC Motor

7.2.2.7.1 Shunt Motor Application

Since shunt motor has approximately constant speed it can be used for lather line-shaft drives, centrifugal pumps, reciprocating pumps, machine tools, blowers, fans, grinders, shapers, wood working machines, small printing presses, etc.

7.2.2.7.2 Series Motor Applications

Since series motor has high starting torque, it can be used for electric traction, tram cars and railway cars, cranes, hoists, elevators, winches conveyors, etc.

7.2.2.7.3 Cumulative Compound Motor Application

Since this motor has high starting torque, it can be used for shears, punches, heavy planers, rolling mills, elevators, conveyors, stamping presses, large printing presses, nine hoists, haulage gears, power fans, etc.

7.3 Transformer

7.3.1 Principle of Operation

Transformer is a static machine which transforms electric power from one circuit to another circuit. It does so without change in frequency. It can increase or decrease the voltage in a circuit with a corresponding decrease or increase in current. If it delivers energy at a higher voltage, it is called *step-up transformer* and if it delivers energy at a lower voltage, it is called *step-down transformer* (Fig. 42). It works on the principle of mutual induction. It consists of two coils which are magnetically coupled through a core. The coil to which alternating voltage is supplied is called *primary winding* and the coil to which load is connected is called *secondary winding* (Fig. 43).

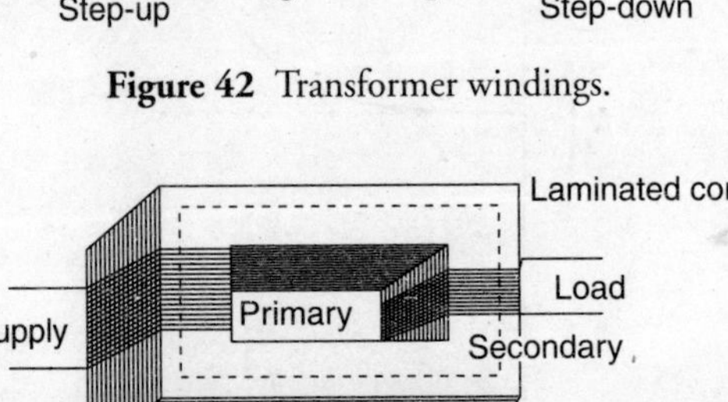

Figure 42 Transformer windings.

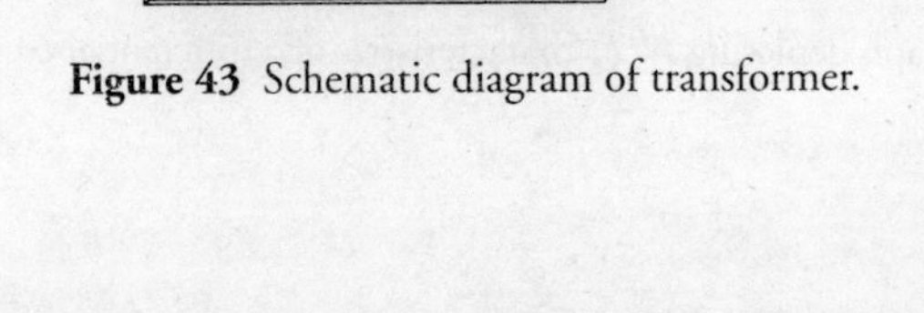

Figure 43 Schematic diagram of transformer.

Whenever the primary winding is given an alternating supply, the alternating flux established in it also links with the secondary winding. According to Faraday's laws of electromagnetic induction, an e.m.f. is induced in the secondary winding, which is called *mutually induced e.m.f.*

7.3.2 Classification of Transformers

Transformers are classified on the basis of:

1. Duty they perform:
- Power transformer – for transmission purpose and distribution.
- Current transformers – instrument transformer.
- Potential transformers – instrument transformer.

2. Construction:
- Core-type transformer.
- Shell-type transformer.
- Berry-type transformer.

3. Voltage output:
- Step-down transformer (higher to lower voltage).
- Step-up transformer (lower to higher voltage).
- Auto transformers (variable from "0" to rated value).

4. Application:
- Welding transformer.
- Furnace transformer.
- Radio transformer.

5. Cooling:
- Dry type transformer: air natural or air last.
- Oil immersed: self-cooled, forced air cooled, water cooled, forced oil cooled.

6. Three-phase transformers:
- Star Star
- Stars–Delta
- Delta–Delta
- Delta–Star
- Open–Delta
- Scott connection

7.3.3 Constructional Details

The various parts of transformer are as follows:

1. **Primary winding:** The winding which is connected to supply is called primary winding.
2. **Secondary winding:** The winding which is connected to load is called secondary winding.
3. **Transformer core:** The two windings of the transformer are magnetically coupled though the core. The core is made up of silicon steel to reduce hysteresis loss. It is also made up of laminated stampings which are insulated from each other by core varnish or oxide layer to reduce the eddy-current loss. The thickness of the lamination varies from 0.35 to 0.5 mm. The two types of transformer cores are as follows:
   - ***Core-type transformer:*** Here the windings surround a considerable part of the core as shown in Fig. 44 and has only core magnetic path. It has two limits for the two windings and is

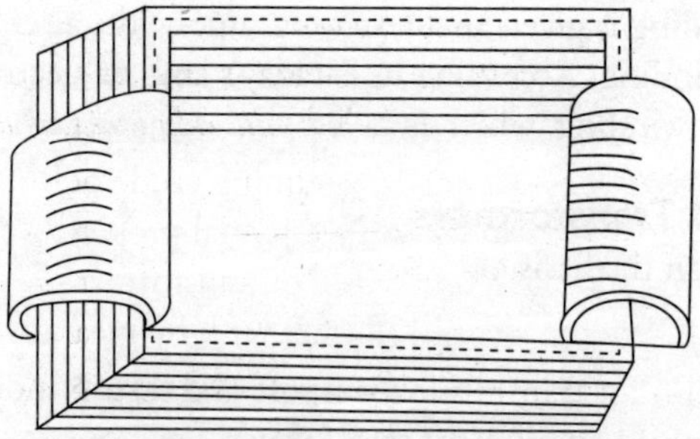

Figure 44 Constructional detail.

made up of two L-type stamping (Fig. 45). The coils used are of cylindrical type and are usually form wound. For transformer of higher rating, stepped core with cylindrical coils are used. For transformer of smaller rating, rectangular coils with core having square or rectangular cross-section is used. Insulating cylinders are used to separate windings from the core and from each other.

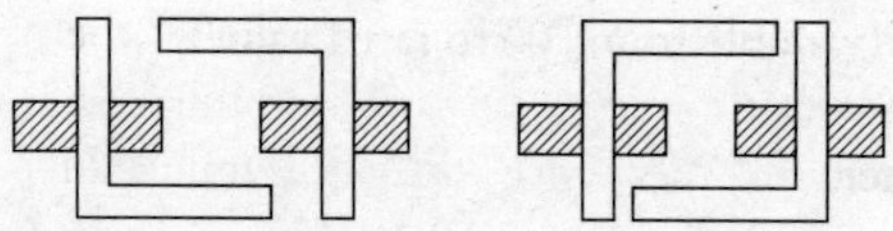

Figure 45 Stampings arrangement.

- ***Shell-type transformer:*** Here the core surrounds considerable part of windings as shown in Fig. 46. The two windings are carried by central limb. It has two parallel paths of magnetic flux. The coils used are of multilayer disc type and are former wound in the foam of pancakes (Fig. 47). Each layer is insulated from other layers by paper.

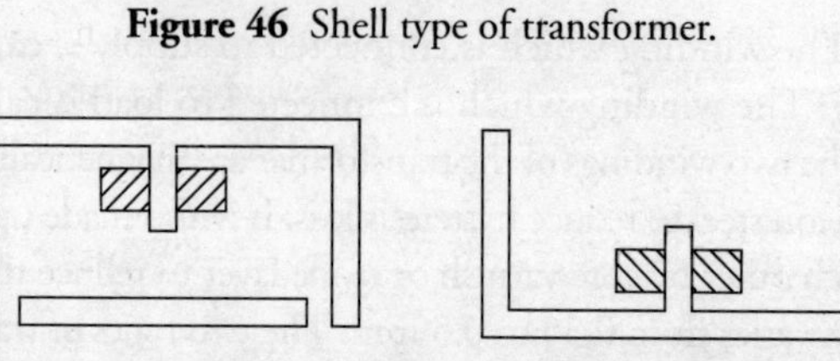

Figure 46 Shell type of transformer.

Figure 47 Distributed form.

4. **Transformer tank:** It is a metallic container used for accommodation assembled core and winding. For better cooling, the surface area of the tank is increased by providing corrugated sheets.
5. **Transformer oil:** It is a mineral oil used for carrying the heat from the core and coils. It also provides additional insulation. It is obtained by refining crude petroleum, High dielectric strength, high flash point, low viscosity are some of the essential properties of transformer oil.
6. **Conservator:** It is a drum mounted on the top of transformer tank and contains transformer oil. The conservator is connected to the transformer tank by means of a pipe. The main purpose of the conservator is to provide space for contraction and expansion of the coil without exposing it to come in contact with the outside atmosphere.
7. **Breather:** A breather is a cylindrical tube containing silica gel impregnated with cobalt. The purpose of the breather is to extract moisture from the air coming inside the transformer during breathing.

7.3.4 E.M.F. Equation of the Transformer

Consider a transformer arrangement as shown in Fig. 48. Let N_1 be the primary number of turns, N_2 the secondary number of turns, ϕ_m the maximum value of flux in the core in weber, A the area of the core in m^2, B_m the maximum value of flux density in the core in Wb/m^2, f the frequency of the AC supply in heat, V_1 the supply voltage across secondary in volts, V_2 the terminal voltage across secondary in volts, L_1 the full load primary current in amperes, L_2 the full load secondary current in amperes, E_1 the e.m.f. induced in the primary in volts and E_2 the e.m.f. induced in the secondary in volts.

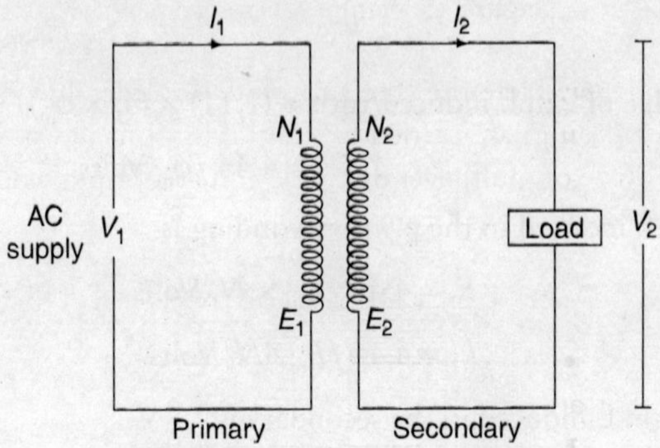

Figure 48 Transformer arrangement.

Since the applied voltage is alternating in nature, the flux established is also an alternating one as shown in Fig. 49. From Fig. 49 it is clear that the flux is attaining its maximum value in one quarter of the cycle, that is, in $T/4$ seconds where "T" is the time period in seconds.

We know that $T = 1/f$, where "f" is the frequency is Hz. Therefore,

$$\text{Average rate of change of flux} = \frac{\phi_m}{1/4f} \text{ Wb/s}$$

$$= 4f \times \phi_m \text{ Wb/s}$$

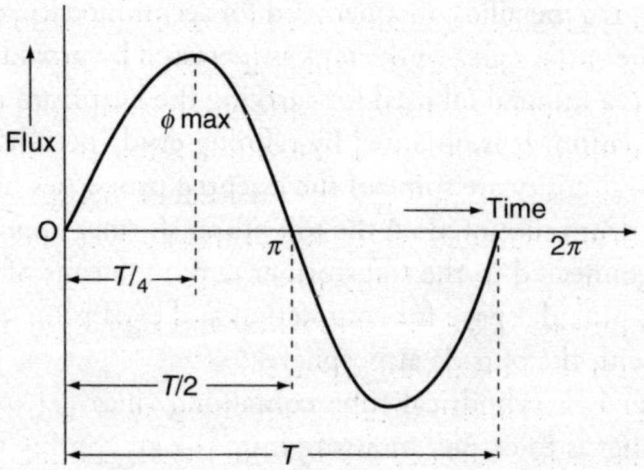

Figure 49 Flux versus time waveform.

If we assume a single-turn coil then according to Faraday's laws of electromagnetic induction, the average value of e.m.f. induced/turn is $4f \times \phi_m$ Volts. We know that

$$\text{Form factor} = \frac{\text{RMS value}}{\text{Average value}} = 1.11 \text{ (since } \phi_m \text{ is sinusoidal)}$$

Therefore,

$$\text{RMS value} = \text{Form factor} \times \text{Average value}$$

Hence

$$\text{RMS value of e.m.f. induced/turn} = (1.11) \times (4f \times \phi_m)$$

$$= 4.44 f\phi_m \text{ Volts}$$

Hence RMS value of e.m.f. induced in the primary winding is

$$E_1 = 4.44 f\phi_m \times N_1 \text{ Volts} \tag{7.2}$$

$$E_1 = 4.44 fB_m AN_1 \text{ Volts} \tag{7.3}$$

Similarly, RMS value of e.m.f. induced in the secondary is

$$E_2 = 4.44 f\phi_m N_2 \text{ Volts} \tag{7.4}$$

$$E_2 = 4.44 f(B_m A)N_2 \tag{7.5}$$

7.3.5 Transformation Ratio (K)

For an ideal transformer,

$$V_1 = E_1$$

$$V_2 = E_2$$

and

$$V_1 I_1 = V_2 I_2$$

$$\Rightarrow \frac{V_2}{V_1} = \frac{I_1}{I_2}$$

$$\Rightarrow \frac{E_2}{E_1} = \frac{I_1}{I_2} \tag{7.6}$$

From Eqs. (7.2) and (7.4), we have

$$\frac{E_2}{E_1} = \frac{N_2}{N_1} \tag{7.7}$$

From Eqs. (7.6) and (7.7), we have

$$\frac{E_2}{E_1} = \frac{N_2}{N_1} = \frac{I_1}{I_2} = K$$

where "K" is called transformation ratio.

Note:

1. If $N_2 > N_1$, that is, $K > 1$, then the transformer is a step-up transformer.
2. If $N_2 < N_1$, that is, $K < 1$, then the transformer is a step-down transformer.

7.3.6 Transformer on No Load

If the primary winding is connected to an alternating voltage and the secondary winding is left open, then the transformer is said to be on no load (Fig. 50). Let the supply voltage be "V_1" volts. This causes a flow of alternating current through the primary winding.

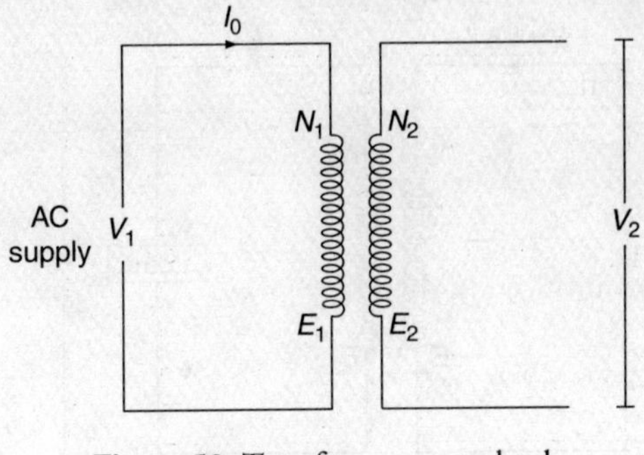

Figure 50 Transformer on no load.

Since secondary winding is open, this current is called no-load primary current (I_0). This "I_0" establishes a flux of "ϕ" weber in the core. "I_0" has two components:

1. Active or working or iron loss component (I_w) which is in phase with "V_1" and supplies the iron loss and negligible amount of primary copper loss:

$$I_w = I_0 \cos \phi_0 \tag{7.8}$$

where $\cos \phi_0$ is the no-load power factor.

2. Reactive or magnetizing or wattless component (I_μ) which is in quadrature with V_1 and its function is to sustain the flux in the core:

$$I_\mu = I_0 \sin \phi_0 \tag{7.9}$$

From Eqs. (7.8) and (7.9) we have

$$I_0 = \sqrt{I_w^2 + I_\mu^2} \tag{7.10}$$

No-load vector diagram is shown in Fig. 51.

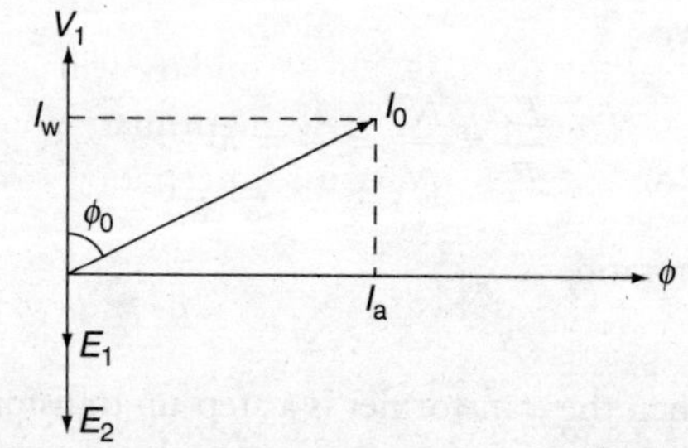

Figure 51 Vector representation of no load.

7.3.7 Transformer on Load

When the secondary winding is connected to a load the transformer is said to be on load (Fig. 52). The phase angle between "V_2" and "I_2" depends on the type of the load.

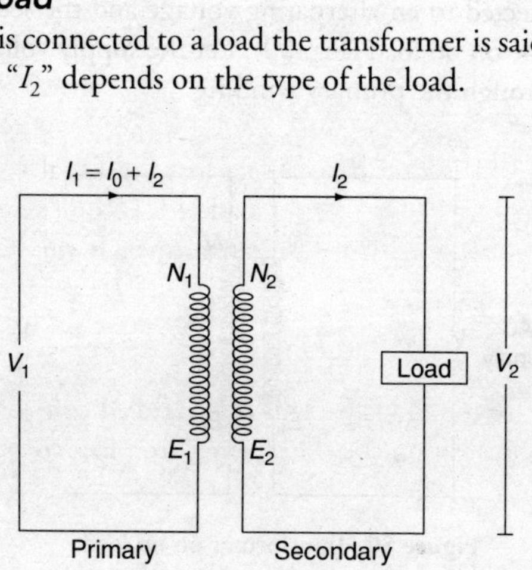

Figure 52 Transformer on load.

When the load is resistive, "I_2" will be in phase with "V_2." When the load is inductive, "I_2" will lag behind "V_2" and when the load is capacitive, "I_2" will lead "V_2." When the secondary circuit is closed, a current "I_2" will flow through the secondary establishing a flux ϕ_2 in it. This ϕ_2 opposes ϕ_1 (flux established by I_0) and weakens it momentarily. Hence E reduces and "V_1" dominates over E_1 and hence causes an additional current ("I_2") to flow through the primary. "I_2" establishes a flux "ϕ_2" which is equal in magnitude to ϕ_1 but opposite in direction. Hence "ϕ_2" and "ϕ_1" cancel each other. Thus when the transformer is loaded,

1. the flux passing through the core is same as that at no load;
2. the total primary current (I_1) will be vector sum of I_0 and I_2.

7.3.8 Transformer Losses

Since transformer is a static device, it has no rotational losses. The losses that occur in the transformer are (a) copper loss and (b) iron loss.

7.3.8.1 Copper Loss

This loss is due to the resistance of the primary and secondary windings. Let I_1 be the primary current in ampere, I_2 the secondary current in ampere, R_1 the primary winding resistance in ohm, R_2 the secondary winding resistance in ohm and K the transformer ratio. Then the total copper loss in the transformer is given by

$$\text{Total copper loss} = I_1^2 R_1 + I_2^2 R_2 \tag{7.11}$$

$$= I_1^2 R_{01} \tag{7.12}$$

$$= I_2^2 R_{02} \tag{7.13}$$

where

$$R_{01} = R_1 + \frac{R_2}{K^2} \tag{7.14}$$

$$R_{02} = R_2 + R_1 K^2 \tag{7.15}$$

Then from Eqs. (7.11) to (7.13) it is clear that copper loss is a variable loss and varies with square of the current and output. This copper loss at quarter load is 1/16th of that at full load. Copper loss can be found by conducting short circuit test on the transformer. If this loss is high it will damage the insulation.

7.3.8.2 Iron Loss

Iron loss remains almost constant for all loads. This is also called core loss or magnetic loss and can be found by conducting open circuit test on the transformer Iron loss consists of the following two types of losses:

1. Hysteresis loss.
2. Eddy-current loss.

7.3.8.2.1 Hysteresis Loss

Since the supply to the primary is alternating, the core gets alternately magnetized and demagnetized causing loss of energy. It is in the form

$$W_h = \eta (B_{max})^{1.6} \times f \times v \text{ Watts} \tag{7.16}$$

where W_h is the hysteresis loss in watts, B_{max} is the maximum value of flux density in Wb/m^2, f is the frequency of magnetic reversal in Hz, v is the volume of the core in m^3 and η is the serenity constant.

This loss can be minimized by selecting a material for the core which has low hysteresis coefficient, for example, silicon steel.

7.3.8.2.2 Eddy-Current Loss

This loss is due to eddy current induced in the transformer core and is seen in the form of heat in the core. It is given by

$$W_e = KB_{max}^2 \ f^2 t^2 \text{ Watts} \tag{7.17}$$

where "t" is the thickness of lamination. This loss can be minimized by using a laminated core.

7.3.9 Efficiency of a Transformer

It is given by the ratio of output to input. Transformer has high efficiency since it has no function and winding loss (static device). Therefore,

$$\text{Efficiency} = \frac{\text{Output}}{\text{Input}} \tag{7.18}$$

However,

$$\text{Input} = \text{Output} + \text{Total losses} \tag{7.19}$$

where total losses are given by

$$\text{Total losses} = \text{Iron loss} + \chi^2(\text{copper loss}) \tag{7.20}$$

where χ is the fraction of the load. Therefore,

$$\text{Efficiency} = \frac{\text{Output}}{\text{Input} + \text{Total losses}} \tag{7.21}$$

Also,

$$\text{Output} = \text{Input} - \text{Total losses} \tag{7.22}$$

and

$$\text{Output in watts} = (\text{Rated KVA}) \times 1000 \times \text{p.f.} \times \chi \tag{7.23}$$

where KVA denotes Kilo Volt Ampere and p.f. denotes power factor. Now

1. $\chi = 1$ for full load;
2. $\chi = 0.75$ for 3/4th full load;
3. $\chi = 0.5$ for 1/2 full load;
4. $\chi = 0.25$ for 1/4 full load.

Therefore,

$$\text{Efficiency} = \frac{\text{Input} - \text{Total loss}}{\text{Input}} = 1 - \frac{\text{Total loss}}{\text{Input}} \tag{7.24}$$

Hence

$$\%\ \text{Efficiency} = \frac{\text{Output}}{\text{Input}} \times 100$$

7.3.10 Application of Transformer

Transformers are used

1. In electrical power engineering for transmission and distribution
2. In communication engineering for impedance transformation, etc.
3. As an instrument for measuring current (CT) and voltage (PT).
4. As a step-down and step-up transformer to get reduced or increased output voltage.
5. In radio and TV circuits, telephone circuits, control and instrumentation circuits.
6. In furnaces and welding.

7.4 Three-Phase Induction Motor

Three-phase induction motor is a workhorse of modern industry. It has a very wide range of application. It has specific advantages like simple construction, low capital cost, ruggedness and absence of commentator problems, reasonably good power factor, sufficiently high efficiency, simple starting arrangement, minimum maintenance, etc. However, its starting torque is somewhat less and as the load increases the speed decreases.

7.4.1 Principle of Operation

The three-phase induction motor consists of a stator (stationary part) and a rotor (rotating part). Whenever a three-phase supply is given to the three-phase winding of the stator, a magnetic flux of constant magnitude but rotating with certain speed is obtained. This speed of the stator magnetic flux is called *synchronous speed* (N_s) given by

$$N_s = \frac{120f}{P} \text{ r.p.m.} \tag{7.25}$$

where N_s is the synchronous speed or speed of the stator magnetic flux; P the number of stator poles; f the frequency of the supply in Hz. For a given motor and supply, "P" and "f" are fixed; hence N_s is fixed. Therefore to reduce ($N_s - N$), "N" has to increase. Thus the rotor starts rotating in the same direction as that of the stator flux.

This rotating flux when passes through the air gap is being cut by the stationary rotor conductors. According to Faraday's laws of electromagnetic induction, the relative speed between the stator (rotating) magnetic flux and rotor (stationary) conductors induces an e.m.f. in the rotor conductors. The magnitude of the induced e.m.f. depends on the relative speed between the stator flux and the rotor conductors. The direction of e.m.f. is given by Fleming's right-hand rule. Since rotor conductors form a loud circuit, the induced e.m.f. causes a current to flow through the rotor conductors whose direction (given by Lenz's law) will be in such a way as to oppose the very cause of its production. The cause of production is the relative speed between the stator flux and the rotor conductor, that is, $(N_s - N)$ where "N" is the speed of the rotor.

Example 1

Consider a 50 Hz, four-pole three-phase induction motor. Initially the rotor is stationary, that is, $N = 0$. We have,

$$N_s = \frac{120f}{P} = \frac{120 \times 50}{4} = 1500 \text{ r.p.m.}$$

Therefore

$$\text{Relative speed} = N_s - N = 1500 - 0 = 1500$$

If the relative speed is opposed by the induced rotor current, then $(N_s - N)$ value has to decrease.

7.4.2 Constructional Details

The main parts of induction motor are as follows:

1. Stator.
2. Rotor.

We will discuss both of them next.

7.4.2.1 Stator

The stator is built up of sheet steel laminations of 0.4–0.5 mm thickness. Each lamination is insulated from one another by means of varnish or oxide coating. In the outer periphery of the hollow cylindrical

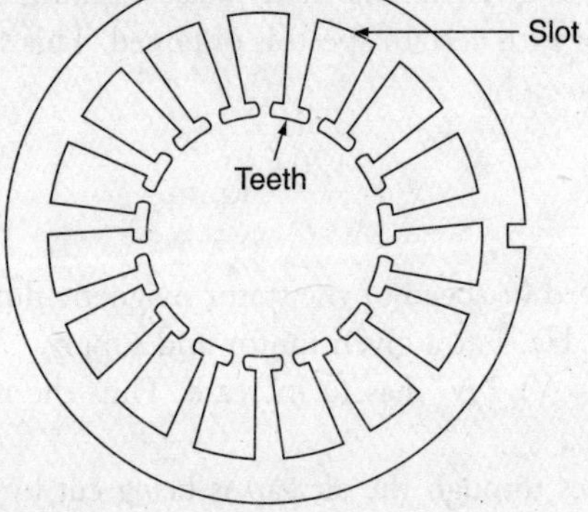

Figure 53 A typical stator stamping.

stator, slots are made to accommodate the stator conductors. The stator is wound for three-phase supply and its connections are brought out to terminal box to the frame. The stator carries three-phase winding and is fed from three-phase supply. It is wound for a definite number of poles, the exact number of poles being determined by the requirements of speed. Greater the number of poles, lesser the speed and vice versa. For cooling purpose radial ventilating ducts are also provided. A typical stator stamping is as shown in Fig. 53.

7.4.2.2 Rotor

There are two types of rotors:

1. Squirrel cage rotor.
2. Phase-wound or slip-ring rotor.

7.4.2.2.1 Squirrel Cage Rotor

It is cylindrical in shape and is built up of sheet steel laminations. The number of ventilating ducts provided is equal to that of the stator. It has slots to accommodate rotor conductors. These rotor conductors or bars are short circuited at the ends of the rotors by leaving the copper rings resembling the cage of a squirrel and hence the name squirrel cage rotor. Since the rotor conductors are permanently short circuited, it is not possible to add any internal resistance in series with the rotor. This type of rotor is cheap and has simple and rugged construction (Fig. 54).

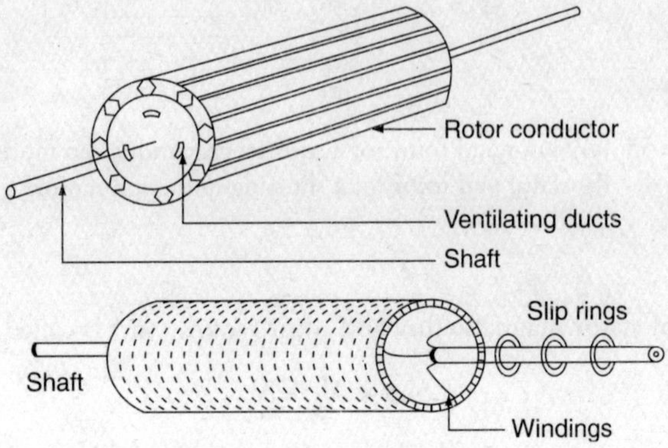

Figure 54 Squirrel cage rotor.

7.4.2.2.2 Phase Wound or Slip-Ring Motor

The rotor winding is wound for the same number of poles and phases as that of stator winding. For a three-phase winding the three phases start at one junction and the other three ends are connected to slip rings mounted on the shaft. To increase the starting torque and to reduce the starting current, starting resistances can be connected in star connection through carlion brushes (which are placed on the slip rings). As soon as the motor achieves its rated speed, the starting resistances can be removed gradually from the rotor circuit and finally slip rings are short circuited.

The disadvantage is that slip-ring rotors are costlier than squirrel cage rotors. The stators and rotors of a squirrel cage and slip-ring induction motor are shown in Fig. 55.

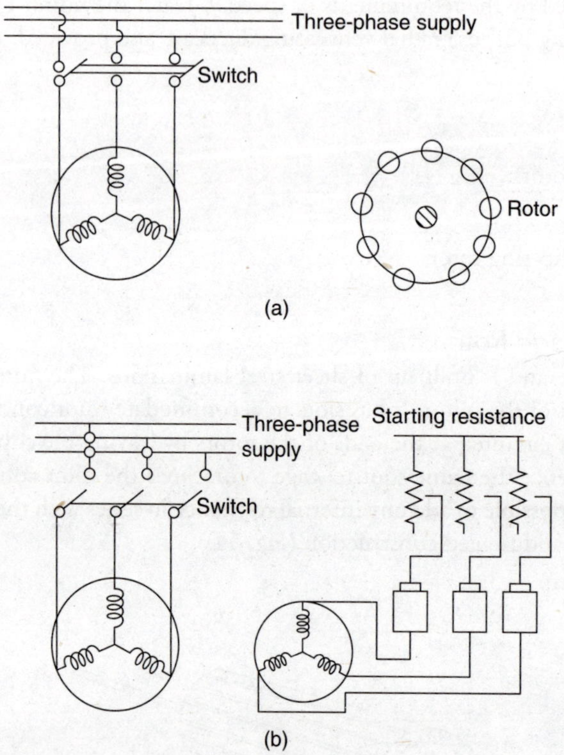

Figure 55 (a) Stator and rotor for a squirrel cage induction motor; (b) stator and rotor for a slip-ring induction motor.

7.4.3 Slip (S)

The difference in speed of stator magnetic flux and rotor (conductor) is called slip. Therefore,

$$\text{Slip} = N_s - N \tag{7.26}$$

where N_s is the synchronous speed or speed of stator magnetic flux and N is the speed of the rotor. It is expressed as the percentage of the synchronous speed. Therefore,

$$\%\ \text{Slip} = \frac{N_s - N}{N_s} \times 100 \tag{7.27}$$

or

$$\%\ S = \frac{N_s - N}{N_s} \times 100 \tag{7.28}$$

This implies

$$S = 1 - \frac{N}{N_s} \quad (7.29)$$

$$\Rightarrow N = N_s(1 - S) \quad (7.30)$$

7.4.4 Frequency of Rotor Current (f′)

When the rotor is stationary the frequency of the rotor current is same as the supply frequency (f). But, when the rotor starts rotating, the frequency of the rotor current depends upon the relative speed. Let the frequency of rotor current be f'. Then we have

$$\text{Realative speed} = N_s - N \quad (7.31)$$

where

$$N_s = \frac{120f}{P}$$

According to definition of f',

$$N_s - N = \frac{120\, f'}{P} \quad (7.32)$$

Dividing Eq. (7.32) by Eq. (7.31), we have

$$\frac{N_s - N}{N_s} = \frac{f'}{N}$$

Therefore, using Eq. (7.29) we get slip

$$S = \frac{f'}{f}$$

$$\Rightarrow f' = S \cdot f \quad (7.33)$$

7.4.5 Expression for Torque

The rotor of a three-phase induction motor has two states:

1. Standstill condition.
2. Running condition (X).

Accordingly its parameters under standstill condition are given by

f: the rotor current frequency
R_2: the rotor resistance/phase
X_2: the rotor reactance/phase
Z_2: the rotor impedance/phase
E_2: the rotor e.m.f./phase

The same parameters under running condition are given by

f': the rotor current frequency
X_1: the rotor reactance/phase
Z_1: the rotor impendence/phase
E_r: the rotor e.m.f./phase
R_2: the rotor resistance/phase (remains same under both the conditions)

We have proved in Eq. (7.33) that

$$f' = S \cdot f$$

where "S" is the slip. Similarly we can write

$$X_r = SX_2 \tag{7.34}$$

$$E_r = SE_2 \tag{7.35}$$

$$Z_r = \sqrt{R_2^2 + X_r^2} \tag{7.36}$$

where X_r denotes rotor reactance per phase and Z_r denotes rotor impedance per phase. We know that in a DC motor the expression for torque is given by $T \propto \phi I_a$. In case of an induction motor,

$$T \propto I_r \cos \phi_r \tag{7.37}$$

where I_r is the rotor current/phase under running condition and ϕ is the stator flux/pole. But

$$I_r = \frac{E_r}{Z_r}$$

$$\cos \phi_r = \frac{R_2}{Z_r}$$

(from impedance triangle shown in Fig. 56.)

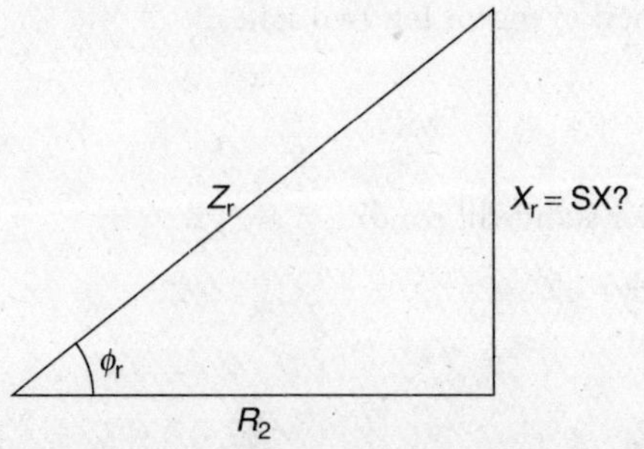

Figure 56 Impedance triangle.

Therefore,

$$T = k\phi \frac{E_r}{Z_r}\frac{R_2}{Z_r}$$
$$= \frac{k\phi E_r \cdot R_2}{Z_r^2}$$

From impedance triangle,

$$Z_r^2 = (R_2)^2 + (SX_2)^2$$

Also, we have $E_r = SE_2$ from Eq. (7.35). Therefore,

$$T = \frac{k\phi \cdot S \cdot E_2 \cdot R_2}{(R_2)^2 + (SX_2)^2} \quad (7.38)$$

This is the expression for torque is

$$\cos\phi_r = \frac{R_2}{Z_r} \quad (7.39)$$

$$\cos\phi_r = \frac{R_2}{\sqrt{(R_2)^2 + (SX_2)^2}} \quad (7.40)$$

7.4.6 Torque–Slip Characteristics

The torque–slip characteristics of three-phase induction motor are a shown in Fig. 57. From this figure it is clear that the curve starts from origin. That is to prove that when $S = 0$, T should be equal to 0. This can be easily verified as follows: From Eq. (7.38).

$$T = \frac{k\phi\, SE_2R_2}{(R_2)^2 + (SX_2)^2}$$

Put $S = 0$. Then

$$T = \frac{0}{(R_2)^2 + 0} = 0$$

Thus when $S = 0$, $T = 0$. Hence the curve starts from the beginning.

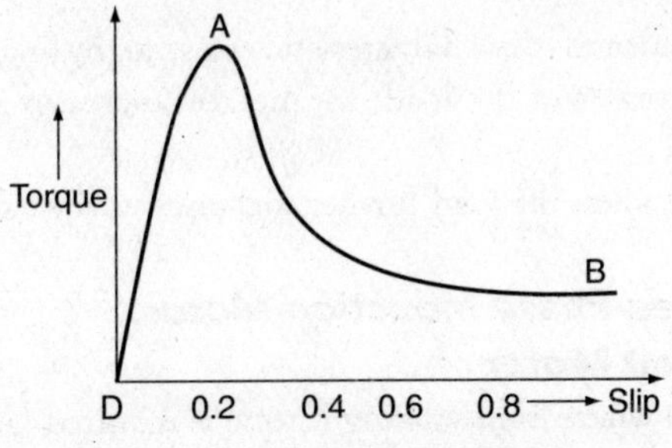

Figure 57 Torque–slip characteristics of three-phase induction motor.

The portion "AB" is a straight line passing through the origin; this can be verified as follows. At speeds close to synchronism, $(N_s - N)/N_s$, that is "S," will be small. Therefore SX_2 is very small compared to R_2; hence can be neglected. Equation (7.38) can be rewritten as

$$T = \frac{k\phi SE_2R_2}{(R_2)^2 + (SX_2)^2}$$

$$= \frac{k\phi \cdot SE_2R_2}{(R_2)^2} \qquad (SX_2 << R_2)$$

This implies

$$T = \frac{k\phi E_2R_2}{(R_2)^2} \cdot S$$

$$\Rightarrow T \propto S$$

$$\Rightarrow T = k'S \tag{7.41}$$

This is equation of straight line passing through the winding, hence, the portion AB. The shape of portion BC can be explained as follows. When "N" is not very close to N_s, $(N_s - N)/N_s$, that is "S," will be large when compared to R_2. Hence "R_2" can be regulated when compared to "SX_2," that is, $Sx_2 >> R_2$. Therefore, Eq. (7.38) can be rewritten as

$$T = \frac{k\phi SE_2R_2}{(SX_2)^2}$$

$$= \frac{k\phi E_2R_2 \cdot S}{(X_2)^2 S^2}$$

$$T = \left(\frac{k\phi E_2R_2}{(X_2)^2}\right)\frac{1}{S}$$

$$\Rightarrow T \propto \frac{1}{S}$$

Thus as slip increases, torque decreases. So the torque–slip characteristic is a rectangular hyperbola.

The points to be noted are as follows:

1. At low values of slip, the torque–slip characteristic is a straight line.
2. As slip increases with increases in the load, torque also increases and becomes maximum when $S = R_2 / X_2$.
3. Beyond the point of T_{max} when the load further increases, torque starts decreasing.

7.4.7 Application of Three-Phase Induction Motor

7.4.7.1 Slip-Ring Induction Motor

These motors are used in places where high starting torque is required (also starting currents should be low) such as pumps, fans, hoists, compressors, lifts, etc.

7.4.7.2 Squirrel Cage Inductor Motor

They find their application under various circumstances as follows.

1. Low starting torque with normal starting current. For example, fans and centrifugal pumps.
2. High torque with low starting current. For example, conveyors, compressor, crushes, agitators, reciprocating pumps.
3. High torque with medium and high slip. For example, sheers, punches, die stamping, etc.
4. Normal torque with normal starting current. For example,, fans, blowers, centrifugal pumps, line shafting, etc.

7.5 Single-Phase Induction Motor

Single-phase induction motor, which are designed to operate from a single-phase supply, are manufactured in a large number of types to perform a wide variety of useful services in home, offices, factories, workshops, in business establishments, etc. Constructionally, this motor is more or less, similar to a polyphase induction motor, except that

1. Its stator is provided with a single phase winding.
2. A centrifugal switch is used in some types of motors, in order to cut out a winding, used only for starting purposes.

It has distributed stator winding and a squirrel cage rotor. When fed from a single-phase supply, its stator winding produces a flux (or field) which is only alternating, that is, one which alternates along one space axis. It is not a synchronously revolving (or rotating) flux, as in the case of a two or a three phase stator winding fed from a two- or three-phase supply. Now, an alternating or pulsating flux acting on a stationary squirrel cage rotor cannot produce rotation (only a revolving flux can). That is why a single phase motor is not self-starting. However, if the rotor of such a machine is given an initial start by hand (or small motor) or otherwise, in either direction, then immediately a torque arises and the motor accelerates to its final speed (unless the applied torque is too high).

7.6 Synchronous Motor

A synchronous motor is electrically identical with an alternator or AC generator. In fact, a given synchronous machine may be used, at least theoretically, as an alternator, when driven mechanically, or as a motor, when driven electrically, just as in the case of DC machines. Most synchronous motors are rated 150 kW and 15 MW and run at speeds ranging from 150 to 1800 r.p.m.

Some characteristics features of a synchronous motor are as follows:

1. It runs either at synchronous speed or not at all, that is, while running it maintains a constant speed. The only way to change its speed is to vary the supply frequency.
2. It is not inherently self-starting. It has to be run upto synchronous speed by some means, before it can be synchronized to the supply.
3. It is capable of being operated under a wide range of power factors, both lagging and leading. Hence, it can be used for power correction purposes, in addition to supplying torque to drive loads.

Summary

In this chapter, different types of electric machines have been discussed. This chapter deals with various fundamental aspects of DC and AC as well as.

1. DC generator is a dynamic electrical machine in which mechanical energy is converted into electrical energy. It operates on the principle based on the Faraday's law of electromagnetic induction. A DC machine that converts electrical energy into mechanical energy is known as DC motor. That is in action the DC motor is reverse of DC generator. The DC motor basically works on the principle that if it is placed is a magnetic field, mechanical force acts on the current-carrying conductor, and as a result the conductor starts rotating in a direction of current and field. This current is given by Fleming left-hand rule.
2. The parallel operation of DC generators gives expressions for sharing of load by different generators, operating in parallel in a power system.
3. Transformer is a static device which transforms electric power from one circuit to another circuit.
4. Three-phase induction motor maintenance is cheaper compared to other types and speed rating. As such, these motors are widely used for most of the industrial applications. In fact, it ranges from a small workshop to a large industry.
5. Synchronous machine is either called as synchronous generator or synchronous motor. When the machine converts mechanical energy into electrical energy which is AC in nature, it is called as an AC generator or alternator. Further, when the alternator operates along with other alternators or with other AC supply, it is called as a synchronous generator. A synchronous motor converts electrical energy which is AC in nature to mechanical energy.

Solved Problems

Problem 1

A 250 V DC shunt motor (Fig. 58) takes 40 A at full load. The resistance of the armature and shunt field windings are 0.5 Ω and 250 Ω, respectively. Determine the back e.m.f. on full load.

Refering

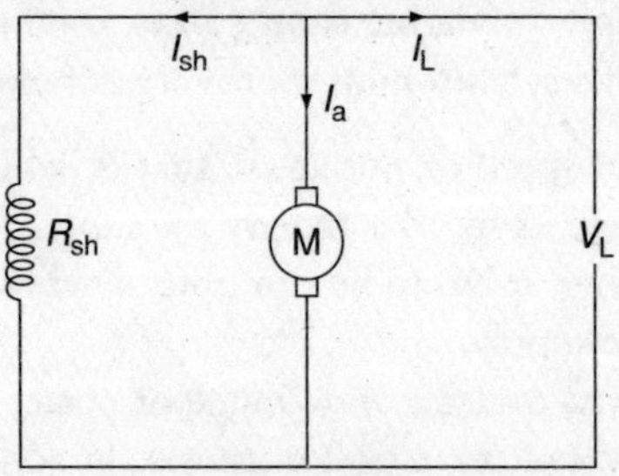

Figure 58 Problem 1.

Solution

Given,

$$R_a = 0.5\Omega$$

$$R_{sh} = 250\Omega$$

$$\text{Line voltage } V_L = 250 \text{ V}$$

$$\text{Line current } I_L = 40 \text{ A}$$

Hence

$$I_{sh} = \frac{250}{250} = 1\text{ A}$$

$$I_a = I_L - I_{sh} = 40 - 1 = 39\text{ A}$$

where I_{sh} denotes shunt current, I_a denotes armature current, I_L denotes load current. Therefore, back e.m.f. is given by

$$E_b = V_L - I_a R_a$$

$$= 250 - 39 \times 0.5$$

$$E_b = 230.5 \text{ V}$$

Problem 2

A 20 kW, 200 kV DC shunt generator has armature and field resistance of 0.5 Ω and 150 Ω, respectively. Determine the total armature power developed (i) when working as a generator delivering 20 kW output and (ii) when working with a motor taking 20 kW.

Solution

Given,

$$\text{Input power } P_L = 20 \times 10^3 \text{ W}$$

$$\text{Input voltage } V_L = 200 \text{ V}$$

$$R_a = 0.52\Omega$$

$$R_{sh} = 150\Omega$$

(i) When the machine acts as a DC shunt generator (Fig. 59):

$$I_{sh} = \frac{V_L}{R_{sh}} = \frac{200}{150} = 1.33$$

$$I_L = \frac{P_L}{V_L} = \frac{20\times10^3}{200} = 100\,\text{A}$$

$$I_a = I_L + I_{sh} = 101.33\,\text{A}$$

Therefore, induced e.m.f. is

$$\begin{aligned} E_g &= V_L + I_a R_a \\ &= 200 + (101.33\times0.05) = 205.06\ \text{V} \end{aligned}$$

Therefore,

$$\begin{aligned} \text{Armature power developed} &= E_g \times I_a \\ &= 205.06\times101.33 \\ &= 20779.388\,\text{W} \end{aligned}$$

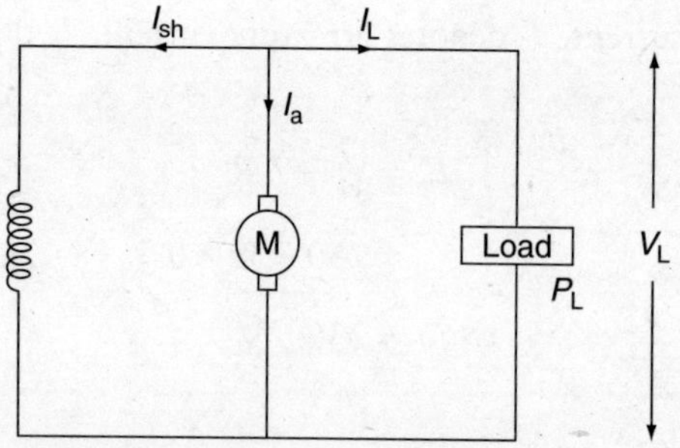

Figure 59 Solution to Problem 2 part (i).

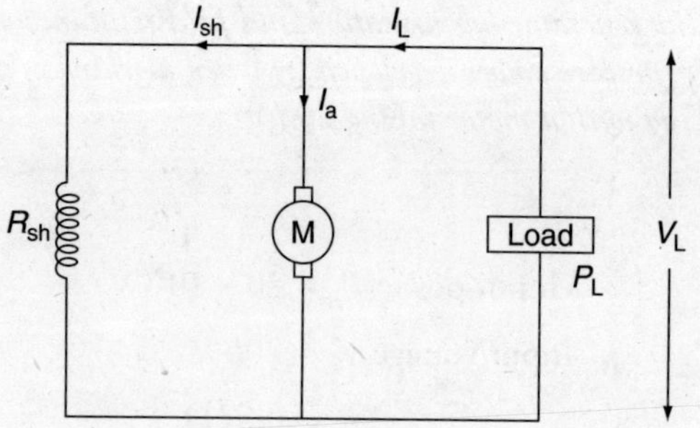

Figure 60 Solution to Problem 2 part (ii).

(ii) When same machine is running as a DC shunt motor (refer to Fig. 60):

$$\text{Input current, } I_L = \frac{20\times10^3}{200} = 100\,\text{A}$$

$$I_{sh} = \frac{V_L}{R_{sh}} = \frac{200}{150} = 1.33$$

$$I_a = I_L - I_{sh} = 98.67 \text{ A}$$

$$E_b = V_L - I_a R_a = 200 - (98.67 \times 0.05) = 195.06 \text{ V}$$

Therefore,

$$\text{Armature power developed} = 19247.2 \text{ W}$$

Problem 3

In Problem 2, if the DC shunt generator runs at a speed of 1000 r.p.m., calculate the speed of the machine when running as a motor, all the data remaining same.

Solution
Given

$$\text{Speed of generator } (N_1) = 1000 \text{ r.p.m.}$$

$$\text{Generated e.m.f. of generator } (E_1) = 205.06 \text{ V}$$

$$\text{Back e.m.f. of motor } (E_2) = 195.06 \text{ V}$$

$$\text{Speed of the motor} = N_2 \text{ r.p.m}$$

We have for two shunt machines,

$$\frac{N_2}{N_1} = \frac{E_2}{E_1}$$

$$N_2 = \frac{E_2}{E_1} N_1$$

$$N_2 = 951.23 \text{ r.p.m.}$$

Problem 4

A 250 V DC series motor has an armature resistance of 0.05 Ω and series field resistance of 0.02 Ω. It is running at a speed of 900 r.p.m. taking a current of 30 A. Determine its speed when it takes a current of 25 A.

Solution
Given,

$$V_L = 250 \text{ V}$$

$$R_a = 0.05 \Omega$$

$$R_{se} = 0.02\,\Omega$$

$$I_L = I_{se} = 30\text{ A} = I_1$$

$$N_1 = 4000 \text{ r.p.m.}$$

Therefore

$$E_1 = V_L - I_a R_a - I_{se} R_{se}$$

$$E_1 = 250 - 1.5 - 0.6$$
$$= 247.9\text{ V}$$

When the same motor takes a current of 25 A, then

$$I_L = I_{se} = I_a = 25\text{ A} = I_2$$

Therefore,

$$E_2 = 250 - 1.25 - 0.5 = 248.25 \text{ V}$$

We have,

$$\frac{N_2}{N_1} = \frac{E_2}{E_1} \times \frac{I_1}{I_2}$$

Therefore,

$$N_2 = N_1 \times \frac{E_2}{E_1} \times \frac{I_1}{I_2}$$

$$= 400 \times \frac{248.25}{247.9} \times \frac{30}{25}$$

$$N_2 = 480.67\text{ r.p.m.}$$

Problem 5

A four-pole wave wound 200 V shunt motor has 700 armature conductors. The resistance of the armature is 0.02 Ω and that of shunt field is 50 Ω. Find the torque developed when it takes a current of 60 A. The flux/pole is 30 mWb.

Solution

Given,

$$V_L = 200 \text{ V}$$

$$I_L = 60 \text{ A}$$

$$R_{sh} = 50\,\Omega$$

$$\phi = 30 \times 10^{-3}$$

$$Z = 700$$

$$P = 4$$

$$A = 2$$

Therefore,

$$I_{sh} = \frac{V_L}{R_{sh}} = \frac{200}{50} = 4\text{ A}$$

$$I_a = I_L - I_{sh} = 60 - 4 = 56\text{ A}$$

Hence,

$$T = 0.159 \times \phi \times Z \times I_a \times \frac{P}{A}\text{ Nm}$$

$$= \frac{0.159 \times 30 \times 10^{-3} \times 700 \times 56 \times 4}{2}$$

$$= 373.96\text{ Nm}$$

Problem 6

A four-pole wave wound DC generator has 50 slots and 24 conductors/slot. The flux/pole is 10 mWb. Determine the induced e.m.f. in the armature if it is rotating at a speed of 600 r.p.m.

Solution

Given,

$$P = 4$$

$$A = 2\text{ (wave wound)}$$

Numbers of slots = 50

Slots/conductor = 24

Total conductor (Z) = 24 × 50

Flux/pole (ϕ) = 10×10^{-3} Wb

Speed of armature (N) = 600 r.p.m.

Therefore, the e.m.f. induced is

$$E = \frac{\phi ZNP}{60A}$$

$$= \frac{(10\times10^{-3})\times(24\times50)\times600\times4}{60\times2}$$

$$= 240\ \text{V}$$

Problem 7

Repeat Problem 6 for lap connection.

Solution

For lap winding, $A = P = 4$. Therefore,

$$E = \frac{\phi ZNP}{60A} = \frac{(10\times10^{-3})\times(24\times50)\times600\times4}{60\times4}$$

$$= 120\ \text{V}$$

Problem 8

A six-pole wound armature has 780 conductors and a flux of 12 mWb. Find the speed of the armature when generated e.m.f. is 400 V. What will be the speed when it is lap wound.

Solution

Given,

$$P = 6$$

Wave wound, $A = 2$

Armature conductors $(Z) = 780$

Flux/pole, $\phi = 12\times10^{-3}$ Wb

$$E = 400$$

$$N = ?$$

Now

$$E = \frac{\phi ZNP}{60A}$$

$$\frac{E60A}{\phi ZP} = N$$

$$\frac{400\times60\times2}{12\times10^{-3}\times780\times6} = N$$

or

$$N = 854.70 \text{ r.p.m.}$$

When it is lap wound, $A = P = 6$. Then

$$N = \frac{400 \times 60 \times 6}{12 \times 10^{-3} \times 780 \times 6}$$
$$= 2564.1 \text{ r.p.m.}$$

Problem 9

A DC shunt generator (Fig. 61) supplies a load of 10 kW at 250 V. Calculate the induced e.m.f. if the armature resistance is 0.5 Ω and shunt field resistance is 100 Ω.

Solution

Given,

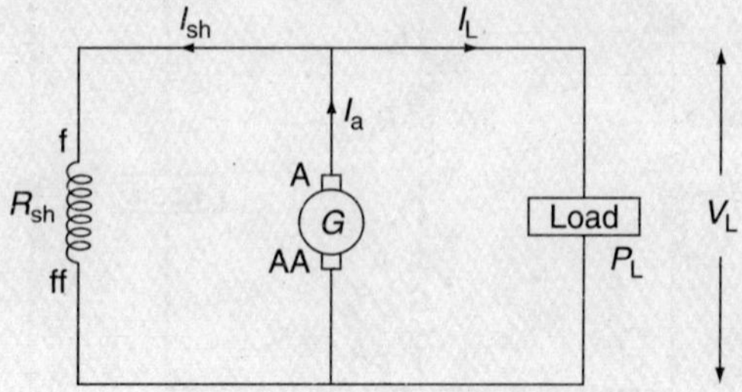

Figure 61 Problem 9.

$$R_a = 0.5 \ \Omega$$
$$R_{sh} = 100 \ \Omega$$
$$V_L = 250\text{V}$$
$$P_L = 10 \times 10^3 \text{ W}$$

Then

$$I_L = \frac{P_L}{V_L} = \frac{10 \times 10^3}{250} = 40 \text{ A}$$

$$I_{sh} = \frac{V_L}{R_{sh}} = \frac{250}{100} = 2.5 \text{ A}$$

Now

$$I_a = I_L + I_{sh} = 40 + 2.5 = 42.5 \text{ A}$$

Therefore induced e.m.f. is

$$\begin{aligned} E_g &= V_L + I_a R_a \\ &= 250 + (42.5) \times 0.5 \\ &= 271.25 \text{ V} \end{aligned}$$

Problem 10

A DC long shunt compound generator (Fig. 62) delivers a load current of 200 A at 500 V. The resistance of the armature, series field and shunt field are 0.03 Ω, 0.015 Ω and 150 Ω. Calculate the e.m.f. induced in the armature. Allow a brush contact drop of IV/brush.

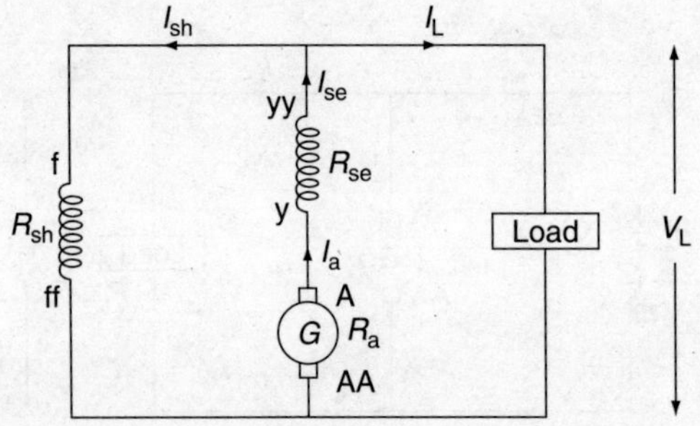

Figure 62 Problem 10.

Solution
Given

$$I_L = 200 \text{ A}$$

$$V_L = 500 \text{ V}$$

$$R_a = 0.03\ \Omega$$

$$R_{se} = 0.015\ \Omega$$

$$R_{sh} = 150\ \Omega$$

$$\begin{aligned} \text{B.C.D.} &= 1 \text{ V/brush} \\ &= 1 \times 2 \\ &= 2 \text{ V} \end{aligned}$$

We have to fing the induced e.m.f. (E_g). Now

$$I_{sh} = \frac{V_L}{R_{sh}} = \frac{50}{150} = 3.33\,\text{A}$$

Hence

$$\begin{aligned} I_a &= I_L + I_{sh} \\ &= 200 + 3.33 \\ &= 203.33\,\text{A} \end{aligned}$$

Also

$$I_{se} = I_a = 203.33\text{ A}$$

So

$$\begin{aligned} E_g &= V_L + I_a R_a + I_{se} R_{se} + \text{B.C.D.} \\ &= 500 + (203.33 \times 0.03) + (203.33 \times 015) + 2 \\ &= 500 + 6.1 + 3.05 + 2 \\ &= 511.15\,\text{V} \end{aligned}$$

Problem 11

Repeat Problem 10 for short shunt compound generator.

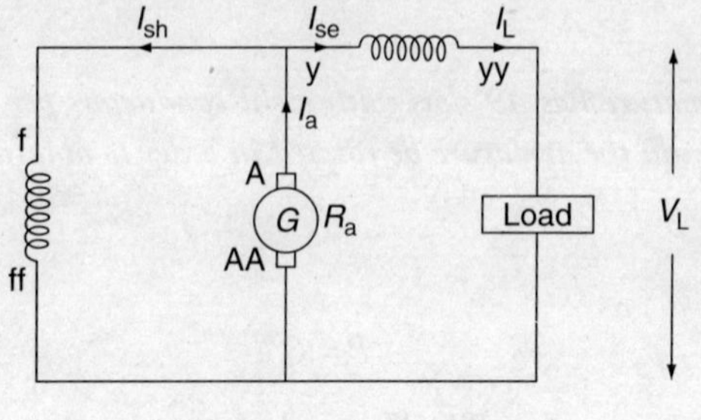

Figure 63 Solution to Problem 11.

Solution

See Fig. 63. Given

$$\begin{aligned} I_L &= 200\,\text{A} \\ V_L &= 500\,\text{V} \\ R_a &= 0.03\,\Omega \\ R_{se} &= 0.015\,\Omega \\ R_{sh} &= 150\,\Omega \\ \text{B.C.D.} &= 1 \times 2 = 2\,\text{V} \end{aligned}$$

Now

$$I_L = I_{se} = 200\ \text{A}$$

$$I_{sh} = \frac{V_2 + I_{se}R_{se}}{R_{sh}}$$

$$= \frac{500 + (200 + 0.015)}{150}$$

$$= 3.353\ \text{A}$$

$$I_a = I_L + I_{sh}$$
$$= 500 + 3.353$$
$$= 503.353\ \text{A}$$

Therefore

$$E_g = V_L + I_a R_a + I_{se}R_{se} + \text{B.C.D.}$$
$$= 500 + (503.353 \times 0.03) + (200 \times 0.015) + 2$$
$$= 500 + 15.1 + 3 + 2$$
$$= 502.1\ \text{V}$$

Problem 12

A six-pole wave wound armature has 49 slots with eight conductors per slot. Its useful flux per pole is 20 mWb. At what speed should the armature be rotated in order to obtain an e.m.f. of 500 V.

Solution
Given

$$P = 6$$

Wave wound, $A = 2$

Slip, $S = 49$

Impedance of the slip, $Z_S = 8$

$\phi = 20\ \text{mWb} = 20 \times 10^{-3}\ \text{Wb}$

e.m.f., $E = 500$ V

Now

$$Z = Z_S \times S$$
$$= 8 \times 49 = 392$$

Therefore speed is

$$N = \frac{60E}{\phi Z} \times \frac{A}{P}$$

$$= \frac{60 \times 500}{(20 \times 10^{-3})} \times 384 \times \frac{2}{6}$$

$$N = 1275.5 \text{ r.p.m.}$$

Problem 13

A 50 KVA, 6000/250 V transformer has 52 turns on the secondary. Determine the primary number of turns. Also calculate primary and secondary currents.

Solution

Given,

$$\text{Rated KVA} = 50$$

$$\text{Primary voltage } E_1 = 6000 \text{ V}$$

$$\text{Secondary voltage } E_2 = 250 \text{ V}$$

We know that

$$\frac{E_2}{E_1} = \frac{N_2}{N_1}$$

$$\Rightarrow N_1 = N_2 \times \frac{E_1}{E_2}$$

$$= \frac{52 \times 6000}{250}$$

$$= 1248 \text{ turns}$$

Also, the secondary current is

$$I_2 = \frac{\text{Rated KVA} \times 1000}{E_2}$$

$$= \frac{50 \times 1000}{250} = 200 \text{ A}$$

Primary current is

$$I_1 = \frac{\text{Rated KVA} \times 1000}{E_1}$$

$$= \frac{50 \times 1000}{6000} = 8.33 \text{ A}$$

Problem 14

A single-phase 50 Hz transformer has 350 turns on the primary and 800 turns on the secondary. The primary voltage is 400 and the cross-sectional area of the core is 750 m². Determine (i) the e.m.f. induced in the secondary and (ii) the maximum value of flux density.

Solution

Given

$$f = 50 \text{ Hz}$$
$$N_1 = 350$$
$$N_2 = 800$$
$$E_1 = 400$$

Now we know that

$$\frac{E_2}{E_1} = \frac{N_2}{N_1}$$

Therefore

$$E_2 = \frac{N_2}{N_1} E_1 = \frac{800}{350} \times 400 = 914.2 \text{ V}$$

Also,

$$E_1 = 4.44 \times f \times B_m A \times N_1$$

Given $A = 75 \times 10^{-4}$ m². Therefore

$$B_m = \frac{E_1}{4.44 \times f \times A \times N_1} \text{ Wb/m}^2$$
$$= \frac{400}{4.44 \times 50 \times 75 \times 10^{-4} \times 350} = 0.686 \text{ Wb/m}^2$$

Problem 15

A 440/200 V transformer takes a current of 0.2 A on open circuit at a power factor of 0.18. Find the magnetizing and iron loss component of current.

Solution

Given

$$E_1 = 440 \text{ V}$$

$$E_2 = 200 \text{ V}$$

$$I_0 = 0.2 \text{ A}$$

Now

$$\cos\phi_0 = 0.18$$

$$\Rightarrow \sin\phi_0 = \sqrt{1-\cos^2\phi_0} = 0.9836$$

Iron loss is given by

$$I_w = I_0 \cos\ \phi_0 = 0.036 \text{ A}$$

and magnetizing loss is

$$I = I_0 \sin\ \phi_0 = 0.1967 \text{ A}$$

Problem 16

Determine the % efficiency of single phase 20 KVA, 2200/20 transformer at full load 0.8 power factor and half full load 0.4 p.f. Given: full load copper loss = 400 W. Iron loss = 350 W.

Solution

$$\text{Rated KVA} = 20$$

$$\text{Iron loss} = 350 \text{ W}$$

$$\text{Full load copper loss} = 400 \text{ W}$$

To find % efficiency at full load 0.8 p.f.
Fraction of the load (X) = 1. Therefore,

$$\begin{aligned}\text{Output} &= \text{Rated KVA} \times 1000 \times \text{p.f.} \times (X)\\ &= 20 \times 1000 \times 0.8 \times 1\\ &= 16000 \text{ W}\end{aligned}$$

$$\begin{aligned}\text{Total loss} &= (\text{Iron loss}) + X^2(\text{Copper loss})\\ &= 350 + (1)^2 \times 400 = 750 \text{ W}\end{aligned}$$

$$\begin{aligned}\text{Input} &= \text{Output} + \text{Total loss}\\ &= 16000 + 750 = 16750 \text{ W}\end{aligned}$$

Therefore

$$\%\eta = \frac{\text{Output}}{\text{Input}} \times 100$$
$$= \frac{16000}{16750} \times 100$$
$$= 95.52$$

To find % efficiency at half full load 0.4 p.f.
Fraction of the load (X) = 0.5. Therefore

$$\text{Output} = 20 \times 1000 \times 0.4 \times 0.5$$
$$= 4000\text{ W}$$

$$\text{Total loss} = 350 + (0.5)^2 \times 400$$
$$= 450\text{ W}$$

$$\text{Input} = 4000 + 450$$
$$= 4450\text{ W}$$

Therefore

$$\%\eta = \frac{\text{Output}}{\text{Input}} \times 100 = 89.88$$

Problem 17

A three-phase, 50 Hz, four-pole induction motor runs at 1460 r.p.m. Calculate the following for this speed

(i) Synchronous speed
(ii) Slip
(iii) Frequency of rotor-induced e.m.f.

Solution
Given

$$\text{Supply frequency, } f = 50\text{ Hz}$$

$$\text{Poles, } P = 4$$

Therefore,

(i) Synchronous speed is given by

$$\text{Synchronous speed, } N_s = \frac{120f}{P}$$

$$= \frac{120 \times 50}{4}$$

$$= 1500 \text{ r.p.m.}$$

(ii) The given speed of the motor, $N = 1460$ r.p.m. Therefore,

$$\text{Slip, } S = \frac{N_s - N}{Ns}$$

$$= \frac{1500 - 1460}{1500}$$

$$= 0.026$$

$$\% \text{ Slip} = 2.66$$

(iii) The frequency of the motor-induced e.m.f. is

$$(f') = S \times f$$

$$= 0.026 \times 50$$

$$= 1.3 \text{ Hz}$$

Problem 18

A six-pole, 50 Hz, three-phase induction motor has frequency of rotor-induced e.m.f. 1.5 Hz. Determine the value of slip and speed of the motor.

Solution

Given, $P = 6$ and $f = 50$ Hz. Therefore, synchronous speed is

$$N_s = \frac{120f}{P}$$

$$= \frac{120 \times 50}{6}$$

$$= 1000 \text{ r.p.m.}$$

Given $f' = 1.5$ Hz. We know

$$f' = Sf$$

Therefore,

$$S = \frac{f'}{f} = \frac{1.5}{50} = 0.03$$

Hence slip is 0.03. The speed of the motor is given by

$$\begin{aligned} N &= N_s(1 - S) \\ &= 1000(1 - 0.03) \\ &= 970 \text{ r.p.m.} \end{aligned}$$

Problem 19

A three-phase induction motor runs almost at 1,000 r.p.m. on no load and 960 r.p.m. on full load when supplied with a 50 Hz three-phase supply. Calculate:

(i) Number of poles.
(ii) Slip at full load.
(iii) Frequency of motor e.m.f.
(iv) Speed of the motor at 8% slip.

Solution

Given, $N_s = 1000$, $N = 960$, $f = 50$ Hz. Now

$$\text{Synchronous speed, } N_s = \frac{120f}{P}$$

(i) Therefore number of poles is

$$\text{Number of poles, } P = \frac{120f}{N_s} = \frac{120 \times 50}{1000} = 6$$

(ii) Full load slip is given by

$$\text{Full load slip, } S = \frac{N_s - N}{N_s} = \frac{1000 - 960}{1000} = 0.04$$

(iii) Frequency of motor-induced e.m.f. is given by

$$f' = S \times f = 0.04 \times 50 = 2 \text{ Hz}$$

(iv) Speed of the motor at 8% slip is given by

$$\begin{aligned} N &= N_s\,(1 - S) \\ &= 1000(1 - 0.08) \\ &= 920 \text{ r.p.m.} \end{aligned}$$

Problem 20

A single-phase transformer has 120 and 60 turns, respectively, in its secondary and primary windings. The respective resistances are 0.233 Ω and 0.067 Ω. Calculate the equivalent resistance of

(i) the primary in terms of the secondary winding.
(ii) the secondary in terms of the primary winding.

Solution

Here,

$$\frac{N_2}{N_1} = n = \frac{120}{60} = 2$$

Also $R_1 = 0.067$ and $R_2 = 0.233$.

(i) Resistance of primary in terms of the secondary: Since resistance of low turns (LT) winding is to be referred to the high-turns (HT) winding, the ratio of transformation $n = 2$; hence

$$n^2 R_1 = R_1' = 4 \times 0.067 = 0.268\ \Omega$$

(ii) Resistance of secondary in terms of primary:

$$R_2' = \frac{R_2}{n^2} = \frac{0.233}{4} = 0.058\ \Omega$$

Multiple Choice Questions

1. A transformer will work on
(a) AC only
(b) DC only
(c) AC as well and DC
(d) none of the above

2. A transformer does not possess ________ changing property.
(a) impedance
(b) voltage
(c) current
(d) power

3. The iron core is used to ________ of the transformer.
(a) increase the weight
(b) provide tight magnetic coupling
(c) reduce core losses
(d) none of the above

4. The no-load ratio of a 50 Hz single-phase transformer is 6000/250 V. The maximum flux in the core is 0.06 Wb. What is the number of primary turns?
(a) 450
(b) 900
(c) 350
(d) 210

5. The stator of a three-phase induction motor produces __________ magnetic field.
(a) steady
(b) rotating
(c) alternating
(d) none of the above

6. In a three-phase induction motor, the rotor speed is __________ the synchronous speed.
(a) greater than
(b) smaller than
(c) equal to
(d) none of the above

7. The reactance of the motor circuit of a three-phase induction motor is maximum at
(a) no load
(b) full-load
(c) half full load
(d) starting

8. Most of single-phase induction motors are __________ machines.
(a) 2-pole
(b) 6-pole
(c) 8-pole
(d) 4-pole

9. A 50 Hz, 4-pole single-phase induction motor will have a synchronous speed of
(a) 1500 r.p.m.
(b) 750 r.p.m.
(c) 1200 r.p.m.
(d) none of the above

10. An alternator is sometimes called __________ generator.
(a) synchronous
(b) asynchronous
(c) Rosenberg
(d) none of the above

11. The number of cycles generated in 6-pole alternator in one revolution is
(a) 3
(b) 6
(c) 50
(d) none of the above

12. High-speed alternators are driven by
(a) diesel engine
(b) hydraulic turbines
(c) steam turbines
(d) none of the above

Review Questions

1. Explain the concept of electrical machines.
2. Mention the classification of AC machines.
3. Define DC machine.
4. Compare between DC machine and AC machine.
5. Write the principle of operation of DC generator.
6. Write the advantages and disadvantages of DC generator.
7. What is a DC motor?
8. Write the working principle and operation of DC motor.
9. Define transformer.
10. Compare between static and dynamic transformers.
11. Mention the advantages and applications of transformer.

12. Explain the torque–slip characteristics of induction motor.

13. Describe the concept and working methodology of synchronous motor.

14. Write the principle of operation of single-phase induction motor.

15. Mention the different types of rotor in three-phase induction motor.

16. Explain the constructional features of DC machines.

17. Explain the construction and operation of three-phase squirrel cage induction motor.

18. Define synchronous motor.

19. Write the short motes on single phase induction motor.

20. List the applications of single phase induction motor.

Problems

1. A 2000/200 V, 20 KVA transformer has 66 turns in the secondary. Calculate
   - (a) Primary turns
   - (b) Primary and secondary full-load currents. Neglect the losses.

2. An ideal 25 KVA transformer has 500 turns on the primary winding and 40 turns on the secondary winding. The primary is connected to 3000 V, 50 Hz supply. Calculate
   - (a) Primary and secondary currents on full load.
   - (b) Secondary e.m.f.
   - (c) The maximum core flux.

3. A single-phase 2200/250 V, 50 Hz transformer has a net core area of 36 cm^2 and a maximum flux density of 6 Wb/m^2. Calculate the number of turns of
   - (a) The primary.
   - (b) The secondary.

4. A 230/2300 V transformer takes no-load current of 5 A of 0.25 power factor lagging. Find
   - (a) Iron loss.
   - (b) Magnetizing loss component of current.

5. A six-pole, three-phase induction motor is connected to 50 Hz supply. If it is running at 970 r.p.m., find the slip.

6. A three-phase induction motor is wound for 4 poles and is supplied from 50 Hz system. Calculate:
   - (a) Synchronous speed.
   - (b) Speed of the motor when slip is 4%.
   - (c) Rotor current frequency when the motor runs at 600 r.p.m.

7. The stator of a three phase induction motor has 6 slots per pole per phase. If supply frequency is 50 Hz, calculate
   - (a) number of stator poles produced and total number of slots on the stator,
   - (b) speed of the rotating stator flux (or magnetic field).

8. A 220 V, 50 Hz, 2-pole, three -phase squirrel cage induction motor develops a torque of 100 N-m at a speed of 1800 r.p.m. If the stator supply frequency is reduced by half, calculate
   - (a) the stator supply voltage required for maintaining the same flux in the machine,
   - (b) the new speed at a torque of 100 Nm.

9. A 4-pole, three-phase induction motor runs at 1200 r.p.m on a 50 Hz supply. Calculate slip speed and slip.

10. A 2 pole, three phase induction motor runs at 2940 r.p.m on a 50 Hz supply. Find the
    (a) Slip.
    (b) Frequency of rotor emf.

11. A 6 pole, three phase induction motor operating on a 50 Hz supply has rotor emf frequency as 2 Hz. Determine
    (a) Slip.
    (b) The rotor speed.

12. The core of 1000 KVA , 11000/550 V, 50 Hz, single phase transformer has a cross section of 20 cm × 20 cm. If the maximum core density is not exceed 1.3 Tesla, calculate
    (a) The number of primary and secondary turns per phase.
    (b) The e.m.f per turn.
    (c) Assume a stacking factor of 0.9.

13. Determine the regulation of a transformer in which Ohmic loss is 1% of the output and the resistance drop 5% of the voltage when power factor is
    (a) 0.8 lag.
    (b) Unity.
    (c) 0.8 lead.

14. Calculate the value of R_0, X_0, R_t and X_t in Fig. 64 for equivalent circuit of a single phase 4 KVA, 200/400 V, 50 Hz transformer of which the following are the test results:
    (a) Open circuit (o.c.) test: 200 V, 0.7 A, 70 W on low voltage primary side
    (b) Short circuit (s.c.) test: 15 V, 10 A, 80 W on HT side

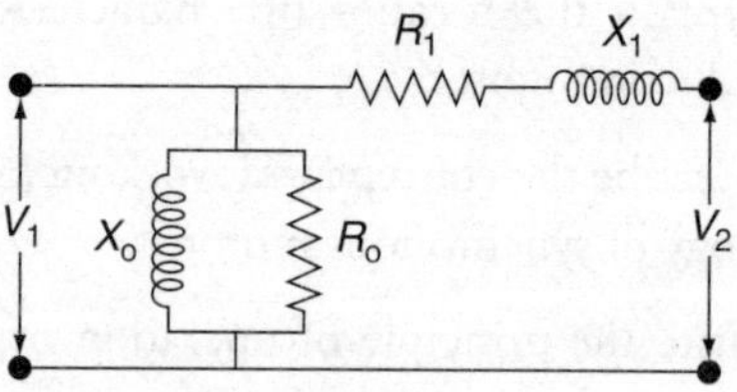

Figure 64 Problem 14.

15. The maximum efficiency of a 500 KVA, 3300/500 V, 50 Hz, single-phase transformer is 97% and occurs at 3/4th full load unity power factor. If the impedence is 10%, calculate the regulation at full load, power factor is 0.8 lag.

16. In a 25 KVA, 2000/200 V transformer the iron and copper losses are 350 and 400 W, respectively. Calculate the efficiency on unity power factor at
    (a) Full load.
    (b) Half load.

    Determine the load for maximum efficiency and the iron and copper losses in this case.

17. Calculate the efficiencies at
    (a) Half.
    (b) Full.
    (c) 1/4 load of a 100 KVA transformer for power factor of unity.

    The copper loss is 1000 W at full load and the iron loss is 1000 W.

18. A 6-pole, two circuit wave-connected armature has 250 conductors and runs at 1200 r.p.m. The electromotive force generated on open circuit is 600 V. Determine the useful flux per pole.

19. An 8-pole lap-connected armature has 960 conductors with a flux of 40 mWb per pole and a speed of 400 r.p.m. Determine the e.m.f. generated.

20. A shunt generator has an induced voltage on open circuit of 127 V. When the machine is on load the terminal voltage is 120 V. Find the load current if the field circuit resistance is 15 Ω and the armature resistance is 0.02 Ω. Ignore armature reaction.

21. Find the useful flux per pole on load of a 250 V, 6-pole shunt motor having wave-connected armature winding with 110 turns. The armature resistance including brush is 0.2 Ω. The armature current is 13.3 A at no load speed of 908 r.p.m.

22. A direct current shunt motor develops 10 bp at 600 r.p.m. when drawing line current of 18 A at 500 V. Find
    (a) The efficiency at this load.
    (b) The useful torque.

23. A series motor of resistance 1 Ω between terminals runs at 800 r.p.m. at 200 V with a current of 15 A. Determine the speed at which it will run when connected in series with a 5 Ω resistance and taking the same current at same supply voltage.

24. The armature of a 2-pole 200 V wave wound generator has 300 conductors and runs at 300 r.p.m. Calculate the useful flux per pole.

25. A 4-pole lap wound DC generator has 230 armature conductors, a flux of 0.04 Wb per pole and runs at 1200 r.p.m. Find the generated e.m.f.

26. A 4-pole lap wound DC generator has 1000 armature conductors, flux of 20 mWb per pole and e.m.f. generated is 400 V. What is the speed of the machine?

27. A 4-pole DC generator has 96 slots and 16 conductors per slot. The flux per pole is 40 mWb and the speed is 960 r.p.m. Find the e.m.f. produced if the machine is
    (a) Wave wound.
    (b) Lap wound.

28. A 2-pole wave wound DC generator has 200 coils of 10 turns each. The speed is 200 r.p.m. and resistance of armature turn is 0.02 Ω. Find
    (a) The e.m.f. produced.
    (b) The resistance of armature winding, if the flux per pole is 0.05 Wb.

Answers

Multiple Choice Questions

1. (a)
2. (d)
3. (b)
4. (a)
5. (b)
6. (b)
7. (d)
8. (d)
9. (a)
10. (a)
11. (a)
12. (c)

Problems

1. **(a)** 660 turns , **(b)** 100 A , **(c)** 10 A
2. **(a)** 8.33 A, 104.125 A, **(b)** 240 V, **(c)** 27 mWb
3. **(a)** 459 turns, **(b)** 52 turns
4. **(a)** 1.25 A, **(b)** 4.84 A
5. 0.03 or 3%
6. **(a)** 1500 r.p.m., **(b)** 1440 r.p.m., **(c)** 30 Hz
7. **(a)** 12 poles, 216; **(b)** 500 r.p.m

8. **(a)** 110 V, **(b)** 2700 r.p.m,
9. **(a)** 300 r.p.m, **(b)** 0.2
10. **(a)** 0.02, **(b)** 1 Hz
11. **(a)** 0.04, **(b)** 960 r.p.m
12. **(a)** 1059 turns, 53 turns , **(b)** 10.387 V
13. **(a)** 3.8%, **(b)** 1%, **(c)** –2.27%
14. **(a)** O.C. test: $R_0 = 571\ \Omega$, $X_0 = 333\ \Omega$, $R_t = 0.8\ \Omega$ and $X_t = 1.269\ \Omega$; **(b)** S.C. $R_t = 0.2\ \Omega$ and $X_t = 0.317\ \Omega$
15. 7.51%
16. **(a)** 97.08%, **(b)** 96.52%; Load for maximum efficiency = 23.38 KVA, losses = 350 W each
17. **(a)** 97.56%, **(b)** 98.04%, **(c)** 97.99%;
18. 0.04 Wb
19. 256 V
20. 342 A
21. $\phi = 24.8$ mWb
22. **(a)** 0.817 or 81.7%, **(b)** 117 Nm
23. 476 r.p.m.
24. $\phi = 1.08$
25. 184 V
26. 1200 r.p.m.
27. **(a)** 1966.08 V, **(b)** 983.04 V
28. **(a)** 666.66 V, **(b)** 10 Ω

8 Measuring Instrument

LEARNING OBJECTIVES

After reading this chapter, you will be able to:

- Understand the importance of analog instruments.
- Understand the concept of permanent magnet type moving coil and moving iron instrument.
- Understand the construction of registering system.
- Understand the construction and operation of megger.

8.1 Introduction

Test and measurements are important in designing, evaluating, maintaining and servicing electrical circuits and equipments. In order to detect electrical quantities such as current, voltage, resistance or power, it is necessary to transform an electrical quantity or condition into a visible indication. This is done with the aid of instruments that indicate the magnitude of quantities by the position of a pointer moving over a graduated scale.

8.2 Classification

Electrical measuring instruments are broadly classified into the following three types:

1. Depending on the quantity measured: for example, voltmeter, ammeter, wattmeter, energy meter, ohmmeter.
2. Depending on the different principles used for their working: for example, moving iron type, moving coil type, dynamometer type, induction type.
3. Depending on how the quantity is measured: for example, deflecting type, integrating type, recording type.

8.3 Principle of Analog Instrument

Analog instrument like ammeter, voltmeter and wattmeter consist essentially of a moving system with a pointer attached to it. For satisfactory operation, the moving system is pivoted by jeweled bearing and is subjected to the following three distinct torques:

1. deflecting torque;
2. controlling torque;
3. damping torque.

8.3.1 Deflecting Torque

The torque that causes the moving system and the pointer of the instrument to get the required deflection is called "deflecting torque." That is, a mechanical force is produced by the current or voltage which causes the pointer to deflect from its zero position. The amount of deflection is proportional to the quantity to be measured.

8.3.2 Controlling Torque

The torque which controls the movement of the moving system and the pointer over the scale is called the "controlling torque." That is, the controlling force acts in opposition to the deflecting force and ensures that the deflection shown on the meter is always the same for a given measured quantity. It also prevents the pointer from always going to the maximum deflection. The controlling torque is usually provided by any one of the following two types:

1. **Spring control:** In this method, two helical springs, spiraled in opposite directions, are used to develop a controlling torque. The use of two springs avoids errors due to temperature variations. The springs are made up of phosphor–bronze. One end of the spring is attached to a stationary point on the body of the instrument whereas the other end is attached to the moving system as shown in Fig. 1. In this type, if T_c is the controlling torque and θ is the angular deflection of the instrument, then

$$T_c \propto \theta$$

$$T_c = C_s\theta \text{ N m} \tag{8.1}$$

   where C_s is the spring constant in Nm per unit deflection.

2. **Gravity control:** Figure 2 shows the arrangement for gravity control. In this method, a small weight W_c, called control weight, is attached to the moving system and produces the controlling torque due to the gravitational force acting on it when there is deflection in the system.

$$T_c = W_c I \sin\theta$$

or

$$T_c \propto \sin\theta$$

   As the movement of the control weight must lie in the vertical plane, the instrument must lie always, mounted with their dials vertical.

8.3.3 Damping Torque

The damping force ensures that the pointer comes to rest in its final position quickly and without undue oscillation. There are three main types of damping used – eddy-current damping, air-friction damping and fluid-friction damping.

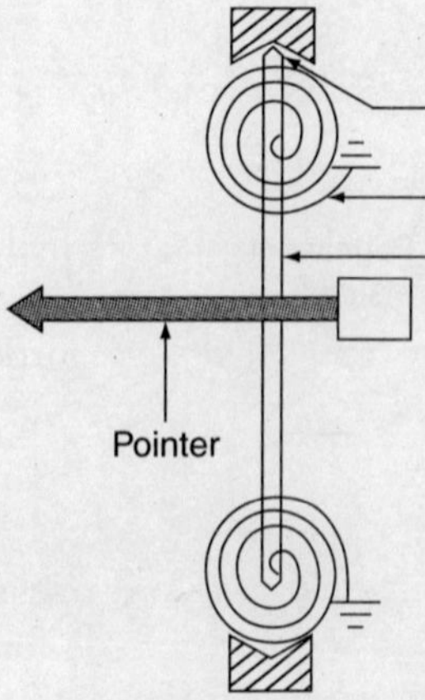

Figure 1 Spring control.

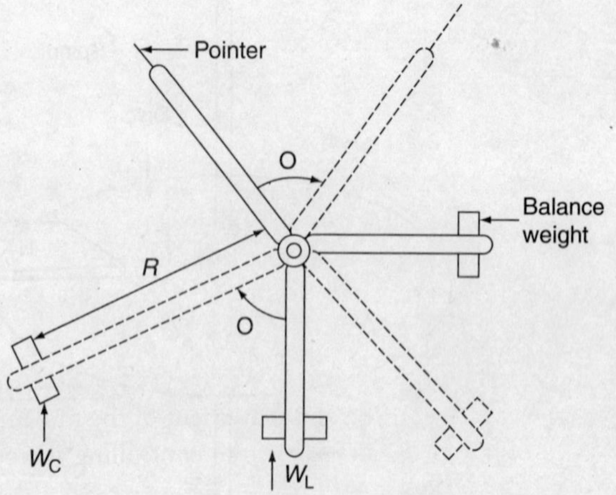

Figure 2 Control weight.

8.3.3.1 Air-Friction Damping

In this method, a light aluminum piston, which is attached to the moving system, moves in an air chamber closed at one end as shown in Fig. 3. There is very little clearance between the piston and the air chamber. When the spindle moves due to a deflecting torque, the piston also moves along with it, say out of the cylinder, the air pressure inside the cylinder falls and thus the outward movement of the piston is opposed by the outside air at atmospheric pressure. Similarly, when the piston is moved into the cylinder, the air trapped inside the chamber is compressed and it opposes the movement. Thus, the section and compression action on the air inside the chamber produces the necessary damping torque and thereby the oscillation of the power is arrested.

8.3.3.2 Eddy-Current Damping

In eddy-current damping, a thin disc of conducting but nonmagnetic material such as aluminum or copper is attached to the instrument spindle as shown in Fig. 4. The disc is allowed to move along

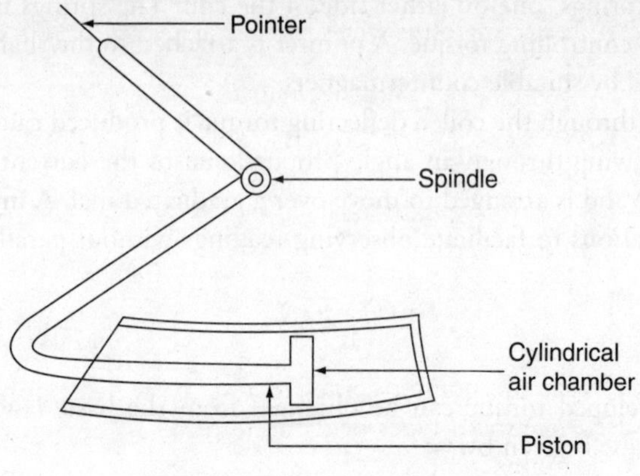

Figure 3 Air-friction damping.

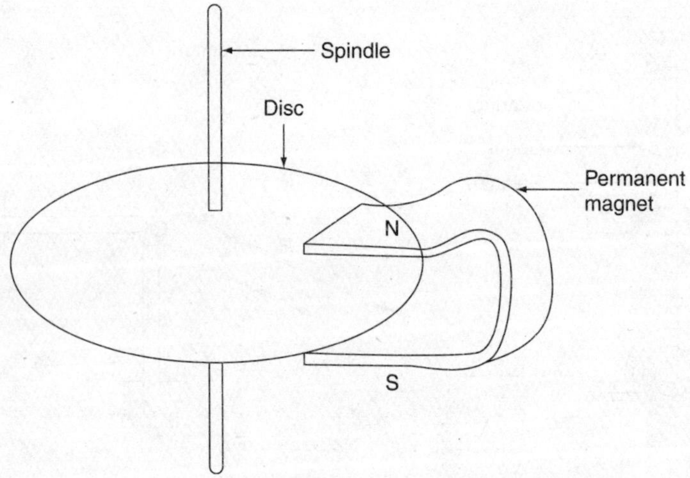

Figure 4 Eddy-current damping.

with the spindle inside the magnetic field provided by a permanent magnet. The movement of the disc causes an eddy e.m.f. to get induced in it causing an eddy current to flow. The flux set up by this eddy current interacting with the magnetic field exerts a force on the disc. The force, according to Lenz's law, opposes the movement of the disc. Thus, necessary damping torque is provided.

8.4 Moving Coil Instruments

8.4.1 Permanent Magnet Type

Figure 5 shows the main parts of a permanent magnet type moving coil instrument, which consists of a coil of very thin insulated copper wire wound on an aluminum former and placed in a field of a strong horseshoe-shaped permanent magnet.

The aluminum former carrying the coil is fitted to the shaft. The shaft ends are housed in jewel bearing to facilitate frictionless movement of the former together with the coil. The current to the coil is fed through two spiral springs, one on either side of the coil. The springs in addition to serving as lead wires also provide the controlling torque. A pointer is attached to the shaft and the entire moving system is carefully balanced by suitable countermagnets.

When a current passes through the coil, a deflecting torque is produced causing the moving system, including the pointer, to swing through an angle proportional to the current. The pointer is usually provided with a knife edge and is arranged to move over a graduated dial. A mirror is also provided on the dial besides the graduations to facilitate observing readings without parallax error. The deflecting torque T_d is given by

$$T_d \propto I \tag{8.2}$$

Torque Equation:

The equation for the developed torque can be obtained from the basic law of the electromagnetic torque. The deflecting torque is given by,

$$T_d = NBAI$$

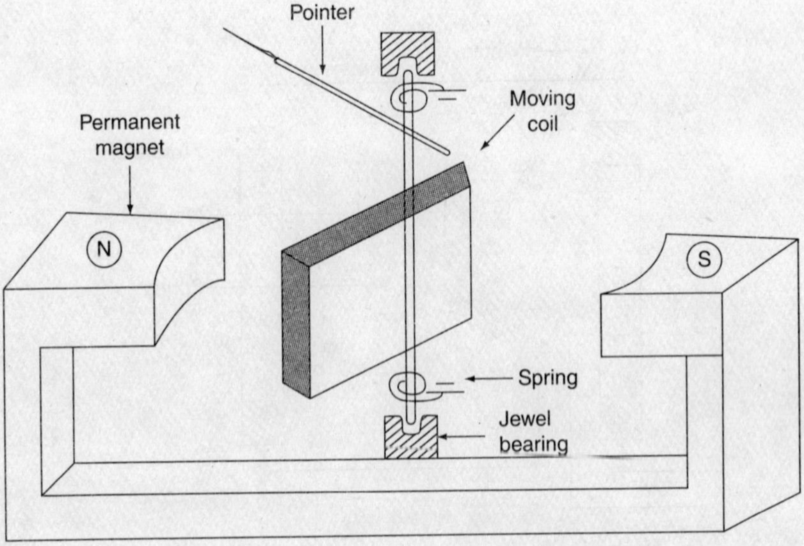

Figure 5 Main parts of a permanent magnet type moving coil instrument.

where T_d is the deflecting torque in Nm; B is the flux density in air gap, Wb/m^2; N is the number of turns of the coil; A is the effective coil area m^2; I = current in the moving coil, amperes. Therefore

$$T_d = GI$$

where

$$G = NBA = \text{constant}$$

The controlling torque is provided by the springs and is proportional to the angular deflection of the pointer.

$$T_c = K\theta$$

where TC is the controlling torque; K is the spring constant, *Nm*/*rad*. For the final steady-state position

$$T_d = T_c$$

$$GI = K\theta$$

$$\theta = (G/K)\, I$$

or

$$I = (K/G)\, \theta$$

Therefore

$$\theta \propto I$$

Thus the deflection is directly proportional to the current passing through the coil. The pointer deflection can therefore be used to measure current. Therefore, scale graduation in moving coil instruments will be uniform throughout the range. The instrument cannot be used is AC circuits.

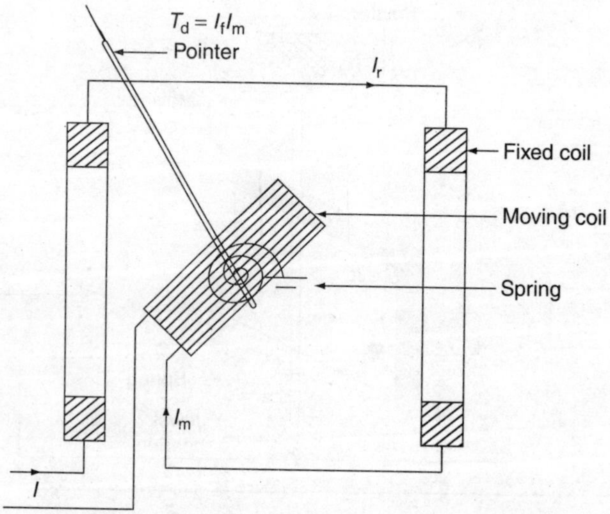

Figure 6 Dynamometer type moving coil instrument.

8.4.2 Dynamometer Type Moving Coil Instrument

In this type, the operating magnetic field is produced by air-cored electromagnets excited by two fixed coils as shown in Fig. 6. In all other aspects, they are similar to the permanent magnet moving coil instrument.

The force on the moving coil is proportional to the flux produced by the fixed coils (and hence the current) through it, I_f, and the current in the moving coil, I_m, the deflecting torque T_d. Mathematically,

$$T_d = I_f I_m$$

Therefore, it is clear from the above equation that the meter deflection is not affected when the currents through the coils are reversed. Thus, this type of instrument can be used in both AC and DC circuits.

8.5 Moving Iron Instrument

There are two types of moving iron instruments, viz., (a) attraction type and (b) repulsion type. We will discuss both of them next.

8.5.1 Attraction Type

In this type of instrument (Fig. 7), a stationary field coil carries the current proportional to the current to be measured in case of ammeter or voltage to be measured in case of voltmeter. The current-carrying coil produces a magnetic field inside it. A small iron piece placed nearby and fitted to the spindle of the instrument gets attracted by magnetic field and turns toward the coil. This causes the spindle and the pointer attached to it to get the required deflection. The pointer moves over a graduated scale.

As the force of attraction due to an electromagnet varies in proportion to square of the flux density, the deflecting torque in this type of instrument is directly proportional to the square of the current through the field coil. Mathematically,

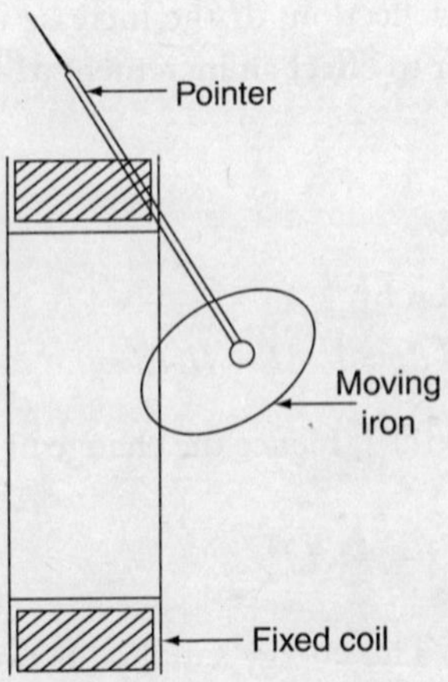

Figure 7 Attraction type moving iron instrument.

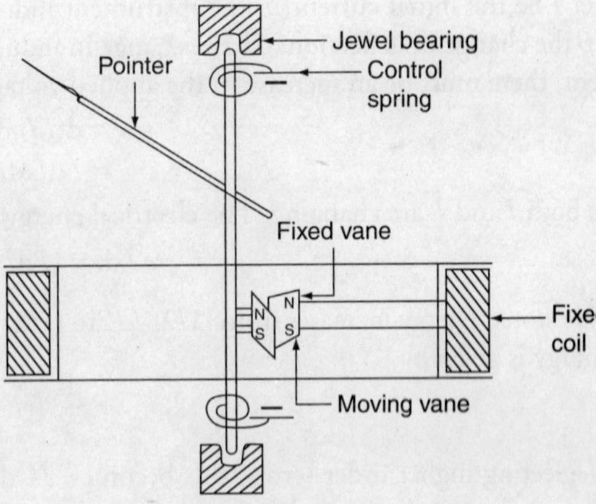

Figure 8 Repulsion type moving iron instrument.

$$T_{\text{d}} \propto I^2$$

Hence, moving iron instruments can be used in both DC and AC circuits.

8.5.2 Repulsion Type

The principle of this instrument is that two iron pieces kept within close proximity in a magnetic field get magnetized to the same polarity and exert a force of repulsion on each other.

As seen in Fig. 8 there are two vanes of iron inside the field coil. One of the vane in rigidly mounted to the frame of the instrument and the other vane is attached to the spindle.

When a current flows through the coil, both the vanes are magnetized with the same polarity and then a force of repulsion between them results. As a result of this force of repulsion, the movable vane rotates causing a corresponding deflection in the pointer. It is evident that in this case too that the deflecting torque is directly proportional to the square of the current:

$$T_{\text{d}} \propto I^2$$

8.5.2.1 Torque Equation of Moving Iron Instruments

Consider a small increment in current supplied to the coil of the instrument. Due to the current let $d\theta$ be the deflection under the deflecting torque T_{d}. Due to such deflection, some mechanical work will be done.

$$\text{Mechanical work} = c$$

There will be change in the energy stored in the magnetic field due to the change in inductance. This is because the vane tries to occupy the position of minimum reluctance hence the force is always in such a direction so as to increase the inductance of coil. The inductance is inversely proportional to the reluctance of the magnetic circuit of coil.

Let I be the initial current; L the instrument inductance; θ the deflection; dI the increase in current; dθ the change in deflection; dL the change in inductance. In order to effect an increment dI in the current, there must be an increase in the applied voltage is given by

$$e = \mathrm{d}(LI)/\mathrm{d}t$$
$$= I\,\mathrm{d}L/\mathrm{d}t + L\,\mathrm{d}I/\mathrm{d}t$$

as both I and L are changing. The electrical energy supplied is given by

$$e\,I\,\mathrm{d}t = (I\,\mathrm{d}L/\mathrm{d}t + L\,\mathrm{d}I/\mathrm{d}t)\,I\,\mathrm{d}t = I^2\,\mathrm{d}L + IL\,\mathrm{d}I$$

The stored energy increases from (1/2) LI^2 to (1/2) $[(L + \mathrm{d}L)\,(I + \mathrm{d}I)^2]$. Hence the change in the stored energy is given by

$$\frac{1}{2}\,[(L + \mathrm{d}L)\,(I + \mathrm{d}I)^2] - \frac{1}{2}\,LI^2$$

Neglecting higher order terms, this becomes, $IL\,\mathrm{d}I + 1/2\,I^2\,\mathrm{d}L$. The energy supplied is nothing but increase in stored energy plus the energy required for mechanical work done:

$$I^2\,\mathrm{d}L + IL\,\mathrm{d}I = IL\,\mathrm{d}I + \frac{1}{2}\,I^2\,\mathrm{d}L + T_\mathrm{d}\,\mathrm{d}\theta$$

$$T_\mathrm{d}\,\mathrm{d}\theta = \frac{1}{2}\,I^2\,\mathrm{d}L$$

$$T_\mathrm{d} = \frac{1}{2}\,I^2\,\frac{\mathrm{d}L}{\mathrm{d}\theta}$$

The controlling torque is given by

$$T_\mathrm{c} = K\theta$$

where K is the spring constant. Now

$$K\theta = \frac{1}{2}\,I^2\,\frac{\mathrm{d}L}{\mathrm{d}\theta}$$

$$\theta = \frac{1}{2}\left(\frac{I^2}{K}\right)\frac{\mathrm{d}L}{\mathrm{d}\theta}$$

Thus the deflection is proportional to the square of the current through the coil. And the instrument gives square law response. Hence the instrument can be used in both DC and AC circuits.

8.6 Use of Voltmeters and Ammeters

8.6.1 Permanent Magnet Type Moving Coil Ammeters

An ammeter, being an instrument to be connected in series with the circuit, must have a low resistance so that it will not cause a voltage change in it. The control springs that take the current to the moving coil are capable of carrying the current to a few milliampere, and in some cases to microamperes. Therefore, if the moving coil instrument has to be used as an ammeter capable of measuring a large current, it should be used in conjunction with suitable shunts as shown in Fig. 9.

A shunt is, essentially, a low resistor usually made of magnesium strips so that its resistance practically remains constant irrespective of temperature changes. It is connected in parallel with the moving coil meter.

Let R_m be the meter resistance, R_sh the resistance of the shunt, I_m the current through the meter and I_sh be the current through the shunt. Then current to be measured is

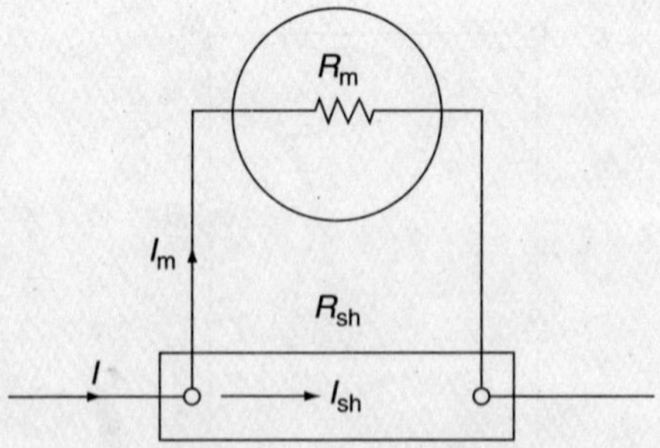

Figure 9 Permanent magnet type moving coil ammeter.

$$I = I_m + I_{sh} \tag{8.3}$$

Also,

$$I_m R_m = I_{sh} R_{sh}$$

Therefore

$$R_{sh} = \frac{I_m}{I_{sh}} \cdot R_m$$

Using Eq. (8.3) we get

$$R_{sh} = \frac{I_m}{(I - I_m)} \cdot R_m \tag{8.4}$$

$$R_{sh} = R_m/[(I/I_m) - 1]$$

The ratio of the total current to the instrument current (i.e., I/I_m) is known as the multiplying factor denoted by η

$$\eta = I/I_m$$

Therefore, shunt resistance,

$$R_{sh} = R_m/(\eta - 1)$$

Equation (8.4) gives the value of shunt resistance to be used in parallel with the meter for measuring a maximum current of I ampere.

8.6.2 Permanent Magnet Type Moving Coil Voltmeters

A voltmeter should have a large resistance so that its power consumption, when connected across the supply line, should be as low as possible. Therefore, a moving coil instrument can be used as a voltmeter by connecting a high resistance in series with the instrument as shown in Fig. 10. The value of the external resistance R_e must be adjusted so that the total resistance limits the current within maximum meter current I_m. That is

$$R_t = (R_m + R_e) = \frac{V}{I_m}$$

where V denotes voltage and R_t denotes total resistance.

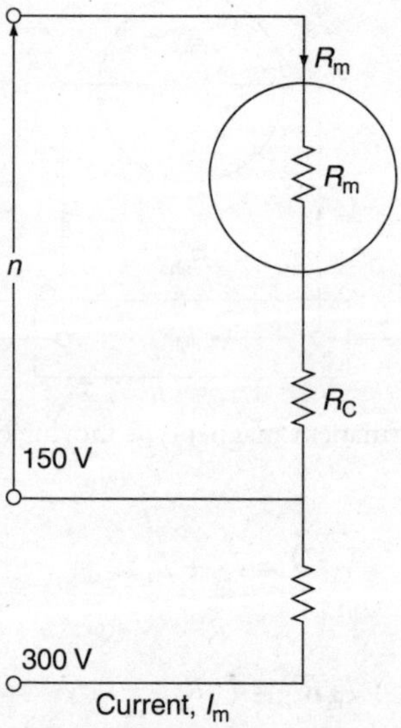

Figure 10 Permanent magnet type moving coil voltmeter.

8.6.3 Dynamic Type Instruments

When used as an ammeter, the stationary coil carries most of the current to be measured and is wound with thick wires. The moving coil carries a definite fraction of the current to be measured as shown in Fig. 11(a). When used as a voltmeter, all the coils carry the same amount of current which is proportional to voltage to be measured. The coil current can be limited by adding a high resistance in series with the instrument coils as shown in Fig. 11(b).

8.6.4 Moving Iron Instruments

Except for the field coil, there is essentially no difference in the construction of the other parts between ammeters and voltmeters. As ammeters are required to be of very low resistance, the exciting coil is wound with wires of large cross-section and contain only few turns. In voltmeters the required amp-turns are produced by passing a small current through a large number of turns. Hence, the coil is wound with thin wires and large number of turns, thereby giving the required higher coil resistance.

8.7 Dynamometer Type Wattmeter

Wattmeter measures the real power, P, in a circuit. In DC circuits $P = VI$ and it may be measured either by wattmeter or by an ammeter and voltmeter and then their product.

In AC circuits the real power at any instance is given as $P = VI$, and the average power is given by $P = VI \cos \phi$ which can be measured by wattmeter only. In AC circuits, both inductance and capacitance having lagging and leading power factor for that purpose the power $P = VI \cos \phi$. The dynamometer type wattmeter measures real power both in AC and DC circuits.

The dynamometer type wattmeter is similar in construction to dynamometer type moving coil instrument discussed in Section 8.3. The two fixed coils carry the current in the circuit and are known as current

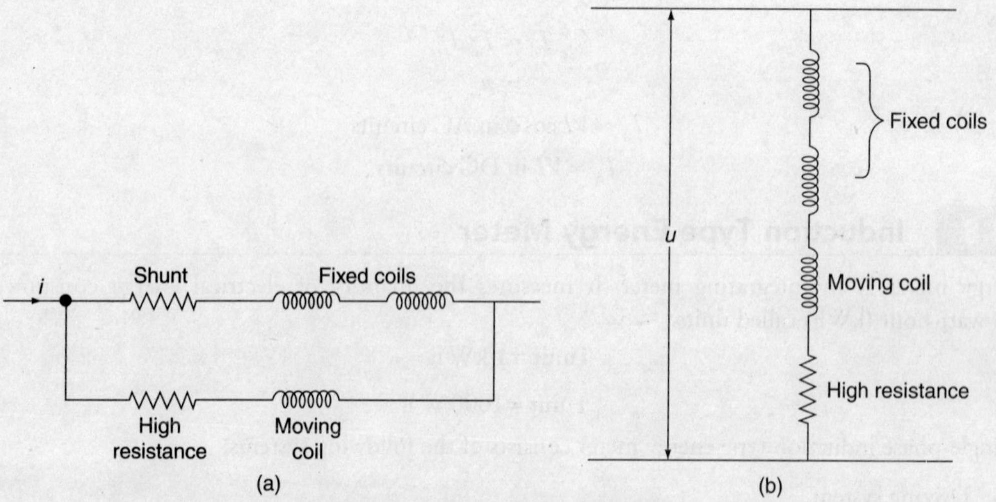

Figure 11 Dynamometer instrument connections: (a) Dynamometer ammeter; (b) dynamometer voltmeter.

coils. The moving coil carries a current proportional to the voltage in the circuit and the coil is called a pressure coil. Figure 12(a) shows the coil connection and Fig. 12(b) shows the schematic connection.

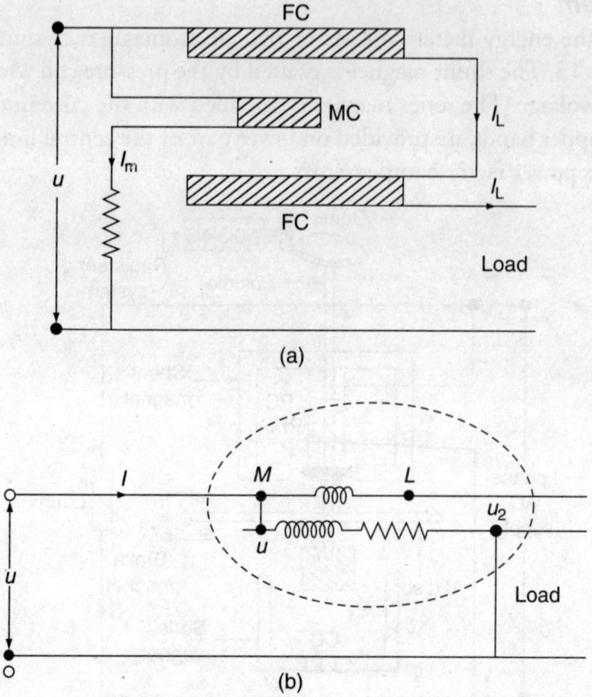

Figure 12 Dynamometer type wattmeter: (a) Instrument coil connections; (b) schematic connection.

Now

$$T_d \propto I_m I_t$$

$$\propto u_t$$

$$T_d \propto VI \cos \phi \text{ in AC circuits}$$

$$T_d \propto VI \text{ in DC circuits}$$

8.8 Induction Type Energy Meter

Energy meter is an integrating meter. It measures the quantity of electrical energy consumed in kilo-watt-hour (kWh) called units.

$$1 \text{ unit} = 1 \text{ kW h} \tag{8.5}$$

$$1 \text{ unit} = 1000 \text{ W h} \tag{8.6}$$

A single-phase induction type energy meter consists of the following systems:

1. Driving system.
2. Moving system.
3. Braking/damping system.
4. Registering system.

We explain each of them next.

8.8.1 Driving System

The driving system of the energy meter consists of two electromagnets, a shunt magnet and a series magnet as shown in Fig. 13. The shunt magnet is excited by the pressure coil which carries current proportional to the supply voltage. The series magnet is provided with the current coil excited by the load current. Short circuit copper bands are provided on lower part of the central limb of the shunt magnet. These loops are called as power factor compensators.

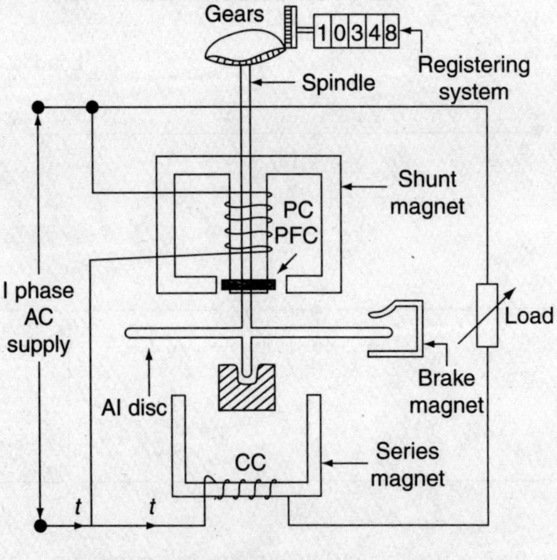

Figure 13 Induction type energy meter.

8.8.2 Moving System

The moving system consists of a thin aluminum disc mounted on the spindle. The disc is placed in the air gap between the shunt and series magnets.

The current proportional to supply voltage in the shunt magnet produces a flux Φ_{sh} which in turn produces an eddy current i_{sh} in the aluminum disc. Similarly the load current in the series magnet sets up a flux which produces an eddy current in the disc. The interaction of these fluxes and eddy current causes a torque developed in the disc and it starts rotating. The torque developed and hence the speed of rotation is proportional to the real power taken by the load circuit.

8.8.3 Braking/Damping System

The braking system consists of a permanent magnet called as brake magnet. It is placed near the edge of the aluminum disc. The brake magnet provides eddy-current damping to the moving system. The amount of braking on damping torque can be adjusted by adjusting the position of the magnet.

8.8.4 Registering System

The spindle is connected through gears to a counting mechanism. This mechanism keeps on registering the number of revolutions of the disc calibrated in kWh energy consumption.

8.9 Megger

Megger is the most portable insulation tester. It is used to measure very high resistances of the order of megaohms. The instrument works on the principle of ratiometer or ohmmeter. The required deflecting torque is produced by both the system voltage and the current. Due to interaction between the magnetic fields produced by the voltage and the current, the deflecting torque is produced. The required coils are so positioned that the deflecting torque is proportional of the ratio *VI*.

8.9.1 Construction

Figure 14 shows the constructional details of a megger which is similar to the permanent magnet type moving coil instrument. It has a horseshoe-shaped permanent magnet.

The moving system consists of two coils, pressure coil PC and current coil CC. The two coils are fitted to the spindle at a fixed angle. A pointer attached to a spindle can move over an Ohmic scale graduated from 0 to ∞ (infinity).

The necessary voltage for resistance measurement (usually 250 or 500 V) is generated by means of a small permanent magnet DC generator provided within the instrument and is hand driven. The pressure coil or control coil is connected across the armature through a current-limiting resistor R_p. The current coil or deflecting coil is connected to the armature through a current-limiting resistor R_c and reset probe terminals.

8.9.2 Operation

With the test probes in open position, if the generator handle is rotated at full speed, the pressure or controlling coil alone gets the current. The current-carrying pressure coil sets itself perpendicular to the magnetic field provided by the permanent magnet and the pointer shows the reading as infinity marked on one end of the scale.

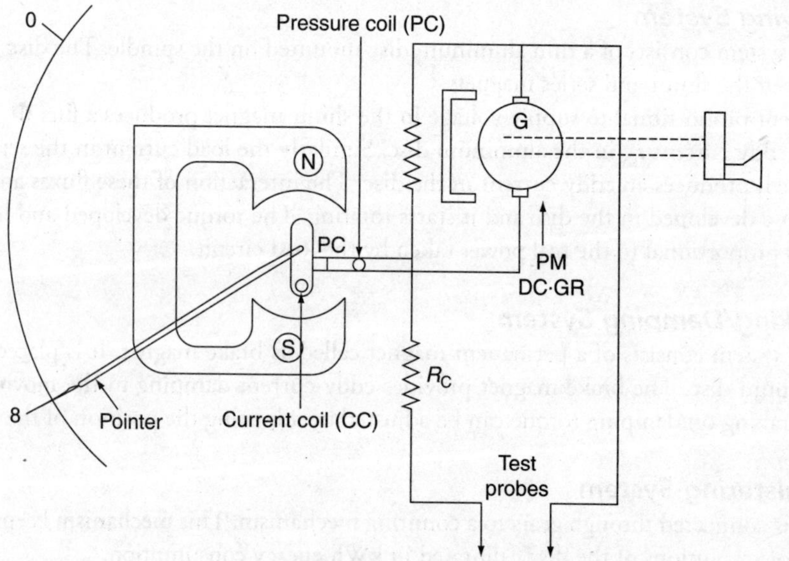

Figure 14 Megger.

When the generator handle is rotated slowly with test probes, if the generator handle is rotated at full speed, the moving system gets a deflection proportional to the ratio of voltage to current in the insulation resistance.

With sufficient insulation resistance available across the test probes, if the generator handle is rotated at full speed, the moving system gets a deflection proportional to the ratio of voltage to current in the insulation resistance.

8.10 Measurement Errors

A measuring instrument is a device used for comparing the unknown quantity with the unit of measurement or a standard quantity. The most important static characteristic of an instrument is its accuracy, which is generally expressed in terms of an error called the static error. Static Errors are classified in to three types, namely,

1. Absolute error.
2. Relative error.
3. Limiting error.

We will discuss the three on them next.

8.10.1 Absolute Error

The value of the unknown quantity obtained on taking measurements with standards and measuring instruments is considered to be its true value. There is always some difference, small or large, between the measured value and the true or exact value of the unknown quantity. The difference between the measured value (A_m) and the true value (A_t) of the unknown quantity is known as the absolute error of measurement, δ_a. Mathematically,

$$\delta_a = A_m - A_t$$

8.10.2 Relative Error

The absolute value of error δa does not indicate precisely the accuracy of measurements. For example, an absolute error of 1 cm is infinitesimal in determining the circumference of earth, negligible in setting out a cricket pitch, but in tolerable in a metre stick. The quality of measurement is, therefore, indicated preferably in terms of relative error. The relative error (t_r) is the ratio of absolute error (δ_a) to the true value (A_t) of the quantity to be measured. Mathematically,

$$t_r = \frac{\delta_a}{A_t} = \frac{A_m - A_t}{A_t}$$

The percentage relative error is expressed as

$$\% t_r = \frac{A_m - A_t}{A_t} \times 100$$

8.10.3 Limiting Error

The manufacturers specify the accuracy of the instruments within a certain percentage of full-scale reading. Components such as resistors, inductors, capacitors are guaranteed to be within a certain percentage of rated value. This percentage indicates the deviations from the nominal or specified value. These deviations are called limiting errors. These are also called guarantee errors.

Example 1

The manufacturer of a certain instrument may specify that the instrument is accurate within ± 1% of full-scale deflection. This means that a full-scale reading is guaranteed to be within ± 1% of a perfectly accurate reading. But for a reading less than full scale, the limiting error increases.

Example 2

Let a resistor be specified by a manufacturer as 4.7 kΩ with a tolerance of ± 5%. Then the actual value of the resistance is guaranteed to be within the limits. Now

$$\begin{aligned} R &= 4.7\ \text{k}\Omega \pm (5\%\ \text{of}\ 4.7\ \text{k}\Omega) \\ &= 4.7\ \text{k}\Omega \pm 0.235\ \text{k}\Omega \\ &= 4.935\ \text{k}\Omega\ \text{and}\ 4.465\ \text{k}\Omega \end{aligned}$$

Thus the actual value with the limiting error can be expressed mathematically as

$$A_a = A_s \pm \delta_a$$

where A_a is the actual value, A_s the specified or rated value, δ_a the limiting error or tolerance.

Relative limiting error is also called fractional error. It is the ratio of the error to the specified magnitude of a quantity. Thus,

$$e = \delta_a / A_s$$

where e is the relative limiting error. From this equation, we can write

$$\delta_a = e \times A_s$$

$$A_a = A_s \pm \delta_a = A_s \pm (e \times A_s) = A_s\,[1 \pm e]$$

The percentage relative limiting error is expressed as

$$\%\, e = e \times 100$$

Summary

1. In this chapter indicating instrument and types of indicating systems were discussed.
2. The instruments which are used to measure the electrical quantities like current, voltage, power, energy, etc. are known as primary instruments. Absolute instruments can be divided broadly on the basis of the value of a quantity measured in terms of the deflection and instrument constant. These instruments are used in standard laboratories and institutions as standardizing instruments.
3. Secondary instruments are those in which the value of a quantity to be measured is determined from the deflection of a pointer as the scale unit of the quantity. These instruments are commonly used in laboratories, institutions, industries and power stations.
4. Indicating instruments are those which indicate the magnitude of the measuring quantity at that time. Recording instruments are those which give the continuous record of variation of the measuring quantity over a specified period of time.
5. Integrating instruments are those which measure the total amount of the electrical quantity supplied over a period of time. Energy meter and ampere hour meter are few examples of integrating instruments.

Solved Problems

Problem 1

The measured and true value of a resistance are 10.26 Ω and 10.23 Ω, respectively. Determine the absolute static error of measurement.

Solution

Given,

Measured value $A_m = 10.26\ \Omega$

True value $A_t = 10.23\ \Omega$

Therefore, the absolute error is

$$\delta_a = A_m - A_t = 10.26 - 10.23$$

$$= 0.03\Omega$$

Problem 2

The measured and true value of a resistance are 18.89 Ω and 17.89 Ω, respectively. Determine the absolute static error of measurement?

Solution

Given,

Measured value $A_m = 18.89\ \Omega$

True value $A_t = 17.89\ \Omega$

Therefore, absolute static error is

$$\delta_a = A_m - A_t = 1.00\ \Omega$$

Problem 3

The measured value of an inductor is 205.3 mH and its true value is 201.4 mH. Determine the relative static error?

Solution

Given,

Measured value $A_m = 205.3 \times 10^{-3}$ H

True value $A_t = 201.4 \times 10^{-3}$ H

Therefore, absolute static error is

$$\delta_a = A_m - A_t = 3.9 \times 10^{-3}\ \text{H}$$

Now, relative static error is

$$\%\ t_r = \frac{\delta_a}{A_t}$$

$$= \frac{3.9 \times 10^{-3}}{201.4 \times 10^{-3}}$$

$$= 0.0194 = 1.94\%$$

where t_r is percentage error.

Problem 4

The measured value of an inductor is 105.3 mH and its true value is 103.2 mH. Determine the relative static error.

Solution

Given,

Measured value $A_m = 105.3\ \text{mH} = 105.3 \times 10^{-3}$ H

True value $A_t = 103.2\,\text{mH} = 103.2 \times 10^{-3}$ H

Therefore, absolute static error is

$$\delta_a = A_m - A_t = 2.1 \times 10^{-3}\ \text{H}$$

Now, relative static error is

$$\%\ t_r = \frac{\delta_a}{A_t} = \frac{2.1 \times 10^{-3}}{10.3.2 \times 10^{-3}}$$

$$= 0.0203$$

$$= 2.03\%$$

Problem 5

Two capacitors $140 \pm 2.3\,\mu F$ and $110 \pm 1.4\,\mu F$ are connected in parallel. Calculate the limiting error of the resultant capacitance in μF and in percent?

Solution

Given,

$$u = 140 \pm 2.3\ \mu\text{F},\ v = 110 \pm 1.4\ \mu\text{F}$$

When two capacitors are connected in parallel, the resultant capacitance is

$$y = u + v = 140 \pm 2.3 + 110 \pm 1.4$$

$$= (250 \pm 3.7)\ \mu\text{F}$$

Therefore, the limiting error is $\pm 3.7\,\mu$F. Relative limiting error is

$$\frac{\delta y}{y} = \pm \frac{3.7}{250} = \pm 0.0148\ \ (\text{since } \delta_a/\text{A}_s)$$

$$= \pm 1.48\%$$

Problem 6

The unknown resistance is determined by Wheatstone bridge which is given by the following expression:

$$R_4 = \frac{R_2 R_3}{R_1}$$

where $R_1 = 85 \pm 0.6\,\%\,\Omega$, $R_2 = 870 \pm 0.7\,\%\,\Omega$, $R_3 = 810 \pm 0.8\%\,\Omega$. Calculate the magnitude of the unknown resistance and the limiting error in percent and in ohm for the unknown resistance?

Solution

The expression for unknown resistance is

$$R_4 = \frac{R_2 R_3}{R_1} = \frac{870 \times 810}{85} = 8290.6\ \Omega$$

Relative limiting error of the unknown resistance is given by

$$\frac{\delta R_4}{R_4} = \pm\left(\frac{\delta R_1}{R_1} + \frac{\delta R_2}{R_2} + \frac{\delta R_3}{R_3}\right) = \pm(0.6 + 0.7 + 0.8)\% = \pm 2.1\%$$

Therefore,

$$\text{Limiting error in ohms} = \pm\frac{2.1}{100} \times 8290.6 = \pm 174.1\ \Omega$$

Problem 7

The full-scale deflection for a moving coil voltmeter occurs at a current of 6 mA. The coil has 120 turns, effective depth 3.5 cm and width 3.0 cm. The controlling torque of the spring is 0.6 g/cm for full-scale deflection. Find the flux density in the air gap.

Solution

Given,

$$\text{Deflecting torque, } T_d = NBIA$$

Therefore,

$$T_d = 120 \times B \times 6 \times 10^{-3} \times (3.5 \times 3) \times 10^{-4} = 7.56 \times 10^{-4}\, B\ \text{N m}$$

Controlling torque,

$$T_c = 0.6\ \text{g cm} = 0.6\ \text{g wt cm} = 0.6 \times 10^{-3} \times 10^{-2}\ \text{kg wt m} = 0.6 \times 9.8 \times 10^{-5}\ \text{N m} = 5.88 \times 10^{-5}\ \text{N m}$$

Now since

$$T_d = T_c$$

Hence,

$$7.56 \times 10^{-4} B = 5.88 \times 10^{-5}$$

$$B = 0.07778\ \text{Wb/m}^2$$

Problem 8

A permanent moving coil instrument has a coil of dimensions 10 mm × 8 mm. The flux density in the air gap is 0.15 Wb/m^2. If the coil is wound for 100 turns, carrying a current of 5 mA then calculate the deflecting torque. Calculate the deflection if the spring constant is 0.2×10^{-6}.

Solution

The deflecting torque is given by

$$T_d = NBAI = 100 \times 0.15 \times (A) \times 5 \times 10^{-3}\ \text{Nm}$$

$$A = \text{Area} = 10 \times 8 = 80\ \text{mm}^2 = 80 \times 10^{-6}\ \text{m}^2$$

Therefore

$$T_d = 100 \times 0.15 \times 80 \times 10^{-6} \times 5 \times 10^{-3} = 6 \times 10^{-6}\ \text{Nm}$$

$$T_d = T_c = K\theta$$

$$6 \times 10^{-6} = 0.2 \times 10^{-6} \times \theta$$

$$\theta = (6 \times 10^{-6})/(0.2 \times 10^{-6}) = 30°$$

Problem 9

The measured and true value of resistance are 4.26 Ω and 3.0 Ω, respectively. Determine the absolute static error of measurement.

Solution

Given,

Measured value, $A_m = 4.26\ \Omega$

True value, $A_t = 3.0\ \Omega$

Therefore, the absolute static error is

$$\delta_a = A_m - A_t$$

$$= 4.26 - 3$$

$$= 1.26\ \Omega$$

Problem 10

The measured value of an inductor is 65.2 mH and its true value is 59.2 mH. Determine the relative static error.

Solution

Given,

Measured value, $A_m = 65.2\ \text{mH}$

True value, $A_t = 59.2\ \text{mH}$

Therefore, the absolute static error is

$$\delta_a = A_m - A_t$$
$$= 65.2\times10^{-3} - 59.2\times10^{-3}$$
$$= 6\times10^{-3}\ \text{H}$$

Now, relative static error is

$$\%\ t_r = \frac{\delta_a}{A_t}$$
$$= \frac{6\times10^{-3}}{59.2\times10^{-3}}$$
$$= 0.1014 = 1.041\%$$

Problem 11

The measured and true value of a resistance are 103.29 Ω and 103.04 Ω, respectively. Determine the absolute static error of measurement.

Solution

Given,

Measured value, $A_m = 103.29\ \Omega$

True value, $A_t = 103.04\ \Omega$

Therefore, the absolute static error is

$$\delta_a = A_m - A_t$$
$$= 103.29 - 103.04$$
$$= 0.05\ \Omega$$

Problem 12

The measured value of an inductor is 2.06 mH and its true value is 1.09 mH. Determine the relative static error.

Solution

Given,

Measured value, $A_m = 2.06\times10^{-3}\ \text{H}$

True value, $A_t = 1.09\times10^{-3}\ \text{H}$

The absolute static error is

$$\delta_a = A_m - A_t = 0.097\times10^{-3}\ \text{H}$$

Now, relative static error is

$$\% \, t_{r} = \frac{\delta_{a}}{A_{t}}$$

$$= \frac{0.97 \times 10^{-3}}{1.09 \times 10^{-3}}$$

$$= 0.0889$$

$$= 8.89\%$$

Problem 13

A 10 A range electrodynamic ammeter is controlled by a spring having a torsion constant of 10^{-7} Nm/degree. The full-scale deflection is 120°. Calculate the inductance of the instrument when measuring a current of 10 A. The initial inductance is 2.5 mH and the change of inductance is linear with deflection of the moving coil.

Solution

Given,

Spring torsion constant $c = 10^{-7}$ Nm/degree

Full-scale deflection current $F = 10$ A

Full-scale deflection $\theta = 120°$

Full-scale deflection torque $T_{d} = c \times \theta$

Therefore,

$$T_{d} = 10^{-7} \times 120 = 12 \times 10^{-6} \text{ N m}$$

Now

$$T_{d} = I^{2} \frac{dM}{d\theta}$$

Therefore,

$$\frac{dM}{d\theta} = \frac{T_{d}}{I^{2}} = \frac{12 \times 10^{-6}}{10^{2}} = 0.12 \ \mu\text{H/rad} \quad (\text{since } dL/2 = dM)$$

Now, change in inductance is

$$dM = 0.12 \times 10^{-6} \times 120 \times \frac{\pi}{180}$$

$$= 0.2512 \ \mu\text{H}$$

Therefore,

$$\text{Total inductance} = \text{Initial inductance} + \text{Change in inductance}$$

$$= 2.5 + 0.2512$$

$$= 2.7512 \ \mu\text{H}$$

Problem 14

The inductance of a certain moving iron ammeter is $(9+50-1/20^2)\,\mu H$ where θ is the deflection in radians from zero position. Determine the scale position in radians for a current of 6 A, if the control spring torque is 1.4×10^{-6} Nm/rad.

Solution

We have,

$$T=\frac{1}{2}I^2\frac{dL}{d\theta}$$

Therefore,

$$\frac{dL}{d\theta}=\frac{d}{d\theta}\left(9+5\theta-\frac{1}{2}\theta^2\right)\times 10^{-6}=(5-\theta)\times 10^{-6}\ \mu\text{H/rad}$$

If the deflection be θ radians for a current of 6 A, it can be written as

$$14\times 10^{-6}\times\theta=\frac{1}{2}\times 6^2\times(5-\theta)\times 10^{-6}$$

$$14\theta=18\,(5-\theta)$$

$$14\theta+18\theta=90$$

Therefore $\theta=2.8125$ rad.

Problem 15

Design a multirange DC milliammeter using a basic movement with an internal resistance $R_m=60\,\Omega$ and a full-scale deflection current $I_m=1\,mA$. The required ranges are as follows: 0–20 mA, 0–60 mA.

Solution

(i) 0–20 mA range: Multiplication factor,

$$\eta=\frac{I}{I_m}=\frac{20}{1}=20$$

Therefore, the resistance of the shunt is

$$R_{sh1}=\frac{R_m}{\eta-1}=\frac{60}{20-1}=3.158\,\Omega$$

(ii) 0–60 mA range: Multiplication factor,

$$\eta=\frac{60}{1}=60$$

Therefore, the resistance is

$$R_{sh2}=\frac{60}{60-1}=1.017\,\Omega$$

Problem 16

A moving coil instruments requires 20 mA and 100 mV for full-scale deflection. What is the value of the shunt required to use the above meter to read upto 10 A?
Data: $I_m = 20\ mA$, $V_m = 100\ mV$, $I = 10\ A$

Solution

$$R_m = \frac{V_m}{I_m} = \frac{100\times10^{-3}}{20\times10^{-3}} = 5\ \Omega$$

$$R_{sh} = \frac{I_m R_m}{(I - I_m)} = \frac{20\times10^{-3}\times5}{(10-20\times10^{-3})}$$

$$R_{sh} = 0.01002\ \Omega$$

Problem 17

A moving coil instrument has 50 Ω coil resistance and gives maximum deflection at 5 mA. Find the external resistance to be connected in series with the instrument so that it reads upto 300 V.
Data:

$$R_m = 50\ \Omega, I_m = 5\ mA = 5\times10^{-3}\ A, U = 300\ V$$

Solution

$$R_1 = (R_m - R_a) = \frac{V_m}{I_m}$$

$$= \frac{300}{5\times10^{-3}}$$

$$R_t = 60000\ \Omega$$

$$R_e = R_1 - R_m = 60000 - 50$$

$$R_e = 59950\ \Omega$$

Multiple Choice Questions

1. An ammeter is ____________ instrument.
- **(a)** an indicating
- **(b)** an integrating
- **(c)** a recording
- **(d)** none of the above

2. Shunts are generally made of
- **(a)** copper
- **(b)** aluminum
- **(c)** silver
- **(d)** magnesium

3. The full-scale deflection current of a moving coil instrument is about
   (a) 50 mA
   (b) 1 A
   (c) 3 A
   (d) 2 A

4. For measuring high values of alternating current with a dynamometer ammeter, we use a
   (a) shunt
   (b) multiplier
   (c) potential transformer
   (d) current transformer

5. A dynamometer type instrument can be used for
   (a) AC work only
   (b) DC work only
   (c) both DC and AC work
   (d) none of the above

6. A dynamometer instrument is chiefly used as a
   (a) DC ammeter
   (b) DC voltmeter
   (c) wattmeter
   (d) none of the above

7. Permanent magnet moving coil instrument can be used for
   (a) AC work only
   (b) DC work only
   (c) both AC and DC work
   (d) none of the above

8. The multiplying power of the shunt of a milliameter is 8. If the circuit current is 200 mA, then the current through the meter is
   (a) 200 mA
   (b) 15 mA
   (c) 1600 mA
   (d) none of the above

9. The pointer of an indicating instrument is generally made of
   (a) copper
   (b) aluminum
   (c) silver
   (d) soft steel

10. The controlling torque of an indicating instrument ______ as the deflection of the moving system increases.
   (a) remains uncharged
   (b) decreases
   (c) increases
   (d) none of the above

Review Questions

1. What is meant by measuring instruments?
2. Mention the different types of torques to which the moving stem of an indicating instrument are subjected.
3. Explain the principle of operation of a permanent magnet moving coil instrument.
4. Describe with the help of a neat sketch the working of a dynamometer type wattmeter.
5. Why cannot a moving coil instrument be used in AC circuits?
6. Explain, with the help of a neat sketch, the principle of working of a single-phase induction type energy meter.
7. Explain how a moving coil instrument can be converted into voltmeters and ammeters.
8. List the merits and demerits of dynamometer type instruments.
9. Explain the purpose of shunts used in ammeter.
10. What are **(a)** indicating instruments and **(b)** recording instruments? Explain.

11. Explain the various methods of damping provided in indicating instruments.

12. Justify the necessity of controlling torque in indicating instruments. Discuss the various methods of producing the same.

13. What is a megger? Explain its working.

14. Describe the construction and operation of a repulsion type moving iron instrument.

15. Describe the construction and operation of an attraction type moving iron instrument.

Problems

1. A moving coil milliammeter with a resistance of 1.6 Ω is connected with a shunt of resistance 0.228 Ω. What will be the current flowing through the instrument if it is connected in a circuit in which a current of 200 mA is flowing?

2. What should be the resistance of moving coil of an ammeter which requires 2.5 mA for full-scale deflection so that it may be used with a shunt having a resistance of 0.0025 Ω for a range of 0–10 A?

3. A moving coil instrument gives a full scale deflection with a current of 40 μA, while the internal resistance of the meter is 500 Ω. It is to be used as a voltmeter to measure a voltage range of 0–10 V. Calculate the shunt resistance needed.

4. An ammeter reads 8.5 A and the true value of the current is 8.3 A. Determine the absolute error and percentage relative error.

5. The change of inductance for a moving iron ammeter is 2 μH/degree. The control spring constant is 5×10^{-7} Nm/degree. The maximum deflection of the pointer is 100°, what is the current corresponding to maximum deflection?

6. The full scale deflecting torque of a 10 A moving iron meter is 4×10^{-5} Nm. Estimate in micro – henry per radian the rate of change of self inductance of the instrument at full scale.

7. A moving coil instrument gives full-scale deflection for a current of 10 mA. Neglecting the resistance of the instrument, calculate the approximate value of series resistance needed to enable the instrument to measure upto **(a)** 20 V, **(b)** 100 V and **(c)** 250 V.

8. A meter of resistance 50 Ω has a full-scale deflection of 4 mA. Determine the value of shunt resistance required in order that the full-scale deflection should be **(a)** 15 mA, **(b)** 20 A and **(c)** 100 A.

9. A meter has a resistance of 40 Ω and registers a maximum deflection when a current of 15 mA flows. Calculate the value of resistance that converts the movement into **(a)** an ammeter with a maximum deflection of 50 A and **(b)** a voltmeter with a range 0–250 V.

10. A voltage of 240 V is applied to a circuit consisting of a 800 Ω resistor in series with a 1.6 kΩ resistor. What is the voltage across the 1.6 kΩ resistor? The potential difference across the 1.6 kΩ resistor is measured by a voltmeter of full-scale deflection of 250 V and sensitivity 100 Ω/V. Determine the voltage indicated.

11. The unknown resistance is determined by Wheatstone bridge and is given by the expression

$$R_x = \frac{R_2 R_3}{R_1}$$

where $R_1 = 90 \pm 0.5\%\ \Omega$, $R_2 = 900 \pm 0.8\%\ \Omega$, $R_3 = 825 \pm 0.6\%\ \Omega$. Determine the magnitude of the unknown resistance?

12. In Problem 11 find the limiting error in percent and in ohms for the unknown resistance R_x.

13. In Problem 11, find the limiting error in ohms.

14. Two capacitors $150 \pm 2.4\ \mu F$ and $120 \pm 1.5\ \mu F$ are connected in parallel. Determine the limiting error of the resultant capacitance **(a)** in mF and **(b)** in percent?

15. The coil of a moving coil ammeter has a resistance of 6.25 Ω and full scale reading of 0.012 A. What would be the full scale reading of the instrument if used as a voltmeter?

16. A single turn 100/5 A, 50 Hz current transformer has a non-inductive burden of 1 Ω. The magnetizing current is 100 A. Calculate **(a)** the current ratio and **(b)** phase angle.

17. A cathode ray oscilloscope is calibrated for 20 V/inch and its sweep speed is set for 6 ms. Three complete sine curves with a distance of 2.25 inch between their upper and lower extremities appear on the screen. Find **(a)** the frequency and **(b)** RMS voltage.

18. In an indicating PMMC type ammeter, the current through the coil for full-scale deflection is 5 mA. Coil resistance is 5 Ω and swamping resistance is 4 Ω. The meter is used for measuring current up to 1 A. Calculate the value of shunt resistance required.

19. In a gravity controlled instrument the controlling weight is 0.005 kg and acts at a distance of 2.4 cm from the axis of the moving system. Determine the deflection in degrees corresponding to deflecting torque of 1.05×10^{-4} kgm.

20. The torque of an ammeter is directly proportional to the current flowing through it. If a current of 10 A causes a deflection of 60°, determine the value of current for a deflection of 40° when the instrument is **(a)** spring controlled **(b)** gravity controlled.

21. A sinusoidal waveform at 10 kHz is applied to the vertical input of a free running oscilloscope. The horizontal sweep speed is set for a full cycle of 0.4 ms. What will be the result for sweep of the beam?

22. A pulse is measured as having a peak amplitude of 18 V. If the vertical dial setting is 10 V/cm, how many centimeters of a signal amplitude can be observed peak to peak?

23. A square wave signal is observed on the oscilloscope to have one cycle measured as 6 cm at a horizontal sweep rate of 30 ms/cm. Find the signal frequency.

24. A wattmeter reads 25.34 W. The absolute error in the measurement is −0.11 W. Determine the true value of power.

25. The measured value of a capacitor is 205.3 μF whereas its true value is 201.4 μF. Determine the relative error.

Answers

Multiple Choice Questions

1. (a) **3.** (a) **5.** (c) **7.** (b) **9.** (b)

2. (d) **4.** (a) **6.** (c) **8.** (b) **10.** (c)

Problems

1. 28.5 mA
2. 10 Ω
3. 249.5 kΩ
4. 2.4%
5. 7.07 A
6. 0.8 μH/radian
7. **(a)** 2 kΩ, **(b)** 10 kΩ, **(c)** 25 kΩ
8. **(a)** 1 13.33 Ω, **(b)** 10.00 mΩ, **(c)** 2.00 mΩ
9. **(a)** 12.00 mΩ in parallel,
   (b) 16.67 kΩ in series
10. 160 V, 156.7 V
11. 8250 Ω
12. ±1.9%
13. ±156.75 Ω
14. **(a)** (270 ±3.9) μF, **(b)** 1.44%
15. 75 mV
16. **(a)** 200.99, **(b)** 5.73°
17. **(a)** 500 Hz, **(b)** 15.9 V
18. 0.04523 Ω
19. 61°
20. **(a)** 6.67 A, **(b)** 7.42 A
21. 0.1 ms
22. 3.6 cm
23. 5.55 kHz
24. 25.45 W
25. 1.94%

9 Domestic Wiring

LEARNING OBJECTIVES

After reading this chapter, you will be able to:

- Understand the factors affecting the choice of wiring system.
- Describe wiring materials.
- Understand different types wiring.
- Understand the layout of simple domestic system.
- Understand the concept of earthing and fuses.

9.1 Introduction

A network of wires connecting various accessories for distribution of electrical energy from the supplier meter board to the numerous electrical energy consuming devices such as lamps, fans and other domestic appliances through controlling and safety devices is known as a wiring system. House wiring (domestic wiring) deals with the distribution system within the domestic premises. The wiring requirements may vary among different consumers. House wiring is generally done for consumption of electrical energy at 230 V, single phase or at 400 V, three phase. In the latter case the total load in the house is expected to be divided among the three phases. An earth wire is also run connecting all the power plugs from where large quantity of electrical energy is tapped by using electrical appliances like heater, electric iron, etc.

This chapter gives information about the wiring practices normally followed for residential places.

9.2 Factors Affecting Choice of Wiring System

The choice of wiring system for a particular installation should be based on technical and economic considerations, both in the content of the wiring system itself and the installation for which it is proposed. In general, the following factors should be considered.

1. **Safety:** First and foremost consideration is safety to a person using electricity against leakage, shock and fire hazard.
2. **Mechanical protection:** The wiring must be protected from mechanical damage during its use.
3. **Permanency:** The wiring must not deteriorate unduly by action of weather, furnes, dampness, etc.
4. **Appearance:** The wiring must have a good appearance.
5. **Durability:** The wiring must be durable. The type of wiring that is suitable for temporary building will definitely be unsuitable for a permanent building.
6. **Accessibility:** In wiring there should be provision for extension, renewal or trouble shooting.

7. **Initial cost:** The wiring system selected should not be prohibitively expensive.
8. **Maintenance cost:** The wiring system selected should have, as far as possible, the lowest maintenance cost.

9.3 Wiring Materials and Accessories

The following are the wiring materials used for domestic wiring.

1. Switches
2. Fuse units
3. Socket outlets
4. Plugs
5. Lamp holders
6. Ceiling roses
7. Mounting blocks
8. Switch board
9. Push-button switches
10. Conduits
11. Teakwood batten
12. Screws
13. Wiring clips and conduit clamps
14. Wires and cables
15. Each leakage circuit breakers
16. Miniature circuit breakers

The accessories used for domestic wiring are as follows:

1. Screw driver;
2. side cutting plier;
3. diagonal cutting plier;
4. long nose plier;
5. slip plier;
6. pocket knife;
7. hammer;
8. woodsaw;
9. hacksaw;
10. chisels;
11. scratch awl;
12. hand drill;
13. portable electric drilling machine;
14. anger bit;
15. center punch;
16. test lamp;
17. wire gange;
18. wire stripper; etc.

9.3.1 Switches

The "ON" and "OFF" electrical connections to the load are done with the help of a switch. When the switch is "ON," the circuit is alive and when the switch is "OFF," the circuit is dead. The usual ratings of switches used for domestic purposes are 5 A and 15 A. In the following, we discuss some of the different types of switches that are commonly used.

9.3.1.1 Tumbler Switches

These switches are mounted on the mounting block or switchboard. They project out of the surface of walls. These switches are either single-way or two-way types. They are manufactured in either complete bakelite or in porcelain base and bakelite cover.

9.3.1.2 Flush Switch

As the name indicates these switches are fixed flush with the wall and do not project out. All modern wiring makes use of only this type of switches. They are available in either single-way or two-way types.

9.3.1.3 Pull Switches

These switches are fixed in the ceiling or raised places and all the live parts are out of the operators. The switch is operated with the help of pulling machines such as hand drills.

9.3.1.4 Grid Switches

Grid switches are similar to tumbler switches and are used in portable machines such as hand drills.

9.3.1.5 Architrave Switches

These types of switches are designed for flush mounting on architraves.

9.3.1.6 Rotary Snap Switches

This type of switch consists of an insulated handle to which blades are fixed. The movement of the handle makes these blades move in a step-wise fashion and make contact with another set of terminals. They are used for speed control and speed reversals of small motors, varying the range of heaters, etc.

9.3.1.7 Main Switches

The purpose of this switch is to control all the subcircuits provided in the building. It is enclosed in an iron box and is called iron clad switch. They are made of cast iron for current ranges of 5, 15 and 30 A. Switches with a current capacity 30 A, 63 A, 100 A 200 A, etc., may be fabricated by steel plates.

9.3.2 Fuse Units

A fuse unit is a safety device, a weak link connected in series with the circuit. The number of fuse units is equal to the number of circuits with which the entire head is splitted. Whenever the current in the circuit exceeds the value of the fuse provided, either due to overload or due to short circuits, the fuse melts, thus opening the circuit and protecting other material in the circuit. Fuse units are to be provided as per IS 2086/1963 and IS 2208/1962 specification. They are available in the two types: wire fuse (of copper–tin or lead) and HRC cartridge fuse.

9.3.3 Push-Button Switches

This type of switch consists of one blade which is given a reducing action by a push button. Its movement is controlled by a cam and a spring. These switches are normally used to control light power circuits.

9.3.4 Socket Outlets

These are of either two-pin or three-pin types and have 2 A, 5 A and 15 A ratings. These are used for temporary electrical connections such as tape recorder, table lamp, etc.

9.3.5 Plugs

The plugs should have an appropriate finger grip for inserting and withdrawing with the sockets. Supply to appliances like radio, table fan, etc., can be fed from socket outlets through plugs and flexible cords.

9.3.6 Lamp Holders

They not only connect the lamps to the supply but also give support to them. Lamp holders are normally available with bakelite exterior with porcelain interior or brass exterior with porcelain interior. They are classified as

1. Bayonet cap lamp holders (for lamps up to 200 W).
2. Screwed lamp holders (for lamps above 200 W).

Bayonet cap lamp holders are further classified as

1. Batten holders and Angle Batten holder for screwing on to a surface rigidly.
2. Pendant holders for suspending the lamps from the ceiling.
3. Fluorescent tube lamp holders.

Types of screwed lamp holders are

1. Edison screw lamp holder (200 W to 300 W).
2. Goliath screw holders (above 300 W lamps).

9.3.7 Ceiling Roses

Ceiling rose consists of two parts, namely, base and cover, both made up of porcelain or bakelite. Tube light, ceiling fan, etc. are connected to supply through flexible cords and ceiling rose. The normal voltage rating of a ceiling rose is 250 V.

9.3.8 Mounting Blocks

They are used to fix ceiling roses, batten lamp holders, surface switches, pull switches, etc.

9.3.9 Switch Boards

They are used for mounting switches, socket outlets, regulators and fuse units. The material used for construction of the board may be iron or teakwood. The top cover may be of a laminated type. They are available in the following sizes: (a) 100 mm × 100 mm, (b) 150 mm × 100 mm, (c) 200 mm × 150 mm and (d) 300 mm × 200 mm.

9.3.10 Teakwood Batten

These are used to lay the wires on the walls. The batten is initially fixed over the wall and thus wires are firmly fixed over the batten by wiring clips. The battens are available in widths of 13 mm, 19 mm, 25 mm, 31 mm, 38 mm, 44 mm, etc.

9.3.11 Conduits

IS 1653/1963 specifies the different types of conduits: metallic such as mild steel or nonmetallic such as polyvinyl chloride (PVC). Metallic conduits are coated with a black enamel or they are galvanized. Conduits are available in different diameters such as 16 mm, 19 mm, 25 mm, 30 mm, 35 mm, etc.

9.3.12 Screws

Screws are made up of brass or iron. They are available in length of 4 mm, 6 mm, 10 mm, 12 mm, 15 mm, 20 mm, etc. The two types are wood screws and metallic screws. Wood screws are used to fix wiring materials like switchboards, teakwood batten mounting blocks, etc. On walls, metallic screws with nuts are used to fix switches, socket outlets, etc. on switch boards.

9.3.13 Wiring Clips and Conduit Clamps

Wiring clips are used to fix insulated wires on teakwood batten. Clips are initially fixed on batten by nails at suitable intervals along its length. They are made up of brass on tinned brass and available in length of 30 mm, 40 mm and 50 mm. Conduit clamps are used to fix the conduit on the walls.

9.3.14 Wires and Cables

The different types are listed below:

1. VIR wires:

Insulation : Vulcanized India rubber (VIR).
Voltage grade : 250 V or 660 V.
No. of cores : Single or twin.
Conductor material : Copper or aluminum.
Conductor size (mm^2) : 1.5, 2, 3, 4, 6, 10, 15, 20, 25.
Use: Suitable for places where acids and alkalies are likely to be present.

2. TRS wires:

Insulation : Tough rubber sheath (TRS).
Voltage grade : 250 V or 660 V.
No. of cores : Single or twin.
Conductor material : Copper or aluminum.
Conductor size (mm^2) : 1.5, 2, 3, 4, 6, 10, 15, 20, 25.
Use: Suitable for places where acids and alkalies are likely to be present.

3. PVC cables:

Insulation : Polyvinyl chloride (PVC).
Voltage grade : 250 V or 660 V.

No. of cores : Single or twin.
Conductor material : Copper or aluminum.
Conductor size (mm^2) : 1.5, 2, 3, 4, 6, 10, 15, 20, 25.
Use: Generally used in all types of wiring.

4. Weather proof (WP) cables:
   Insulation : Vulcanized India rubber.
   Voltage grade : 250 V or 660 V.
   No. of cores : Single or twin.
   Conductor material : Aluminum.
   Conductor size (mm^2) : 1.5, 2, 3, 4, 6, 10, 15, 20.

5. Sheathed Cables:
   Specification : IS 692/1957.
   Insulation : Paper or PVC insulated with lead alloy sheath with or without steel armouring.
   Voltage grade : 660 V, 1100 V.
   No. of cores : 2, 3, 3.5 or 4.
   Conductor material : Copper or aluminum.
   Conductor size (mm^2) : 4, 16, 25, 35, 50, 185, 300, 500.
   Use: For service connection and outdoor fittings connections.

6. Flexible cord:
   Insulation : VIR and/or PVC.
   Voltage grade : 250 V or 660 V.
   No. of cores : 2, 3 or 4.
   Conductor material : Copper or aluminum.
   Conductor size (mm^2) : 0.4, 0.6, 1, 2, 3, 4.
   Use: For taking supply from socket outlets, ceiling roses, etc.

9.3.15 Earth Leakage Circuit Breaker

This can be optionally fitted in the main supply circuit to open the circuit automatically in the event of any supply leakage in any part of the circuit. Under healthy condition in a single-phase AC circuit, the current through the line wire and neutral wire are equal and opposite. As and when any leakage occurs, there exists a difference in the current in the two wires and this will make the circuit breaker to trip off, opening the supply and thus safeguarding the users against electric shock.

9.3.16 Miniature Circuit Breakers

Miniature circuit breakers (MCBs) are equivalent to have a switch and fuse in a single unit. They are normally used to make and break the supply in subcircuits. In addition to the above purpose, it also protects the circuits against overloading or short-circuiting. They are available in single-, two-, three- or four-pole configuration with current ratings 5 A, 16 A, 32 A, etc. As and when the circuit current exceeds the capacity of MCB, it gets tripped off.

9.4 Types of Wiring

The type of wiring to be selected for a particular place of use is based on several factors such as durability, safety, mechanical protection, appearance, environmental condition, accessibility for extension, cost, etc. The various type of wiring in practice are as follows:

1. Cleat wiring.
2. Wooden casing and capping wiring.
3. Batten wiring.
4. Conduit wiring.
5. Lead sheathed wiring.

9.4.1 Cleat Wiring

This system of wiring is used for temporary shelters and farmhouses. Porcelain or seasoned teakwood cleats are made of two halves, base and top. The cleats are available as two-way, three-way, four-way, etc. according to the number of grooves provided. The base cleats are fixed on wooden plugs fitted on walls or other supporting surface. Spacing between adjacent cleats vary from 30 cm to 60 cm. After laying the wires on the grooves, top cleats are fixed to the base cleats, thus holding the wires firmly. In this cleat wiring, the vulcanised India rubber conductors are supported in porcelain cleats.

9.4.2 Wooden Casing and Capping Wiring

This type of wiring makes use of wooden casing with grooves and cap made from first-class seasoned teakwood or any other approved wood. VIR or PVC wires are used. The casing is initially fitted on the walls. The wires are laid in such a way that wires of opposite polarity are laid in different grooves. The casing is then covered at the top by means of capping.

9.4.3 Batten Wiring

This type of wiring is most commonly used since it is cheap and can be quickly installed. The wooden battens are fixed to the walls or ceiling by means of wooden plugs with wood screws. The wooden screws should be fixed on the batten maintaining a gap not exceeding 75 cm. The minimum width of the batten is 13 mm for two wires. The width increases with increase in the number of wires. Wires are firmly fixed over the batten by means of wiring clips. Wires are firmly fixed. Clips are initially hailed on the batten at a gap of 10 cm in the case of horizontal run and 15 cm in the case of vertical run.

9.4.4 Conduit Wiring

The advantages of this type of wiring are as follows: it provides good mechanical protection, it provides good electrical protection, it is fire proof, it is damp-proof, and is durable. When the conduits are buried inside the walls and ceilings they give good appearance. Conduits may also be fixed on walls and ceiling by clamps. The required numbers of wires are then drawn through the conduits and connections are then given.

9.4.5 Lead Sheathed Wiring

Sheathed wiring is carried out on places exposed to the sun, rain and other damp conditions, provided the joints are exposed. Sheathed cables may be directly clamped on walls or ceilings. Armored sheath cables may be laid underground for outdoor connections.

9.5 Rules for Wiring

According to IS specifications, the following rules have to be kept in mind while wiring:

1. The total lighting load in a subcircuit should not be more than 100 W. The maximum power load in a subcircuit should not exceed 2000 W.
2. Every fitting or appliance must be controlled by a switch.
3. Every socket outlet must also be controlled by a switch.
4. The switch should be on the line conductor.
5. The incandescent lamp should be hanged at least 2.5 m above the floor level.
6. All ceiling fans should be hanged at least 2.75 m above the floor level.
7. Power wiring should be separated from light wiring.
8. Every subcircuit must have a separated distribution fuse.
9. The switches and starters of the motors should be easily accessible.
10. All the metal coverings, frames, etc. should be earthed.

9.6 Specification of Wires

The wire employed for wiring of buildings may be divided into different groups according to

1. Conductors used;
2. number of cores used;
3. voltage grading and
4. type of insulation used.

According to the conductor material used in cables, these may be divided into two classes:

1. Copper conductor cables.
2. Aluminum conductor cables.

According to the number of cores used in the cable, the cable may be divided into the following classes:

1. Single-core cables;
2. twin-core cables;
3. three-core cables;
4. two cores with earth continuity conductor cables, etc.

According to voltage grading, the cables may be divided into two classes:

1. 250/440 V cables;
2. 650/1,100 V cables.

According to type of insulation, the cables are of the following types:

1. Vulcanized Indian rubber (VIR) cables.
2. Tough rubber sheathed (TRS) or cab type sheathed (CTS) cables.
3. Lead sheathed cables.
4. Polyvinyl chloride (PVC) cables.
5. Weather proof cables.
6. Flexible cables.

9.7 Two-Way and Three-Way Control of Lamps

The lamp circuits used for house wiring are quite simple and they are generally controlled from one point, such as room lighting, bathroom lighting, lavatory lighting, etc. Note that in staircase wiring, it is necessary to control the lamp circuit from two points: at the top and bottom of the staircase. Similarly, in halls, corridors or bedrooms, it may be necessary to control the lamp from more than two points. This is explained in the following.

9.7.1 Two-Way Control of Lamp (Staircase Lighting)

For the lamps to be controlled by two points, it is necessary to use two-way switches. The wires used between the switches are called as strap wires.

The necessary rigged-up circuit is shown in Fig. 1. In this figure S_1 and S_2 are the two-way switches connected in series with the lamp filament and the mains. The control of the lamp by the two switches is shown in Table 1 in Chapter 9.

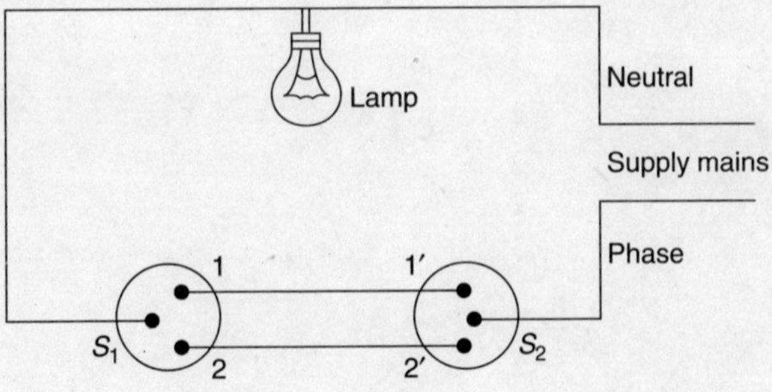

Figure 1 Staircase lighting circuit.

Table 1 Staircase switch position

Switch S_1 Position	*Switch S_2 Position*	*Lamp Condition*
1	1	ON
1	2	OFF
2	2	ON
2	1	OFF

9.7.2 Three-Way Control of Lamps (Corridor Lighting)

In such a method of control the lamp can be controlled independently from three places. This finds applications in corridor lighting, bedroom lighting, godowns, cellars, etc. In such a circuit we require two two-way switches and an intermediate switch. The circuit diagram and the state of the lamp at different positions (see Fig. 2) are explained as follows.

The intermediate switch can be operated in two modes: (a) straight (S) and (b) cross (C). The various positions of switches and the state of the lamp are given in Table 2.

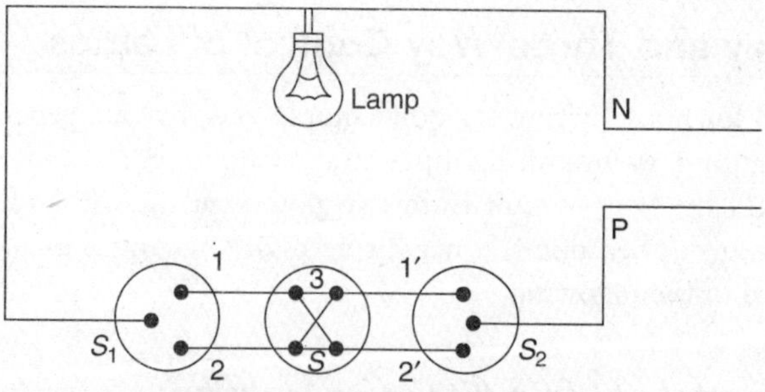

Figure 2 Three-way control lamp circuit.

Table 2 Three-way control lamp condition

Intermediate Switch Position	*Switch S_1 Position*	*Switch S_2 Position*	*Lamp Condition*
Straight (S)	1	1′	ON
	1	2′	OFF
	2	1′	OFF
	2	2′	ON
Cross (C)	1	1′	OFF
	1	2′	ON
	2	1′	ON
	2	2′	OFF

9.8 Fluorescent Tube Wiring

Fluorescent tube has filaments on either ends. The filaments are coated with electron-emitting material. The inside wall of the tube has a phosphorous coating which converts ultraviolet radiation into visible light. A ballast is used to give a transient high voltage so as to initiate the electron movement. It is an iron-cored coil having high inductance. Other accessories are a bimetallic starter and capacitors. Figure 3 shows the connection diagram.

When the supply switch S is closed, a current initially passes through the ballast, the two filaments of the tube and the starter. The bimetallic strip in the starter, after getting heated up, suddenly opens the circuit. Due to the high inductive property of the ballast, a transient high voltage appears across the two filaments, causing the initial electron flow to take place between the filaments.

The lamp starts glowing and now the current from the supply passes through the ballast, the two filaments and the mercury vapour inside the tube. A capacitor C across the supply is to improve the power factor of the circuit and another capacitor across the starter is to suppress radio interference.

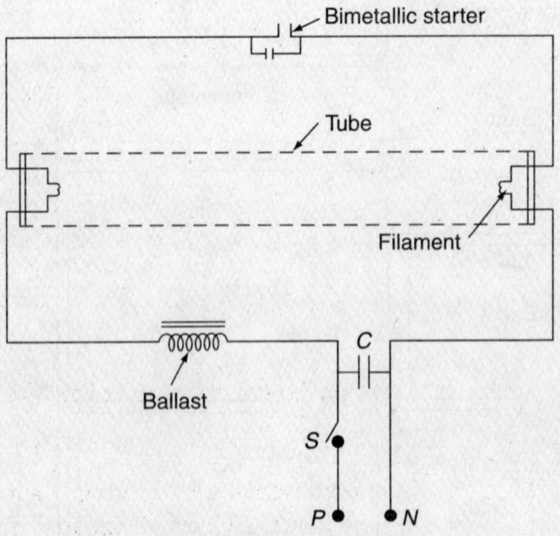

Figure 3 Fluorescent tube wiring circuit.

9.9 Simple Domestic Layout

Figure 4 shows the layout for a residential building. The supply is taken from Electricity Board (EB) distribution lines through series mains. It goes to energy meter, the main switch and then to the distribution board. The supply to the entire building may be divided into three circuits.

Circuit 1:

1. Study room: 2 Lights and 1 Fan
2. Verandah: 2 Lights
3. Bed: 1–2 Lights and 1 Fan
4. Toilet: 1–1 Light and 1 Exhaust fan
5. Bath: 1–1 Lights and 1 Water heater

Circuit 2:

6. Hall: 3 Lights, 2 Fans, 1 TV
7. Dining: 2 Lights, 1 Fan, 1 Refrigerator

Circuit 3:

8. Toilet: 2–1 Lights and 1 Exhaust fan
9. Bath: 2–1 Lights and 1 Water heater
10. Bed: 2–2 Lights and 1 Fan

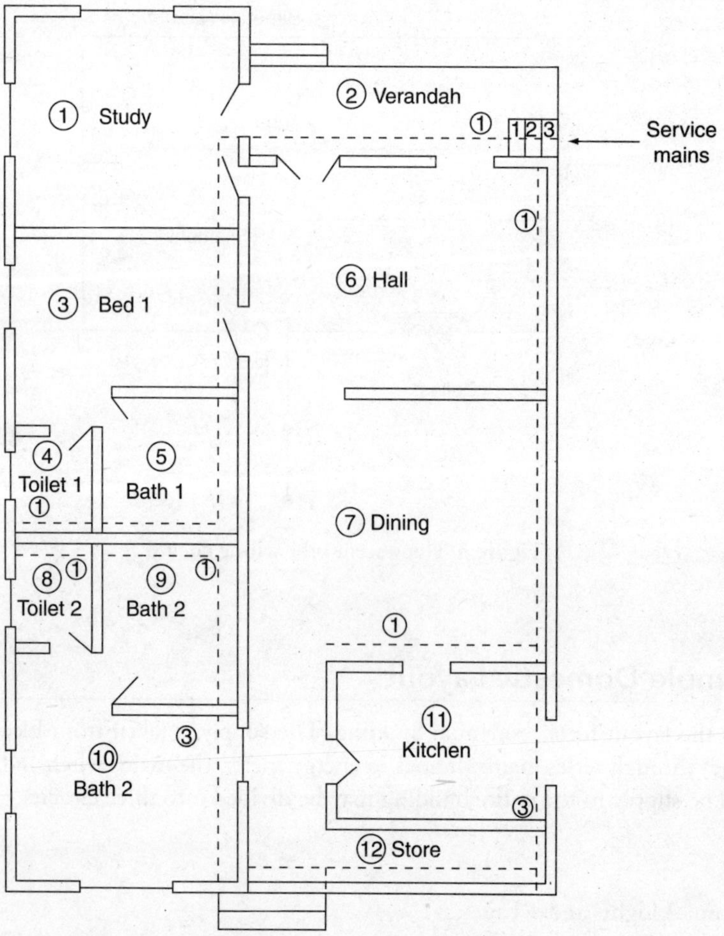

Figure 4 Simple domestic layout.

9.10 Earthing

Earthing generally means connecting the body of the equipments, appliances, etc. to the earth so as to ensure at all times that an immediate and safe discharge of electric current to the earth should take place in the event of any leakage, fault, etc. The earth is considered to be at zero potential. This arrangement is to provide enough safety to the user from electric shock.

All metallic part of every electrical wiring systems such as metallic conduits, metallic sheath steel armoring of cables, metallic frames, iron lad switches, household appliances, motor body, starting gear frame, earth point of socket outlet, etc. should be earthed by running one continuous earth bus (thick base copper wire/strip) throughout the wiring layout. No joints are permitted in the earth bus. Only firm connections are recommended.

The two ends of the earth bus are to be connected to two separate earth pits installed just outside the building with suitable spacing. If sufficient spacing is not available one earth pit may be used.

Two different earthing systems are practiced:

1. Pipe earthing.
2. Plate earthing.

9.10.1 Pipe Earthing

Galvanized iron pipe of 38 mm diameter, 2 m length with 12 mm holes on its surface is buried vertically into the ground. Alternate layers of powdered charcoal and salt are arranged 15 cm around the pipe. A funnel with a mesh is connected at the top of the pipe and water is poured in it periodically for maintaining a good earth connection.

9.10.2 Plate Earthing

In this system, in addition to the arrangement listed in pipe earthing, an additional earth plate is provided at the bottom of the earth rod. The size of the plate is 60 cm × 60 cm × 6.3 cm if it is made of galvanized iron or 60 cm × 60 cm × 3.15 cm if it is made of copper. Alternate layers of powdered charcoal and salt are arranged both below and top of the earth plate.

Summary

In this chapter basis of domestic wiring and system of distributional electrical energy are discussed. The main concept of electrical installation will depend to a large extent on the conditions of service. While selecting a particular system of wiring for an electrical installation, the following points must be given consideration:

1. Life of installation.
2. Future of extensions.
3. Construction of buildings.
4. Five haryards or other special condition.
5. Corrosions fames.
6. Dampness.

There are various systems by which power is distributed for domestic purposes:

1. Single-phase two-wire system.
2. Three-phase three-wire system.
3. Three-phase four-wire system.

The most commonly used system for the domestic purposes is the two-wire system or the three-phase four-wire system. The loads on the three phases are balanced by connecting various consumers or groups of consumers to different phases of three-phase four-wire supply.

Multiple Choice Questions

1. __________ is used to make or break the electric circuit.
   (a) Switch
   (b) Lamp
   (c) Earthing
   (d) Fuse

2. __________ is used to hold the lamp required for lighting purposes.
   (a) Ceiling roses
   (b) Adopter
   (c) Socket outlets
   (d) Lamp holders

3. ________ are used to provide a tapping to the lamp holder through the flexible wire.
   (a) Lamp holder
   (b) Ceiling roses
   (c) Socket outlets
   (d) Flush switch

4. Push-button switch consists of ________ blade only.
   (a) two
   (b) one
   (c) three
   (d) four

5. ________ are generally connected to the mains of Earth.
   (a) Switch
   (b) Lamp holder
   (c) Earthing
   (d) Socket outlets

Review Questions

1. Mention the different types of materials used for electrical wiring.
2. List the various accessories used for electrical wiring.
3. Describe the operation of a low-pressure fluorescent lamp with a circuit diagram.
4. Give particulars about the different kinds of wires and cables used for a domestic wiring.
5. Draw the wiring diagram for controlling one lamp from two different places.
6. What are the various types of wiring used for residential buildings?
7. What are the advantages of earthing?
8. Discuss the safety measures used in a domestic wiring.
9. Draw a neat wiring diagram for a staircase for lighting and explain its working.
10. Present a simple domestic layout and explain how wiring is carried out is it.
11. Explain the following types of wiring systems: **(a)** batten and **(b)** conduit.
12. List the various types of a switches used in a domestic wiring system.
13. What are the various types of lamp holders?

Answers

Multiple Choice Questions

1. (a) **2.** (d) **3.** (b) **4.** (b) **5.** (c)

10 Power System

LEARNING OBJECTIVES

After reading this chapter, you will be able to:

- Study the hydro, thermal and nuclear power stations in detail.
- Understand the concept of transmission.
- Compare the overhead and underground cables.

10.1 Introduction

Energy exists in different forms in nature. Electrical energy is one of the most popular form of energy. This is because compared to the other forms of energy, it possesses the following unique advantages:

1. It may be carried in far greater quantities and to far greater distances at high efficiency.
2. The methods of control are extremely elastic and convenient.
3. Since it can easily be changed to other forms like heat, light, mechanical energy, chemical energy, etc., so it is the highest grade of energy since.

Day-by-day, the demand of electrical energy is increasing due to increasing population, industrial sectors, shops and supermarkets. Electrical energy is generated from different sources of energy such as solar, wind, water, biomass and nuclear.

Figure 1 shows a single-line diagram of a typical alternating current (AC) power system. In the generating stations, AC generators generate electrical energy at 11 kV (6.6 kV/33 kV), the transformer steps up the voltage to 230 kV (110 kV/132 kV/400 kV) at which long distance transmission takes place.

The receiving station steps down the voltage to 33 kV and feeds the substations where the system voltage is further stepped down to 3.3 kV (6.6 kV). High tension distribution starts from the substations and bulk consumers are provided the electrical energy at this voltage level. A consumer has his own transformation station. There are other transformer stations maintained by the system authority where the voltage is stepped down to 415 V/240 V (415 V) between lines and 240 V between line and neutral at which other consumers draw the electric energy to their loads.

High transmission (HT) may be carried out by high voltage direct current (HVDC) system also. Several generating stations may be interconnected aiming at good grid connections for the integrated process of continuous generation transmission and distribution.

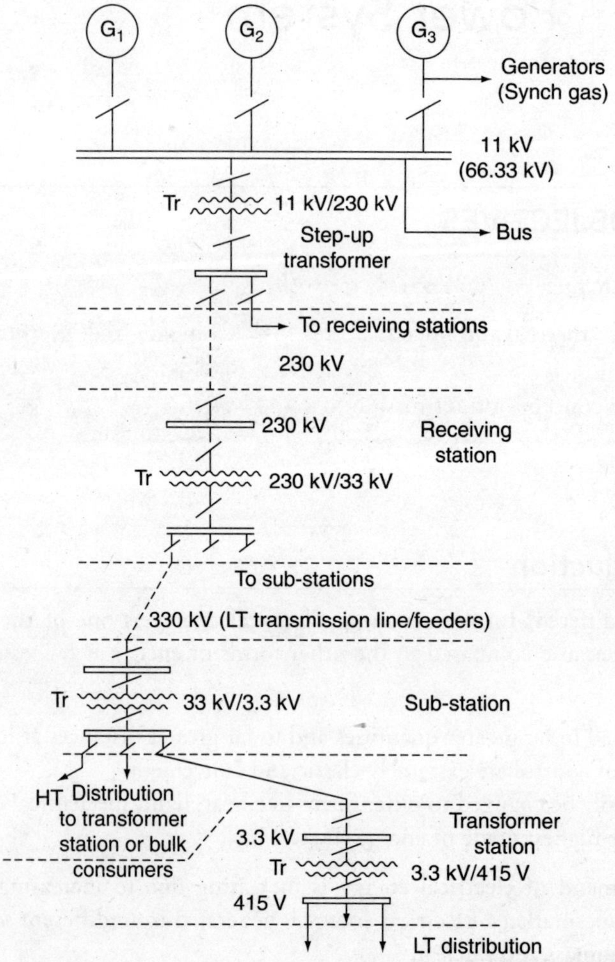

Figure 1 AC power system layout.

The electrical power is generated by two main sources, namely

1. Renewable energy sources.
2. Non-renewable energy sources.

Some of non-renewable energy sources are mentioned below.

10.2 Types of Power Stations

Electrical energy in bulk amount is generated in the following three types of generating stations:

1. Hydro-power stations.
2. Thermal-power stations.
3. Nuclear-power stations.

10.2.1 Hydro-Power Stations

Hydroelectric power plants utilize the potential energy of water at a high level to produce electrical energy. These plants are generally located in hilly areas where dams can be built conveniently and large reservoirs can be obtained. Besides generating electric energy they also aid in river flood control; storage of water for irrigation, drinking, etc.

10.2.1.1 Layout of Hydro-Power Stations

The main features of a typical hydro-power station are as follows:

1. The catchment area, reservoir and dam.
2. The pipelines, tunnels and surge tank.
3. The turbines.
4. The electrical plant.
5. The discharge arrangements and tail race.

Figure 2 shows the typical layout of hydro-power stations. The dam is constructed to create the required water level, and water from catchment area collects at the back of the dam to form a reservoir. From the dam, the water is allowed to flow to the power house usually by two or more pipelines. In case an intervening mountain is present, tunneling is employed. Prior to the valve house, a surge tank is provided. In the event of load being suddenly thrown off and water turbines likely to be shut off by closing the valves, the sudden pressure built up in the waterway can be instantly dissipated by the rise in water level in this tank. The potential energy of water is converted into mechanical energy by water turbines. The choice of turbine depends on the water head. For high head, above 200 m, Pelton wheel is used. For medium heads, that is, 70–200 m, Francis turbines are used. Thaplan and propeller turbines are used for low heads, that is, less than 70 m. A synchronous generator coupled with the water turbine converts mechanical energy into electrical energy.

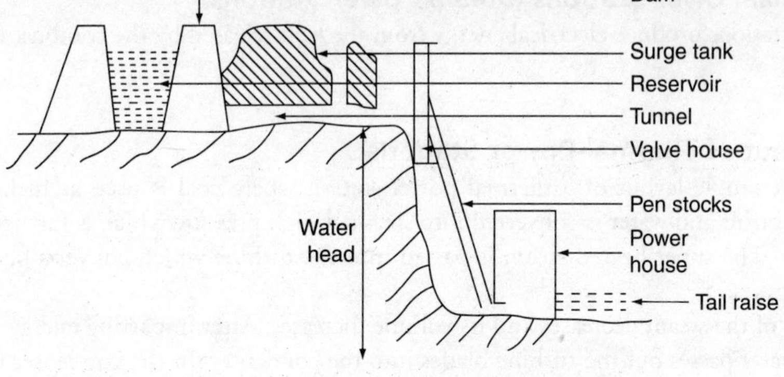

Figure 2 Hydro-power station layout.

10.2.1.2 Derivation to Calculate the Firm Capacity and Yearly Gross Output

Let

$$\text{Weight of the water } W = \text{Volume of water} \times \text{Density}$$

We already know that electrical energy generated per hour (or) power produced per hour is given by

$$W \times H \times \eta_{\text{overall}} \text{ in kWh (kilo Watt hour)}$$

where H is the water head in meters, W is the weight of water available in kg, overall efficiency (η_{overall}) is equal to (heat equivalent of electrical output)/(heat of water). This gross plant capacity is

$$\text{Firm capacity} = \text{Plant efficiency} \times \text{Gross plant capacity}$$

$$\text{Yearly gross output} = \text{Firm capacity} \times \text{Hours in a year}$$

where

$$\text{Hours in a year} = \text{Number of days} \times \text{Total hour per day} = 365 \times 24 = 8760$$

10.2.1.3 Advantages and Disadvantages of a Hydro-Power Station

Advantages:

1. It requires no fuel as water is used for the generation of electrical energy.
2. It is quite neat and clean as no smoke or ash is produced.
3. It is simple in construction.
4. It requires less maintenance.
5. It is robust and has a longer life.

Disadvantages:

1. It involves high capital cost bue to construction of dam.
2. It requires high cost of transmission lines as the plant is located in hilly areas which are quite away from the consumers.

10.2.2 Thermal-Power Stations (Steam Power Stations)

Thermal power stations produce electrical energy from the heat released by the combustion of the coal, oil or natural gas.

10.2.2.1 Layout of Thermal-Power Stations

Figure 3 shows a simple layout of a thermal power station where coal is used as fuel. In the boiler house the fuel is burnt and water is converted into steam at high pressure which is further super heated in a super heater. The super heated steam is passed into the turbine which converts heat energy into mechanical energy.

The pressure of the steam decreases and its volume increases. After imparting energy to the turbine rotor, the cold water passes out the turbine blades into the condenser. In the condenser the cold water

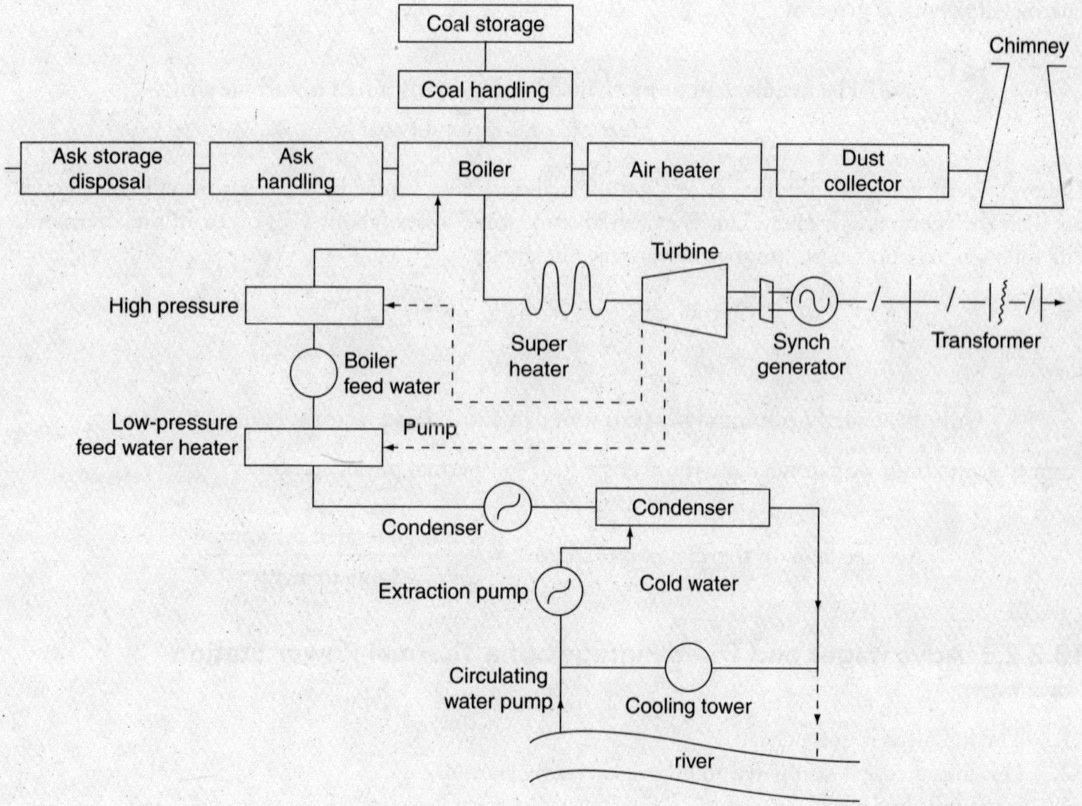

Figure 3 Thermal power station layout.

is circulated with help of a pump which condenses the low-pressure wet steam. This condensed water is further supplied to low-pressure water heater where the low-pressure steam supplied from the turbine causes the temperature of the feed water to increase. It is again heated in the high-pressure heater where high-pressure steam is used for heating. The method of taking out steam from turbines for feed water heating is called *bleeding of turbines,* which increases the overall efficiency of the boiler. The steam turbine acts as the prime-mover of the synchronous generator which converts mechanical energy into electrical energy.

10.2.2.2 Derivation to Calculate the Average Load

Electrical efficiency is the ratio of heat equivalent of electrical output to the heat of combustion of coal. It is also known as overall efficiency and is given by

$$\eta_{\text{Electrical}} = \frac{\text{Heat equivalent of electrical output}}{\text{Heat of combustion of coal}}$$

Thermal efficiency is given by

$$\eta_{\text{Thermal}} = \frac{\text{Heat equivalent of mechanical energy transmitted to turbine shaft}}{\text{Heat of combustion of coal}}$$

The overall efficiency of a thermal power station is about 29%. It may be seen that overall efficiency is less than the thermal efficiency. This is expected since some losses (about 1%) occur in the alternator. The following relation exists among the various efficiencies.

$$\text{Overall efficiency} = \eta_{\text{Electrical}} \times \eta_{\text{Thermal}}$$

Now

$$\text{Units generated per annum} = \text{Maximum demand} \times \text{Load factor} \times \text{Hours in a year}$$

From this equation, we can calculate the average load on thermal power station as

$$\text{Average load on thermal power station} = \frac{\text{Units generated per annum}}{\text{Hours in a year}}$$

10.2.2.3 Advantages and Disadvantages of a Thermal-Power Station

Advantages:

1. The fuel used is quite cheap.
2. Less initial cost as compared to other generating stations.
3. It requires less space.
4. The cost of generation is less.

Disadvantages:

1. It pollutes the atmosphere due to the production of large amount of smoke and fumes.
2. It is costlier in running cost.

10.2.3 Nuclear-Power Stations

Nuclear-power plants convert nuclear energy into electrical energy. Here, the heat energy required to produce steam is obtained by nuclear fission carried out in a nuclear reactor. The two materials which undergo nuclear fission are Uranium (U_{235}) and Plutonium (PU_{239}). These are the heaviest materials existing on Earth. Heat energy obtained from 1 kg of uranium is equivalent to that obtained from burning 36000 tonnes of coal.

10.2.3.1 Layout of Nuclear-Power Station

Figure 4 shows the typical layout of nuclear-power station. A nuclear power station consists of a steam turbine, synchronous generators, condense feed water heater, etc. as a thermal power station. However, here the heat is produced by nuclear fission carried out in the reactor house. The heat is taken to the heat exchanger by suitable gas or liquid. The steam at the required pressure and the temperature is produced in the heat exchanger and passed into the steam turbine.

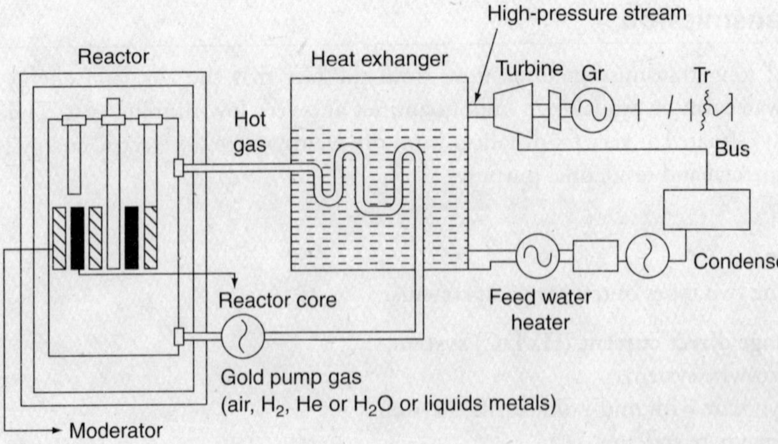

Figure 4 Layout of a nuclear-power station.

10.2.3.2 Derivation to Calculate the Output Power (P)

We have

$$\text{Number of atoms released per hour} = \frac{\text{Energy received per hour}}{\text{Energy released per fission}}$$

Now 1 gram-atom (i.e., 235g for $_{92}U^{235}$) has Avogadro's number = 6.023×10^{23} atoms. Therefore

$$\text{Mass of uranium fissioned per hour} = \frac{235}{6.023 \times 10^{23}} \times \text{Number of atoms fissioned per hour}$$

From this we can calculate the energy released per second or power output P is

$$P = \text{Energy released per fission} \times \text{Fission rate in Watt}$$

10.2.3.3 Advantages and Disadvantages of a Nuclear-Power Station

Advantages:

1. The amount of fuel required is quite small. Therefore, there is a considerable saving in the cost of fuel transportation.
2. It requires less space.
3. It has low running charges.
4. This types of plant is very economical for producing bulk electric power.
5. It ensures reliability of operation.

Disadvantages:

1. The fuel used is expensive and is difficult to recover.
2. The capital cost on a nuclear plant is very high.
3. The disposal of the by-products, which are radioactive, is a big problem. They have either to be disposed off in a deep trench or in a sea away from sea-shore.
4. Maintenance charges are high.

10.3 Transmission

The necessity of long transmission lines arose from the fact that the electrical energy obtained by harnessing the water can be available in bulk quantities at a very low running cost. The hydro-power stations are always situated at very far distances from the plains where the electric energy is required for industrial, commercial and residential purposes.

10.3.1 Types

Basically, there are two types of transmission systems:

1. **High voltage direct current (HVDC) system**
   - DC two-wire system.
   - DC two-wire with mid-point earthed system.
   - DC three-wire system.
2. **High voltage alternating current (HVAC) system**
   - Single-phase, AC, two-wire system.
   - Single-phase, AC, two-wire with mid-point earthed system.
   - Single-phase, AC, three-wire system.
   - Two-phase, AC, three-wire system.
   - Two-phase, AC, four-wire system.
   - Three-phase, AC, three-wire system.
   - Three-phase, AC, four-wire system.

The advantages of HVDC system are as follows:

1. Only two conductors are required.
2. Potential stress produced on the insulation by the DC system is approximately 70% of AC effective voltage.
3. Inductance, capacitance, phase displacement and surge problems are not there.
4. No stabilizers are required.
5. Economical use of underground cable is possible.

The advantages of HVAC system are as follows:

1. It is possible to generate electric energy at very high voltage, at 33 kV in AC as compared to 11 kV in DC.
2. The AC voltage can be stepped-up by transformers.
3. Stepping down the voltage at the substation is easier and cheaper.
4. Static substation using transformer are much efficient.
5. Maintenance of AC substation is easier and cheaper.

10.3.2 Effect of Increase in Voltage

By suitable derivation it may be proved that

1. For the same power to be transmitted, the line loss is inversely proportional to square of voltage and power factor.

2. For the same power to be transmitted with the same line loss, the weight of the conductor materials is inversely proportional to square of voltage and power factor.
3. For the same power to be transmitted, the line efficiency will increase with square of increase in voltage and power factor.
4. For the same amount of power, the voltage regulation of transmission line decreases with increase in voltage in square proportion and with increase in power factor.
5. Cost for insulation, line supports, switch gears, etc. increase with increase in system voltage.

The following are the voltage level employed at different parts of power system:

1. **For generation:** 6.6, 11, 13.2 or 33 kV.
2. **For HT transmission:** 110, 132, 230 or 400 kV.
3. **For LT transmission:** 11, 33 or 66 kV.
4. **For HT distribution:** 3.3, 6.6 or 11 kV.
5. **For LT distribution:** 415 V, 3 phase, 240 V, 1 phase.

10.4 Distribution

Distribution system consists of the following three elements:

1. **Feeders:** Feeders are conductors of large current carrying capacity, supplying power to the distributors at the feeding points.
2. **Distributors:** Distributors are conductors from which power is tapped off for supply to the consumers.
3. **Service mains:** They are the small cables taking power from distributor to the consumer premises.

10.4.1 Types

Distribution system is classified into the following types:

1. **Radial system:** Figure 5 shows the typical arrangement in a radial system. Here, the distributor is connected to the supply system on one end only by the feeder.

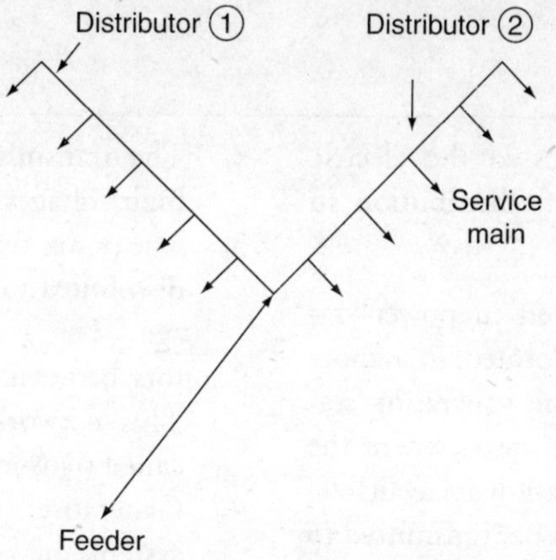

Figure 5 Radial system.

2. **Ring main system:** Figure 6 shows the arrangement in ring main system. In this system, the distribution starting from point A, after covering different areas B, C, D and E finally ends up with starting point A itself. The supply is given to the distributor by the feeder at point A. Ring main system is more reliable than radial system.

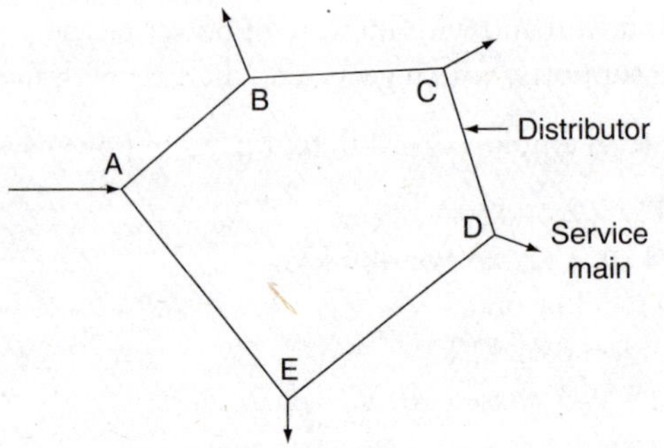

Figure 6 Ring main system.

10.4.2 Comparison of OHL with UGC

The advantages of overhead line (OHL) are as follows:

1. Transmission and distribution are cheaper.
2. High voltage can be employed easily.
3. Easy accessibility for extension and repairs.

The advantages of underground cable (UGC) are as follows:

1. Cable transmission and distribution are not subjected to supply interruption caused by lightening, birds and other severe weather conditions.
2. It reduces accidents caused by the breaking of conductors.
3. The use of cable does not spoil the beauty of the cities.

Summary

In this chapter we have discussed the electric power generation, transmission, distribution in step-by-step procedure.

1. Electric power is generated in power stations which are mostly located in remote places. This is because the generating stations are installed in the areas, where the resources for power generation are available. The power generated is to be transmitted to different load centers.
2. The transmission of electricity is done at high voltages.
3. The power received at the load center is to be distributed to the consumers at normal voltage. There is a complex network of conductors between power stations and consumer. This network is divided into major parts called transmission and distribution systems.
4. Generation, transmission and distribution systems are known as electric supply system or electric power supply system.

Solved Problems

Problem 1

A large hydro electric power station has a head of 324 m and an average flow of 1370 m^3/s. The reservoir has an area of 6400 km^3. Calculate the available hydraulic power.

Solution
Given

$$\text{Average flow } (Q) = 1370 \text{ m}^3/\text{s}$$

$$\text{Head } (H) = 324 \text{ m}$$

Therefore available hydraulic power,

$$P = 9.8\, Q\, H$$

where Q is average flow, H is the water head. Therefore

$$P = 9.8 \times 1370 \times 324 \text{ kW}$$

$$= 4350 \times 10^3$$

$$= 4350 \text{ MW}$$

Problem 2

The relation between water evaporated (M kg), coal consumption (C kg), and energy generated (in kilo Watt hour, kWh) for 8 hour shift in a thermal power plant is given by

$$M = 15000 + 10 \text{ kWh}$$

$$C = 5000 + 5 \text{ kWh}$$

Calculate to what limiting value does the water evaporation per kg of coal consumed approach as the station output increases.

Solution
Weight of water evaporated per kg of coal consumption for eight hour shift is

$$\frac{M}{C} = \frac{15000 + 10 \text{ kWh}}{5000 + 5 \text{ kWh}}$$

As the station output increases towards infinity, the ratio M/C approaches a limiting value given by

$$10/5 = 2 \text{ kg}$$

Problem 3

A thermal-power station has an overall efficiency of 20% and 0.6 kg of coal is burnt per kW of electrical energy generated. Calculate the calorific value of fuel.

Solution

Let x kcal/kg be the calorific value of fuel.
Heat produced by 0.6 kg of coal = 0.6 x kcal
Heat equivalent of 1 kWh = 860 cal
Now,

$$\text{Efficiency} = \frac{\text{Electrical output in heat unit}}{\text{Heat of combustion}}$$

$$0.2 = \frac{860}{0.6x}$$

Therefore

$$x = 7166.67 \text{ kcal/kg}$$

Problem 4

A hydro electric generating station is supplied from a reservoir of capacity 5×10^6 cubic meters at a head of 200 m. Find the total energy available in kWh if the overall efficiency is 75%.

Solution

Weight of water available is

$$\begin{aligned} W &= \text{Volume of water} \times \text{Density} \\ &= (5\times10^6)\times1000 \\ &= 5\times10^9 \text{ kg} \\ &= 5\times10^9 \times 9.81 \text{ N} \end{aligned}$$

$$\begin{aligned} \text{Electrical energy available} &= W \times H \times \eta_{\text{overall}} \\ &= 5\times10^9 \times 9.81 \times 200 \times 0.75 \text{ Watt per hour} \\ &= \frac{(5\times10^9 \times 9.81)\times(200)\times0.75}{3600\times1000} \text{ kWh} \\ &= 2.044\times10^6 \text{ kWh} \end{aligned}$$

Problem 5

An atomic power reactor can deliver 300 mW. If due to fusion of each atom of $_{92}U^{235}$, the energy released is 200 MeV, calculate the mass of uranium fissioned per hour.

Solution

Energy received from the reactor = 300 mW = 3×10^8 W (or J s^{-1})
Energy received/hour = $(3 \times 10^8) \times 3600 = 108 \times 10^{10}$ J

$$\text{Energy released/fission} = 200\,\text{MeV} = 200 \times 10^6 \times 1.6 \times 10^{-12}\,\text{J}$$
$$= 3.2 \times 10^{-11}\,\text{J}$$

$$\text{Number of atoms fissional per hour} = \frac{108 \times 10^{10}}{3.2 \times 10^{-11}} = 33.75 \times 10^{21}$$

Now 1 gram-atom (i.e., 235 kg) has 6.023×10^{23} atoms. Therefore

$$\text{Mass of uranium fissional per hour} = \frac{235}{6.023 \times 10^{23}} \times 33.75 \times 10^{21}$$
$$= 13.17\,\text{g}$$

Problem 6

A thermal power plant spends Rs 25 lakhs in one year as was consumption. The coal has heating value of 5000 kcal/kg and cost Rs 500 per tonne. If the thermal efficiency is 35% and electrical efficiency is 90%. Find the average load on the power plant?

Solution

Given

$$\text{Overall efficiency, } \eta = 0.35 \times 0.9 = 0.315$$

$$\text{Coal consumption in 1 year } = (25 \times 10^5) + 500 = 5000 \text{ tonne} = 5 \times 10^6\,\text{kg}$$

$$\text{Heat of combustion} = \text{Coal consumption per year} \times \text{Heating value of coal}$$
$$= 5 \times 10^6 \times 5000 = 25 \times 10^9\ \text{kcal}$$

$$\text{Heat output} = \eta \times \text{Heat of combustion}$$
$$= 0.315 \times 25 \times 10^9$$
$$= 7.875 \times 10^9\ \text{kcal}$$

$$\text{Annual unit generated (kwh)} = \frac{7.875 \times 10^9}{860}$$

Therefore

$$\text{Average load on the power station} = \frac{7.875 \times 10^9}{860 \times 8760}$$
$$= 1047.51 \text{ kW}$$

Problem 7

A thermal power station has an overall efficiency of 27% and 0.8 kg of coal is burnt per kWh of electrical energy generated. Calculate the calorific value of fuel.

Solution

Let x kcal/kg be the calorific value of fuel.
Heat produced by 0.8 kg of coal = 0.8 x kcal
Heat equivalent of 1 kWh = 860 cal
Now,

$$\text{Efficiency} = \frac{\text{Electrical output in heat unit}}{\text{Heat of combustion}}$$

$$0.27 = \frac{860}{0.8x}$$

Therefore

$$x = 3981.48 \text{ kcal/kg}$$

Problem 8

It has been assessed in a hydro plant that a minimum run off 95 m^3/s will be available with a head of 40 m. Find out the firm capacity and yearly gross output. Assume the efficiency of the plant to be 73%.

Solution

Given

$$\text{Weight of water available, } W = \text{Volume of water} \times \text{Density}$$
$$= 95 \times 1000 = 95000 \text{ kg/s}$$

$$\text{Water head, } H = 40 \text{ m}$$
$$\text{Work done/second} = W \times H$$
$$= 95000 \times 9.81 \times 40$$
$$= 37278 \times 10^3$$
$$= 37278 \text{ kW}$$

This is gross plant capacity. Now

$$\text{Firm capacity} = \text{Plant efficiency} \times \text{Gross plant capacity}$$

$$= 0.73 \times 37278 \times 10^3$$

$$= 27212 \text{ kW}$$

$$\text{Yearly gross output} = \text{Firm capacity} \times \text{Hours in a year}$$

$$= 27212 \times 10^3 \times 8760$$

$$= 238 \times 10^6 \text{ kWh}$$

Problem 9

Water for a hydro-power plant is obtained from a reservoir with a head of 80 m. Calculate the electrical energy generated per hour per cubic metre of water if the thermal efficiency be 0.78 and electrical efficiency 0.83.

Solution

Given

$$\text{Water head } H = 80 \text{ m}$$

$$\text{Discharge } Q = 1 \text{ m}^3/\text{s}$$

$$\eta_{\text{overall}} = 0.78 \times 0.83 = 0.6474$$

Weight of water available per second is

$$W = Q \times 1000 \times 9.81 = 9810 \text{ N}$$

Power produced is thus given by

$$W \times H \times \eta_{\text{overall}} = 9810 \times 80 \times 0.6474$$

$$= 508 \times 10^3$$

$$= 508 \text{ kW}$$

Problem 10

A river-based hydro plant has its capacity as "firm capacity" when it operates at the peak part of the load curve, the load factor here being 15%. If the rated installed capacity of the generator in order to operate the plant at the base part of the load curve is $= 10 \times 10^3$ *kW, calculate the average flow.*

Solution

Given

Rated installed capacity of the plant $= 10 \times 10^3$ kW

Head $H = 50$ m

Efficiency $\eta = 0.8$

Now

$$15\% \text{ load factor } = 10 \times 10^3 \times 0.15 \text{ kWh}$$

which for one week becomes

$$10 \times 10^3 \times 0.15 \times 168 \text{ kWh}$$

Now

$$P = 9.81\ QH\ \eta = 392.4 \text{ kW}$$

Total units generated per week = 392.4 kW.

Then,

$$10 \times 10^3 \times 0.15 \times 168 = Q \times 392.4 \times 168$$

Hence average flow, $Q = 3.823$ cubic m/s.

Multiple Choice Questions

1. In India, the transmission of electric power is done by
(a) 3-phase-3wire system
(b) 3-phase, 4-wire system.
(c) 1-phase system
(d) none of the above

2. When transmission voltage is increased, the line losses are
(a) decreased
(b) increased
(c) same
(d) none of the above

3. A 3-wire DC distribution system makes available
(a) one voltage
(b) two voltages
(c) three voltages
(d) none of the above

4. The base load power station supplies load __________ in a day.
(a) 10 hours
(b) 16 hours
(c) 24 hours
(d) 12 hours

5. An isolator is designed to open a circuit under
(a) full load
(b) normal conditions
(c) no load
(d) none of above

6. The thermal efficiency of a steam power station is about
(a) 28%
(b) 69%
(c) 80%
(d) 75%

7. As a matter of economy, voltage for power transmission should be
(a) low
(b) medium
(c) high
(d) none of the above

8. The most commonly used material for insulator of over head lines is
   (a) porcelain
   (b) glass
   (c) steatite
   (d) none of the above

9. A line which connects distributor to substation is called
   (a) service mains
   (b) distributor
   (c) feeder
   (d) none of the above

10. The cheapest plant in operation and maintenance is
    (a) steam power plant
    (b) nuclear power plant
    (c) hydro electric plant
    (d) diesel power plant

Review Questions

1. State the different systems of power transmission.
2. Compare the merits and demerits of underground cables and overhead lines.
3. With the help of a neat sketch explain the various components of a nuclear power stations.
4. Discuss the working of hydro-electric power stations with a suitable layout diagram.
5. Explain the general layout of generation transmission and distribution of electric power system.
6. Why is high voltage adopted for transmission? Explain.
7. Distinguish the different types of distribution system.
8. Discuss the working of a thermal power station with suitable layout diagram.

Problems

1. What is the power output of a $_{92}U^{235}$ reactor if it takes 30 days to use up 2 kg of fuel? Given that energy released per fission is 200 MeV and Avogadro's number = 6.023×10^{26} per kilomole.
2. Water for a hydro-electric station is obtained from a reservoir with a head of 100 m. Calculate the electrical energy generated per hour per cubic meter of water if the hydraulic efficiency be 0.86 and electrical efficiency be 0.92.
3. A thermal power station spends Rs 30 lakhs per annum for coal used in the station. The coal has a calorific value of 5000 kcal/kg and costs Rs 300 per tonne. If the station has thermal efficiency of 33% and electrical efficiency of 90%, find the average load on the station.
4. A thermal power plant spends Rs 15 lakhs in one year as coal consumption. The coal has heating value of 5000 kcal/kg and costs Rs 250 per tonne. If the thermal efficiency is 40% and electrical efficiency is 95%, find average load on power plant?
5. Prepare a rough heat balance sheet for a steam power station consuming 1 kg of coal per kW output. Take calorific value of coal as 500 0kcal. Assume boiler efficiency as 85% and electrical efficiency as 95%.

6. It has been assessed in hydel power plant that a minimum run-off of 85 m^3/s will be available with a head of 50 m. Find out the **(a)** firm capacity and **(b)** yearly gross output.

7. Find available continuous power of hydroelectric station from following data: Catchment area (A) = 100 sq km. Annual rainfall (F) = 2000 mm. Effective head (h) = 400 m. Yield factor to allow run off (K) = 60%. Efficiency of plant (η) = 70.

8. A river-based hydel plant has its capacity as "firm capacity" when it operates at the peak part of load curve. The load factor here being 25%. If rated installed capacity of generator head and plant efficiency is 5 mW, 40 m and 0.7, respectively, calculate the minimum flow of river water in order to operate the plant at base part of load curve.

9. A thermal power station has an overall efficiency of 10% and 0.6 kg of coal is burnt per kW of electrical energy generated. Calculate the calorific value of fuel.

10. An atomic power reactor can deliver 300 MW. If due to fission of each atom of ${}_{92}U^{235}$ the energy released is 250 mW, calculate the mass of fissioned per hour.

Answers

Multiple Choice Questions

1.	(a)	**3.**	(b)	**5.**	(c)	**7.**	(c)	**9.**	(c)
2.	(a)	**4.**	(c)	**6.**	(a)	**8.**	(a)	**10.**	(a)

Problems

1. 63.2 mW
2. 775 kWh
3. 1971 kW
4. 378.3 kW
5. Heat balance sheet: boiler loss = 0.8721 kW, turbine loss = 3.854 kW; total loss = 4.7261. Output: 1 + 4.7261 = 5.7261. Input: 5.7261. Heat balance is observed as input = output + losses.
6. **(a)** 42271.73 kW, **(b)** 370×10^6
7. 10550 kW
8. $Q = 9.1015$ cubic m/s
9. 143.4 kcal/kg
10. 16.85 kg

11 Essentials of an Electronic Circuit

LEARNING OBJECTIVES

After reading this chapter, you will be able to:

- Understand the essentials of passive electronic circuit.
- Understand the concepts of resistor, inductor and capacitor.
- Understand the concept of chokes.

An electronic circuit essentially consists of two types of components, namely,

1. Active components.
2. Passive components.

Voltage and current sources, semiconductors and vacuum devices come under the category of active. In addition to these components, passive components such as resistors, capacitors and inductors are also necessary to control electronic circuits. This chapter deals with passive components while the theory, operation and application of semiconductor devices are covered in Chapter 13.

Passive components are capable of controlling the current, rate of change of current and rate of change of voltage in an electronic circuit. Further signals of desired shapes can also be derived using proper combination of these elements. However, for signal generation in addition to passive components, active components are also required.

11.1 Resistors

Any material that offers resistance to the flow of electric current is called resistor. A resistor is capable of conducting current in both directions. But while doing so, a fraction of the energy is converted into heat energy. In electronic circuits, resistances are mainly used for limiting the current flow through the devices. The specifications of a resistor include in Ohmic value, power rating, tolerance and working voltage.

11.2 Fixed Resistors

A fixed resistor is one whose value cannot be changed by the user. It could be any one of the following types.

1. Carbon composition resistors.
2. Thin film resistors.
3. Wire wound resistors.

11.2.1 Carbon Composition Resistors

These resistors are made up of finely divided carbon or graphite, mixed with a power insulated material as a binder. The resistance element is usually enclosed in a plastic case for mechanical strength and insulation. The two ends of the carbon resistance elements are fixed to metal caps with loads of tinned copper wire. Carbon composition resistors are available with Ohmic values ranging from 1 Ω to 20 MΩ. The size and the power ratings of these resistors are small. Quarter watt and half watt resistances are very commonly used in electronic circuits. The cross-sectional view of a carbon composition resistor is shown in Fig. 1. The resistance values are indicated by circular rings of different colors which follow a specific color code.

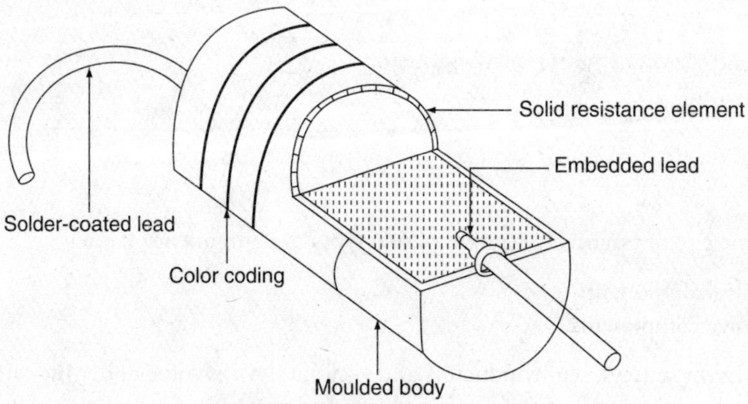

Figure 1 Carbon composition resistor.

11.2.2 Thin Film Resistors

These resistors are manufactured by depositing a very thin layer of conducting material on an insulating rod, tube or plate made of ceramic or glass. There are two types of thin film resistors, namely,

1. Carbon film resistor.
2. Metal film resistor.

Figure 2 illustrates a thin film resistor.

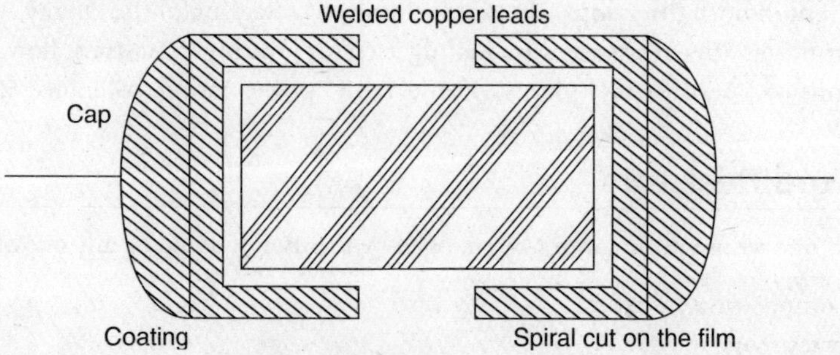

Figure 2 Thin film resistor.

11.2.2.1 Carbon Film Resistors

These resistors are manufactured by depositing pure carbon on a ceramic substance by thermal decomposition of hydrocarbons at 1273 K. Lead wires are attached to the film at each end and the whole assembly is surrounded by an insulating material. The sheet resistance of the film is controlled by the carbon film thickness.

These resistors are commercially available in the range of 10 Ω to 10 MΩ with the power rating available upto 5 W. Carbon film resistors are less noisy in comparison to carbon composition resistors. These resistors are used in measuring instruments where close tolerance is required. They also have better stability.

11.2.2.2 Metal Film Resistors

A common type of axial resistor today is referred to as a metal film resistor. Metal electrode leadless face (MELF) resistors often use the same technology, but are a cylindrically shaped resistor designed for surface mounting. Note that other types of resistors (e.g., carbon composition) are also available in MELF packages. Metal film resistors are usually coated with nickel chromium (NiCr), but might be coated with any of the cermet materials listed above for thin film resistors. Unlike thin film resistors, the material may be applied using different techniques than sputtering (though that is one such technique). Also, unlike thin film resistors, the resistance value is determined by cutting a helix through the coating rather than by etching. (This is similar to the way carbon resistors are made.) The result is a reasonable tolerance (0.5, 1, or 2%) and a temperature coefficient that is generally between 50 and 100 ppm/K. Metal film resistors possess good noise characteristics and low non-linearity due to a low voltage coefficient. Also beneficial are the components: efficient tolerance, temperature coefficient and stability.

11.2.3 Wire-Wound Resistors

This type of resistor is obtained by tightly winding a known length of wire on a cylindrical ceramic core. The material of the wire is chosen from high resistivity alloys like Unichrome and Manganin. The wire is normally applied with a coating of insulating material like enamel. Resistors of different values are obtained by varying the length or thickness of the wire or both. The end leads are so surrounded to end card and the whole assembly is coated with a positive material such as enamel. A wire-wound resistor is shown in Fig. 3.

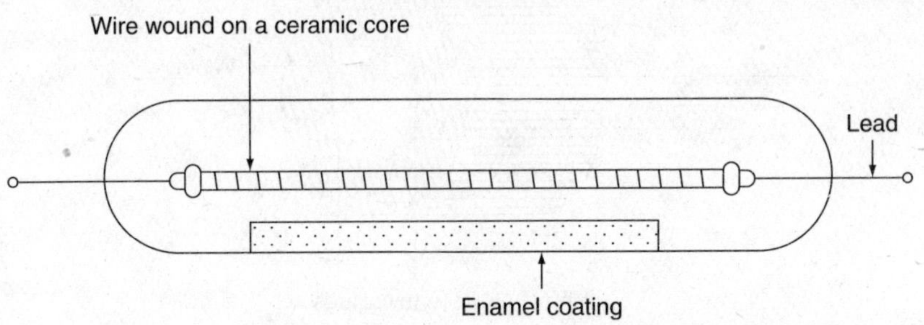

Figure 3 Wire-wound resistors.

Wire-wound resistors are available with resistance values ranging from 0.1 Ω to 0.1 MΩ. They are preferred when resistors of high power rating (10 W to 75 W) are required.

11.3 Variable Resistors

Variable resistors are those whose resistance values can be adjusted between zero and maximum value. They are immediately useful in adjusting voltages and currents. Volume control in radio and brightness control in television are examples of the applications of variable resistors. Variable resistors can be classified into

1. Wire-wound variable resistors.
2. Carbon composition potentiometers.

11.3.1 Wire-Wound Variable Resistors

The wire-wound variable resistors are further classified into the following three types:

1. Wire-wound rotary variable type resistors.
2. Wire-wound solenoid type resistors.
3. Helical-wound variable resistors.

11.3.1.1 Wire-Wound Rotary Variable Resistors

These types of resistors are popular in TV and radio receivers for volume control. On a strip of Bakelite former, a nichrome wire is wound as shown in Fig. 4. The strip is then bent to form an arc which is fixed in a Bakelite mould with end plates and screws as shown in the figure. High power dissipation aluminum former with an oxide coating is preferred.

11.3.1.2 Wire-Wound Solenoid Type Resistors

The wire-wound solenoid type resistor makes use of copper in oxidized form as the resistance winding. The winding is of circular or hexagonal shape wound on a ceramic or steel former. The commercial wire wound solenoid resistor, is depicted in Fig. 5.

This type of resistor is also referred to as rheostat. The resistance of a rheostat is changed by varying the position of the sliding brush, contact. Copper or graphite can be used as the brush material.

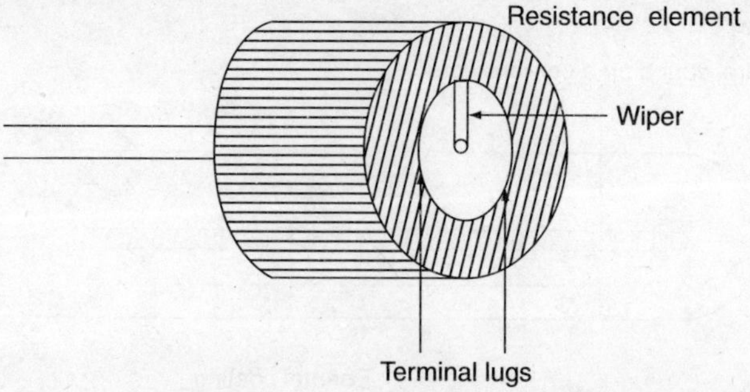

Figure 4 Wire-wound rotary variable resistor.

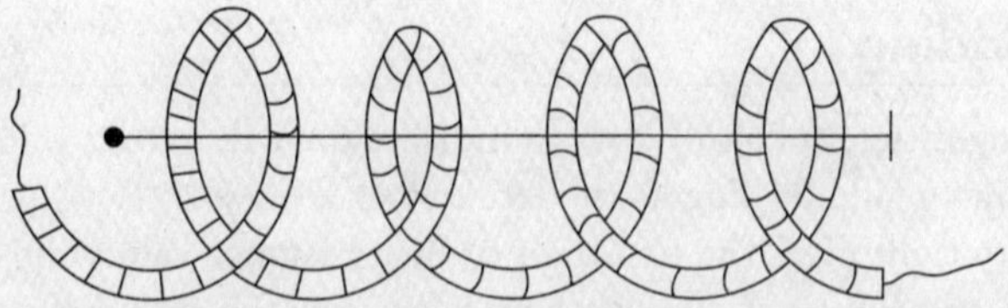

Figure 5 Wire-wound solenoid type resistor.

The rheostat is capable of withstanding temperature up to 300°C. These are used in industries and laboratories for obtaining the desired voltage drop.

11.3.1.3 Helical-Wound Variable Resistors

These resistors are obtained by winding the resistance element over a helical former of circular or rectangular cross-section. These resistors are highly linear, possess very large angle of rotation and high revolution. They are available in the resistance range of 1 Ω to 10 kΩ. The power ratings can go up to 200 W.

11.3.2 Carbon Composition Potentiometers

Carbon composition potentiometers can be classified into the following two types:

1. **Molded type:** In this type, the carbon composition mixture is molded into a cavity over a plastic base, or a mixture of carbon is coated on a ring of insulating material. The resistance can be changed by varying the movable contact brush made of carbon. This type of potentiometer finds applications in industries and defence equipments.
2. **Coated film type:** The coated film type potentiometers find their use in Television. They affect brightness and contrast control. They are also used in radio receivers and measuring instruments. The coated film potentiometer is obtained by coating mixture of carbon and silica with a binder on a ring of insulating material. The film surface is hardened to minimize the abrasiveness of the film due to sliding.

The external and internal view of carbon composition potentiometers is shown in Fig. 6.

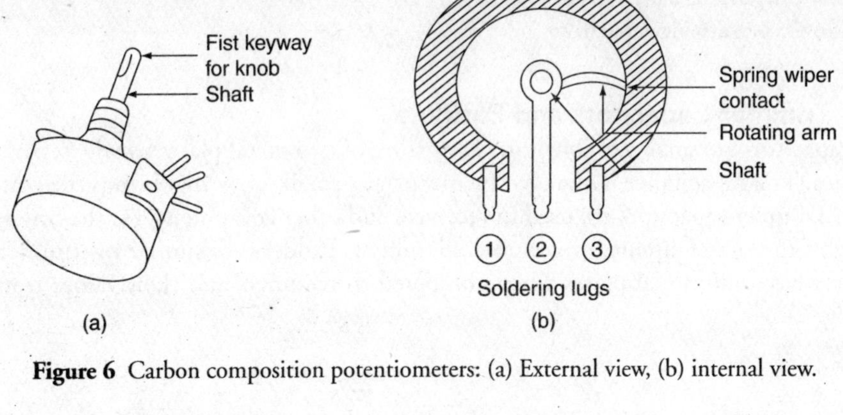

Figure 6 Carbon composition potentiometers: (a) External view, (b) internal view.

11.4 Color Coding

Resistors are coded using different color bands to indicate their resistance values and tolerance. There are four color bands drawn side by side at the left end of a resistor. Reading the first band close to the left edge from left to right gives the first digit of the resistance value, the next band indicates the second digit, the third band gives the number of rows after the two digits, while the fourth band gives the tolerance value. Figure 7 shows the color codes and tolerance values for various colors. For example, a resistor with following color bands Red-Red-Orange-Gold will have 2200 Ω tolerance resistance.

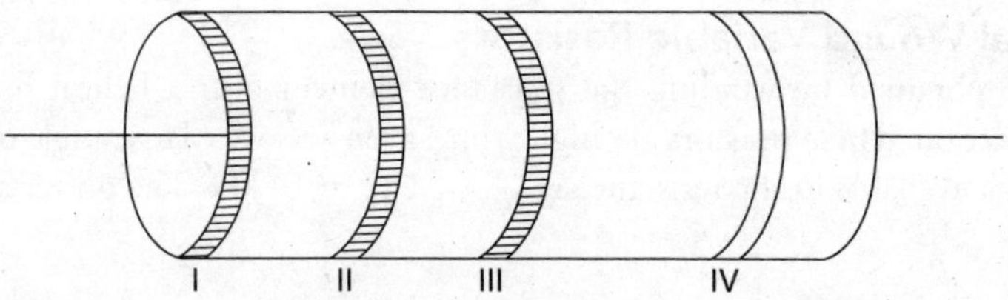

Figure 7 Color coding.

11.5 Capacitors

The function of a capacitor is to store electric charge. Capacitors vary in shape, size and design depending on intended use. Capacitors are generally classified into the following two types:

1. Variable capacitors.
2. Fixed capacitors.

11.5.1 Variable Capacitors

Variable capacitors are so named because of their ability to vary their capacitance values. The different types of variable capacitors are as follows:

1. Trimmer capacitor and padders.
2. General-purpose variable capacitors.
3. Precision type variable capacitors.

11.5.1.1 Trimmer Capacitors and Padders

Trimmer capacitors are small variable units consisting of two metal plates usually separated by thin piece of mica. The capacitance is changed by means of a small screw that brings the plates closer to each other. Trimmer capacitors are used in receivers and other equipments for the fine adjustments of capacity of the tuned circuit for accurate alignment. Padders are similar to trimmer but larger in size. They have additional plates when compared to trimmers and their values from 10 pF to 500 pF (Fig. 8).

Figure 8 Trimmers and padders: (a) Cormin type, (b) mine type.

11.5.1.2 General-Purpose Variable Capacitor

The rotor stator capacitor comes under this type. If consists of two sets of metal plates. The fixed plates are connected together and form the stator. The movable plates are connected together to the shaft in such way that the rotor plates mesh with stator plates. There is no electric contact between the two sets of plates. Air is the dielectric medium. When the plates are fully meshed, the effective area is at its greatest and hence maximum capacitance results. Conversely, when the plates are out of mesh, the effective area is gently reduced and minimum capacitance results. Theoretically, an infinite number of continuous variables capacitance can be realized between the limits of maximum and minimum capacitances.

Two rotor-stator capacitances are shown in Fig. 9 along with their symbols. The split-stator capacitor is similar to the capacitor used for tuning most (older) types of radio receivers. The midgot single capacitor is shown in the figure is used for tuning high-frequency circuits. The dotted lines joining the symbol in figure indicate that the capacitors are gauged together. That is, both the rotors move simultaneously because they are attached to a common shaft.

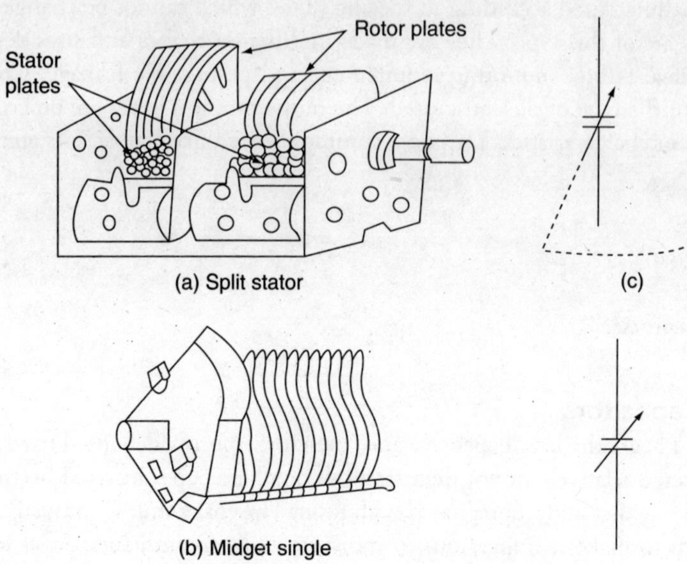

Figure 9 Rotor-stator capacitances.

11.5.1.3 Precision Variable Capacitor

The precision variable capacitor comprising a stator and a rotor with aluminum plates is shown in Fig. 10.

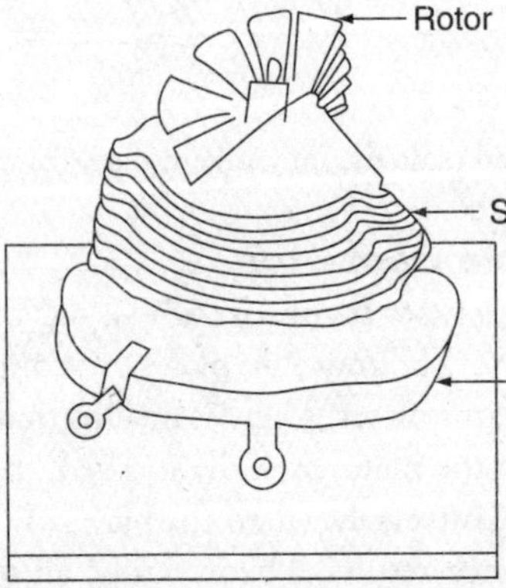

Figure 10 Precision variable capacitor.

Quartz is used to insulate the light alloy and plates which are mounted on a duralium shaft. The capacitance value ranges from 40 pF to 1000 pF. The drive to the moving assembly is provided by a spring loaded worm wheel.

11.5.2 Fixed Capacitors

Fixed capacitors are manufactured according to specific values which cannot be changed. Most capacitors used in industries are of this type. They are made in different shapes and sizes depending on the required capacity, voltage rating, mounting requirement and permissible leakage. Customarily, they are named according to the type of dielectric used. The most common kinds are described below with emphasis on their salient characteristics. The most commonly used fixed capacitors are:

1. Paper capacitors.
2. Mica capacitors.
3. Ceramic capacitors.
4. Plastic capacitors.
5. Electrolytic capacitors.

11.5.2.1 Paper Capacitors

Paper capacitors (Fig. 11) are the least expensive and, therefore, most widely used fixed capacitors. The dielective is Kraft paper, a relatively heavy, high strength Sulphate paper inserted between thin aluminum foil plates walled together and impregnated with resin. The entire unit is encapsulated in a tubular plastic or metal sheath to make it impervious to moisture and contaminants. Axial leads are usually brought out of each end. Typical capacities range from about 0.0001 μF to 2 μF with voltage rating ranging from 200 V to 600 V.

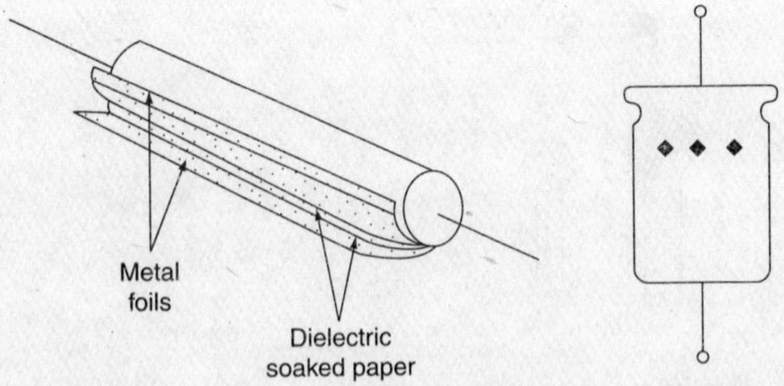

Figure 11 Paper capacitor.

11.5.2.2 Mica Capacitors

Mica capacitors (Fig. 12) find application when capacitors of high voltage ratings are required. Mica is one of the best insulator with low loss. Radio transmitters use these capacitors since the voltage rating and current rating and current may go as high 30 kV and 100 A. These units are structurally quite large and capacities in excess of 0.05 μF are uncommon. A variation of this kind of capacitor is the silver mica capacitor. A thin layer of silver is deposited on the surface of the mica, and resulting capacitor has excellent stability and tolerance properties. These capacitors are used principally for temperature compensation in radio frequency tuned circuits owing to their predictable and linear temperature–capacitance (TC) variations.

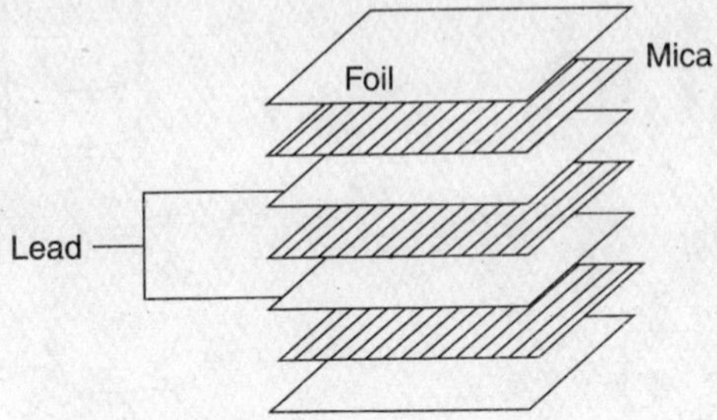

Figure 12 Mica capacitor.

11.5.2.3 Ceramic Capacitors

Ceramic capacitors consist of a ceramic disc with silver electrodes attached to each flat surface. Leads are bonded to these electrodes to provide connection to the unit. Figure 13 shows the details of a monolithic ceramic capacitor. Ceramic capacitors are used in application with requirements ranging from low frequency and audio to high, very high and ultra high frequency through 1000 MHz. Dielectric materials for ceramic capacitors are made from mixture of barium and strontium titanates, blended with rare earth and other additives for improving electrical characteristics.

Figure 13 Ceramic capacitors: (a) Disc type; (b) tubular type.

11.5.2.4 Plastic Capacitors

Plastic films such as polysters or Mylar have largely replaced paper as a dielectric for general capacitors. Plastic is much denser than paper and contamination due to foreign particles is absent. Plastic can withstand higher temperatures and is more stable than paper. Plastic film and paper capacitors are generally constructed in same manner. The details of constructions of a plastic film extended foil capacitor are shown in Fig. 14.

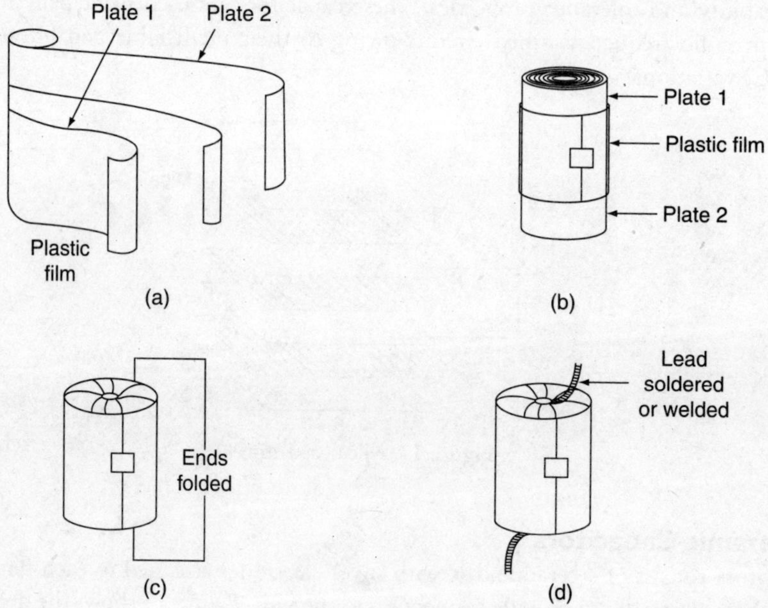

Figure 14 Plastic capacitors.

11.5.2.5 Electrolytic Capacitor

Aluminium electrolytic capacitors are used when a large amount of capacitance is needed and the leakage current is not critical. Their capacities range from a few microfarads to hundreds or thousands of farads. Generally, they are not rated higher than 450 V in values upto several hundred microfarads. The voltage ratings tend to decreases as the capacitance increases because thinner dielectrics are used in larger capacitances.

In its simplest form, the electrolytic capacitors are constructed similar to paper capacitors. The construction of an electrolytic capacitor is shown in Fig. 15.

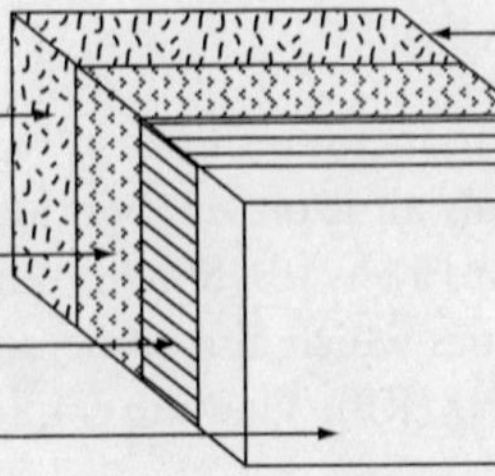

Figure 15 Electrolytic capacitor.

The electrolytic capacitor consists of a roll of aluminum foil coated on one side with aluminum oxide, the aluminum being the positive plate and the oxide the dielectric. A layer of porous paper or gauge saturated with an electrolytic is placed over the aluminum oxide. Another layer of aluminum without coating is the placed over this layer which serves as the negative plate. In most cases, the negative plate is connected directly to the aluminum contained. Since the dielectric film is extremely thin, very large capacitors are possible.

In electrolytic capacitors, tantalum is used in place of aluminum when the size of the capacitor is a consideration. In tantalum capacitors, a dielectric oxide film – tantalum pentoxide – can be electrolytically formed on a tanalum metal. These capacitors are superior to aluminum capacitors.

There are two types of tantalum electrolytic capacitors:

1. Wet type.
2. Dry type.

11.6 Dissipation Factor

The dissipation factor gives the relation between the power wasted and power stored in a capacitor.

$$\text{Dissipation factor} = \frac{\text{Power wasted per cycle}}{\text{Power stored per cycle}}$$

where power storage is due to dielectric loses and current leakage.

11.7 Inductors

An inductor is a passive component that is capable of storing energy. It always opposes the change in the current flowing through it by producing a self-induced e.m.f. There are two type of inductors:

1. Fixed type inductors.
2. Variable type inductors.

11.7.1 Fixed Inductors

In fixed inductors, the inductance value remains unchanged. They are further classified as

1. Air-core inductors.
2. Iron-core inductors.
3. Ferrite-core inductors.

11.7.1.1 Air Core Inductors

The shape and symbol of an air-core inductor is shown in Fig. 16. Air-core inductors are most widely used in communication circuits. Since, only air is present within the coils, they have very low values of inductance ranging from a fraction of μH to a few μh. Sometimes the coil of wire is wound over non-magnetic materials like ceramic rods and plastics which also come under the category of air-core inductors. They are used at frequencies upto 200 MHz (RF). They are employed mainly in radio and TV equipments.

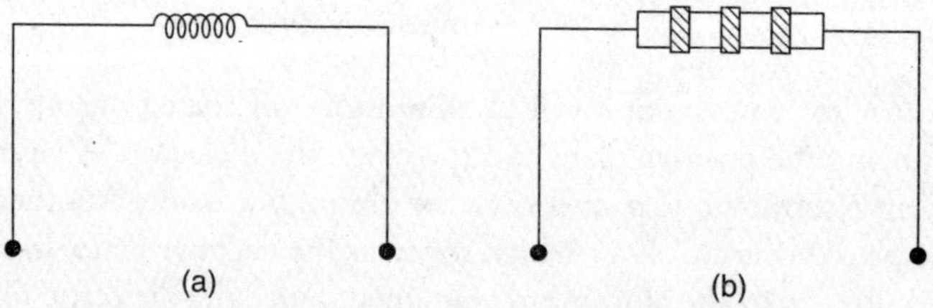

Figure 16 Air-core inductors: (a) Symbol; (b) shape.

11.7.1.2 Iron-Core Inductors

When a coil is wound over a solid or laminated iron core, an iron-core inductor is obtained. Usually the iron core is laminated by passing together thin laminated sheets to reduce eddy current loss. Iron-core inductors are used in low-frequency circuits such as filter circuits in power supplies, chokes in fluorescent tubes (Fig. 17).

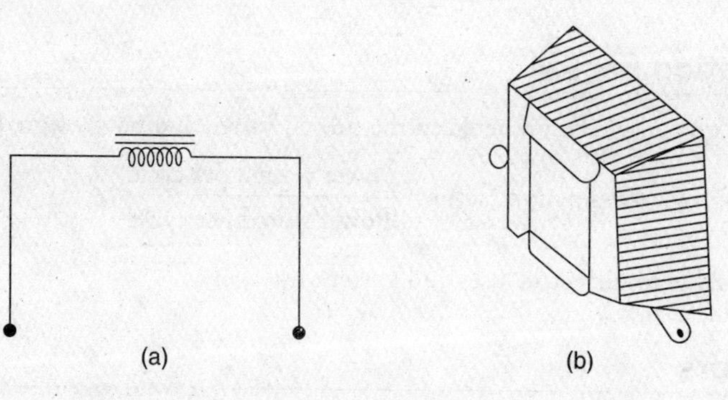

Figure 17 Iron-core inductors: (a) Symbol; (b) shape.

11.7.1.3 Ferrite-Core Inductors

An iron-core inductor is not suitable for high-frequency applications because of increase in hysterics and eddy current loss. This difficulty is overcome by using ferrite material as core. Ferrite-core inductors are shown in Fig. 18. A ferrite is basically insulator having high permeability. Ferrites are non-metallic compounds.

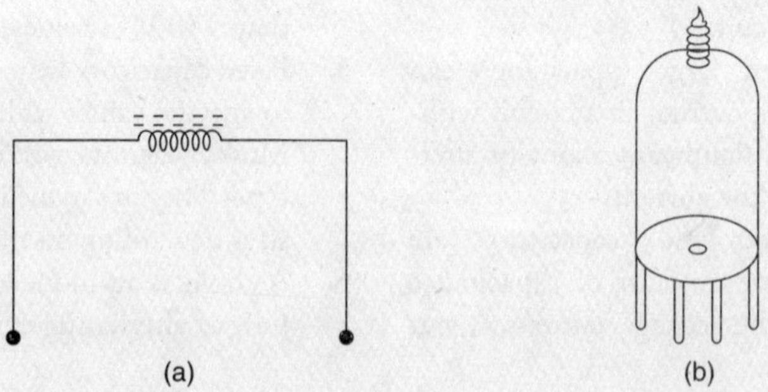

Figure 18 Ferrite-core inductors: (a) Symbol; (b) shape.

The values of ferrite inductors range from a few micro Henry to few milli Henry. The typical applications of ferrite-core inductors are in the following circuits:

1. RF chokes.
2. Switching mode power supply.
3. Filters and communication equipments.

11.7.2 Variable Inductor

Variable inductors are those inductors whose inductance can be varied to a desired value. This can be made either by moving the coil in and out of the core or by varying the air gap in the core gap. It can also be done by varying the number of turns coil used in the inductors.

11.8 Chokes

A choke is an inductance coil to restrict the flow of alternating current chokes may be classified.

1. **Filter choke:** They are usually laminated core inductors. These are used for smoothing pulsating current output from a rectifier.
2. **Audio frequency choke:** These are used to provide high impedance to audio frequencies. Their size and inductance values are less than that of filter choke.
3. **Radio frequency choke:** Chokes having still lesser inductance value are used as radio frequency chokes. They help to block the radio frequency.

Summary

This chapter discussed about the passive electronic components: resistors, inductors, capacitors.

1. Resistors in series have a common current and share the total voltage and power in proportion to the resistance value. Resistors in parallel have a common voltage, and share the total current and power in proportion to conductance value.

2. An inductor with a large inductance value can absorb voltage in a circuit without contributing significant about of alternating current to the circuit.
3. A capacitor with large capacitance can absorb alternating current in a circuit without contributing significant about of alternating voltage in the current.
4. Aluminum electrolytic capacitors are used when a large amount of capacitance is needed and the leakage current is not critical. Their capacities range from a few microfarads to hundreds or thousands of farads. Generally, they are not rated higher than 450 V in values.
5. Fixed capacitors are manufactured according to specific values which cannot be changed. Most capacitors used in industries are of this type. They are made in different shapes and sizes depending on the requirement.
6. A choke is an inductance coil to restrict the flow of alternating current.

Solved Problems

Problem 1

Determine the resistance of 1200 m of copper cable having a diameter 12 mm if the resistivity of copper is 11.7 × 10⁻⁸ Ωm.

Solution

Cross-sectional area of cable,

$$A = \pi r^2 = \pi\left(\frac{12}{2}\right)^2$$
$$= 36\pi\ \text{mm}^2$$
$$= 36\pi \times 10^{-6}\ \text{m}^2$$

Therefore, resistance is

$$\frac{\rho l}{A} = \frac{(11.7\times 10^{-8}\ \Omega\,\text{m})(1200\,\text{m})}{(36\pi \times 15^{-6}\ \text{m}^2)}$$
$$= 0.180\,\Omega$$

Problem 2

The resistance of 1.5 km of wire of cross-sectional area 0.17 mm² is 150 Ω. Determine the resistivity of the wire.

Solution

We know that resistance is given by

$$R = \rho l / A$$

So resistivity

$$\rho = \frac{RA}{l} = \frac{(150\,\Omega)(0.17\times10^{-6}\,\text{m}^2)}{(1500\,\text{m})}$$

$$= 0.017\,\mu\Omega\text{m}$$

Problem 3

Calculate the cross-sectional area, in mm², of a piece of copper wire 40 m in length and having a resistance of 0.25 Ω. Take the resistivity of copper as 0.02×10^{-6} Ωm.

Solution

We know that resistance is given by

$$R = \rho l/A$$

So cross-sectional area is

$$A = \rho l / R$$

$$= \frac{0.02\times10^{-6}\times40}{0.25}$$

$$= 3.2\,\text{mm}^2$$

Problem 4

A direct current of 4 A flows in to a previously unchanged 20 μF capacitor for 3 ms. Determine the voltage between the plates.

Solution

Given that

$$I = 4\text{ A}$$

$$C = 20\ \mu\text{F} = 20 \times 10^{-6}\text{ F}$$

$$t = 3\text{ ms} = 3 \times 10^{-3}\text{ s}$$

Now

$$Q = It = 4 \times 3 \times 10^{-3}\text{ C}$$

Therefore voltage is given by

$$V = \frac{Q}{C} = \frac{4\times3\times10^{-3}}{20\times10^{-6}} = 600\text{ V}$$

Problem 5

A 5 μF capacitor is charged so that the voltage between its plates is 800 V. Calculate how long the capacitor can provide an average discharge current of 2 mA.

Solution
It is given that

$$C = 5\ \mu\text{F} = 5 \times 10^{-6}$$

$$V = 800\,\text{V}$$

$$I = 2\ \text{mA} = 2 \times 10^{-3}\text{A}$$

Now

$$Q = CV = 4 \times 10^{-3}\ \text{V}$$

Also,

$$Q = It$$

Thus

$$t = \frac{Q}{I} = \frac{4\times10^{-3}}{2\times10^{-3}} = 2\,\text{s}$$

Therefore, the capacitor can provide on average discharge current of 2 mA for 2 s.

Problem 6

Determine the resistance of a metal tube in terms of the external diameter D = 10 cm, the internal diameter d = 9 cm, the length l and the resistivity ρ = 0.017. Calculate the resistance of a copper tube 0.5 cm thick and 2 m long.

Solution
We know that

$$\begin{aligned} R &= \frac{\rho l}{A} \\ &= \frac{\rho l}{(\pi/4)(D^2 - d^2)} \\ &= \frac{0.017\times 2}{(\pi/4)(10^2 - 9^2)\times 100\ \text{mm}^2} \\ &= 23.14\ \mu\Omega \end{aligned}$$

Problem 7

A capacitor having a capacitance of 80 μF is connected above a 500 V DC supply. Calculate the charge.

Solution

Using the formula

$$Q = CV$$

we get the charge as

$$\begin{aligned}\text{Charge} &= (80\times10^{-6})\ \text{F}\times 500\ \text{V}\\ &= 0.04\ \text{C}\\ &= 40\ \text{mC}\end{aligned}$$

Problem 8

Three capacitors have capacitances of 2, 4, 8 μF, respectively. Find the total capacitance when they are in parallel.

Solution

Total capacitance when in parallel is equal to the sum of individual capacitances. Therefore

$$\begin{aligned}C &= C_1 + C_2 + C_3\\ &= 2+4+8\\ &= 14\ \mu\text{F}\end{aligned}$$

Problem 9

Three capacitors have capacitances of 2, 4, 8 μF, respectively. Find the total capacitance when they are in series.

Solution

Total capacitance when they are in series is given by

$$\begin{aligned}\frac{1}{C} &= \frac{1}{C_1} + \frac{1}{C_2} + \frac{1}{C_3}\\ &= 0.5+0.25+0.125\\ &= 0.875\end{aligned}$$

Therefore

$$C = 1.14\ \mu\text{F}$$

Problem 10

Calculate the capacitance of two metal plates of area 30 m^2 and separated by a dielectric 2 mm thick and of relative permittivity 6. If the electric field strength in the dielectric is 500 V/mm, calculate the total charge on each plate.

Solution

Capacitance is given by

$$C = \frac{\varepsilon_0 \varepsilon_r A}{d}$$
$$= \frac{8.854 \times 10^{-12} \times 6 \times 30}{2 \times 10^{-3}}$$
$$= 0.797 \times 10^{-6} \text{ F}$$
$$= 0.797 \ \mu\text{F}$$

The voltage is given by

$$V = E \times d$$
$$= 500 \times 2$$
$$= 1000 \text{ V}$$

Hence the charge is

$$Q = CV$$
$$= (0.797 \times 10^{-6}) \times (1000)$$
$$= 0.797 \text{ mC}$$

Problem 11

A mica dielectric parallel-plate capacitor has 21 plates each having an effective area of 5 cm and each separated by a gap of 0.005 mm. Find the capacitance. Take the relative permittivity of mica as 6.

Solution

Capacitance

$$C = (n-1)\frac{\varepsilon_0 \varepsilon_r A}{d}$$
$$= (21-1)\frac{(8.854 \times 10^{-12}) \times 6 \times (5 \times 10^{-4})}{0.005 \times 10^{-3}}$$
$$= 0.1062 \times 10^{-6} \text{ F}$$
$$= 0.1062 \ \mu\text{F}$$

Problem 12

Three capacitors 2 μF, 4 μF, 6 μF are connected in series to a 220 V DC supply. Find
(a) Total capacitance.
(b) Charge on each capacitor.

Solution

(a) The total capacitance is given by

$$\frac{1}{C_T} = \frac{1}{2} + \frac{1}{4} + \frac{1}{6} = \frac{6+3+2}{12} = \frac{11}{12}$$

Therefore

$$C_T = \frac{12}{11} = 1.091\ \mu\text{F}$$

(b) In series connection, charge on each capacitor is the same. Therefore

$$\begin{aligned} Q &= C_T\ V \\ &= (1.091 \times 10^{-6}) \times 220 \\ &= 240 \times 10^{-6}\ \text{C} \end{aligned}$$

Problem 13

A 5 μF capacitor is charged to a potential difference of 100 V and then connected in parallel with an uncharged 3 μF capacitor. Calculate the p.d across the parallel capacitors.

Solution

Charge on 5 μF capacitor is given by

$$Q = CV = (5 \times 10^{-6}) \times 100 = 0.0005\ \text{C}$$

When the capacitors are connected in parallel, the total capacitance is

$$C_T = 5 + 3 = 8\ \mu\text{F}$$

Potential difference across parallel capacitor is given by

$$\frac{Q}{C_T} = \frac{0.0005}{8 \times 10^{-6}} = 62.5\ \mu\text{F}$$

Multiple Choice Questions

1. The unit of resistivity is
   (a) ohms
   (b) ohm millimeter
   (c) ohm meter
   (d) ohm/meter

2. The resistance of a 2 km length of cable of cross-sectional area 2 mm^2 and resistivity of 2×10^8 Ωm is
   (a) 0.02 Ω
   (b) 20 Ω
   (c) 0.02 mΩ
   (d) 200 Ω

3. A color coding of red–violet–black on a resistor indicates a value of
   (a) 27 Ω ± 20%
   (b) 270 Ω
   (c) 270 Ω ± 20%
   (d) 27 Ω ± 10%

4. A resistor market as 47 kΩ indicates a value of
   (a) 47 Ω ± 20%
   (b) 4.7 KΩ ± 20%
   (c) 0.47 Ω ± 10%
   (d) 4.7 KΩ ± 2%

5. The charge of 10 pF capacitor when the voltage applied to it is 10 kV is
   (a) 100 μC
   (b) 0.1 C
   (c) 0.1 μC
   (d) 0.01 μC

Review Questions

1. List a few passive circuit components and active circuit components.
2. Distinguish between resistors and other passive circuit components, namely, inductors and capacitors.
3. Name different types of potentiometers. Explain carbon potentiometer in detail.
4. What is color coding? Explain.
5. What is meant by tolerance of resistors?
6. List the various capacitors used in electronic circuits.
7. What are polarized and non-polarized electrolytic capacitors?
8. Compare the constructional features and characteristics of the following capacitors: (a) paper, (b) mica, (c) ceramic, (d) plastic.
9. What is the application of the chokes?
10. Give an account of fixed inductors.
11. Classify the inductors and explain briefly.
12. State and explain the various losses of a capacitor.
13. What are the specifications of a capacitor?
14. State the factor affecting the capacitance of a capacitor.
15. Write short notes on variable capacitors.
16. Write the applications of inductors.
17. Define the dielectric strength when referred to a capacitor.

18. Describe briefly polarized and non-polarized electrolytic capacitors.

19. Explain the dissipation factor of a capacitor.

20. Explain the active components in detail.

Problems

1. Calculate the resistance of 2 km length of aluminum overhead power cable if the cross-sectional area of the cable is 100 mm^2. Take the resistivity of aluminum to be 0.03×10^{-6} Ωm.

2. The resistance of 1.5 km of wire of cross-sectional area 0.17 mm^2 is 150 Ω. Determine the resistivity of the wire.

3. Find the charge on 10 μF capacitor when the applied voltage is 250 V.

4. Determine the voltage across a 100 pF capacitance to charge it with 2 μF.

5. The charge on the place of a capacitor is 6 mC when the potential between them is 52.4 kV. Determine the capacitance of a capacitor.

6. Calculate the capacitance of an air insulated capacitor with 13 plates, each 10 cm^2. The distance between the plates being 2 mm.

7. A direct current of 3 A flows into a previously uncharged 20 μF capacitor for 2 ms. Determine the voltage between the plates.

8. A transformer has primary winding having 1250 turns and secondary winding having 740 turns. The length of the magnetic path is 27.5 cm and cross-sectional area of core is 5.5 cm^2. Assume $\mu_r = 750$. Find the self-inductance of (**a**) primary coil and (**b**) secondary coil.

9. Determine the parallel resistance between the points *A* and *B* of the circuit shown in Fig. 19.

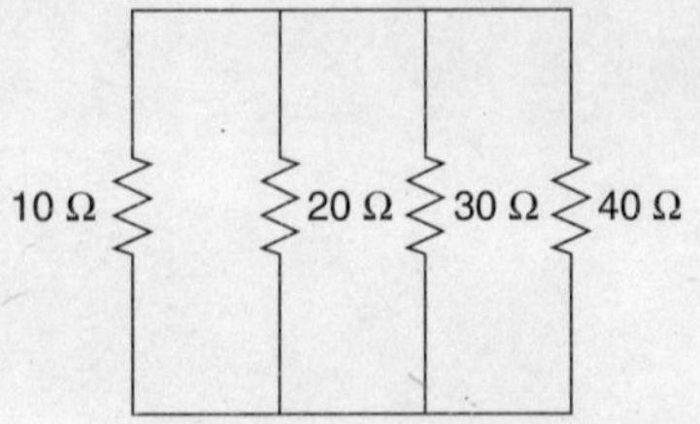

Figure 19 Problem 9.

10. Two inductances $L_1 = 30$ mH and $L_2 = 60$ mH are connected in parallel. Find equivalent inductance.

11. Two capacitances $C_1 = 0.1$ μF and $C_2 = 0.3$ μF are connected in parallel. Find equivalent capacitance.

Answers

Multiple Choice Questions

1. (c)
2. (b)
3. (c)
4. (d)
5. (b)

Problems

1. 0.6 Ω
2. 0.017 μΩm
3. 2.5 mC
4. 2 kV
5. 114.5 nF
6. 530 pF
7. 300 V
8. (**a**) 2.6773 H, (**b**) 0.9383 H
9. 4.8 Ω
10. 20 mH
11. 0.4 μF

12 Basic Semiconductor Devices and Circuits

LEARNING OBJECTIVES

After reading this chapter, you will be able to:

- Understand the structure of atoms.
- Able to describe what is happening with electrons in an electric current.
- Understand what a semiconductor is, in particular silicon and germanium semiconductors.
- Understand N-type and P-type materials.
- Able to analyze the current flow through both a half-wave rectifier and a full-wave rectifier.
- Have knowledge of the construction of a bipolar junction transistor.
- Understand the use of bipolar transistors in common-base and common-emitter circuits.
- Interpret the static characteristics for transistor circuits.
- Have the skills to draw a load line for a transistor amplifier circuit.
- Understand how low field effect transistors operate.
- Describe the advantages of FETs over bipolar transistors.

Electronics is a branch of science and engineering which deals with the study of current conduction phenomena in vaccum or gas or semiconductor and the utilization of the devices based on these phenomena. Vaccum tubes and gas tubes were the dominant electronic devices before the advent of the transistor in the year 1948. Owing to the growth of electronics during this period many important semiconductor devices were invented and new branches of electronic engineering and technology came into existence.

Miniaturization of electric equipment with the introduction of integrated circuit technology had also gained momentum during the last three decades. Electronics today finds wide ranging applications in various fields such as communication, control and automation of industrial processes computer, medicine, defence, consumer electronics, etc.

Germanium, silicon and gallium arsenide are some of the important semiconductor materials forming the backbone of solid state electronics. A semiconductor device like the transistor is credited with small size, long life, low operating voltage and low cost.

This chapter deals with the principle of operation, characteristics and application of few basic semiconductor devices like junction diodes, transistors, field effect transistor (FET), etc.

12.1 Classification of Materials

Materials such as copper, silver and aluminum, known as conductors, are good conductors of electricity while materials such as glass, rubber, and porcelain, collectivity known as insulators, are poor conductors of electricity. Semiconductors such as germanium, silicon and gallium arsenide have their conductivity ranging between these two extremes. The differences in conducting properties of these three classes of materials are found to arise from the different energy band structures they posses. A typical energy band diagram depicting an energy band structure is shown in Fig. 1.

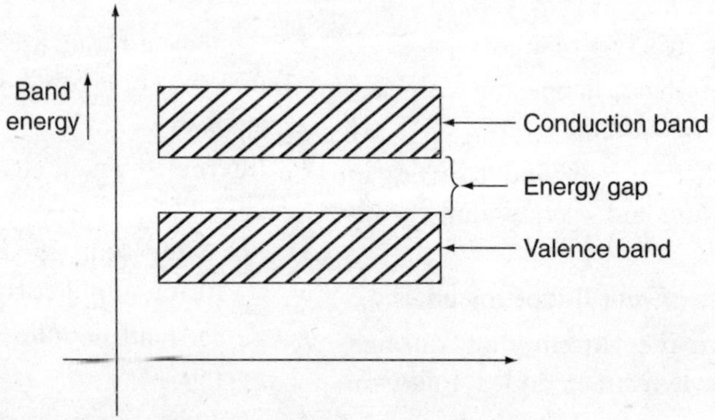

Figure 1 Typical energy band diagram.

12.1.1 Energy Band Structures of Conductors, Insulators and Semiconductors

In this section we present some important terminologies and band structures

1. **Valence electrons:** Electrons present in the outermost orbit of an atom are called valence electrons.
2. **Energy band:** The range of energies possessed by electrons in an atom is known as the energy band. The electron energy is expressed in electron volts (eV). One electron volt is equivalent to 1.6×10^{-19} Ws.
3. **Valance band:** The range of energies possessed by valence electrons is known as valance band.
4. **Conduction (free) electrons:** The electrons that are free to move and thereby conduct current in a material are called the conduction (free) electrons.
5. **Conduction band:** The range of energies possessed by conduction electrons is known as conduction band.
6. **Forbidden band:** The gap between conduction band and valence band expressed in eV is known as forbidden band. It is also called forbidden gap.
7. **Conductor:** Energy band diagram of a good conductor is shown in Fig. 2. Here valence band and conduction band overlap indicating the absence of forbidden band. Therefore, at room temperature, valence electrons can become free. For example, one electron/atom becomes free in the case of copper in this way. Hence, plenty of electrons are available for conduction. Application of small electric field results in the movement of free electrons constituting an electric current. Resistivity in this case is very low.

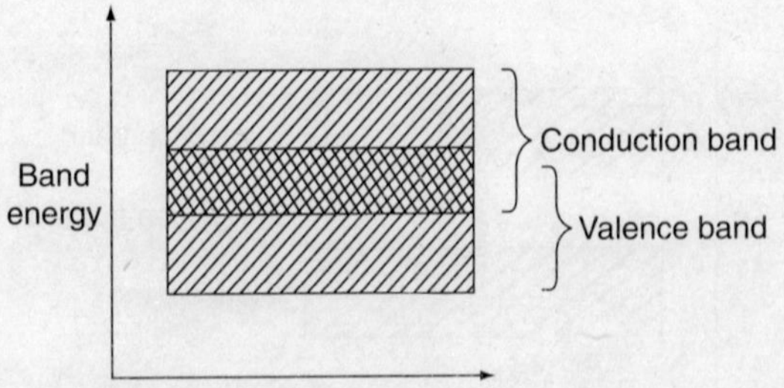

Figure 2 Energy band diagram: Conductor.

8. **Insulator:** The energy band diagram of an insulator, shown in Fig. 3, exhibits filled valence band, an empty conduction band and a large band gap separating them. All the valence electrons are lightly bound to their parent atoms. The large band gap ensures that no electrons at room temperature can jump from valence band to conduction band. Thus, insulator behaves as a bad conductor. Resistivity is very high in this case. However, application of a very strong electron field results in the flow of small current. This is because, valence electrons are set free by the large force exerted by this field. These electrons constitute the current in the outer circuit. Under this condition, the insulator is said to have "broken" down, that is, it has become conductive.

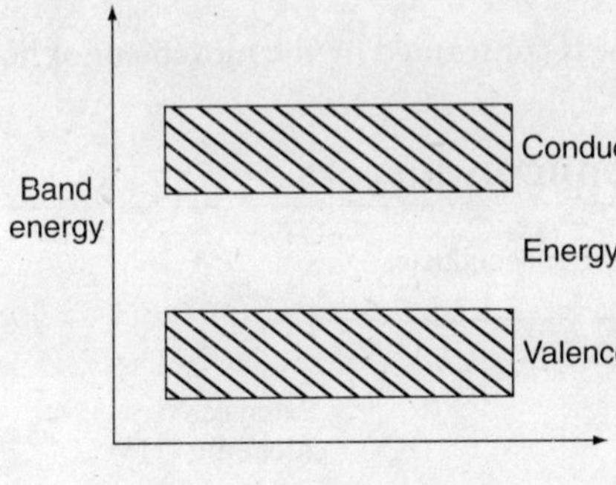

Figure 3 Energy band diagram: Insulator.

9. **Semiconductor:** Semiconductors such as germanium and silicon posses almost filled valence band and nearly empty conduction band, with a very small energy gap (1.1 eV in the case of silicon) separating them as shown in Fig. 4. This narrow band gap implies that only a small amount of energy is needed to take the valence electrons to the conduction band. Therefore, at room temperatures a few electrons on acquiring thermal energy can get themselves detached from parent atom and jump into the conduction band to take part in current conduction. The number of electrons getting into the conduction band in this manner increase with temperatures. However, the semiconductor behaves like an insulator at absolute zero temperature, as the conduction band is then completely empty. The forbidden energy gap for germanium is 0.7 eV and for silicon it is 1.1 eV. We will discuss properties and classification of semiconductors in the following sections.

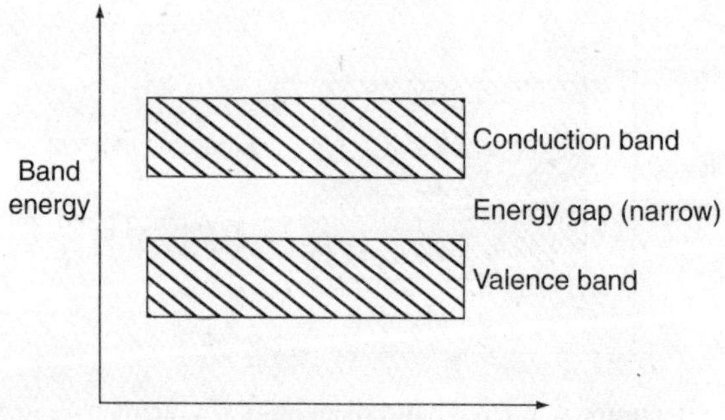

Figure 4 Energy band diagram: Semiconductor.

12.2 Properties of Semiconductors

Germanium and silicon are semiconductors which posses the following properties:

1. They are tetravalent, that is, they have four valence electrons.
2. They have negative temperature coefficient, that is, the resistivity decreases as the temperature increases.
3. Their conductivities lie between that of conductor and insulator at room temperatures. The conductivity can be improved by the addition of suitable impurities. It also increases with incidence of light.
4. The current in the semiconductor is constituted by the movement of holes and electrons.

12.3 Classification of Semiconductors

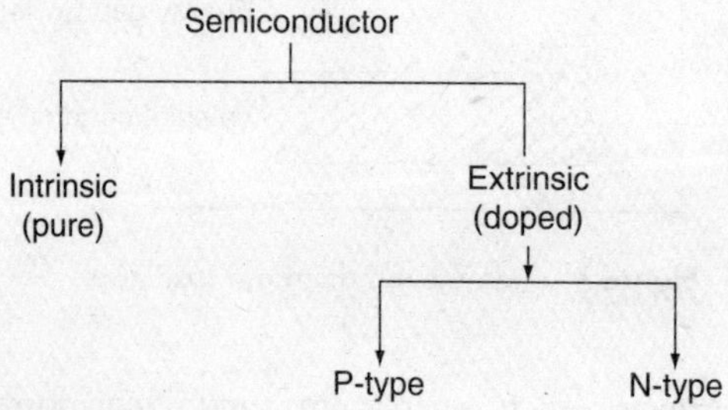

Semiconductors can be classified as

1. Intrinsic semiconductors.
2. Extrinsic semiconductors:
   - P-type.
   - N-type.

12.3.1 *Intrinsic Semiconductors*

The semiconductor in its purest form is called intrinsic semiconductor. The covalent bond arrangement of pure germanium is shown in Fig. 5. For these semiconductors a few free electrons can be traced in the conduction band even at room temperatures, due to breaking of covalent bonds. They are the valence electrons which acquire enough thermal energy (0.7 eV and above for germanium, 1.1 eV and above for silicon) to crossover to the conduction band. As the electron leaves the valence band, a vacant space or void is created. This void is termed as hole; it means the absence of electron. Holes, therefore, are treated as positive charges since electrons are negative charges. Thus, every electron that crosses into the conduction band leaves a hole in the valence band creating what is called a hole–electron pair. When subjected to an electric field, hole–electron pairs drift, that is, holes drift to the negative terminals of the supply and electrons to its positive terminal as shown in Fig. 6. Therefore; the total current inside the semiconductor is the sum of the currents due to electrons and holes. However, the current in the external wires is only due to electrons. This type of conduction is called intrinsic conduction.

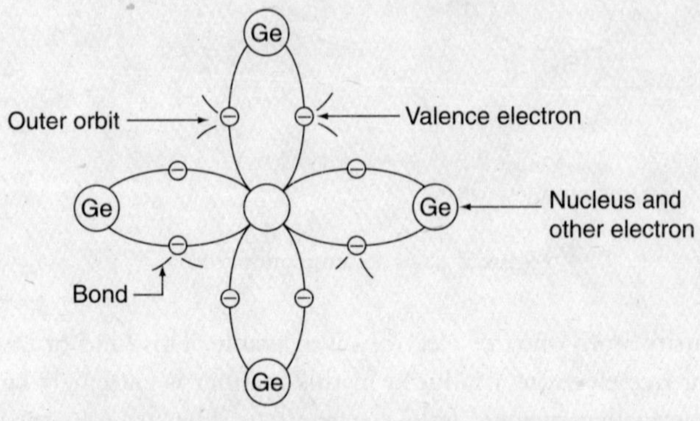

Figure 5 Covalent bond arrangement of pure germanium.

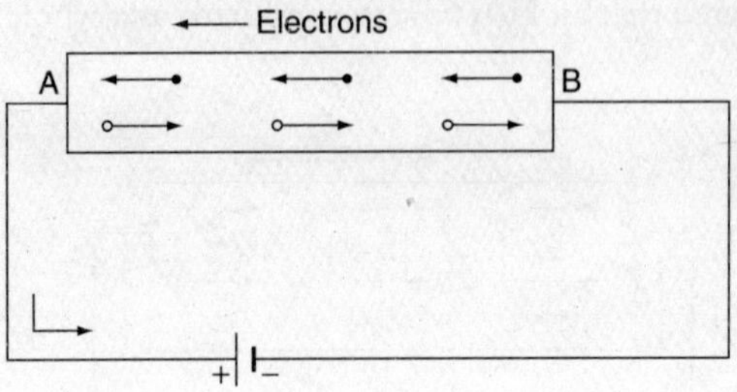

Figure 6 Conduction in intrinsic semiconductor.

12.3.2 Extrinsic Semiconductors

Intrinsic semiconductor has little current conduction capability at room temperatures. To be useful as a semiconductor device its conduction property has to be vastly improved. This can be achieved by adding suitable impurities to the plain semiconductor. The process of adding impurities to the semiconductor is called *doping*, the impurities being referred to as *doping agents* or *dopants*. Aluminum and boron are trivalent doping agents, whereas phosphorus and antimony are pentavalent doping agents. The doped semiconductor or extrinsic semiconductor is either N-type or P-type depending on the valency of doping agents.

12.3.2.1 N-Type Semiconductor

Adding of pentavalent impurity to intrinsic silicon results in N-type semiconductor. It is a covalent bonding arrangement between one phosphorus atom (pentavalent impurity) and four silicon atoms. Fifth valence electron of phosphorus which is unable to enter into covalent bonding with silicon is almost free and can cross over to conduction band very easily at room temperature (Fig. 7).

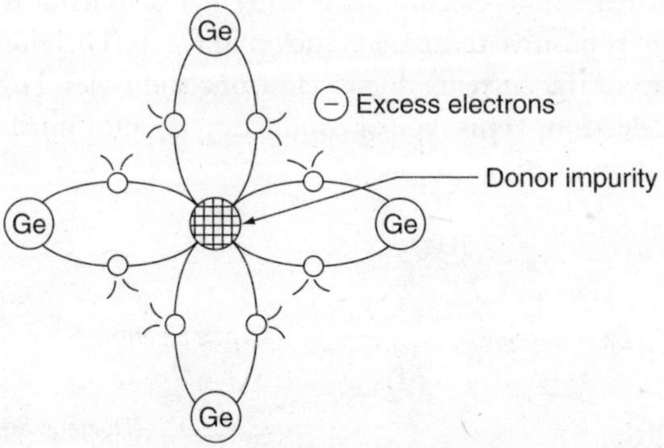

Figure 7 N-type semiconductor.

Thus, for every impurity atom one free electrons is available. This kind of impurity is called *donor impurity*. The number of free electrons produced in this manner is extremely large in comparison to the holes belonging to thermally generated hole–electron pair. The current contribution in an N-type semiconductor shown in Fig. 8 is predominantly by these free electrons, that is, negative charges and hence the name N-type. Here negative electrons are mobile charges and positive donor ions are fixed charges. It should be remembered that an N-type semiconductor as a whole is electrically neutral.

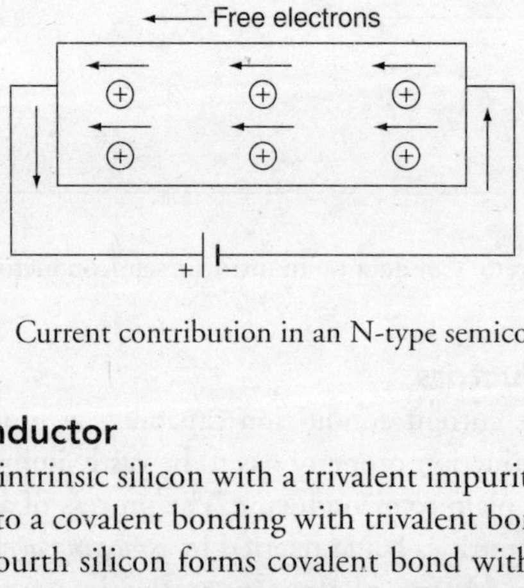

Figure 8 Current contribution in an N-type semiconductor.

12.3.2.2 P-Type Semiconductor

This is obtained by doping the intrinsic silicon with a trivalent impurity such as boron. Silicon having four valence electrons enters into a covalent bonding with trivalent boron. While doing so, three electrons of boron are used. The fourth silicon forms covalent bond with three electrons of boron. The fourth silicon electron is unable to complete the bonding, as boron cannot provide an electron from its side. The formation of covalent bonds between silicon and trivalent boron impurity is depicted in Fig. 9. The absence of electron at "x" gives rise to a hole. Thus, every boron impurity atom contributes one hole. This type of impurity is referred to as *acceptor impurity*. The holes which are the positive ions outnumber the electrons, hence the semiconductor is called P-type semiconductor. Current conduction in this type of semiconductor, shown in Fig. 10, is predominantly by holes. Holes are mobile charges and negative acceptor ions are fixed charges.

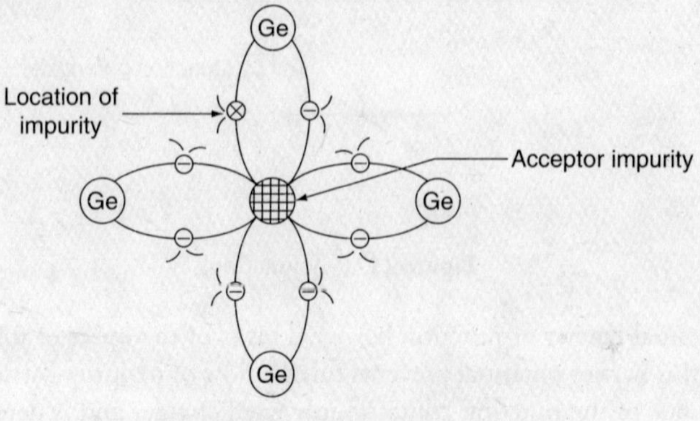

Figure 9 P-type semiconductor.

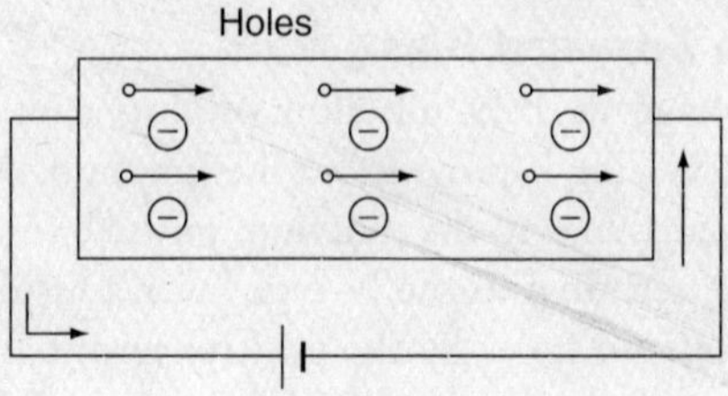

Figure 10 Current contribution in a P-type semiconductor.

12.4 P-N Junction

P-N junction (Fig. 11) is the boundary region between N-type material on the right and P-type material on the left of a continuous and single crystalline structure. They are usually manufactured by planer diffusion process. The study of P-N junction is more important as it forms the basic building block of diodes and other more versatile semiconductor devices like transistors.

No sooner than a P-N junction is formed, the highly concentrated holes in the P-region move towards the N-region. This process is called *diffusion*. During the diffusion process, electrons and holes recombine in the neighboring areas of the junctions and annihilate one another. In fact, recombination leaves open fixed charges, that is, positive donor ions in the N-region and negative acceptor ions in the P-region, lying adjacent to the junction.

Further diffusion of holes and electron is prevented by the positive and negative fixed charges. Positive charges on N side repel holes that come from "P" side to "N" side and similarly, negative charges on P side repel free electrons that come from the "N" side to "P" side. Thus a barrier is set up against the further movement of carriers.

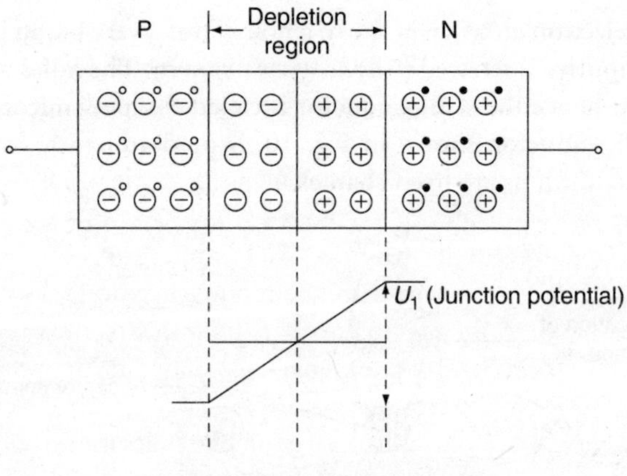

Figure 11 P-N junction.

This is called potential barrier or junction barrier. This is of the order of 0.1 V to 1.3 V. The electronic field set up by this barrier potential prevents further flow of majority carriers across this junction. The region on either side of the junction contains only fixed charges and is depleted of holes and electrons. Hence this region is also called *depletion region.*

12.4.1 P-N Junction with Forward Bias

Application of an external voltage to P-N junction produces interesting effects on the junction depending on the nature of the biasing it provides to the junction. A forward-biased P-N junction is shown in Fig. 12. To apply forwards bias to the junction, positive terminal of the battery is connected to the P-type layer and negative terminal to the N-type layer. During the forward bias, electrons get repelled by the negative terminal and holes by the positive terminal of the battery. As a result, both move towards the junction, thus reducing the potential barrier. A small forward voltage (0.1 V to 0.3 V) is sufficient to wipe off this potential barrier by which a low resistance path is established in the circuit. Hence, considerable amount of forward current results. Therefore, in forward biasing current conduction is due to majority carriers only.

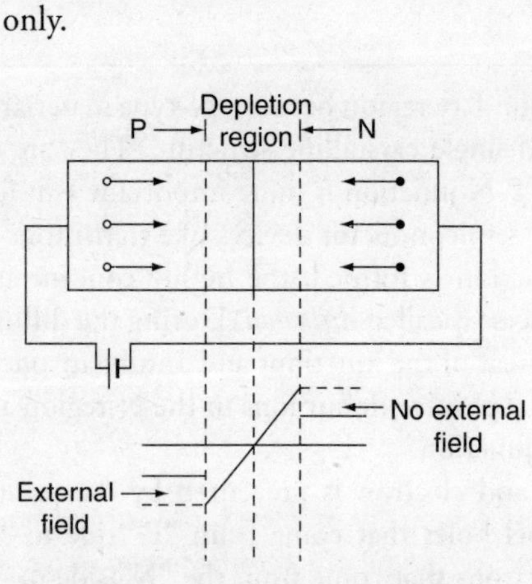

Figure 12 Forward-biased P-N junction.

12.4.2 P-N Junction under Reverse Bias

During the reverse bias, positive terminal of the battery is connected to N-type and negative terminal to P-type. Hence electrons get attracted by the positive terminal of the battery and holes to its negative terminal. As a result, there is a widening of the battery accompanied by an increase in barrier potential. This increased barrier potential presents the flow of majority carriers across the junction. Therefore, there should not be any current in the external circuit.

However, in practice a small current of the order of a few microamperes flows in the external circuit under reverse-biased condition. This is caused by the following process. Electrons forming the covalent bonds of the semiconductor atoms absorb energy mainly due to the rise in temperature and free themselves from the covalent bonds. As a result, hole–electron pairs are produced under the reverse-biased condition. Holes in the P-region drift towards the negative terminal of the battery and electrons towards the positive terminal. But electrons in the P-region and holes in the N-region being the majority carriers easily drift across the junction under the barrier potential.

12.5 P-N Junction Diode

It is a two-terminal device that makes use of the "valve" nature of the junction by allowing the current to flow in one direction and blocking it in the other direction. The circuit symbol of the device is shown in Fig. 13. The unidirectional conduction property of the device makes it useful in the rectifier circuits that are meant for connecting the alternating current into direct current.

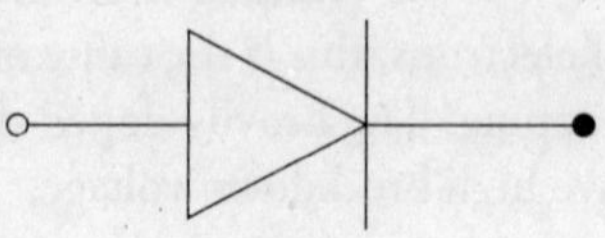

Figure 13 Circuit symbol of a P-N junction.

12.5.1 V–I Characteristics of Junction Diode

V–I characteristics are plotted in the first and third quadrants of the graph (Fig. 14). The characteristic in the first quadrant is called the *forward characteristics* as it represents the relationship between voltage and current forward-biased condition.

The forward current is very small for low initial forward voltages. Once the forward voltages exceeds the barrier potential, the diode current starts increasing. The voltage at which the current starts rising rapidly is called cut-in or threshold voltage (V_r). For forward voltages less than V_r barrier, potential exists and it blocks majority carrier flow. For forward voltages greater than V_r, barrier potential disappears and majority carriers start crossing the junction in large numbers constituting relatively large amount of current flow in the outer circuit.

The volt–ampere characteristics of the diode under reverse-biased condition is shown in third quadrant (Fig. 14). The current under reverse-biased condition is of the order of microamperes and is constituted by minority carriers.

As the reverse-biasing voltage is gradually increased, the current remains almost constant until the biasing voltage attains a particular high value called *breakdown voltage*. At this voltage, the current suddenly and sharply rises due to the onset of phenomenon called *avalanche breakdown.* At very high reverse

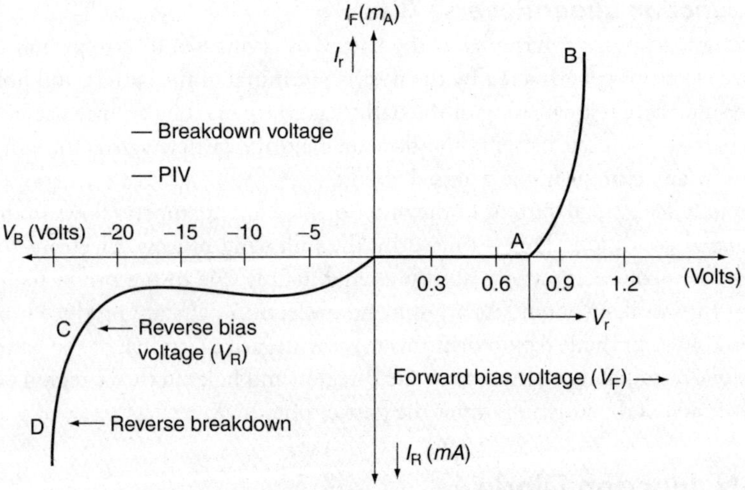

Figure 14 *V–I* characteristics of junction diode.

voltage, the electron velocities shoot up under the influence of large potential gradients. Hence their kinetic energies attain large values. When they collide, they knockout electrons from the semiconductor atoms. The knocked out electrons also join the colliding electrons and further collisions take place. This results in avalanche multiplication of electrons; this is the cause of the sudden rise in diode current. The breakdown depends on the level of doping. The heavily doped diode will have low breakdown voltage while the lightly doped diode will have high breakdown voltage.

12.5.2 Diode Ratings

The satisfactory operation of the diode is possible only when following parameters do not exceed their maximum values:

1. Maximum forward current.
2. Peak inverse voltage – the maximum negative voltage the diode experiences when it is operating.
3. Maximum power rating.

12.6 Rectifiers

The circuits which are used to connect alternating voltage to unidirectional voltage are called *rectifiers.* They perform this function by using diodes. As studied, junction diodes can conduct current only in one direction. This property of the diode is made use of in designing various rectifier circuits.
The rectifier circuits are classified as follows:

1. Half-wave rectifier.
2. Full-wave rectifier.
3. Bridge rectifier.

12.6.1 Half-Wave Rectifier

The half-wave rectifier circuit in its simplest form is shown in Fig. 15. Its input and output waveforms are given in Fig. 16.

Only the positive half of the input sine wave is available as output (V_o). Since the diode can conduct only during the positive half, output voltage is a pulsating unidirectional voltage while the input is a sine wave. Sometimes, it is a necessary to use a transformer as shown in Fig. 17. The transformer provides the necessary isolation between the load circuit and supply. More importantly, the turns ratio of the transformer can be chosen so that for any desired DC output voltage can be obtained. During the negative half cycle of the supply voltage, the diode is reverse-biased and therefore it cannot conduct. No output voltage is available over as the load during this period.

The average value of load current is equal to the average value of load voltage divided by load resistance. A measure of purity of the DC output voltage is given by the term *ripple factor*. This is defined as the

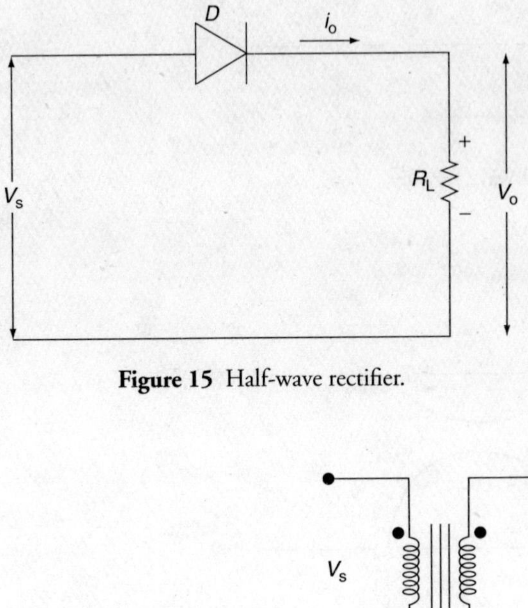

Figure 15 Half-wave rectifier.

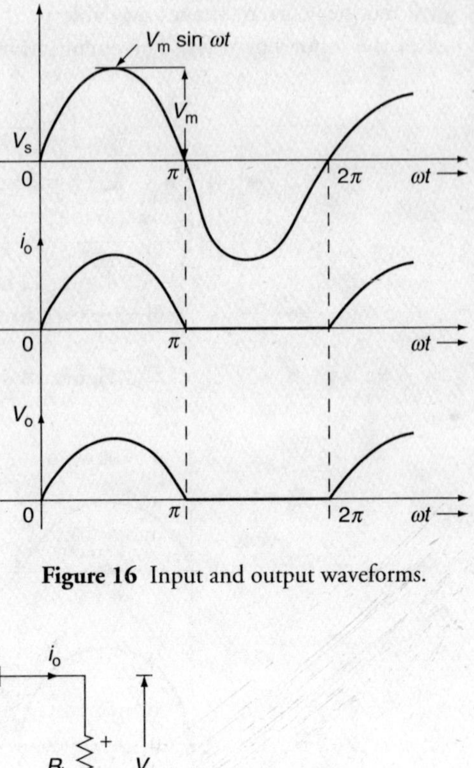

Figure 16 Input and output waveforms.

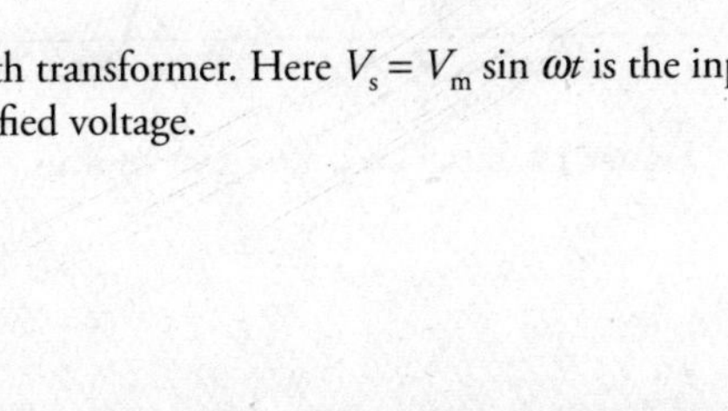

Figure 17 Half-wave rectifier with transformer. Here $V_s = V_m \sin \omega t$ is the input voltage, i_o is the rectified current, V_o is the rectified voltage.

ratio of root mean square (r.m.s.) value of the AC component in the DC wave to the average value of DC wave.

$$\text{Ripple factor} = \frac{\text{RMS Value of AC component}}{\text{Average value of DC wave}}$$

It is 1.21 for half-wave rectifier. Another quantity of interest is rectification efficiency which is defined as

$$\text{Rectification efficiency} = \frac{\text{DC power output}}{\text{AC power input}} = \frac{P_{DC}}{P_{AC}}$$

Rectification efficiency for half-wave rectifier is 40.6%. Excessive ripple factor and low rectification efficiency makes the half-wave rectifier circuit unsuitable for rectification of large amounts of power.

$$\text{Peak inverse voltage} = V_m \text{ for this rectifier}$$

12.6.2. Full-Wave Rectifier

Full-wave rectifier current makes available to the load resistance R_L both the positive half and negative half of the input sine wave. The circuit arrangement and waveforms are given in Figs. 18 and 19,

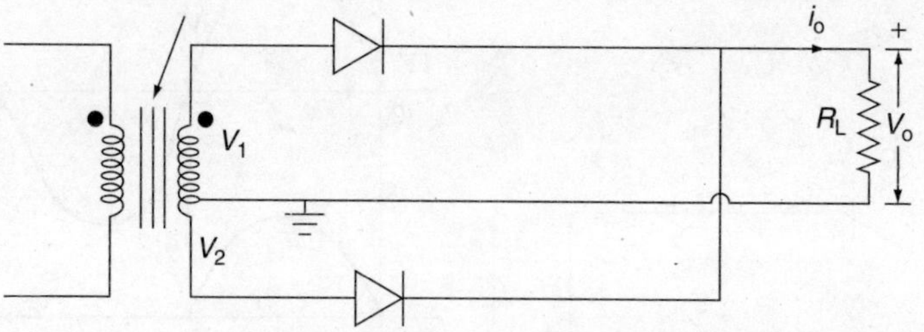

Figure 18 Full-wave rectifier.

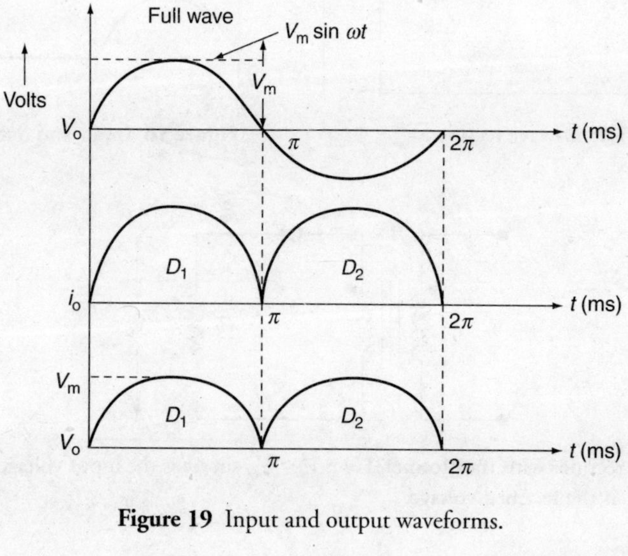

Figure 19 Input and output waveforms.

respectively. During positive half cycle, the diode D_1 gets forward-biased and diode D_2 gets reverse-biased. D_1, therefore, conducts during this period and positive half of the input wave appears across the load. During negative half cycle, diode D_2 gets forward-biased and diode D_1 gets reverse-biased. The current flows through D_2 and hence negative half cycles of the input wave appears across the load. During both the half-cycles current through the load is in the same direction producing two voltage pulses across the load for each cycle of the input wave. The output voltage in this case is twice the output voltage of half-wave rectifier. But the ripple factor reduces to 0.48 from 1.21, indicating higher degree of purity for the output voltage. The rectification efficiency increases to 81.2% which is twice that of half-wave rectifier. However peak inverse voltage is twice that of single phase of half-wave rectifier.

12.6.3 Bridge Rectifier

This is most frequently used circuit for electronic DC power supplies. Unlike the full-wave rectifier, the bridge rectifier does not require a transformer for its operation. However, it can be used for obtaining the desired output voltage and isolation. The circuit configuration and waveforms for this rectifier are given in Figs. 20 and 21, respectively. During the positive half cycle of the input voltage, the diodes D_1 and D_2 conduct whereas during the negative half cycle, diodes D_3 and D_4 conduct. The current flow through the load is unidirectional as shown in Figs. 20 and 21 during positive and negative half cycles.

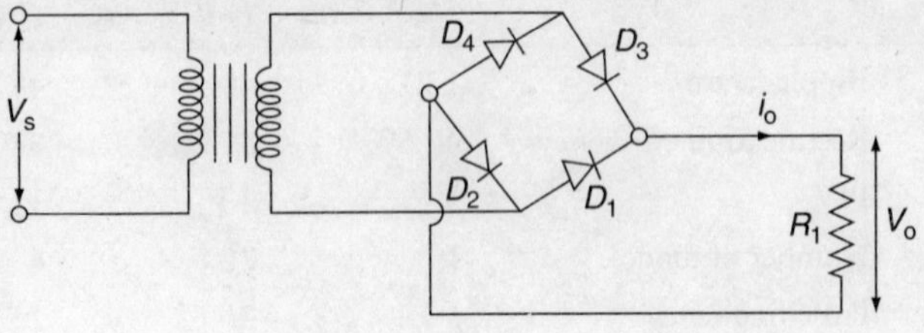

Figure 20 Bridge rectifier.

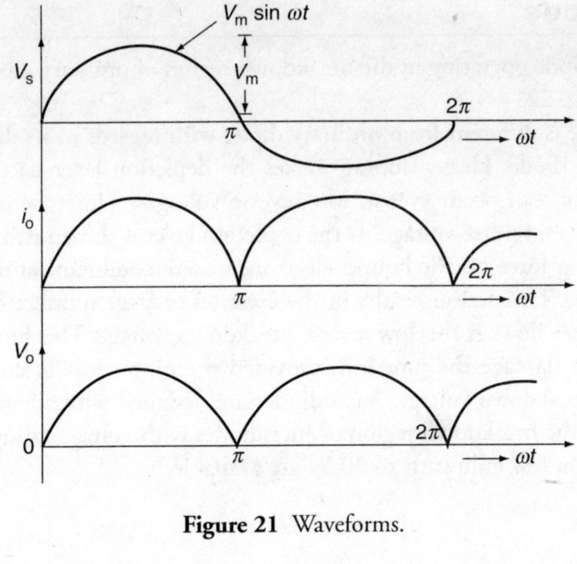

Figure 21 Waveforms.

The polarity of the output voltage remains the same during both half-cycles, positive at the top and negative at the bottom. The bridge rectifier has the following merits:

1. There is no need for a center-tapped transformer as is the case with full-wave rectifier.
2. The peak inverse voltage (PIV) rating of diode is less. It is actually half the rating required for full-wave rectifier for the same output voltage.
3. For the same output voltage, the transformer secondary voltage is only 50% of the value required in the case of full-wave rectifier.
4. It is suitable for high voltage application.

It should be noted that ripple factor and rectification efficiency are same for both full-wave rectifier and bridge rectifier. Thus, a bridge rectifier is economical in compassion to full-wave rectifier.

12.6.4 Comparison of Rectifier Circuits

A comparison of various kinds of rectifiers is given in Table 1.

Table 1 Compression of various rectifier circuits

S. No.	*Parameter*	*Type of Rectifier*		
		Half-Wave	*Full-Wave*	*Bridge*
1.	Ripple factor	1.21	48	48
2.	Rectification effeciency	40.6%	81.2%	81.2%
3.	PIV	V_m	$2\ V_m$	V_m
4.	Number of diodes	1	2	4
5.	Pulse frequency	f	$2f$	$2f$

12.7 Zener Diode

Zener diode is a special diode operating in the breakdown region of ordinary diode. Circuit symbol for Zener diode is given is Fig. 22.

However, Zener diode is different from ordinary diode with regards to its doping level. It is heavy doped than the ordinary diode. Heavy doping makes the depletion layer narrow and consequently, breakdown of the function can occur ever at low reverse voltages. This type of breakdown is called *Zener breakdown*. For a given reverse voltage, as the depletion layer is thin, a strong electric field is produced. This exerts a strong force on the bound electrons of semiconductor atoms pulling them away from their covalent bonds. This action results in the creation of large number of electron–hole pairs. Thus, a large reverse current flows at this low reverse breakdown voltage. This breakdown phenomenon is reversible and does not damage the junction, provided the power loss is small. A properly doped diode will have a sharp breakdown voltage. Such diodes are designed with adequate power dissipation capabilities to operate in the breakdown region. Zener diodes with voltage ratings from 3 V to 100 V, and power dissipation from few milliwatts to 50 W are available.

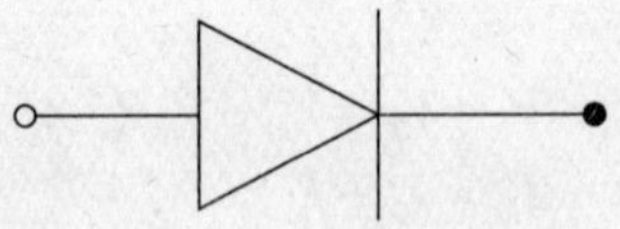

Figure 22 Circuit symbol of Zener diode.

12.7.1 *V–I* Characteristics

V–I characteristics of Zener diode are depicted in Fig. 23. This operation of Zener diode is not different from the ordinary diode. This is revealed from its forward characteristics which are same as that of ordinary diode. Breakdown occurs at point "*x*" of reverse saturation characteristic. Magnitude of this breakdown voltage depends on the degree of doping. Lightly doped diode will have higher breakdown voltage. This breakdown voltage can be selected by the amount of doping. At higher breakdown voltage, avalanche breakdown (explained in Section 12.5.2) also comes into play enhancing the conductivity of the function. It should be emphasized that "Oa" the breakdown region of the Zener diode characteristic is its region of operation.

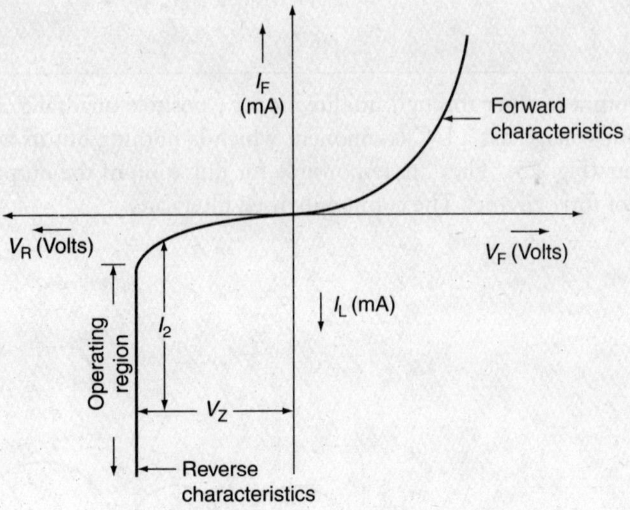

Figure 23 *V–I* characteristics of Zener diode.

12.7.2 Zener Diode as Voltage Regulator

The sharp breakdown characteristics of the Zener diode are useful in devices where it can serve as voltage difference or voltage regulator (Fig. 24). A simple Zener diode maintains a constant voltage across the load resistance R_L even when:

1. The load current (I_L) varies over a small range.
2. The input voltage (Vs) varies over a small range.

The relevant equations are,

$$V_o = V_Z = V_{in} - V_{in}R_{se}$$
$$I_{in} = I_Z + I_L$$
$$V_o = I_o R_L$$

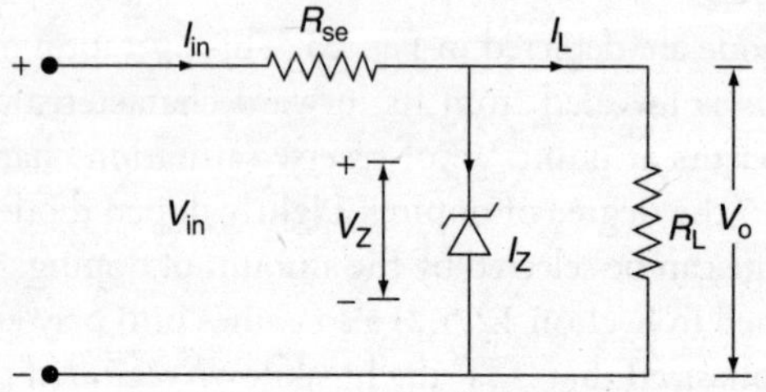

Figure 24 Voltage regulator circuit.

12.8 Filters

The output voltage from a rectifier through unidirectional (positive quadrant or negative quadrant) is pulsating. The output voltage has a DC component which is nothing but its average value in addition to AC components (Fig. 25). They are responsible for pulsation of the output voltage. These are minimized by means of filter circuits. The commonly used filters are:

1. Capacitor filter.
2. Induction filter.
3. LC filter.
4. Π filter.

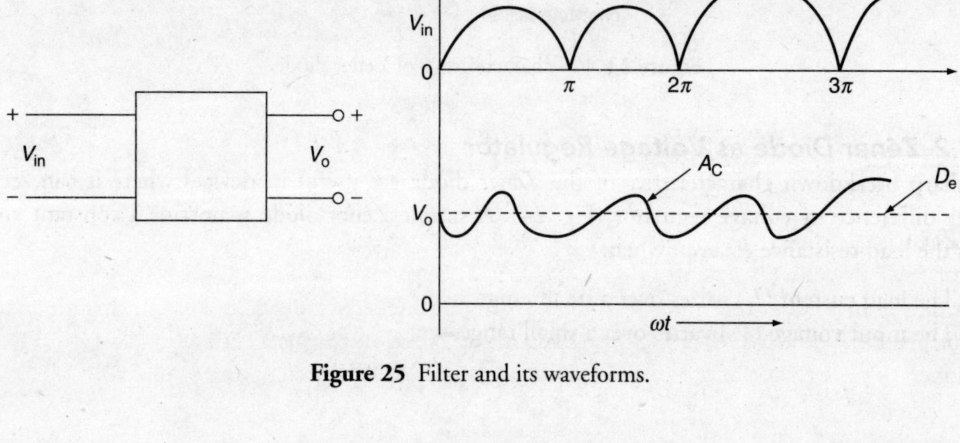

Figure 25 Filter and its waveforms.

12.8.1 Capacitor Filters

The capacitor filter is formed by connecting capacitor directly across the load of a bridge rectifier circuit (Fig. 26). The output voltage waveforms with and without capacitor are shown in Fig. 27. The output voltage is nearly a constant and is equal to the maximum value of the supply voltage. The capacitor gets charged during the period when rectifier diodes conduct. When the diodes are off capacitor discharges through the load. The current though the rectifier diode is pulsating. The capacitor filter performs better at lighter loads. As the load current increases, ripple increases. Hence it is essentially a voltage filter.

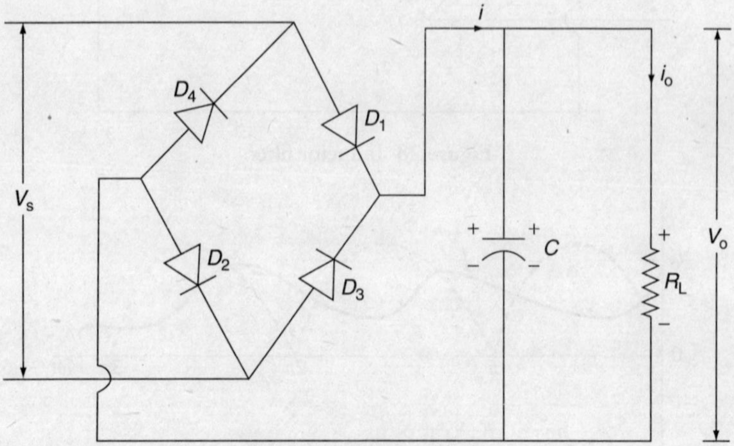

Figure 26 Capacitor filter.

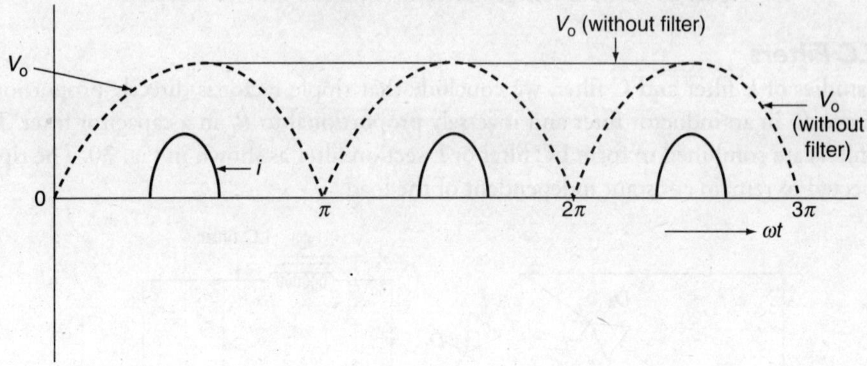

Figure 27 Output voltage waveforms with and without capacitor.

12.8.2 Inductor Filters

Inductor filter is formed by connecting a choke in series with a load as shown in Fig. 28. It is well known that inductor has the inherent property of opposing the changes in current. Introduction of inductor in series with the load will have a smoothing effect on the load current as shown in Fig. 29. The induction filter performs well under heavy loads. It is essentially a current filter.

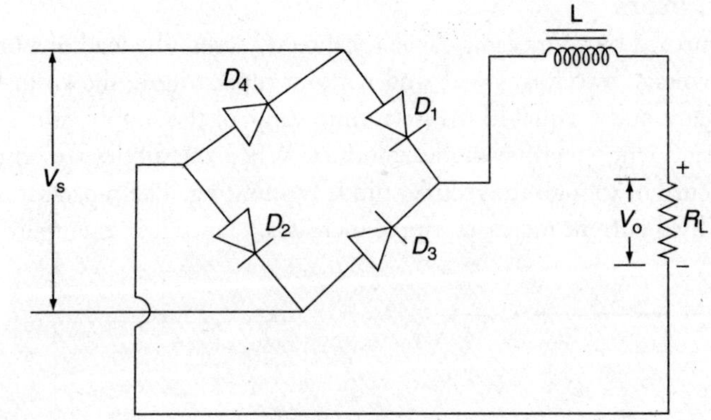

Figure 28 Inductor filter.

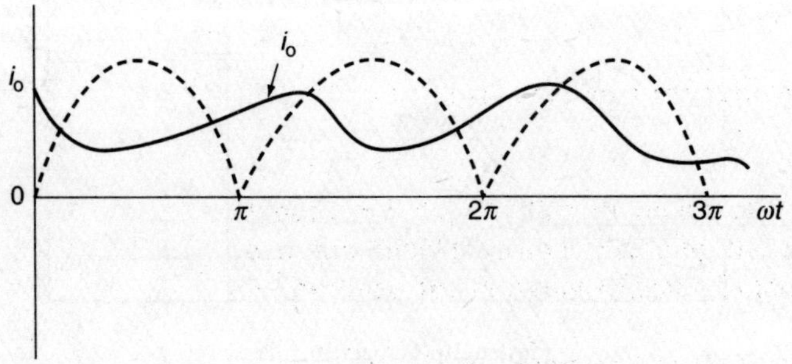

Figure 29 Output voltage waveforms with and without inductor.

12.8.3 LC Filters

From the studies of L filter and C filter, we conclude that ripple factor is directly proportional to the load resistance R_L in an inductor filter and inversely proportional to R_L in a capacitor fitter. Therefore, these two filters are combined to form LC filter or L section filter as shown in Fig. 30. The ripple factor can be expected to remain constant independent of the load.

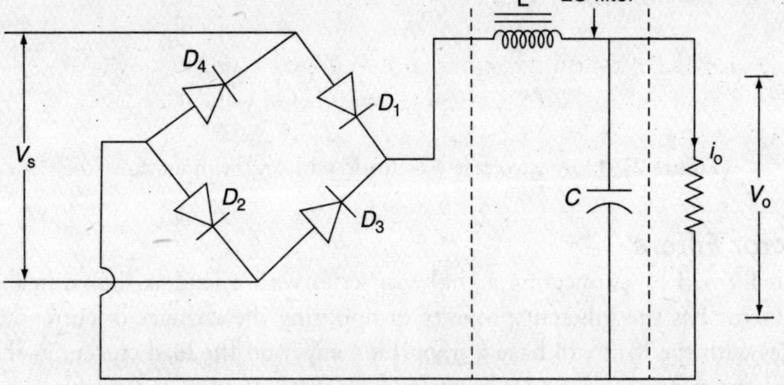

Figure 30 LC filter.

12.8.4 Π or CLC Filters

A full-wave rectifier with Π filter is shown in Fig. 31. A resistor can replace an inductor for lighter loads. The ripple factors for various types of filters are given in Table 2. Here

$$\omega = 2\pi f$$

where f is supply frequency.

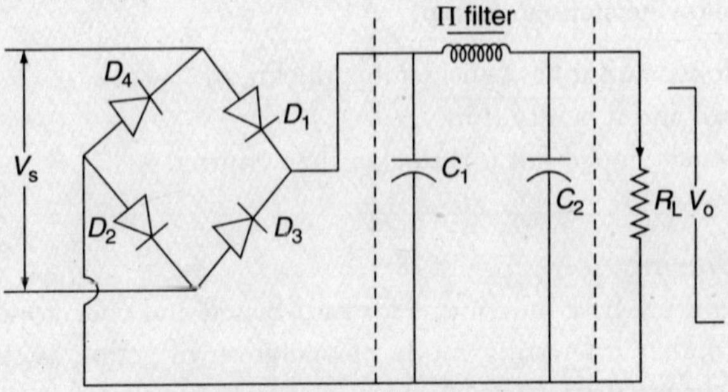

Figure 31 Π or CLC filters.

Table 2 Ripple factors for various types of filters

S. No.	*Type of Filter*	*Ripple Factor*
1.	C	$1/4\sqrt{3}fRLC$
2.	L	$R_L/(3\sqrt{2}\,\omega L)$
3.	LC	$1/(6\sqrt{2}\,\omega^2 LC)$
4.	CLC or Π	$\dfrac{\sqrt{2}}{\omega^3 C_1 C_2 L R_L}$

12.9 Bipolar Transistor

12.9.1 Constructional Details

A bipolar or junction transistor is a three-layer, two-junction device obtained by sandwiching a layer of doped semiconductor between two layers of oppositely doped semiconductors. With N-type and P-type semiconductors available, it is possible to arrive at two types of transistors, namely, PNP and NPN transistors. The transistors and their circuit symbol are given in Fig. 32.

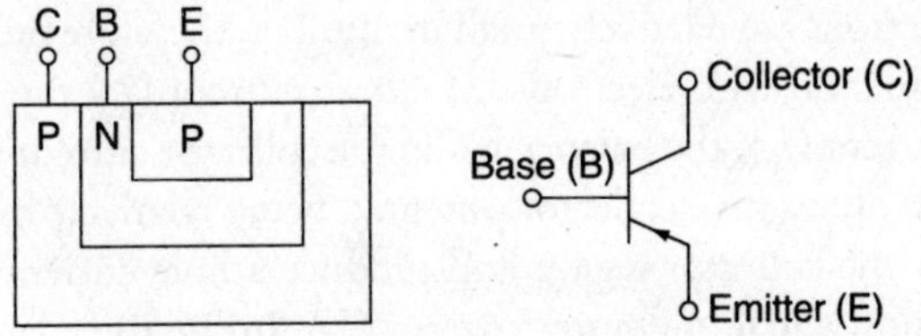

Figure 32 (a) Transistor; (b) circuit symbol.

Three regions of the transmitter are emitter, base and collector, deneted by letters E, B, C respectively The arrow on the connection to the emitter indicates the conventional current direction of current flow. The junctions between emitter and base and between collector and base are referred to as emitter–base junction and collector–base junction, respectively. These P-N junctions exhibit the diode characteristics. It is important to note that the E, B and C regions do not have the same thickness, and are not doped to the same level. This is necessary for the transistor to be used as an active device for improved transistor action. The following steps are taken:

1. Emitter is heavy doped and it has considerable thickness.
2. Base is lightly doped and it is very thin.
3. Collector is moderately doped but it is thicker than emitter.

12.9.2 Transistor Biasing

For normal operation, emitter–base junction is forward-biased and collector–base junction is reverse-biased. Transistor has another interesting mode of operation of acting as switch. If both the junctions are forward-biased then transistor acts like a closed switch. On the other hand, if both the junctions are reverse-biased, then transistor acts like an open switch. The transistor biasing for NPN and PNP transistor for normal operation is shown in Fig. 33.

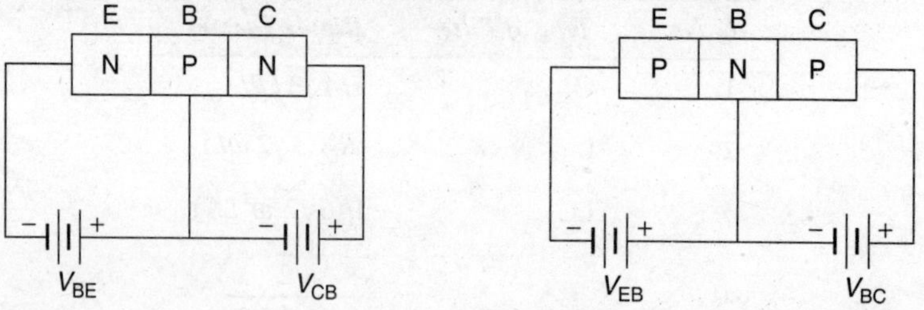

Figure 33 Transistor biasing for (a) NPN and (b) PNP transistor for normal operation.

12.9.3 Working of PNP Transistor

A properly biased PNP transistor for a normal operation in shown in Fig. 34. When emitter–base junction is forward-biased, significant current will flow through it once the base to emitter voltage exceeds the threshold value. In this P-type emitter, holes are the majority carriers. The forward biasing causes the holes to diffuse from emitter into the base for the same reason. The electrons from the base region diffuse across the forward-biased junction (E-B) to the emitters. The total current across the junction is due to the movement of these two types of carriers. At the emitter–base junction, some of the holes and electrons combine. The electrons are relatively small in number, therefore only a few majority carriers are lost by the way of recombination with electrons. The base current (I_B) produced by this action is only about 2% of the emitter current (I_E), the balance being the collector current (I_C). Since the base region is very thin and the attraction of negative collector potential being relatively high, almost all the holes that diffuse into the base, enter the collector region and produce a hole current in the collector. Therefore, collector current (I_C) is almost equal to emitter current (I_E). In the three layers of PNP transistor, holes contribute current flow. However, in the other circuit, the current flow remains only electronic.

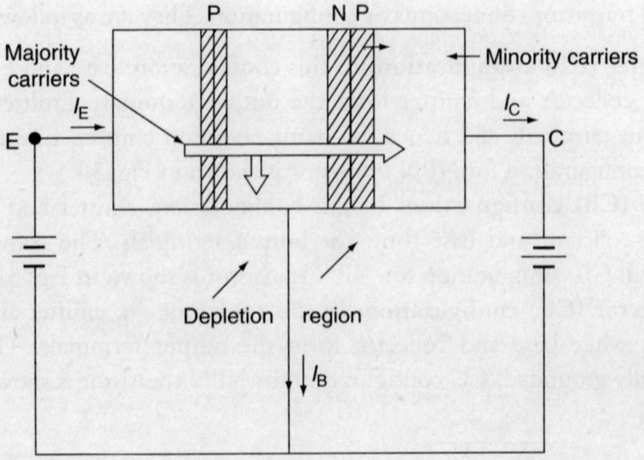

Figure 34 A properly biased PNP transistor for a normal operation.

In the case of NPN transistor (Fig. 35), when the emitter–base junction is forward-biased, the electrons diffuse into the base region. Simultaneously, a small number of holes from the base diffuse into the emitter region. Some of the electrons recombine with the holes in the base region producing a base current (I_B). As the collector–base junction is reverse-biased, minority carriers, namely electrons, in the base region and the holes in the collector region can drift across the reverse-biased collector–base junction, thus constituting the collector current (I_C).

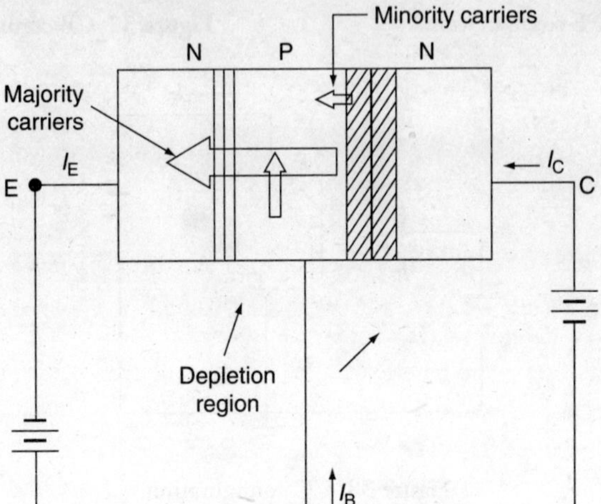

Figure 35 NPN transistor.

For both NPN and PNP transistors

$$I_E = I_B + I_C$$

where I_E is the emitter current, I_B is the base current, I_C is the collector current.

12.9.4 Transistor Configuration

There are three useful transistor connections or configuration. They are as follows:

1. **Common–emitter (CE) configuration:** In this configuration, base and emitter form the input terminals while collector and emitter form the output terminals. Emitter is common to both input and output terminals and hence the name common emitter. Emitter terminal is usually grounded. CE configuration for NPN transistor is shown in Fig. 36.
2. **Common–base (CB) configuration:** In this configuration, emitter and base constitute input terminals while collector and base form the output terminals. The common terminal base is usually grounded. CB configuration for NPN transistor is shown in Fig. 37.
3. **Common–collector (CC) configuration:** In this configuration, emitter and collector constitute input terminals while base and collector form the output terminals. The common terminal collector is usually grounded. CC configuration for NPN transistor is shown in Fig. 38.

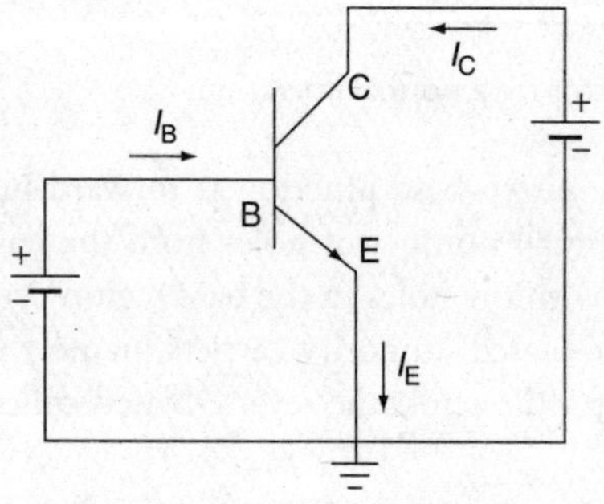

Figure 36 CE configuration.

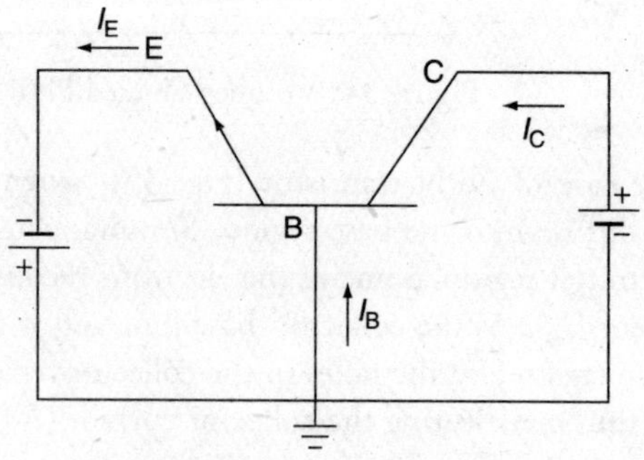

Figure 37 CB configuration.

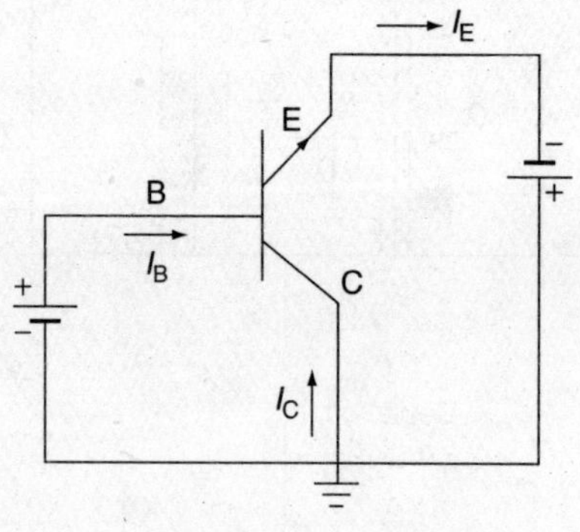

Figure 38 CC configuration.

12.9.5 Static Transistor Characteristic

Common–emitter (CE) configuration is the most popular one. Since it posses the properties suitable for wide ranging applications for CE configuration, voltage and current gains are high, input resistance is low and output resistance is high. In this section, input and output characteristics of the CE transistors are studied (Table 3).

Table 3 Comparison of transistor configurations

Configuration	*Input Resistance*	*Output Resistance*	*Current Gain*	*Voltage Gain*
Common emitter (CE)	Low	High	High	High
Common base (CB)	Very low	Very high	< 1	High
Common collector (CC)	High	Low	High	< 1

12.9.6 Current Amplification Factor (α)

In the CB configuration of a transistor, I_E is the input current and I_C is the output current. The current amplification factor α is given by the ratio of collector current I_C to emitter current I_E. Therefore

$$\alpha = \frac{I_C}{I_E}$$

12.9.7 Current Simplification Factor (β)

In the CE configuration of a transistor, I_B is the input current and I_C is the output current. The current simplification factor is given by

$$\beta = \frac{I_C}{I_B}$$

12.9.8 Relation Between α and β

In a transistor.

$$I_E = I_B + I_C$$

Dividing by I_C on both sides of the equation we get

$$\frac{I_E}{I_C} = \frac{I_B}{I_C} + 1$$

$$\frac{1}{\alpha} = \frac{1}{\beta} + 1$$

$$= \frac{1+\beta}{\beta}$$

Hence we get

$$\alpha = \frac{\beta}{1+\beta}$$

Also

$$\beta = \frac{\alpha}{1-\alpha}$$

12.9.9 Applications

1. Transistors are used in amplifier circuits to amplify weak signals.
2. They are used in oscillators as active devices.
3. Transistors can also be used as switches in digital circuits, and power transistors are used in power control circuits.

12.10 Field Effect Transistor

In bipolar transistor, current conduction occurs due to majority and minority carriers. However, conduction in field effect transistors is solely by majority carriers and therefore it is monopoler. Also the current conduction is controlled by an electric field in this device. There are two types of FET, namely,

1. **Junction FET (JFET):**
   - N-channel JFET with electrons as majority carriers.
   - P-channel JFET with holes as majority carriers.
2. **Metal oxide semiconductor FET (MOSFET):**
   - Enhancement MOSFET.
   - Depletion MOSFET.

Also, it should be noted that FET is a voltage-controlled device like tubes, unlike the transistors which are current-controlled devices.

12.10.1 Juncton FET (JFET)

12.10.1.1 N-Channel JFET

1. It consists of an N-type silicon bar having ohmic contacts at its terminals (Fig. 39).

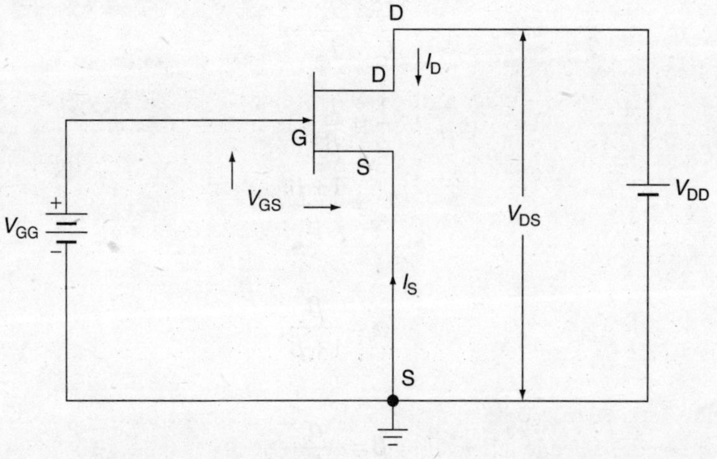

Figure 39 N-channel JFET with an N-type silicon bar.

2. These terminals are source and drain like emitter and collector of bipolar transistor.
3. A heavily doped P-type silicon layer is diffused on the two sides of the bar called "Gate: like base of the transistor.
4. The N-type silicon bar forms a conducting channel for minority carriers.

The construction of an N-channel JFET is shown in Fig. 40.

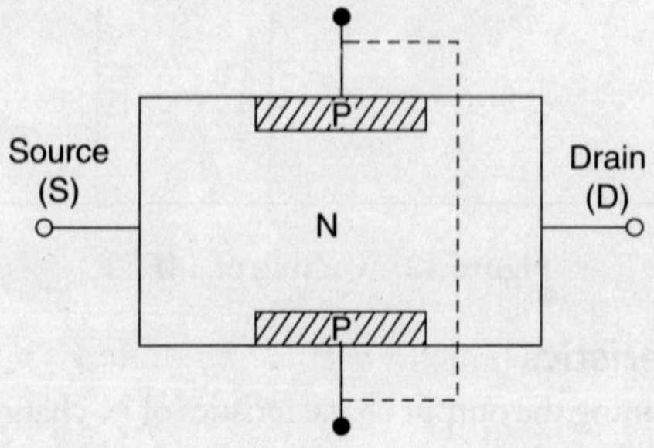

Figure 40 Construction of an N-channel JFET.

The electrons enter the bar at the source terminal and leave the bar at the drain terminal. The gate is used to control the flow of electrons from source to drain.

12.10.1.2 P-Channel JFET

Holes are the majority carriers of P-channel JFET. The constructional details of the JFET are shown in Fig. 41. The holes, being the majority carriers, enter the P-type silicon bar, at the source terminal and leave the bar at the drain terminal. The heavily doped N-regions on both sides of the bar function as gate terminals controlling the movement of holes from source to drain.

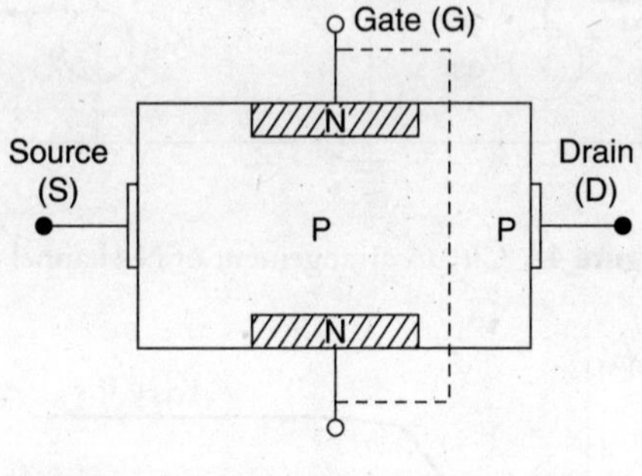

Figure 41 Construction of P-channel JFET.

12.10.1.3 Principle of Working of JFET

An N-channel JFET with positive drain to source voltage V_{DS} is shown in Fig. 42. As a result of V_{DS}, an electron current flows from source to drain. This drain current I_D flowing in the channel between two P-regions can be controlled by controlling the width of the channel. This is accomplished by V_{GS} which can be easily varied. The operation and control of P-channel JFET follow the same lines as that of N-channel JFET. However, the polarity of the biasing voltage, V_{DS} and V_{GS} are reversed.

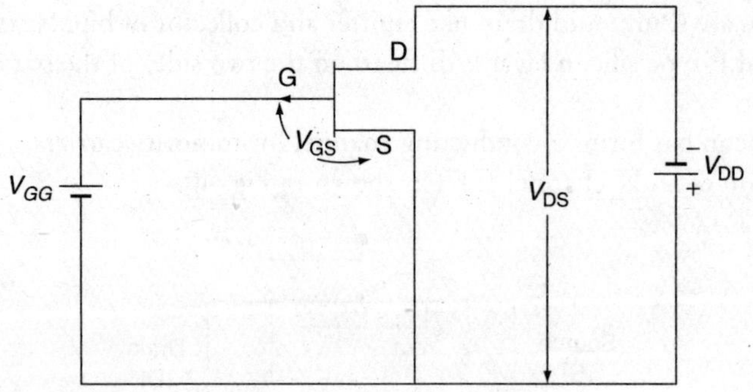

Figure 42 Working of a JFET.

12.10.1.4 JFET Characteristics

The circuit arrangement for obtaining the output characteristics of N-channel JFET is given in Fig. 43. These characteristics plotted in Fig. 44 relates drain current (I_S) to drain to source voltage (V_{DS}) to different values of gate to source voltage (V_{GS}). Keeping V_{DS} constant at 0, V_{DS} is increased in steps. I_D increases linearly with V_{DS} initially, but tends to become constant when V_{DS} attains a voltage called "pinch off" voltage. Further increase in $V_{DS,}$ do not cause I_D to increase. This is due to the fact that for V_{DS} above pinch-off voltage depletion layers formed at the junction expand and almost touch each other making the channel narrow (Fig. 45).

For negative values of $V_{RS,}$ the curves shift downwards because of the fact that pinch-off occurs for the lower values of V_{DS}.

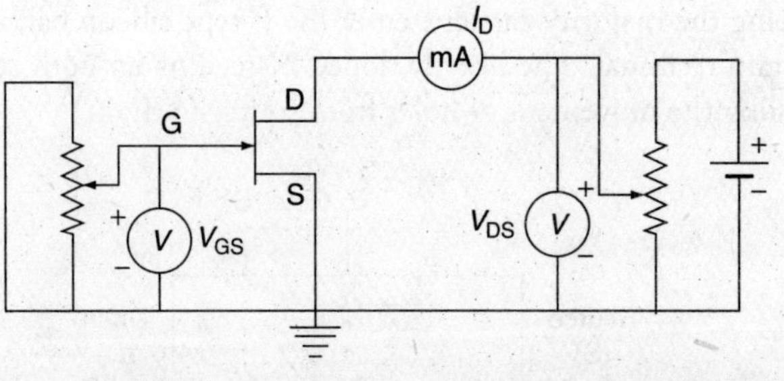

Figure 43 Circuit arrangement of N-channel JFET.

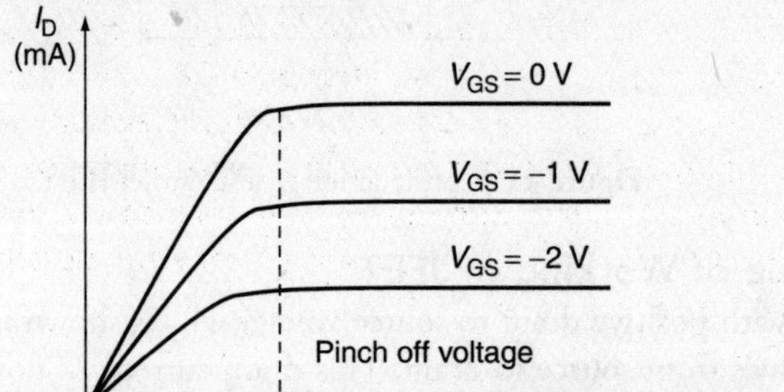

Figure 44 Drain characteristics of JFET.

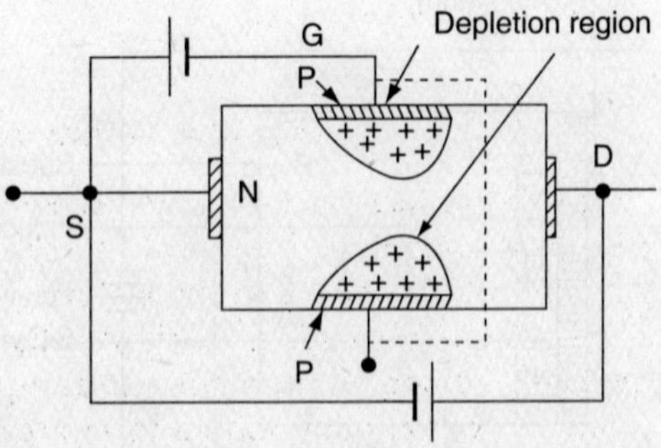

Figure 45 Formation of depletion region in JFET.

12.10.1.5 Advantages and Applications of JFET

The following are the advantages of JFET:

1. It has much higher input impedance than BJT enabling high degree of isolation between input and output.
2. FET is less noisy than BJT.
3. FET is easier to fabricate and are particularly suitable for ICs.
4. FET is normally less sensitive to temperature.

The following are the application of JFET:

1. FET is preferred in phase-shift oscillation
2. They can also serve as buffers bridging amplifier stages.

12.11.2 Metal Oxide Semiconductor FET (MOSFET)

The MOSFET is an FET in which the metal gate is insulated from the semiconductor channel by a very thin oxide layer. MOSFET is also called insulated gate FET (IGFET). The major difference between JFET and MOSFET is that MOSFET has much larger input resistance than JFET. The other important feature of MOSFET is its low power consumption.

12.11.2.1 Enhancement MOSFET

Two heavily doped N^+ regions are diffused over a lightly doped P-type silicon substrate (Fig. 46). One N^+ region is called the source while the other one is called drain. A thin insulating layer of SiO_2 is grown over the surface of the structure and holes are drilled into this oxide layer for making contacts with the source and the drain. Also, a thin layer of metal aluminum is formed over the SiO_2. This metal layer covers the entire channel region. The ohmic contact made with this metal layer forms the gate terminal. The metal area of the gate in conjunction with the insulating oxide of SiO_2 and the semiconductor channel forms a parallel plate capacitor. The SiO_2 layer is responsible for high input resistance.

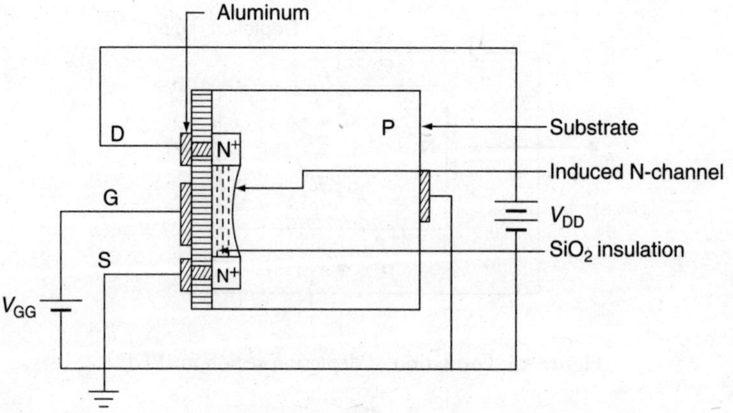

Figure 46 Enhancement MOSFET.

Working Principle

If a positive voltage is applied to the gate with respect to the grounded substrate, positive charge on *G* induces an equal negative charge on the substrate side between the source and the drain regions. Thus an electric field is produced between the source and drain regions. The negative charge of electrons, which are the minority carriers in P-type substrate, forms an inversion layer. As positive voltage of the gate increases, the conductivity increases and current flows from the source to the drain through the induced channel. Thus, the drain current is enhanced by the positive gate voltage. This can be clearly seen from the volt–ampere characteristic shown in Fig. 47.

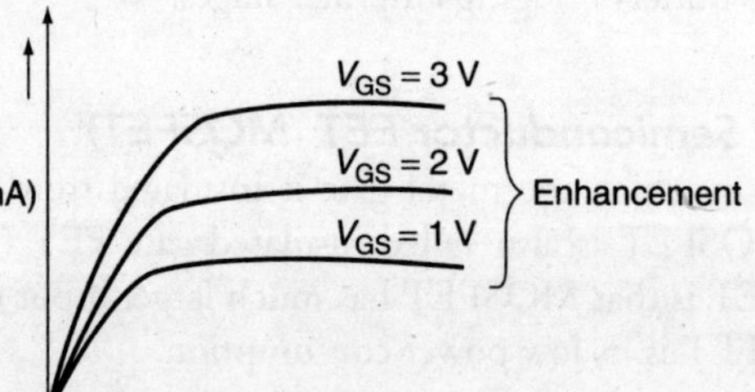

Figure 47 Volt–ampere characteristics of enhancement MOSFET.

12.11.2.2 Depletion MOSFET

The constructional details of N-channel depletion MOSFET is shown in Fig. 48. Here N-channel is diffused between source and drain. With $V_{GS} = 0$, if the drain is kept at higher potential with respect to the source, the electrons flow through N-channel from *S* to *D*. If the gate voltage is made negative, positive charge consisting of holes is induced in the channel through SiO_2 of the gate channel capacitor.

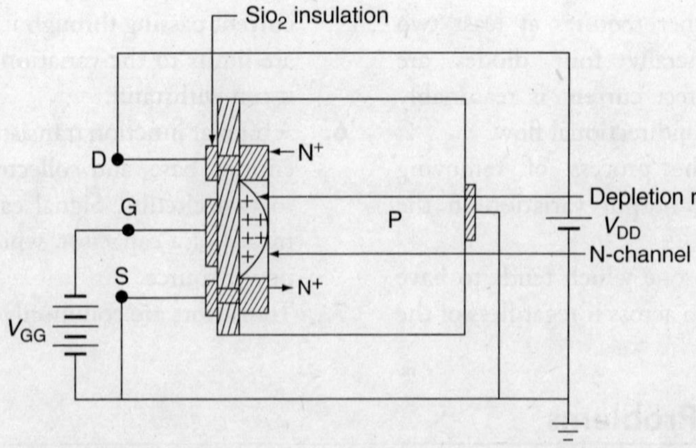

Figure 48 Constructional details of N-channel depletion MOSFET.

The introduction of the positive charge causes depletion of the mobile electrons in the channel. Thus a depletion region is produced in the channel. The shape of the depletion region depends on V_{GS} and V_{DS}. The channel as a result will appear wedge-shaped. When V_{DS} is increased, I_D also increases and it becomes constant at a certain value of V_{DS} called *pinch-off voltage*. Since the current in the FET is due to majority carriers, only induced positive charge makes the channel less conductive because of the reduction in the number of electrons. Therefore I_D drops as V_{GS} is made more negative. It is, thus, clear that N-channel is modulated for higher and lower conductivities by changing the polarity as well as magnitude of V_{GS}. Drain characteristics of MOSFET are shown in Fig. 49.

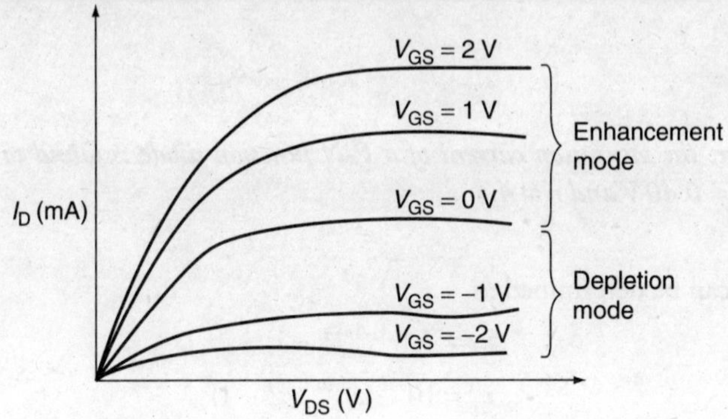

Figure 49 Drain characteristics of depletion MOSFET.

Summary

In this chapter, we have discussed about the following points:

1. A rectifier circuit or network is generally used to convert alternating current to direct current.
2. A single diode can only provide half-wave rectification. It is not efficient and the direct current can only be supplied half the time.

3. A full-wave rectifier requires at least two diodes and generally four diodes are involved. The direct current is reasonably consistent in its unidirectional flow.
4. Smoothing is the process of removing the worst of the output variation in the current.
5. A Zener diode is one which tends to have the same volt drop across it regardless of the current passing through it. In practice, there are limits to the variation of current which it can withstand.
6. A bipolar junction transistor has three layers: emitter, base, and collector. It is connected to two circuits. Signal can be removed by means of a capacitor, which blocks the DC power source.
7. Transistors are commonly used as switches.

Solved Problems

Problem 1

At room temperature, the saturation current in the P-N junction diode is found to be 10^{-14} A. Find the diode current if $V_D = 0.60$ V and $n = 1$.

Solution

The diode current can be determined as

$$i_D = I_S(e^{V_D/nV_T} - 1)$$

$$i_D = 10^{-14}(e^{0.60/0.026} - 1) = 0.105 \text{ mA}$$

Problem 2

At room temperature, the saturation current of a P-N junction diode is found to be 10^{-15} A. Find the diode current if $V_D = 0.40$ V and $n = 4$.

Solution

The diode current can be determined as

$$i_D = I_S(e^{V_D/nV_T} - 1)$$

$$i_D = 10^{-15}(e^{0.40/0.104} - 1)$$

$$= 1.335 \text{ mA}$$

Problem 3

The saturation current in the P-N junction diode is found to be 10^{-8} A. At room temperature, determine the diode current if $V_D = 0.70$ V and $n = 2$.

Solution

The diode current can be determined as

$$i_D = I_S(e^{V_D/nV_T} - 1)$$
$$i_D = 10^{-8}(e^{0.70/0.052} - 1)$$
$$= 3.77\,\text{mA}$$

Problem 4

Figure 50 shows a circuit where a diode is connected with a voltage source through a resistor. Determine the diode current if $V_D = 0.60$ V. Assume negligible diode resistance.

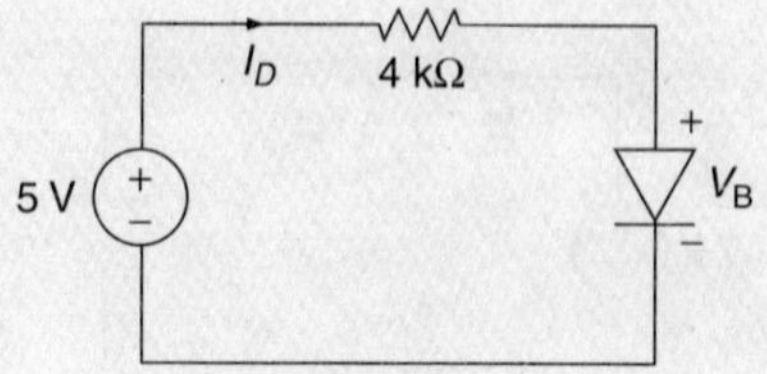

Figure 50 Problem 4.

Solution

We have

$$I_D = \frac{5 - 0.6}{4000}$$
$$= 1.1\,\text{mA}$$

Problem 5

Figure 51 shows a circuit where a diode is connected with a voltage source through a resistor. Determine the diode current if $V_D = 0.40$ V. Assume negligible diode resistance.

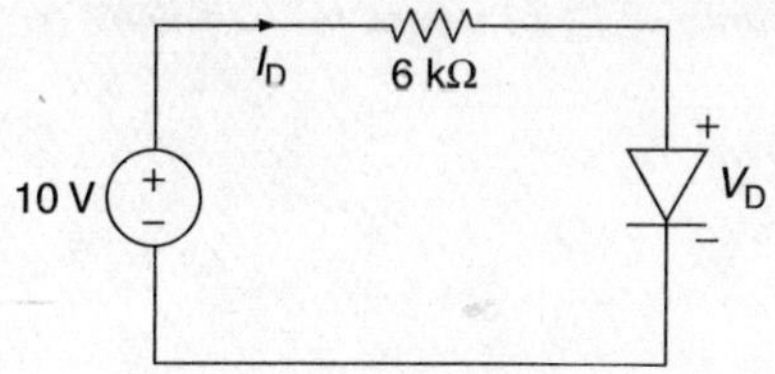

Figure 51 Problem 5.

Solution

$$I_D = \frac{V_{ss} - V_D}{R}$$

$$= \frac{10 - 0.4}{6000}$$

$$= 1.6 \text{ mA}$$

Problem 6

Figure 52 shows a circuit where a diode is connected with a voltage source through a resistor. Determine the diode current if $V_D = 0.60$ V and $r_D = 15\ \Omega$.

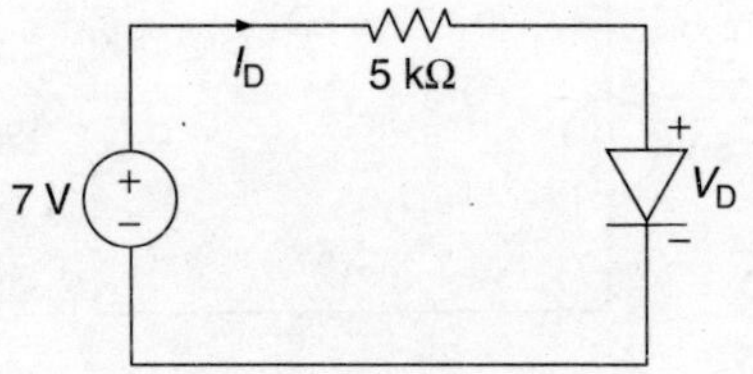

Figure 52 Problem 6.

Solution
We have

$$I_D = \frac{V_{ss} - V_D}{R}$$

$$= \frac{7 - 0.60}{5000} = 1.28 \text{mA}$$

Problem 7

A diode having an internal resistance of 15 Ω is used for half-wave rectification for the motor control. Determine the

(i) maximum value of the current.
(ii) average value of the current.
(iii) r.m.s. value of the current.
(iv) DC output power.

Consider the expression of the input voltage is $V = 45 \sin \theta$ V and the load resistance is 900 Ω.

Solution

We have $V_m = 45$ V.

(i) Maximum value of the current is given by

$$I_m = \frac{V_m}{r_d + R_L} = \frac{45}{15+900} = 0.049 \text{ A}$$

(ii) The average value of the current is given by

$$I_{DC} = \frac{I_m}{\pi} = \frac{0.049}{\pi} = 0.0245 \text{ A}$$

(iii) r.m.s. value of the current is given by

$$I_{rms} = \frac{I_m}{2} = \frac{0.049}{2} = 0.0245 \text{ A}$$

(iv) DC output power is given by

$$P_{DC} = I_{DC}{}^2 R_L = 0.015^2 \times 900 = 0.20 \text{ W}$$

Problem 8

A diode having an internal resistance of 25 Ω is used as half-wave rectifier for the motor control. Determine the

(i) maximum value of the current.
(ii) average value of the current.
(iii) r.m.s. value of the current.
(iv) DC output power.

Consider the expression of the input voltage is $V = 45 \sin \theta V$ and the load resistance is 800 Ω.

Solution

We have $V_m = 45$ V.

(i) Maximum value of the current is given by

$$I_m = \frac{V_m}{r_d + R_L} = \frac{45}{25+800} = 0.0545 \text{ A}$$

(ii) The average value of the current is given by

$$I_{DC} = \frac{I_m}{\pi} = \frac{0.0545}{\pi} = 0.0173 \text{ A}$$

(iii) r.m.s. value of the current is given by

$$I_{rms} = \frac{I_m}{2} = \frac{0.0545}{2} = 0.0272 \text{ A}$$

(iv) DC output power is given by

$$P_{DC} = I_{DC}^2 R_L = 0.025^2 \times 800 = 0.5 \text{ W}$$

Problem 9

A full-wave rectifier with two diodes is used in the field circuit of the DC motor. The internal resistance of each diode is 15 Ω. The r.m.s. value of the secondary voltage (center top to each end) is 45 V. Determine the

(i) maximum current.
(ii) direct component of the current.
(iii) r.m.s. value of the current.
(iv) DC output power.

Consider the load resistance is 700 Ω.

Solution

We have $V_m = 45$ V.

(i) Maximum value of the current is given by

$$I_m = \frac{V_m}{r_d + R_L} = \frac{45}{15+700} = 0.062 \text{ A}$$

(ii) The direct component of the current is given by

$$I_{DC} = \frac{2I_m}{\pi} = \frac{2\times 0.062}{\pi} = 0.039 \text{ A}$$

(iii) r.m.s. value of the current is given by

$$I_{rms} = \frac{I_m}{\sqrt{2}} = \frac{0.062}{\sqrt{2}} = 0.0245 \text{ A}$$

(iv) DC output power is given by

$$P_{DC} = I_{DC}^2 R_L = 0.015^2 \times 900 = 0.20 \text{ W}$$

Problem 10

A full-wave rectifier with the two diodes is used in the field circuit of the DC motor. The internal resistance of each diode is 10 Ω. The r.m.s. value of the secondary voltage (center top to each end) is 40 V. Determine the

(i) maximum current.
(ii) direct component of the current.
(iii) r.m.s. value of the current.
(iv) DC output power.

Consider the load resistance is 800 Ω.

Solution

We have $V_m = 40$ V.

(i) Maximum current is given by

$$I_m = \frac{V_m}{r_d + R_L} = \frac{40}{10+800} = 0.049 \text{ A}$$

(ii) Direct component of the current is given by

$$I_{DC} = \frac{2I_m}{\pi} = \frac{2 \times 0.049}{3.14} = 0.031 \text{ A}$$

(iii) r.m.s. value of the current is given by

$$I_{rms} = \frac{I_m}{\sqrt{2}} = \frac{0.049}{\sqrt{2}} = 0.034 \text{ A}$$

(iv) DC output power is given by

$$P_{DC} = I_{DC}^{\,2} R_L = 0.010^2 \times 800 = 0.08 \text{ W}$$

Problem 11

The Zener diode shown in Fig. 53 is ideal. Find whether it is correctly biased or not.

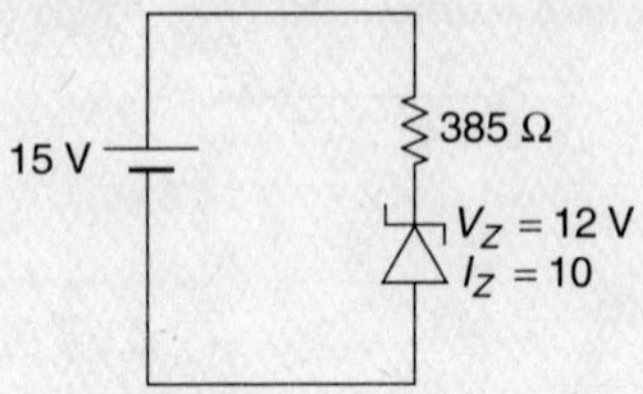

Figure 53 Problem 11.

Solution

We have

$$I_Z = \frac{15-12}{385} = 7.792 \times 10^{-3} \text{ A}$$
$$= 7.792$$

Since $I_Z < (I_Z)_{max}$, the Zener diode is properly biased.

Problem 12

Find whether the Zener diode shown in Fig. 54 is correctly biased. Find I_Z and the power dissipated by the diode where $(I_Z)_{max} = 10\,mA$ and $V_Z = 10\,V$.

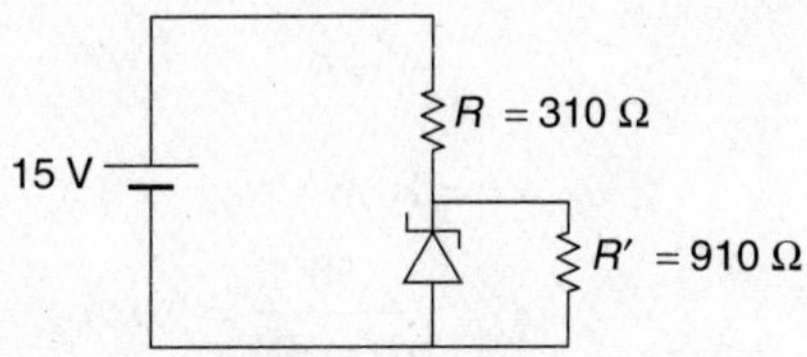

Figure 54 Problem 12.

Solution

Voltage drop across $R = 310\ \Omega$ is

$$(15-10)\text{ V}$$

Current through $R = 310\ \Omega$ is

$$I = 5/310 = 16.13\text{ mA}$$

Current through $R' = 910\Omega$ is

$$I = 10/910 = 10.98\text{ mA}$$

Current through Zener diode is

$$(16.13 - 10.98)\text{ mA} = 5.15\text{ mA}$$

Thus the Zener diode is correctly biased.

Problem 13

Calculate the Zener current if the load resistance R_L (Fig. 55) is (i) 30 kΩ, (ii) 10 kΩ.

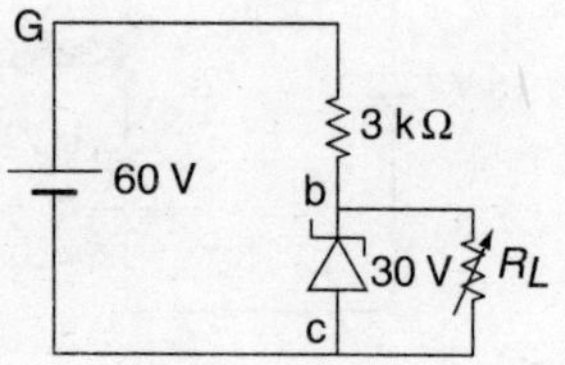

Figure 55 Problem 13.

Solution

(i) When $R_L = 30\text{ k}\Omega$.

$$V_{DC} = 30\text{ V}$$

$$V_R = V_{ab} = 60 - 30 = 30\text{ V}$$

$$I = \frac{3}{3\times 10^3} = 10\text{ mA}$$

$$I_L = \frac{30}{30\times 10^3} = 1\text{ mA}$$

Therefore

$$I_Z = I - I_L$$
$$= (10-1)\text{ mA}$$
$$= 9\text{ mA}$$

(ii) When $R_L = 30\text{ k}\Omega$.

$$I = 10\text{ mA}$$

$$I_L = \frac{30}{10\times 10^3} = 3\text{ mA}$$

Therefore

$$I_Z = I - I_L = 10 - 3\text{ mA} = 7\text{ mA}$$

Problem 14

Calculate the Zener current if the load resistance R_L is (i) 50 kΩ, (ii) 20 kΩ.

Solution

(i) When $R_L = 50\ \text{k}\Omega$.

$$V_{bc} = 50\ \text{V}$$

$$V_R = V_{ab} = 60 - 50 = 10\ \text{V}$$

$$I = \frac{50}{3\times 10^3} = 16$$

$$I_L = \frac{50}{50\times 10^3} = 1$$

Therefore

$$I_Z = I - I_L$$

$$= (16 - 1)\,\text{mA}$$

$$= 15\,\text{mA}$$

(ii) When $R_L = 20\ \text{k}\Omega$

$$I = 16\,\text{mA}$$

$$I_L = \frac{50}{16\times 10^3} = 3\,\text{mA}$$

Therefore

$$I_Z = I - I_L = 13\,\text{mA}$$

Multiple Choice Questions

1. Electronics is the branch of engineering which deals with flow of electrons through
- **(a)** semiconductor
- **(b)** gas
- **(c)** vacuum
- **(d)** all of these

2. The metal from which the electrons are emitted is called
- **(a)** an emitter
- **(b)** a cathode
- **(c)** an emitter or a cathode
- **(d)** none of those

3. The process of conversion AC into DC is called
- **(a)** amplification
- **(b)** modulation
- **(c)** rectification
- **(d)** factorization

4. The addition of impurity atoms to a pure semiconductor makes it an
- **(a)** extrinsic semiconductor
- **(b)** intrinsic semiconductor
- **(c)** both (a) and (b) are correct
- **(d)** none of these

5. The semiconductor diode can be used for
- **(a)** half-wave rectifier
- **(b)** full-wave rectifier
- **(c)** both (a) and (b) are correct
- **(d)** none of these

6. In a full-wave rectifier, the diodes conduct for
- **(a)** one-half cycle
- **(b)** full cycle

(c) alternate half cycle
(d) none of these.

7. The maximum efficiency of a half-wave rectifier is
(a) 33.33%
(b) 40.6%
(c) 50%
(d) 68%

8. The arrow in the transistor symbol indicates the direction of current in
(a) base
(b) collector
(c) emitter
(d) none of these.

9. When C_B and E_B junctions get forward-biased, then the transistor is said to be
(a) active
(b) saturated
(c) cut-off
(d) unsaturated

10. The transistors are
(a) low-voltage and low-current devices
(b) high-voltage and high-current devices
(c) low-voltage and high-current devices
(d) only low-current devices

11. The material which is not a semiconductor is
(a) selenium
(b) gallium-arsenide
(c) silica
(d) carborundum

12. In an intrinsic semiconductor,
(a) there are no holes in the material
(b) the number of holes is too small
(c) electrons in the material are neutralized by holes
(d) there are no electrons in the material

13. If the collector material of a transistor is of high resistivity, the avalanche breakdown voltage would be
(a) high
(b) low
(c) independent of resistivity
(d) constant

14. In a MOSFET, if gate voltage V_{GS} is increased then
(a) channel conductivity increases
(b) depletion region decreases
(c) channel conductivity decreases
(d) channel current increases

15. Which of the following transistor can be used in enhancement mode?
(a) UIT
(b) JFET
(c) MOSFET
(d) NPN transistor

Review Questions

1. What is semiconductor?
2. What is energy band diagram? What is the significance of band gap?
3. How does the energy band structure of a semiconductor differ from that of a conductor and an insulator?
4. What is an intrinsic semiconductor?
5. How can the conductivity of an intrinsic semiconductor be improved?
6. Define doping and dopant.
7. What is an N-type semiconductor?
8. Define ripple factor and transform utilization factor.
9. Compare single-phase full-wave rectifier with single-phase bridge rectifier and bring out the merits of the latter.
10. What is Zener breakdown? How does it differ from avalanche breakdown?

11. How does a Zener diode differ from an ordinary diode?

12. Compare FET with BJT?

13. Explain the construction of an N-channel FET.

14. What is pinch-off voltage?

15. Explain the working of depletion type MOSFET.

16. Explain the use of filters.

17. What are the different types of filters?

18. Compare *C* filter with *L* filter.

19. What is Π filter?

20. Write the expression for Π filter ripple factor.

Problems

1. A transistor operates with a collector current of 100 mA and an emitter current of 102 mA. Determine the value of the base current.

2. A bipolar transistor has a common-emitter current gain of 125. If the transistor operates with a collector current of 50 mA, determine the value of the base current.

3. A bipolar junction transistor operates with a collector current of 120 mA and a base current of 50 mA. What will be the value of the common emitter current gain?

4. A field-effect transistor operates with a drain current of 100 mA and a gate source bias of −1 V. The device has a g_{fb} forward transfer conductance value of 0.25. If the bias voltage decreases to -1.1 V, determine
   (a) The change in drain current.
   (b) The new value of drain current.

5. An amplifier has a current gain of 40 and a voltage gain of 30. Determine the power gain.

6. A load of resistance 100 Ω receives a DC power of 100 W from a full-wave rectifier. Determine the DC output voltage.

7. Consider a full-wave rectifier having r.m.s. voltage as 100.02 V and output voltage as 100 V . Calculate the ripple voltage.

8. A full-wave rectifier with the two diodes is used in the field circuit of the DC motor. The internal resistance of each diode is 20 Ω. The r.m.s. value of the secondary voltage (center top to each end) is 30 V. Determine the **(a)** maximum current, **(b)** direct component of the current, **(c)** r.m.s. value of the current, **(d)** DC output power. Consider the load resistance is 800 Ω.

9. A transistor operates with a base current of 2 mA and a collector current of 26 mA. Determine the value of the emitter current.

10. An amplifier has a current gain of 60 and a voltage gain of 100. Determine the power gain.

11. At room temperature, the saturation current in the P-N junction diode is found to be 10^{-10} A. Find the diode current if $V_D = 0.3$ and $n = 4$.

12. Figure 56 shows a circuit where a diode is connected with a voltage source through a resistor. Determine the diode current, if $V_D = 0.60$ V. Assume negligible diode resistance.

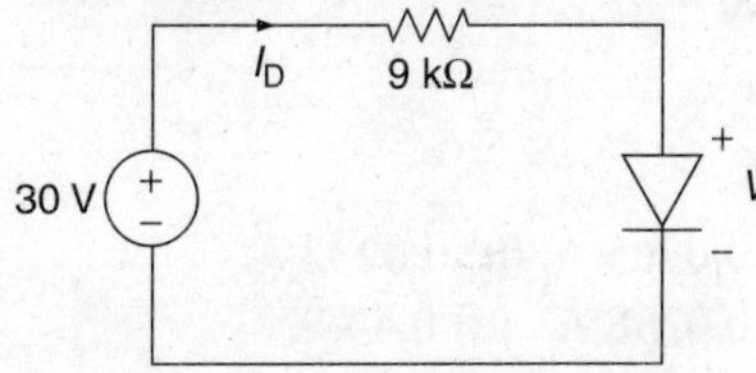

Figure 56 Problem 12.

13. A diode having an internal resistance of 20 Ω is used for half-wave rectification for the motor control. Determine the
(a) maximum value of the current.
(b) average value of the current.
(c) r.m.s. value of the current.
(d) DC output.

Consider the expression of the input voltage is $V = 40\sin\theta$ V and the load resistance is 500 Ω.

14. A full-wave rectifier with two diodes is used in the field circuit of the DC motor. The internal resistance of each diode is 20 Ω. The r.m.s. value of the secondary voltage is 80 V. Determine **(a)** the maximum current, **(b)** direct component of the current, **(c)** r.m.s. value, **(d)** DC output. Consider the load resistance is 650 Ω.

15. The Zener diode shown in Fig. 57 is ideal. Find whether it is correctly biased or not.

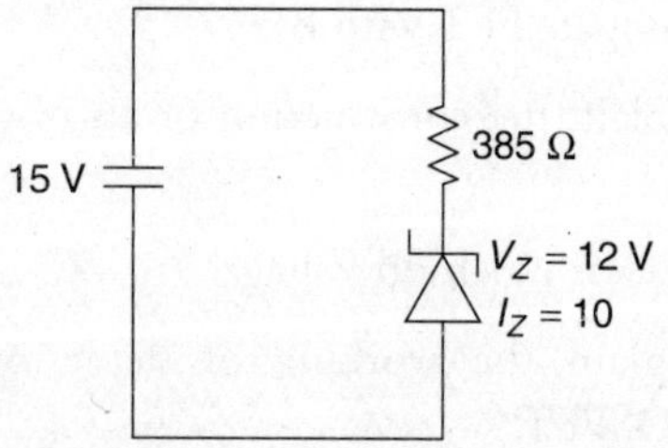

Figure 57 Problem 15.

16. A bipolar transistor has a common-emitter current gain of 100. If the transistor operates with a collector current of 30 mA, determine the value of the base current.

17. Calculate the Zener current if the load resistance R_L is **(a)** 50 kΩ and **(b)** 120 kΩ.

Answers

Multiple Choice Questions

1. (d) **2.** (c) **3.** (c) **4.** (a) **5.** (c) **6.** (c) **7.** (b) **8.** (c) **9.** (b) **10.** (a) **11.** (a) **12.** (b) **13.** (a) **14.** (c) **15.** (c)

Problems

1. 2 mA
2. 400 μA
3. 2.4
4. **(a)** −25 mA, **(b)** 75 mA
5. 1200
6. 100 V
7. 2 V
8. **(a)** 0.0365 A, **(b)** 0.0232 A, **(c)** 0.0258 A , **(d)** 0.43 W
9. 28 mA
10. 6000
11. 1.197 mA
12. 29.9 mA
13. **(a)** 0.076 A, **(b)** 0.024 A, **(c)** 0.038 A, **(d)** 0.2 W
14. **(a)** 0.119 A, **(b)** 0.075 A, **(c)** 0.084 A, **(d)** 0.26 W
15. Zener diode is correctly biased
16. 300 μA
17. **(a)** 9 mA, **(b)** 3 mA

13 Special Semiconductor Devices

LEARNING OBJECTIVES

After reading this chapter, you will be able to:

- Understand how silicon controlled rectifier operates.
- Understand the action of TRIAC.
- Appreciate how injunctions transistor is applicable as a relaxation oscillator.
- Understand the concept of photoconductive cell, LED, photodiode.
- Estimate quiescent operating conditions and gain from transistor characteristic and other data.

It is the field of science and engineering which deals with the study of electron devices and their utilization. An electron device is one in which movement of electrons takes place through vacuum or gas or semiconductor. Electrons have a very wide range of applications in various fields such as instrumentation, process control, medical sciences, defence, communication systems, generation of power, etc. Initially vacuum tubes dominated the field of electronics, and later transistors created revolution in the electronic industry. In this chapter various semiconductor devices, their working principle, characteristics and application are dealt.

13.1 Silicon Controlled Rectifiers (SCRs)

Silicon controlled rectifier (SCR) was invented in the year 1953. SCR is the most important family member of thyristor. It is the power semiconductor device which can handle currents of several hundred amperes and voltages of several thousand volts. Because of its compactness, reliability and low loss, it is employed in almost all the power-controlled devices. The other members of thyristor family include TRIAC, DIAC, silicon controlled switch, programmable unijunction transistor, etc.

13.1.1 Construction Details

The SCR is a four-layer, three-terminal device with junctions J_1, J_2, J_3 as shown in Fig. 1(a). The three terminals are anode, cathode, and gate. The construction of SCR shows that the gate terminal is kept near the cathode. In terms of their lateral dimensions, thyristors are the largest semiconductor devices made. A complete silicon wafer as large as 10 cm in diameter may be used to make a single high power thyristor. The circuit symbol is shown in Fig. 1(b).

SCR is a rectifying device like diode. However, it requires the application of trigger pulse to switch into conduction. When its anode to cathode voltage is positive, the device conducts once the trigger pulse is applied. This is called turn ON of SCR. Once the device is ON, gate loses its control over it.

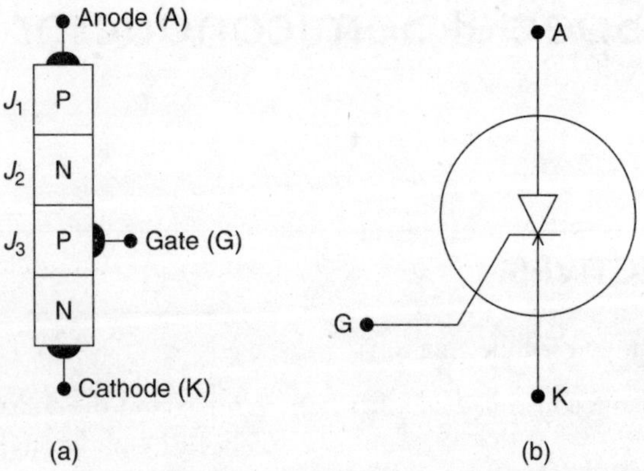

Figure 1 SCR: (a) Structure; (b) circuit symbol.

To switch OFF the thyristor, the current flowing in the device has to be brought down to zero. This is done either by applying negative voltage to the thyristor or by increasing the resistance in series with the thyristor. This process is called "turn OFF."

13.1.2 Characteristics of SCR

A typical *V–I* characteristics of a thyristor are shown Fig. 2. In the reverse direction, the thyristor appears similar to a reverse-biased diode which conducts very little current until avalanche breakdown occurs. In the forward direction the thyristor has two stable states or modes of operation

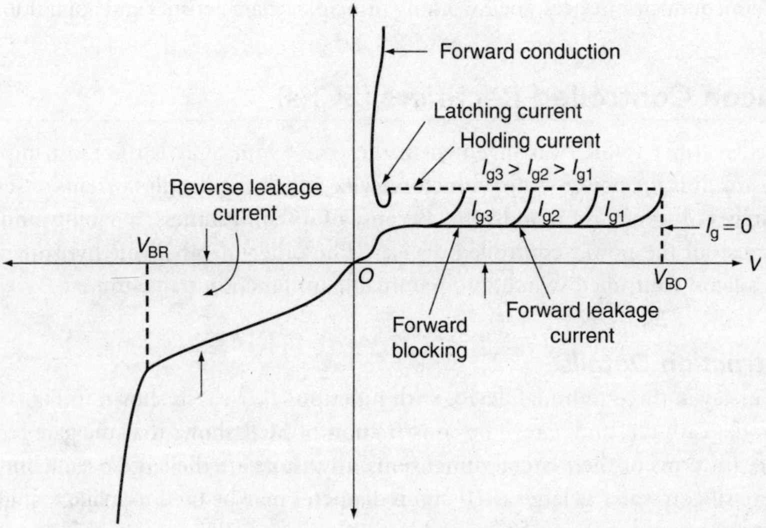

Figure 2 Characteristics of SCR. Here V_{BO}: forward breakover voltage; V_{BR}: reverse breakover voltage; I_g: gate current.

that are connected together by an unstable mode that appears as a negative resistance on the *V–I* characteristics. The low-current, high-voltage region is the forward-blocking state or the OFF state and the low-voltage, high-current mode is the ON state. For the forward-blocking state, the quantity of interest is the forward-blocking voltage V_{BO} which is defined for zero gate current. If a positive gate current is applied to a thyristor then the transition or breakover to the ON state will occur at smaller values of anode to cathode voltage as shown. Although not indicated, the gate current does not have to be a DC current but instead can be a pulse of current having some minimum time duration. This ability to switch the thyristor by means of a current pulse is the reason for widespread application of the device.

However, once the thyristor is in the ON state, the gate cannot be used to turn the device OFF. The only way to turn OFF the thyristor is for the external circuit to force the current through the device to be less than the holding current for a minimum specified time period.

13.1.2.1 Holding Current I_H

After an SCR has been switched to the ON state, a certain minimum value of anode current is required to maintain the thyristor in this low impedance state. If the anode current is reduced below the critical holding current value, the thyristor cannot maintain the current through it and reverts to its OFF state. Usually I_H is associated with turn OFF of the device.

13.1.2.2 Latching Current I_L

After the SCR has switched ON, there is a minimum current required to sustain conduction. This current is called the latching current. It is associated with turn ON and is usually greater than holding current.

13.1.3 Application of SCR

1. It is used in AC to DC converters.
2. It is used in AC voltage controllers.
3. It is used in DC to AC invertors.
4. It is used in HVDC transmission UPS and static circuit breakers.

13.2 Triode for Alternating Current (TRIAC)

Unlike SCR, which is a unidirectional switching device, TRIAC is a bidirectional switching device with three terminals. It is used for controlling alternating current in a load. The basic structure and its circuit symbol for a TRIAC as shown in Fig. 3(a) and (b), respectively TRIAC is functionally equivalent to two SCRs connected in antiparallel as shown in Fig. 3(c). The characteristics of a TRIAC are given in Fig. 4.

TRIAC are used extensively in residential lamp dimmers, heat control and for speed control of single-phase induction motor.

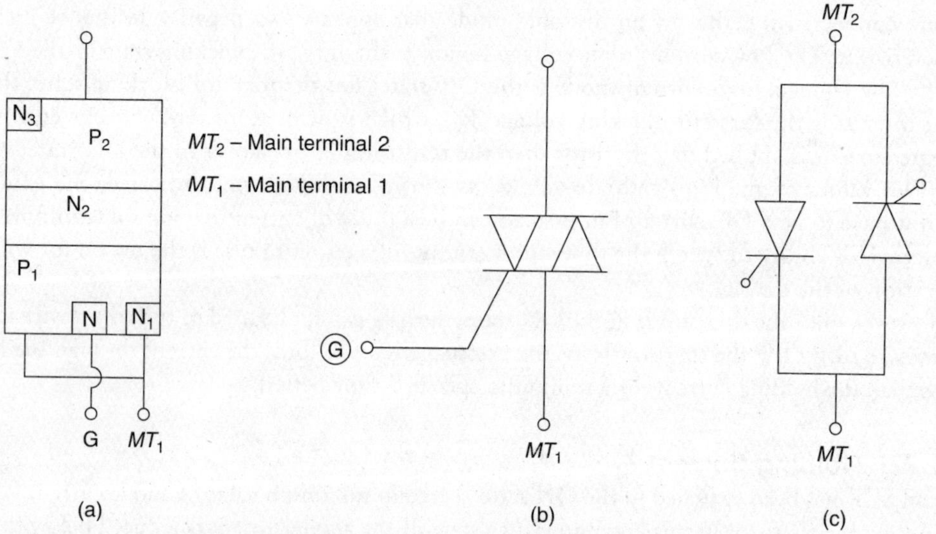

Figure 3 TRIAC: (a) Structure, (b) symbol (c) two SCR are in antiparallel.

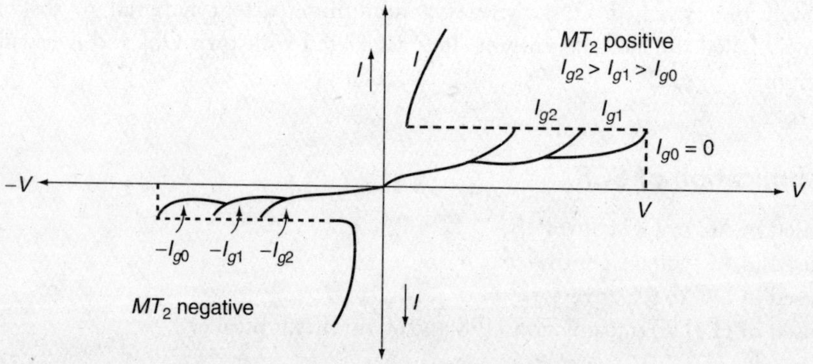

Figure 4 *V–I* characteristics of a TRIAC.

13.3 Unijunction Transistor (UJT)

UJT is an n-type silicon bar in which p-type emitter is embedded. It is a three-terminal semiconductor switching device commonly called as unijunction transistor (UJT) as it has only one junction. A UJT is made up of an n-type silicon base to which a p-type emitter is embedded (Fig. 5). R_{B1} and R_{B2} are the internal resistance from base B_1 and B_2 to point A. Its symbolic representation is given Fig. 6 and its equivalent circuit is shown Fig. 7.

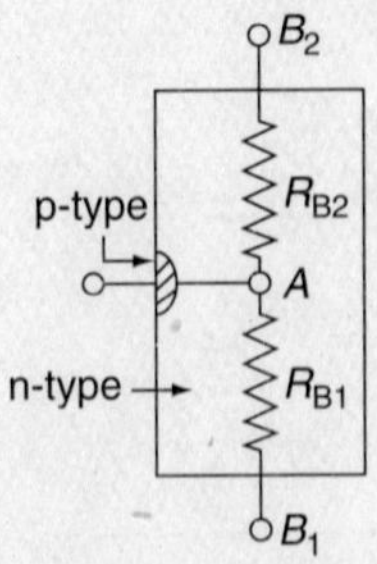

Figure 5 A UJT.

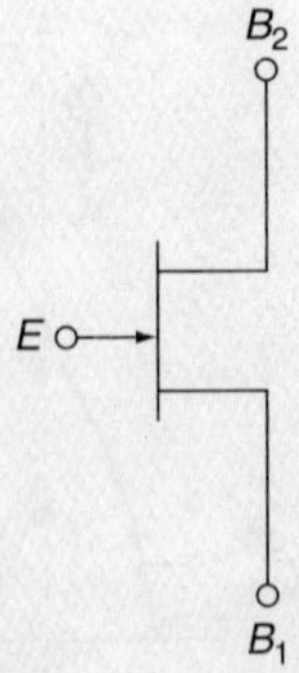

Figure 6 Symbolic representation of a UJT.

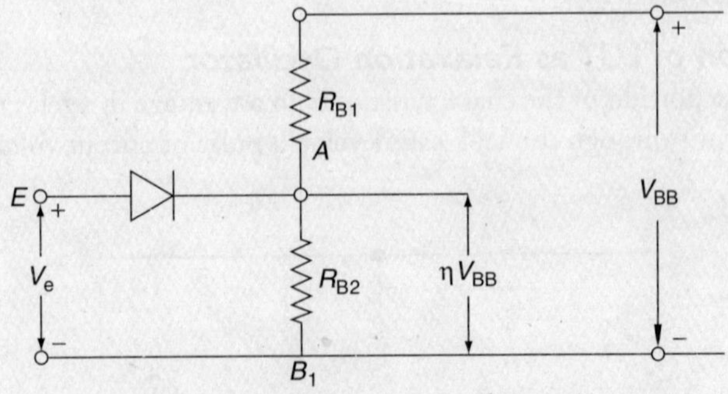

Figure 7 Equivalent circuit symbol of UJT.

When a voltage V_{BB} is applied across the base terminal B_1 and B_2, the potential of point A with respect to B_1 is given by

$$V_{AB1} = \frac{R_{B1}}{R_{B1} + R_{B2}} V_{BB}$$

$$= \eta V_{BB}$$

where

$$\eta = \frac{R_{B1}}{R_{B1} + R_{B2}}$$

is called intrinsic stand-off ratio. Typical values of η are 0.051 to 0.82. The UJT characteristics are shown in Fig. 8. Interbase resistance ($R_{BB} = R_{B1} + R_{B2}$) is 9, the order of 5–10 kΩ. As the input voltage V_E is increased the p-n diode is reverse-biased till the peak point P is reached. It is because $V_E < \eta V_{BB}$. At P, the diode starts conducting since $V_E > \eta V_{BB}$ and holes are injected into the n-layer supply. Due to this negative resistance property, UJT can be employed in a variety of applications, such as in saw-tooth wave generator, pulse generator, switching, and timing and phase control circuits.

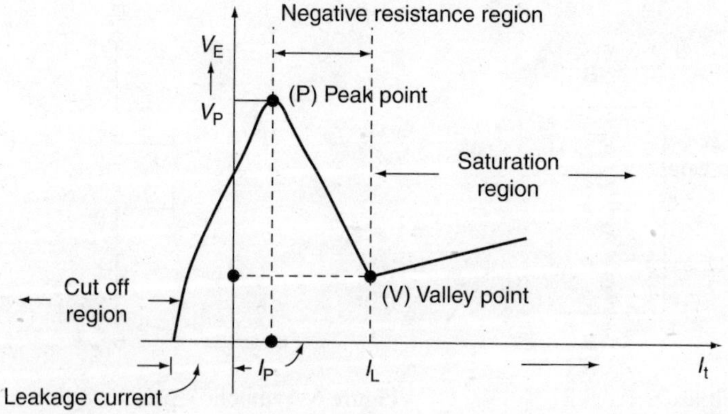

Figure 8 UJT characteristic.

13.3.1 Application of UJT as Relaxation Oscillator

The negative resistance portion of the characteristics is an advantage in a relaxation oscillator where the discharging of a capacitor through the UJT can develop a pulse of output voltage. A practical oscillator is shown in Fig. 9.

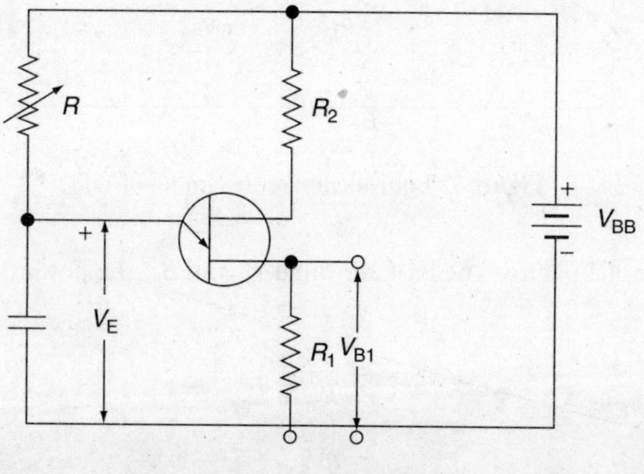

Figure 9 UJT relaxation oscillator.

Capacitor C charges through R from V_{BB} until V_E reaches V_P and the UJT turns ON. When V_E drops to about 2 V, the emitter ceases to conduct and the UJT switches OFF allowing C to charge again. The period of oscillation is given by

$$T = \frac{1}{f}$$

$$= RC \ln \frac{1}{1-\eta}$$

$$= 2.3RC \log_{10} \frac{1}{1-\eta}$$

The input and output voltage waveforms are shown in Fig. 10.

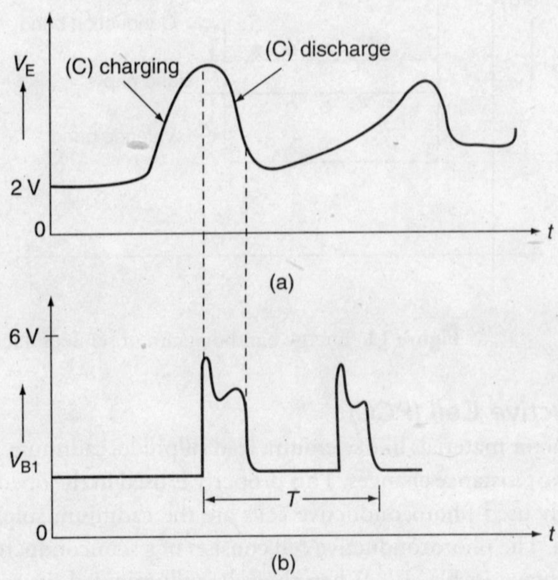

Figure 10 Voltage waveform: (a) Input waveform; (b) output waveform.

13.4 Opto-Electronic Devices

When light radiation falls on certain semiconductors their conductivity increases. This phenomenon is called photoconductive effect. Light radiation is nothing but a stream of photons with each photon having energy of

$$E = hf$$

where $h = 6.626 \times 10^{-34}$ Js is Planck's constant and f is the frequency of the incident light.

Consider the energy band diagram of a semiconductor shown in Fig. 11. If the energy received from the light radiation is greater than the band gap energy, this results in the formation of electron–hole pairs, electrons in the conduction band and holes in the valence band. The increased number of electrons in the conduction band and holes in the valence band improves the conductivity of the semiconductor. Such a semiconductor is called photoconductor or photoresistor. For photoconduction effect to take place in an intrinsic semiconductor, it is important that the photon energy exceeds its band gap energy. This leads to the calculation of the minimum radiating frequency f_C as

$$f_C = \frac{E_g}{h}$$

where E_g is gap energy.

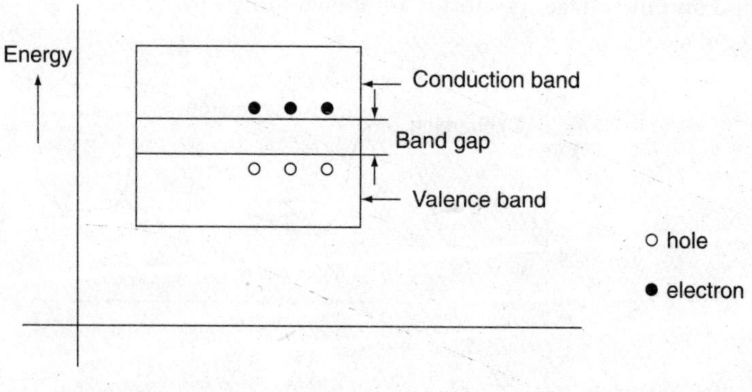

Figure 11 Energy band diagram of semiconductor.

13.4.1 Photoconductive Cell (PCC)

When certain semiconductor materials like selenium, lead sulphide, cadmium sulphide, etc. are exposed to radiations, their specific resistance changes. This property is used in the operation of photoconductive cell. The most commonly used photoconductive cells are the cadmium sulphide cells as they possess very high dark/light ratio. The photoconductive cell consists of a semiconductor material with two electrodes attached to it as shown in Fig. 12. When the cell is illuminated, its resistance decreases, circuit current increases, and a large ratio of dark to light resistance can be obtained by selecting suitable shape of semiconductor material. The photoconductive cells are used for detecting the location of air crafts and ships with the help of the radiations coming out from their exhausts.

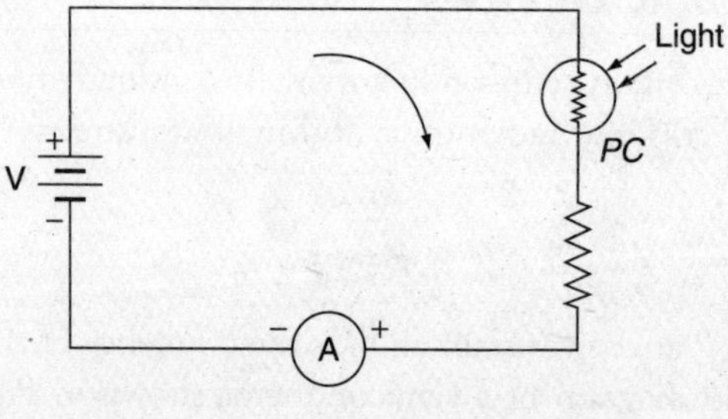

Figure 12 Photoconductive cell.

13.4.2 Photodiode

It is also known as photodetector. A p-n junction diode under reverse-biased conditions carries reverse saturation current. This is due to flow of minority carriers which are nothing but the hole–electron pairs generated by the thermal energy which is absorbed by the crystal at room temperature. When the thermal energy is increased by increasing the temperature, the reverse saturation current increases. If, now the p-n junction is irradiated by light, instead of thermal energy, the reverse current through the diode increases. The circuit symbol and the biasing are shown in Fig. 13. For maximum activity, sometimes a lens is placed on the junction to focus the incident light. The changes in the output voltage are proportional to the incident light. The output signal is amplified and sent for further processing.

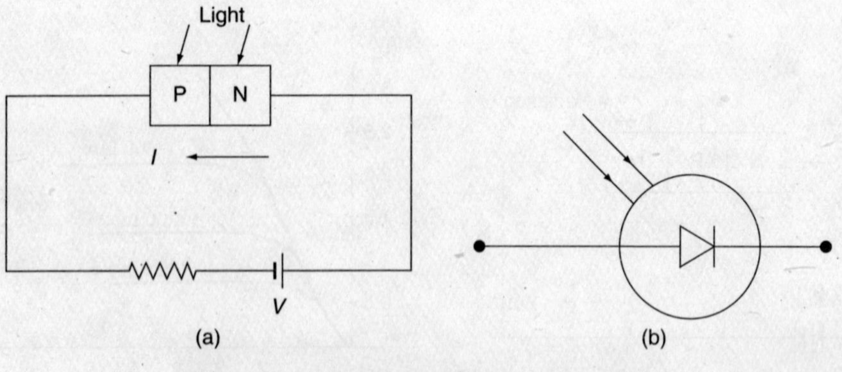

Figure 13 (a) Photodiode; (b) its circuit symbol.

The characteristics of photodiode are shown in Fig. 14. The reverse current increases in direct proportion to the level of illumination. Even when no light is applied, there is minimum reverse leakage current, called dark current, flowing through the device. Photodiodes are therefore used as light detectors, demodulators and encoders. Besides these application, photodiodes are also used in optical communication system, switching circuits, in industries, process control and automation, instrumentation and communication.

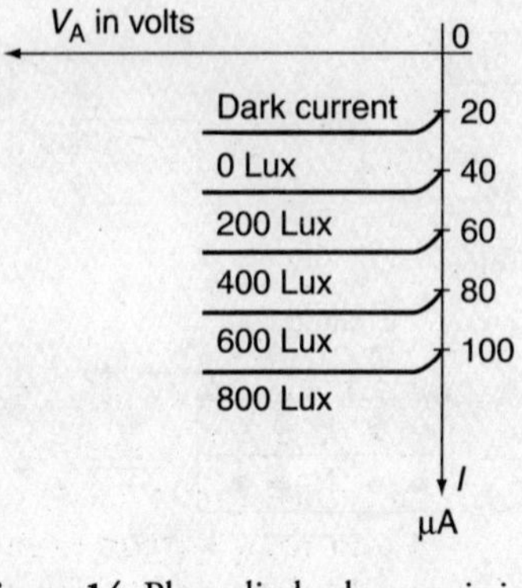

Figure 14 Photodiode characteristics.

13.4.3 Phototransistor

The current produced by a photodiode is very low. Hence this current should be amplified for the use in control circuits. Any bipolar junction transistor is affected by light on its collector–base junction. Phototransistor is constructed to optimize its light activating characteristics. It is an npn transistor with the collector–base junction reverse-biased as shown Fig. 15. A line is built into the package to direct the radiation as collector–base junction. Generally only two heads from the emitter and collector are used. In this device a base current is generated by the light incident in the base–collector junction.

The current in phototransistor is mainly dependent on the intensity of light entering the lens and it is affected by the voltage applied to the external circuit. Figure 16 shows as graph of collector current I_C as a function of collector emitter voltage V_{CE} and as a function of illumination H.

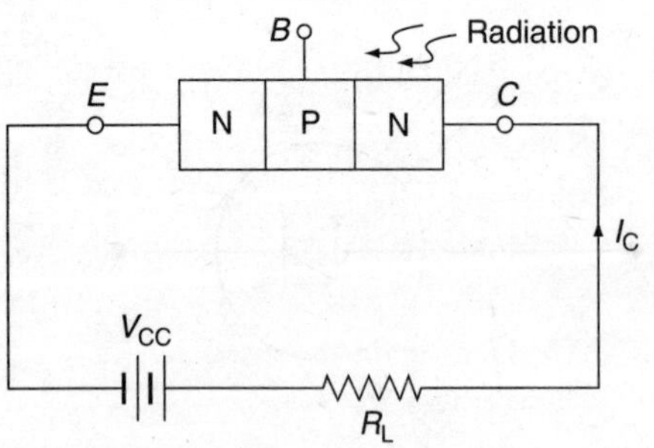

Figure 15 Phototransistor.

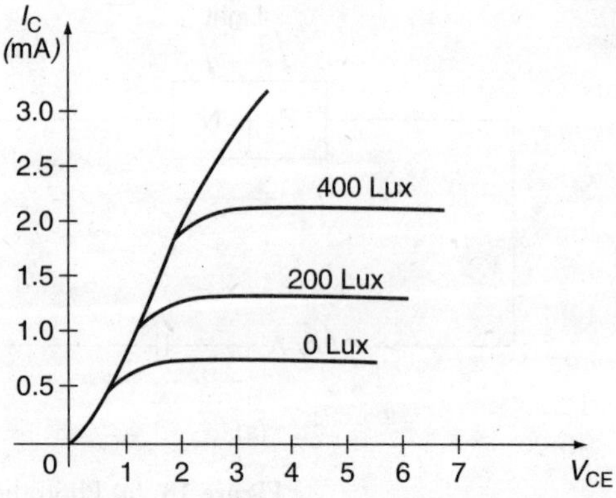

Figure 16 Phototransistor characteristics.

13.4.4 Light-Emitting Diode (LED)

The photodiode and phototransistor produce current when their light-sensitive junction is exposed to radiation. The LED, on the other hand, emits light when it conducts current in forward direction. This phenomenon is called electroluminescence. Figure 17 shows the circuit of the LED, symbol of the LED and construction of surface emitting LED.

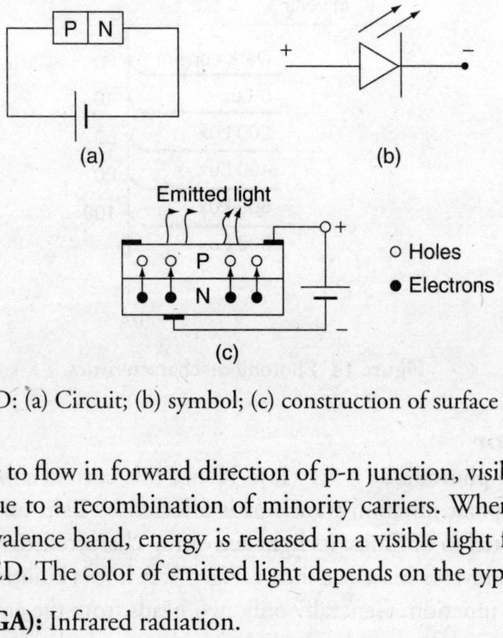

Figure 17 LED: (a) Circuit; (b) symbol; (c) construction of surface emitting LED.

When a current is made to flow in forward direction of p-n junction, visible light is emitted from the depletion region. This is due to a recombination of minority carriers. When an electron falls from the conduction band into the valence band, energy is released in a visible light form. Such a light-emitting diode is referred to as an LED. The color of emitted light depends on the type of material used for LED:

1. **Gallium Arsenide (GA):** Infrared radiation.
2. **Gallium Phosphate (GP):** Red or green.
3. **Gallium Arsenide Phosphate (GAP):** Red or yellow.

LEDs operate at levels ranging from 1.5 V to 3.3 V drawing current of tens of mileamperes. They are used in burglar alarms, calculators, digital computers and optical communication system.

Summary

In this chapter, we have discussed some of the points mentioned below.

1. SCR and its related devices collectively known as thyristors have revolutionized industrial power control applications.
2. UJT is used in wave generation and in trigger circuits for thyristors.
3. The minimum forward voltage at which SCR will conduct even if the gate circuit is open (I_{go}) is known as breakover voltage.
4. The minimum value of the anode current required for the device to conduct is known as holding current.
5. Latching current is the minimum value of the anode current required for the SCR to stay in conduction even when the gate signal is removed.

Multiple Choice Questions

1. An emitter is generally called
   (a) anode
   (b) cathode
   (c) grid
   (d) drain

2. SCR is the __________ rating.
   (a) voltage
   (b) current
   (c) voltage and current
   (d) power

3. The transistors are
   (a) low voltage and low current devices
   (b) high voltage and high current devices
   (c) low voltage and high current devices
   (d) only low current devices

4. A LED is basically a ________ p-n junction.
   (a) forward-biased
   (b) reverse-biased
   (c) lightly doped
   (d) heavily doped

5. The other name for photodiode is
   (a) photoconductive cell
   (b) phototransistor
   (c) photodetector
   (d) light-emitting diode

6. The arrow in a transistor symbol indicates the direction of current in
   (a) base
   (b) collector
   (c) emitter
   (d) none of these

7. The device possessing the highest sensitivity is a
   (a) photoconductive cell
   (b) phototransistor
   (c) photovoltaic cell
   (d) n photodiode

8. The forward resistance of a diode will be
   (a) very high
   (b) small
   (c) infinity
   (d) zero

9. SCR was invented in the year
   (a) 1946
   (b) 1952
   (c) 1953
   (d) 1956

10. Which of the following transistor circuits has the lowest voltage gain?
   (a) Common-base
   (b) Common-collector
   (c) Common-emitter
   (d) None of these

Review Questions

1. What are thyristors? What is an SCR?
2 Compare diode and SCR.
3. Draw the symbol of an SCR.
4. What are the applications of an SCR?
5. What is TRIAC?
6. Draw the symbol of TRIAC.
7. What is a photodiode?
8. Explain the working of LEDs.
9. Draw the UJT oscillator circuit and explain its operation.
10. List the applications of photodiode.
11. Explain the operation of the photo-conductive cell (PCC).
12. Define LED.
13. Define photoconductive effect.
14. Define forward-blocking region of an SCR.
15. Write the application of LED.
16. Mention the types of SCR.
17. How many junctions are available in SCR?
18. Explain the operation of unijunction transistor.
19. Draw and explain the operation of a light emitting diode.
20. Distinguish between photodiode and phototransistor.

Answers

Multiple Choice Questions

1. (b) **2.** (c) **3.** (a) **4.** (a) **5.** (c) **6.** (c) **7.** (b) **8.** (b) **9.** (c) **10.** (b)

14 Transducers

LEARNING OBJECTIVES

After reading this chapter, you will be able to:

- Explain about the basic concepts of classification of transducers.
- Understand LVDT.
- Understand Hall effect transducer.

An instrumentation system generally consists of three components, namely, an input device, a signal conditioning device and an output device. The input device with transducer in it senses the quantity under measurement and changes it to a proportional electrical signal. In the signal conditioning device, the signal is amplified, filtered or otherwise modified to a level accepted by the output device. The output device may be an inductor or a stored medium. The input measurement quantities for most instrumentation systems are non-electrical. The non-electrical quantity is converted to an electrical signal by a device called transducer. A transducer is an element which converts one form of energy into another. In practice, transducers convert non-electrical signals to electrical signals. This chapter describes a few important transducers such has linear variable differential transformer (LVDT), strain gauge, etc.

14.1 Basic Requirements of Transducers

For satisfactory operation, a transducer has to meet the following requirements.

1. **Linearity:** The input and output characteristics of a transducer should be linear.
2. **Accuracy:** The transducer output should conform to the actual value to measured.
3. **Sensitivity:** The transducer should be highly sensitive if its resolving power is to be high.
4. **Ruggedness:** The transducer should be rugged enough to withstand over loads and shocks.
5. **Frequency response:** The transducer gain should remain constant over the measurement range of frequencies. Also, it should respond to its input changes as fast as possible ensuring good dynamic response.
6. **Repeatability:** The transducer should produce identical output signals when the same input signal is applied at different times under the same environment conditions.

14.2 Classification of Transducers

Basically there are two types of transducers: Active transducers and passive transducers. A list of active and passive transducers is given in Table 1.

Table 1 Active and passive transducer

Active Transducers	*Passive Transducers*
1. Thermocouple	1. Resistance transducer thermistor, photo conductive cell resistance strain gauge
2. Piezoelectric transducer	2. Inductance transducer LVDT
3. Photovoltaic cell	3. Capacitance transducer
4. Moving coil generator	4. Photoemissive cell

1. **Active transducers:** An active transducer converts energy directly from one form to another. It does not require any external power source or excitation for this conversion. Hence they are also referred to as self-generating transducers.
2. **Passive transducers:** These types of transducers drive the power required for energy conversion from an external source. Hence they are also referred to as externally powered transducers.

Thermocouple is an active transducer, since it converts heat energy to electrical energy directly, while strain gauge is passive transducer (sensor). Strain caused in the body, in which the strain gauge is embedded, due to application of stress to the body leads to the change in resistance of the strain gauge wire. This results in the change in the output voltage of the Wheatstone bridge with strain gauge forming one of its forearms. It is to be noted that the battery source is required for its operation.

It is important that a distinction be made between transducers and sensors. The word "transducer" refers to the whole assembly of the device while "sensor" is a part of the transducer and it responds directly to the quantity to be measured. For example, microphone is a transducer and its diaphragm is its sensor.

14.3 Displacement Transducers

Displacement is normally defined as a change in position (location) of some object with reference to a fixed point. It can be along a straight line (linear) or rotational (angular). This importance of displacement transducers which produce electrical signals in proportion to the mechanical displacements can be well understood by the fact that many physical variables are first converted to mechanical displacement and then to the electrical signals using suitable transducers.

For example, for the measurement of pressure, mechanical elements shown in Fig. 1 are normally used to convert the applied pressure into mechanical displacement. The displacement so produced is converted into electrical signal with the help of suitable transducers. These mechanical elements, such a diaphragm, bellows, etc., are also called as force summing devices.

Transducers like resistance strain gauge, capacitance transducers, LVDT, etc. are used to convert mechanical displacement to electrical signals.

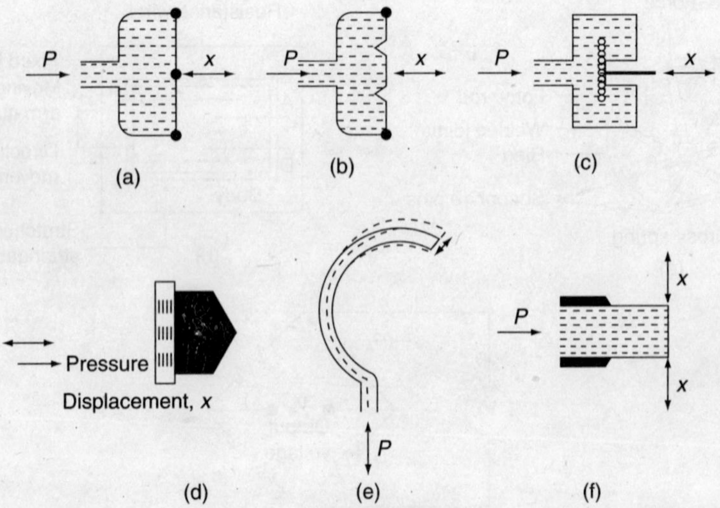

Figure 1 Force summing devices: (a) Flotcircular diaphragm; (b) corrugated (annular); (c) capsule diaphragm; (d) bellow; (e) circular bounded tube; (f) straight tube.

14.3.1 Electrical Strain Gauges

It a metal conductor is elongated or compressed due to a tensile or compression stress, its resistance changes because of the charges in the following dimensions: length (L) and area of cross-section (A). It a wire elongates from L to $L + \Delta L$ under the influence of tensile stress, its resistance increases from R to $R + \Delta R$. The strain produced by stress is given by elongation/length, that is, $(\Delta L / L)$. The sensitivity of a strain gauge is described in terms of a characteristics variable called gauge factor G. It is actually the ration of the unit change in resistance to unit change in length. Mathematically, it is given by

$$G = \frac{(\Delta R / R)}{(\Delta L / L)} = \frac{\Delta R / R}{\text{Strain}}$$

For better performance, gauge factor G should be high. In addition, the strain gauge wire should be made of a material which has low temperature coefficient of resistance. There should be a linear relationship between the stress and the strain.

The strain gauge is used to measure strain and thereby compute the stress caused due to strain. Resistance strain gauges are available in two forms: unbonded strain gauge and bonded strain gauge. We will discuss them next.

14.3.1.1 Unbonded Strain Gauge

The unbonded strain gauge is depicted in Fig. 2. The arrangement shown in the figure is capable of detecting displacement as low as 50 μm. It consists of four sapphire posts holding equal lengths of tungsten platinum wire with the diameter of 25 μm. The unbonded strain gauge is connected to form a Wheatstone bridge.

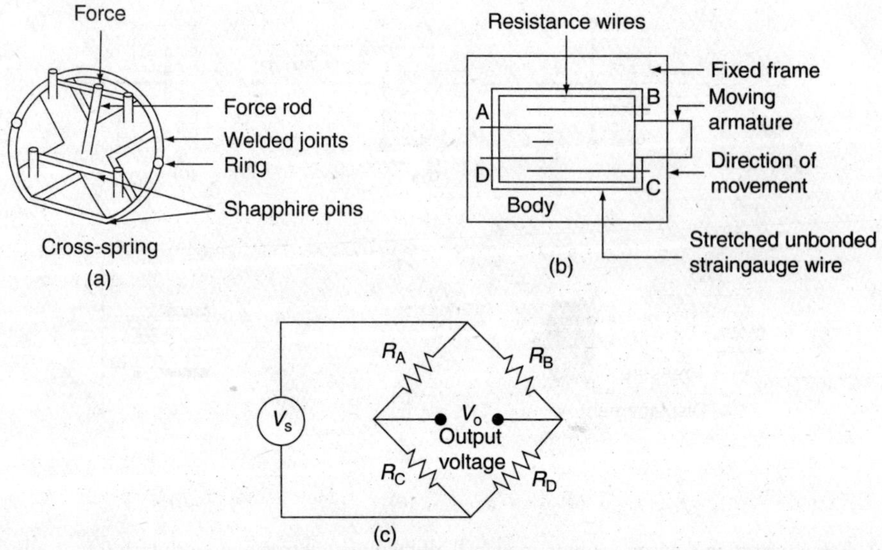

Figure 2 Unbonded strain gauge.

On no load condition, the bridge is in balanced state and hence there is no output voltage. However, when the load is applied to the strain gauge, the armature moves in the direction shown in Fig. 2(b). This results in the coils getting stressed differently, some coils get elongated and the remaining get compressed. This leads to an unbalanced Wheatstone bridge and hence an output voltage in produced. Thus, the external force causes variation in the resistance of wires unbalancing the bridge and causing in the output voltage. This output voltage is proportional to the pressure or load applied.

14.3.1.2 Bonded Strain Gauge

It consists of grid of thin wire made of nichrome, platinum or constantan of about 25 μm diameters. This grid is connected to a carrier or base. The base may be made of paper, Bakelite or Teflon sheet. The carrier is bonded to the structure under study by means of an adhesive material.

Here again the strain gauge wire forms one of the four arms of the Wheatstone bridge. When the structure is stressed by means of a force, the gauge wire is subjected to a strain causing a change in its resistance. This affects the balance of the bridge circuit and produces an output voltage which is, thus, proportional to the applied force.

The most commonly used bonded strain gauges are as follows:

1. Linear strain gauge.
2. Rosette strain gauge.
3. Torque gauge.
4. Helical gauge.

These are shown in Fig. 3.

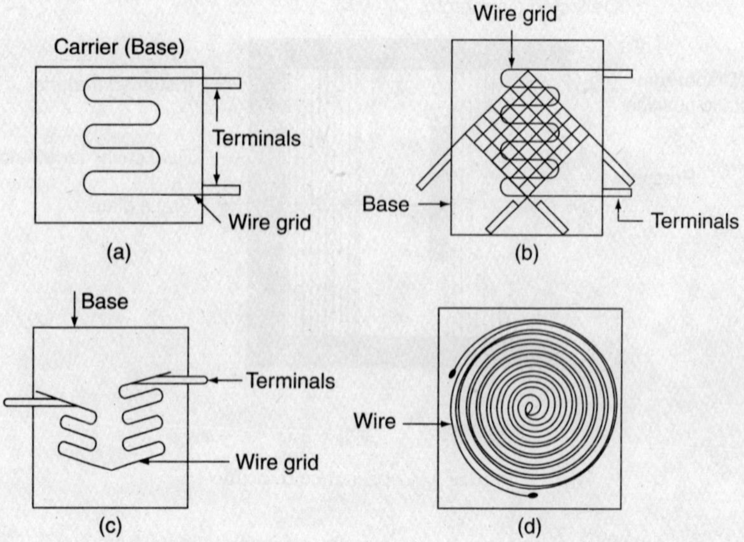

Figure 3 Bonded strain gauges: (a) Linear strain gauge, (b) Rossette strain gauge, (c) torque gauge, (d) helical gauge.

14.3.2 Capacitance Transducer

The capacitance of a parallel plate is expressed as

$$C = \varepsilon_0 \varepsilon_r \frac{A}{d} d$$

where A is the area of plates in meter, d the distance between the plates in meter, ε_0 the dielectric constant (permittivity) of free space, ε_r the relative dielectric constant of material. From above, it is found that the capacitance is directly proportional to "A" and inversely proportional to "d." Therefore, any variation in "d" leads to variation in "C." The capacitance transducer shown in Fig. 4 makes use of this principle.

The capacitance transducer consists of a diaphragm, a static plate and a dielectric between their two plates, consulting a variable capacitor. On the application of force, the diaphragm gets deflected which causes the change in the distance between the plates. This results in change of capacitance, which can be measured by means of suitable AC bridge. Alternatively, an oscillator can be used to measure the change of capacitance, as this affects the oscillator frequency. Indirectly, it is a measure of applied force.

The capacitor transducer can measure static and dynamic changes. The drawback of this transducer is its sensitivity of the temperature variations.

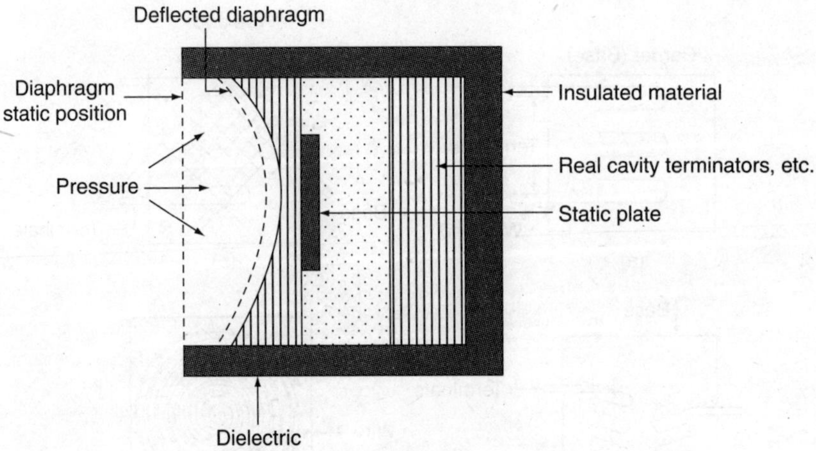

Figure 4 Capacitance transducer.

14.3.3 Linear Variable Differential Transducer (LVDT)

The linear variable differential transducer is the most widely used inductive transducer. It is a costly displacement transducer with excellent performance. The circuit diagram as shown in Fig. 5.

The LVDT consists of a primary coil (*P*), two identical secondary coils (S_1 and S_2) and a rod-shaped magnetic core (*C*) at the center. The primary coil is connected to an alternating current source.

The displacement to be measured is applied to the arm attached to the core. An alternative current is fed into the primary coil and hence voltages V_{01} and V_{02} are induced in the secondary coils. As these coils are connected in series opposition, the output voltage $V_o = V_{01}-V_{02}$. If the core is placed ideally at center position or null position, the output voltage in $V_o = 0$, as the source voltage $V_{01} = V_{02}$. An upward displacement of the core from its null position will make V_o positive as $V_{01} > V_{02}$. On the other hand, the downward movement of the core from its null position will make $V_{02} > V_{01}$ resulting in the negative output voltage V_o. Thus, the voltage available across the secondary output terminals is a measure of displacement from its null position.

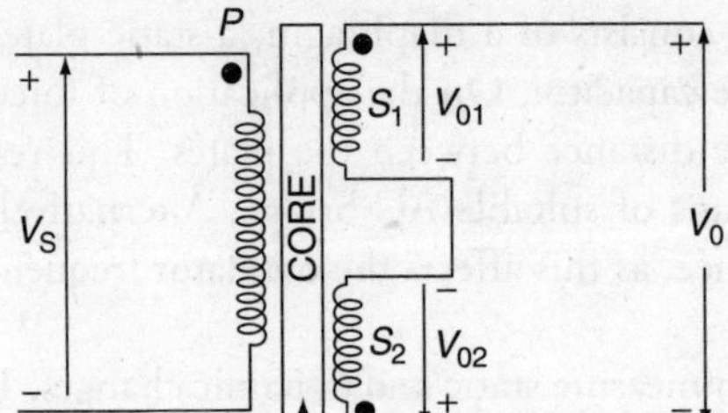

Figure 5 Circuit diagram of an LVDT.

The sign of the output voltage (i.e., whether V_o is positive or negative) represents the direction of movement of the core from its null position. The characteristics of LVDT is depicted in Fig. 6. The output voltage is linear over a considerable range but becomes flat at both the ends.

Advantages:

1. LVDT has a rugged construction and can withstand high degree of shock and vibration.
2. It does not possess any sliding contacts and therefore has less friction with less noise.
3. It possess high sensitivity.
4. It also posses high and continuous resolution.
5. The output voltage is practically linear over a considerable range.
6. It has excellent repeatability.

Disadvantages:

1. LVDT is sensitive to stray magnetic field.
2. For appreciable differential output, relatively large displacement is required.
3. The mass of core and frequency of the applied voltage limit the dynamic response of the transducer.

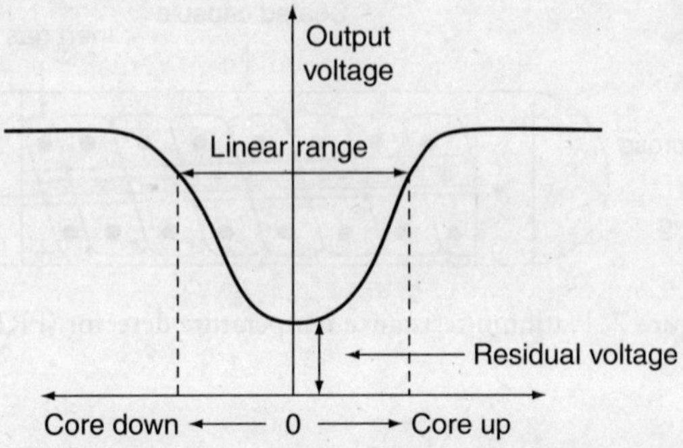

Figure 6 Graph depicting characteristics of LVDT.

14.4 Temperature Transducers

Temperature transducers are normally employed to monitor, record, and control temperatures in industrial processes. The following are the different types of temperature transducers:

1. Resistance temperature detectors (RTD).
2. Thermocouples.
3. Thermistors.

14.4.1 Resistance Temperature Detectors (RTD)

In RTDs, the property of change in electrical resistance due to the change in temperature is employed in the measurement of temperature. The platinum resistance thermometer is widely used in temperature measurement between the boiling point of oxygen (−182.96°C) and the boiling point of antimony

(630.74°C). The fabrication of a strain-free platinum resistance thermometer is described as follows: A helically spiraled platinum wire is wound over a mica cruciform as shown in Fig. 7. The two pieces of mica all normally notched of receive the platinum wire. The winding is annealed and the value of the resistance is adjusted. Normally, two gold leads are welded at each end of the platinum winding. The assembly is then fitted in a glass tube (made of borosilicate glass). The leads are insulated with glass tubing spaced with washers of mica. The assembly is normally evacuated at one-half atmosphere of dry air and sealed. Such a construction is usable over the temperature range of −190°C to 500°C. The protecting tubing for PTRD is made of pyrex glass or mica depending on temperature. Platinum is good for industrial environment due to its stability and ability to withstand high temperatures. A PTRD can show an accuracy of ±0.1°C for a resistance temperature detector. The variation of resistance with temperature is typically given as

$$RT_2 = RT_1[1 + \alpha(T_2 - T_1)]$$

where RT_2 is the resistance of the thermocouple at temperature T, RT_1 is the resistance of the thermometer at temperature T_1 (the reference temperature), α is the temperature coefficient of resistance, T_2 is the operating temperature.

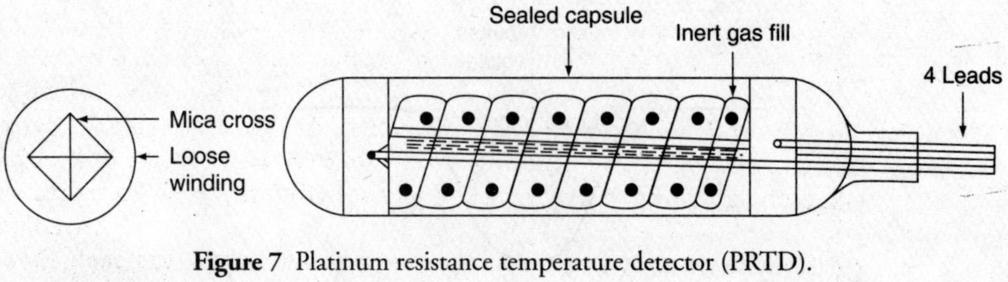

Figure 7 Platinum resistance temperature detector (PRTD).

14.4.2 Thermocouples

A thermocouple is a device for measuring temperature. It has a junction between two dissimilar metals of semiconductor that generate a small voltage, typically in the millivolt range, with a coefficient of about 50 µV/°C.

A typical thermocouple circuit for temperature measurements is shown in Fig. 8. It consists of two junctions, namely, reference junction and sensing junction. Each junction is made by welding two dissimilar metals together. They are maintained at different temperatures. Usually the reference junction is maintained at fixed temperature and the output voltage depends on temperature of the sensing junction. As only a relatively small output voltage of order of millivolts is obtained from this arrangement, it is necessary to amplify the output for calibration and measurement.

If the sensing junction of a thermocouple is kept at a temperature of T_2°C and if the temperature of the reference junction is T_1°C, then e.m.f. (E) of thermocouple can be approximately given as

$$E = \alpha_0(T_2 - T_1) + \alpha_1(T_2^2 - T_1^2)$$

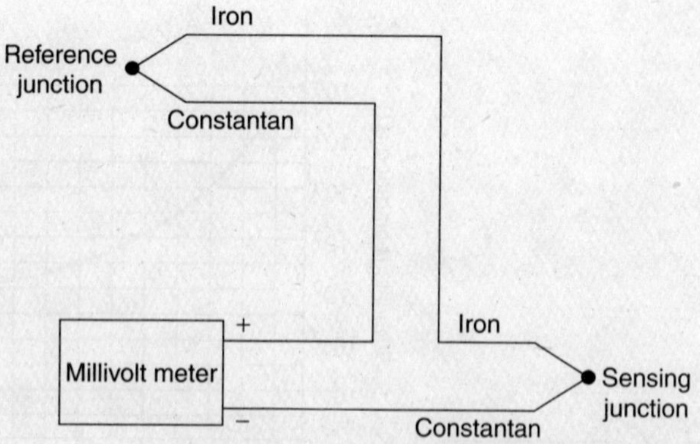

Figure 8 A thermocouple circuit.

where α_0 and α_1 are thermocouple constants. The copper–constant thermocouple can be used for the temperature measurement over a range of −184°C to 400°C and chromel–constant on thermocouple junction can be used for temperature measurement over a range of 0°C to 982°C.

14.4.3 Thermistors

Thermistor is a semiconductor device having negative temperature coefficient of resistance. The symbol and characteristics of this element are shown in Fig. 9. Materials employed in the manufacturing of thermistor include oxides of cobalt, nickel, copper, iron and manganese. The thermistor has very high coefficient of resistance, of the order of 3% to 5% per degree Celsius, making it highly suitable temperature transducer. The resistance at any temperature T, can be theoretically estimated from the equation:

$$R_T = R_0 \exp \beta \left(\frac{1}{T} - \frac{1}{T_0} \right)$$

where R_T is the thermistor resistance at temperature T (K), R_0 the thermistor resistor at temperature T_0 (K), β the percent constant determined by calibration. At high temperatures, the equation reduces to

$$R_T = R_0 \exp \left(\frac{\beta}{T} \right)$$

An observation of the resistance temperature characteristics is shown in Fig. 9. The characteristics curve is highly non-linear and the resistance drops from 5000 Ω to 10 Ω for an increasing temperature from 20°C to 100°C. This property of a thermistor finds application in industrial control mechanisms. Thermistor can be used to measure temperature, fluid flow pressure and also as a compensation element in electronic circuits.

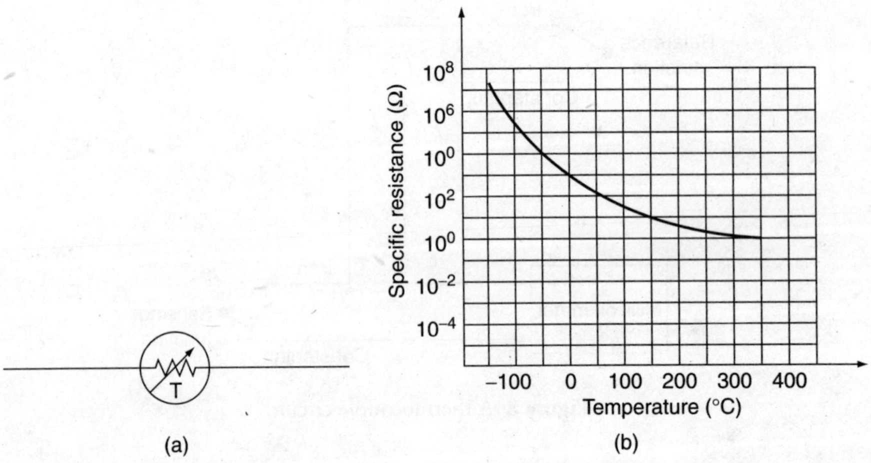

Figure 9 Thermistor: (a) Symbol; (b) graph showing the characteristics.

14.5 Piezoelectric Transducers

If the dimensions of asymmetrical crystalline materials such as quartz rochelle salt and barium titrate are changed by the application of mechanical force, the crystal produces an e.m.f. This property is used in piezoelectric transducers.

In a piezoelectric crystal a surface deformation will take place under the action of tension, compression and shear. This will develop a surface charge that is directly proportional to force exerted. Thus charge Q developed on the surface when subjected to a pressure difference $(P_1 - P_2)$ is given by

$$Q = K(P_1 - P_2)\,A$$

where $(P_1 - P_2)$ is the pressure exerted, A is the area and K is the crystal charge sensitivity. The voltage developed across the device is

$$V = \frac{Q}{C}$$

$$= \frac{K(P_1 - P_2)A}{C}$$

$$= K(P_1 - P_2)td$$

where diameter d is in meter and t is the thickness. The crystal sensitivity of quartz is

$$K = 2.210^{-9} \text{ Coulombs N}$$

or

$$K = 0.05 \text{ Vm/N}$$

Pressure spikes may sometimes damage the transducers. These transients are caused by pumps and values. A snubber in the form of a restriction in the pressure post prevents the transducer from such spikes. This transducer has very high frequency response and is, therefore, used in high-frequency accelerometers. It is an active transducer. A piezoelectric transducer is shown in Fig. 10.

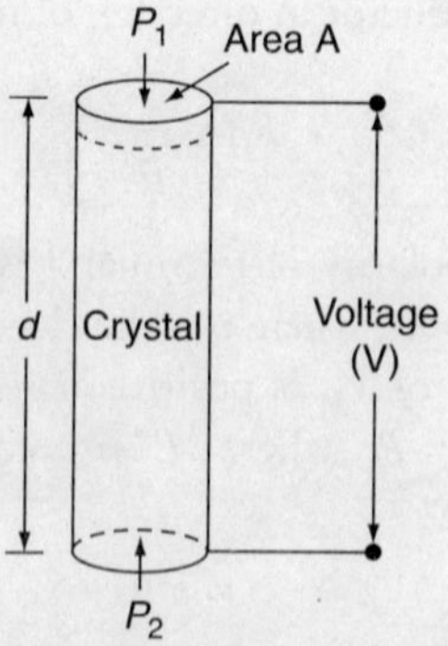

Figure 10 Piezoelectric transducer.

14.6 Hall Effect Transducer

When a transverse magnetic field "*B*" is applied to a metal of a semiconductor carrying a current *I*, an electric field *E* is induced in the direction perpendicular to both *I* and *B*. This phenomenon is called as the Hall effect.

The principle of Hall effect can be explained using an arrangement depicted in Fig. 11. An electron travelling with velocity *V* in magnetic field will experience a force *F* equal to

$$F = eE = BeV$$

In an n-type semiconductor, electrons are majority carriers. When such a semiconductor-carrying current 1 is placed in magnetic field *B* as shown in Fig. 11, a half voltage V_H will appear across the semiconductor because of the accumulation of electrons on one side. The half voltage across surfaces 1 and 2 is given by

$$V_H = \frac{R_H}{t} BI$$

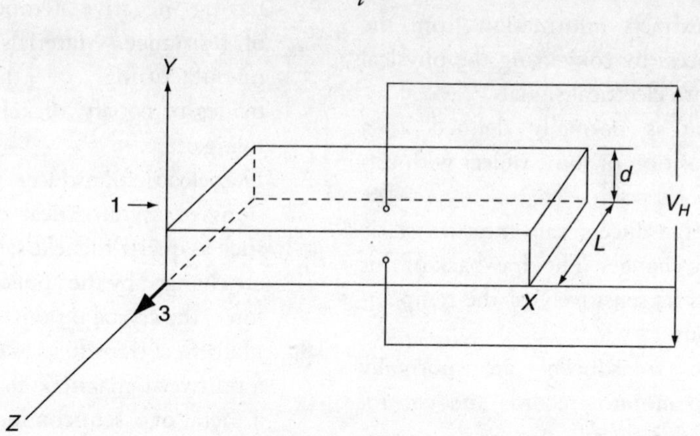

Figure 11 Hall effect.

where t is the thickness of the semiconductor in direction of magnetic field and Hall coefficient

$$R_H = \frac{1}{\rho}$$

where ρ is the charge density. If the polarity of terminal 2 is positive with respect to terminal 1, the semiconductor must be n-type and $p = ne$, where n is the electron concentration.

On the other hand, if the polarity of V_H is positive at terminal 1 with respect to terminal 2, the semiconductor must be p-type and $p = Pe$, where "P" is hole concentration. The conductivity σ and mobility μ are related by equation

$$\sigma = p\,\mu$$

$$\mu = R_H\,\sigma$$

The Hall effect is used to find whether a semiconductor is n-type or p-type and to determine the carrier concentration. The mobility μ can also be used to determine the carrier concentration, calculated with simultaneous measurement of conductivity σ.

Advantages:

1. Hall effect transducers are non-contact devices with high resolution and small size.

Applications:

1. It is used in the measurements of velocity, rpm, sorting limit sensing.
2. It is also used for non-contact current and magnetic field measurements.

Summary

In this chapter, we have discussed the following points:

1. A transducer serves as a vital link between the physical process and its instrumentation system. It extracts information from the physical process by converting the physical parameter into electrical signals.
2. Displacement is normally defined as a change in position of some object with reference to a fixed point.
3. Capacitive transducers can measure static and dynamic changes. The drawback of this transducer is its sensitively of the temperature variations.
4. Temperature transducers are normally employed to monitor records and control temperature in industrial process.
5. Thermocouple is a device for measuring temperature. It has a junction between two dissimilar metals of semiconductors that generates a small voltage, typically in millivolt range.
6. Thermistor is a semiconductor device having negative temperature coefficient of resistance. Materials employed in the manufacturing of thermistor include oxides of cobalt, nickel, copper, iron and manganese.
7. Piezoelectric transducer in which the dimensions of asymmetrical crystalline materials such as quartz rochelle salt and barium titrate are changed by the application of mechanical force, the crystal produces an e.m.f.
8. Hall effect transducer is defined as the when a transverse magnetic field "B" is applied to a metal of a semiconductor carrying a current I, an electric field E is induced in the direction perpendicular to both I and B.

Multiple Choice Questions

1. Dummy strain gauges are used for
   (a) compensation of temperature changes
   (b) increasing the sensitivity of bridge in which they are included
   (c) compensating for different expansion
   (d) calibration of strain gauges

2. Low of intermediate temperature in thermocouples allows them to
   (a) reference junction compensation
   (b) r measurement without disturbing the circuit conditions
   (c) extend wires of materials other than one used for making thermocouples
   (d) both (b) and (c)

3. Capacitive transducer are normally used for
   (a) static measurement
   (b) dynamic measurement
   (c) both static and dynamic measurements
   (d) transient measurements

4. An LVDT
   (a) exhibits linear characteristics up to a displacement of ±5mm
   (b) has a linearity of 0.05%
   (c) has an infinite is of order of 40 V/mm
   (d) all of these

5. Thermocouples are
   (a) passive transducers
   (b) active transducers
   (c) both (a) and (b)
   (d) output transducers

6. The temperature transducers exhibit non-linear behavior. The order in which they exhibit non-linearity (highest to lowest) is
   (a) Thermocouples, RTD, thermistors
   (b) Thermistors, thermocouples, RTD
   (c) RTD's, thermocouples, thermistors
   (d) Thermistors, RTD, thermocouples

7. A capacitive transducer working on the principle of change of capacitances with change of displacement, exhibits non-linear characteristics. The response of these transducers can be made linear by using
   (a) different arrangements
   (b) use of an OPAMP
   (c) working then over a small displacement range
   (d) all of these

8. Variations in very fine thickness, say of paper, are to be monitored continuously in a paper industry. The most suitable transducer will be
   (a) strain gauge
   (b) LVDT
   (c) diaphragm
   (d) capacitive transducer

9. A thermistor
   (a) exhibits only a negative change of resistance with increase in temperature
   (b) exhibits only a positive change of resistance with increase in temperature
   (c) can exhibit either a negative in positive change of resistance with increase of temperature depending upon the type of material
   (d) none of these

10. Which of the following statements is wrong for thermocouple measuring instruments?
    (a) They read rms values.
    (b) If calibrated on DC, they cannot be used for AC signals.
    (c) They are incapable of standing any overloads.
    (d) Their calibration does not change with time or temperature.

11. The output of VDT is in the form of
(a) pulses
(b) high-frequency signals
(c) rotary movement of core
(d) linear displacement of core

12. LVDT is a ______________ transducer.
(a) eddy current
(b) resistance
(c) magnetostiction
(d) inductive

13. ______________ is a low cost thermocouple.
(a) iron–constantan
(b) chromel–constantan
(c) rhodium–indium
(d) tungsten–rhenium

14. Which of the following is an analog transducer?
(a) Strain gauge
(b) Thermistor
(c) LVDT
(d) All of the above

15. LVDT possesses the following advantage:
(a) Excellent repeatability
(b) Infinite resolution
(c) High linearity of output
(d) All of the above

Review Questions

1. What is a transducer?
2. Briefly describe any one of the displacement transducers.
3. What are the basic requirements of a transducer?
4. Discuss the working principle of an inductive transducer.
5. Explain in detail the working of a linear variable differential transducer.
6. Describe the concept of piezoelectric transducer.
7. Explain the Hall effect.
8. Mention the applications of Hall effect.
9. What is meant by gauge factor of a strain gauge?
10. How is a thermocouple used for temperature measurement?
11. What is thermocouple?
12. Explain the principle of operation of a resistance thermometer.
13. Explain the concept of capacitive transducer.
14. Briefly describe about the piezoelectric transducer.
15. Define strain gauge.
16. What is the difference between bonded and unbounded strain gauges?

Answers

Multiple Choice Questions

1. (a)	**5.** (b)	**9.** (c)	**13.** (a)
2. (d)	**6.** (b)	**10.** (b)	**14.** (d)
3. (b)	**7.** (d)	**11.** (d)	**15.** (d)
4. (d)	**8.** (b)	**12.** (d)	

15 Linear and Digital Integrated Circuits

LEARNING OBJECTIVES

After reading this chapter, you will be able to:

- Understand the operation of operational amplifiers.
- Determine the component values for inverting, non-inverting and summing operational amplifier circuits.
- Be aware of integrated circuit (IC) timer.
- Understand the concept of Boolean algebra.
- Appreciate the use of universal gates and flip-flops.

15.1 Integrated Circuits

An integrated circuit (IC) is an electronic assembly built in such a way that all the components in the circuit are fabricated together in a small container called "chip." These chips have become very important in the computer industry. These are used in home appliances, communications, control systems, etc. A small IC chip can be made to hold thousands of devices and components. In an IC, the devices and components are formed at the same time and once formed they cannot be accessed individually. This arrangement is in contrast to that of discrete circuit in which individual resistors, diodes, capacitors, transistors and other components are interconnected extremely to form a working circuit. Discrete circuits are, therefore, large in size and costly. Specific codes are assigned to linear ICs, like LM, LH, CA, etc.

The following are the advantages of ICs over discrete components:

1. They have high reliability.
2. Generally their size is small and the weigh less.
3. The power consumption is low.
4. They are available at low cost.
5. They can be replaced easily.

The following are the disadvantages of ICs over discrete components: Limited range of values for the passive components.

1. High grade of portable North pole (PNP) unit is not easily possible.
2. Coils or inductors cannot be fabricated.
3. Limitations in the building of inductors and transformers.
4. As IC is small in size, it is made to dissipate large amount of power. This offers severe limitations to the current-carrying capacity of the IC.

15.1.1 Classification of ICs

On the basis of functions and signals, ICs are classified as follows:

1. **Linear ICs:** It includes operational amplifier, voltages regulators, voltage comparators and timers. These handle analog or continuous signals.
2. **Digital ICs**: These include logic gate, flip-flops, counters, memory chips, microprocessor, etc. These operate on signals representing binary values 0 or 1.

On the basis of fabrication process used, ICs can be classified as follows:

1. **Monolithic ICs:** Monolithic is a Greek word meaning one stone. In monolithic type, all the devices and components are fabricated together on a small piece of silicon material.
2. **Hybrid ICs:** Thin film and thick film ICs are the two broad categories of hybrid ICs. Passive component values (resistance and capacitance) can be trimmed to precision in the case of hybrid ICs. Hybrid ICs use combinations of integrated and discrete circuit components.

Monolithic ICs are more common, and the next section we describe the formation of monolithic IC.

15.1.2 Monolithic ICs

The following stages are involved in manufacturing ICs of monolithic type:

1. Silicon wafer preparation.
2. Epitaxial and oxide layer growth.
3. Photolithography.
4. Diffusion.
5. Aluminum metallization.

We will discuss each of them next.

15.1.2.1 Silicon Wafer Preparation

The starting material for integrated circuits is a slice of single crystal silicon, which is normally the Metallurgical Grade Silicon (MGS). By the use of chemical vapor deposition techniques (CVD) and hydrogen reduction method, the Electronic Grade Silicon (EGS) is obtained. The EGS is a polycrystalline material of high purity and it is used as the raw material for the preparation of single crystal silicon. The EGS has very small impurity level and is a polycrystalline material.

15.1.2.2 Epitaxial Layer Growth

It is a chemical reaction in which silicon is precipitated from a gaseous solution and made to grow on the silicon wafer as a thin film of single crystal silicon. Here a thin p-type or n-type epitaxial layer is grown on the substrate. Epitaxial layers are controlled for their thickness and resistivity.

15.1.2.3 Oxide Layer Growth

In order to prevent contamination of the epitaxial layer, pure oxygen is blown over the surface to grow an oxide layer of about 0.00005 mm thickness over the surface. This layer of silicon dioxide (SiO_2) seals the surface and prevents further chemical reaction. This is also called passivation. The arrangement of the apparatus to carry out oxidation as well as epitaxial layer growth is shown Fig. 1. The silicon wafers are kept in a quartz boat and inserted into a quartz tube.

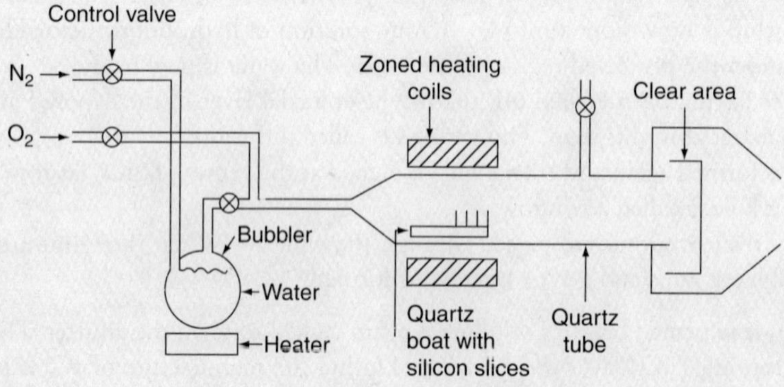

Figure 1 Arrangement for epitaxial and oxide layer growth.

15.1.2.4 Photolithography

The entire top surface of silica is now covered with a film of photo-sensitive emulsion such as Kodak photoresist (KPR). This material is soluble in certain solvents unless it is polymerized. Polymerization takes place when KPR is exposed to ultraviolet light. A photographic mask containing electrical isolation pattern is placed over the silica and then exposed to UV light when only the transparent areas get polymerized. The unexposed areas are then dissolved in a chemical solution to remove silicon oxide layer. The various stages involved in the development of NPN transistor IC are as follows (Fig. 2):

1. A p-type silicon crystal, 2 to 5 cm in diameter and several centimeters long, is formed and sliced into several thin wafers.
2. A thin layer of n-type semiconductor is formed over p-type substrate by epitaxial growth.

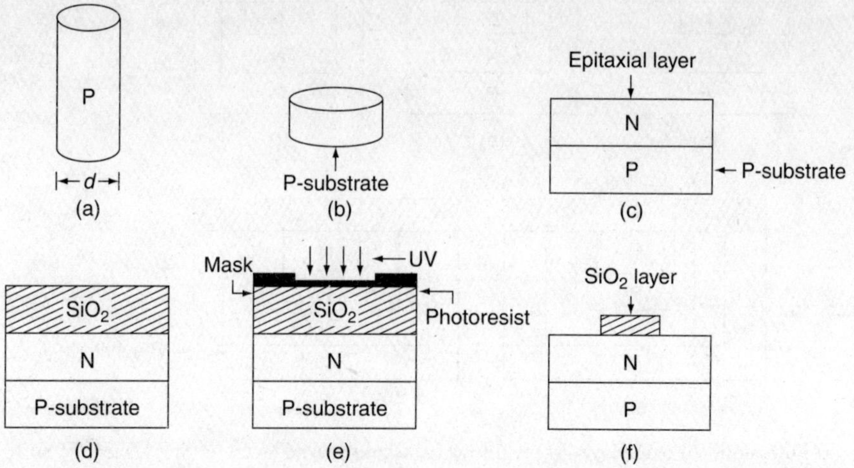

Figure 2 Various stages involved in the development of NPN transistor IC.

3. A thin layer of SiO_2 is formed on the epitaxial layer. Then the entire top surface area of the silica is covered with a thin layer of photoresist.
4. The mask is placed in contrast with the oxide layer. Now, when exposed to UV light, the photoresist under the exposed region of the mask gets polymerized, whereas the remaining area is unaffected. The chip is now dipped into an etching solution of hydrofluoric acid. This removes the oxide layer under the protected region of the wafer. The wafer is now ready.
5. Part of SiO_2 layer having been etched off, the n-type epitaxial layer in the exposed areas is converted to p-type material in this strip. This process is called diffusion.
6. The oxide layer is formed again and then a hole is etched at the centre of SiO_2 layer to expose the n-type layer. This hole is called a window.
7. To get the base, trivalent atoms are passed through the window. When these impurities diffuse into the epitaxial layer, an island *p*-type material is formed.

The SiO_2 layer (Fig. 3) is reformed before a window is again etched to form the emitter. The structure is then passivated by forming SiO_2 layer on the wafer. During the manufacture of IC, it is necessary to deposit a thin layer of metal on the silicon to form the aluminum interconnections using vacuum position technique.

Transistors, diodes, capacitors and resistors can be easily integrated on a chip in a similar manner.

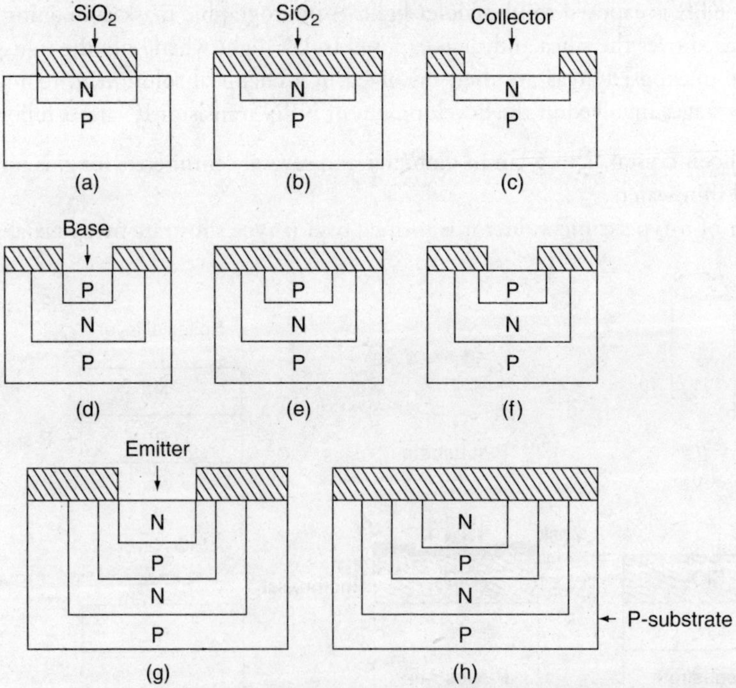

Figure 3 (a) Diffusion, (b) oxide layer, (c) window, (d) island p-type layer, (e) SiO_2 layer, (f) SiO_2 layer is reformed, (g) SiO_2 layer is again reformed before the emitter, (h) SiO_2 layer formed on the wafer.

15.2 Linear Integrated Circuits

15.2.1 Operational Amplifier

An operational amplifier (OPAMP) is a direct-coupled amplifier with high gain in which feedback is added to control its overall response characteristics. It is used to perform wide variety of functions; it forms a basic building block of many electronic circuit applications. An OPAMP is so-called because it is used to carry out division, integration, differentiation, etc. Circuit symbol for an OPAMP is shown in Fig. 4. It has two inputs – an inverting input (marked –) and a non-inverting input (marked +) – and an output. The signs (+) and (–) indicate the polarity of the gain offered to the signals at that input with respect to ground. The terminal V^+ indicates the supply voltage terminal to which the positive polarity of the battery or power supply is indicated and its negative is grounded. V^- indicates another supply voltage terminal to which a second battery is connected with its negative to V^- and the positive of the battery is connected to the ground.

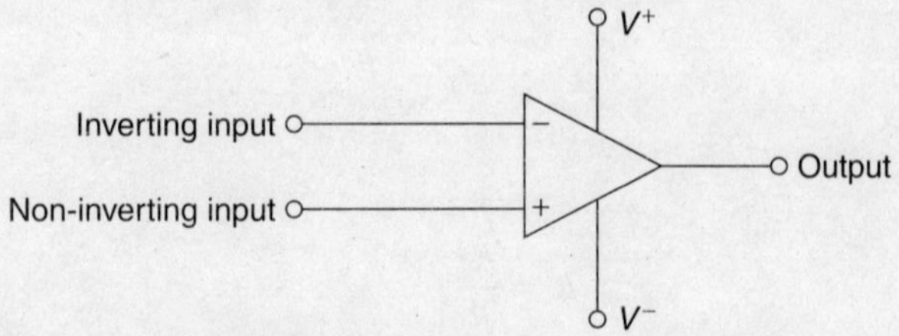

Figure 4 Circuit symbol for an OPAMP.

15.2.2 Characteristics of an Ideal OPAMP

1. Infinite voltage gain.
2. Infinite band width.
3. Infinite input impedance.
4. Zero output impedance.

15.2.3 Application of OPAMP

15.2.3.1 Inverting Amplifier

An inverting amplifier circuit using an OPAMP is shown in Fig. 5. The input V_{in} is given to inverting input terminal of the OPAMP through a resistance R_i. The feedback resistance R_f is connected between output terminal and the inverting input. The non-inverting input terminal is grounded. Now

$$\text{Current, } i = \frac{V_{in}}{R_i} = \frac{-V_{out}}{R_f}$$

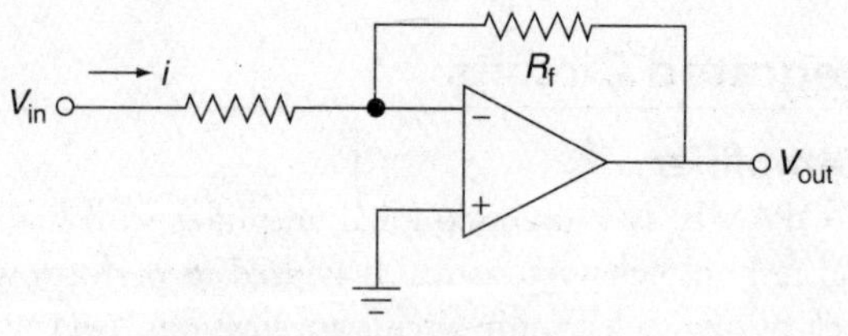

Figure 5 Inverting amplifier.

As input impedance $R_i = \infty$, therefore

$$\text{Gain} = \frac{V_{out}}{V_{in}} = \frac{-R_f}{R_i}$$

$$V_{out} = \left(\frac{-R_f}{R_i}\right)V_{in}$$

Hence

$$V_{out} = -KV_{in}$$

where

$$K = \frac{R_f}{R_i}$$

From the above relationship between V_{out} and V_{in}, it may be concluded that V_{out} depends on K for a given V_{in}. The inverting amplifier can also be considered as a sign charger, that is, if $R_i = R_f$, then $V_{out} = -V_{in}$ as $K=1$. Note that output voltage is the negative of the input voltage.

15.2.3.2 Non-Inverting Amplifier

The circuit arrangement using the OPAMP is shown in Fig. 6. It can be shown that

$$V_{out} = A'(V_{in} - fV_{out})$$

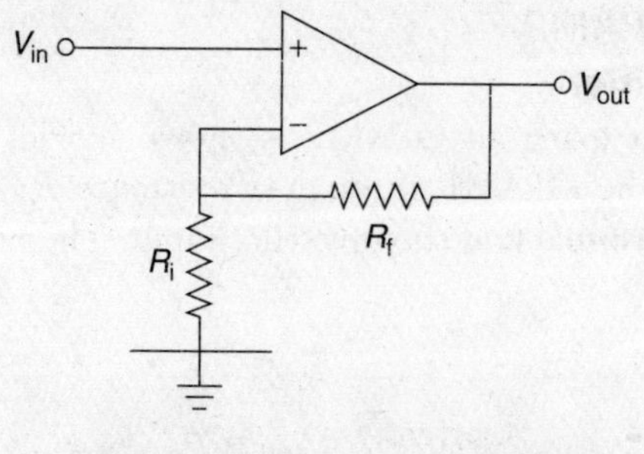

Figure 6 Non-inverting amplifier.

where

$$f = \left(\frac{R_i}{R_i + R_f} \right)$$

and A is open-loop gain. Now

$$A' = \frac{V_{out}}{V_{in}} = \frac{1}{f + (1/A)}$$
$$\cong \frac{1}{f} = \frac{R_i + R_f}{R_i}$$

Hence

$$V_{out} = \left(1 + \frac{R_f}{R_i} \right) V_{in}$$
$$V_{out} = K'V_{in}$$

As the constant K' is positive, the output voltage will be of the same sign as the input voltage.

15.2.3.3 Differential Amplifier

The circuit diagram of this amplifier using OPAMP is shown in Fig. 7. Note that the inputs are given to both inverting an non-inverting terminals. The output will be the difference between these two inputs. Hence, this can also be called a subtractor.

The output voltage V_{out} consists of two components: V_{01} due to input V_{in1} and V_{02} due to input V_{in2}. Now

$$V_{01} = -V_{in1} \frac{R_f}{R_i}$$

$$V_{02} = +V_{in\,2} \frac{R_f}{R_i + R_f} \cdot \frac{R_i + R_f}{R_i}$$
$$= +V_{in\,2} \frac{R_f}{R_i}$$

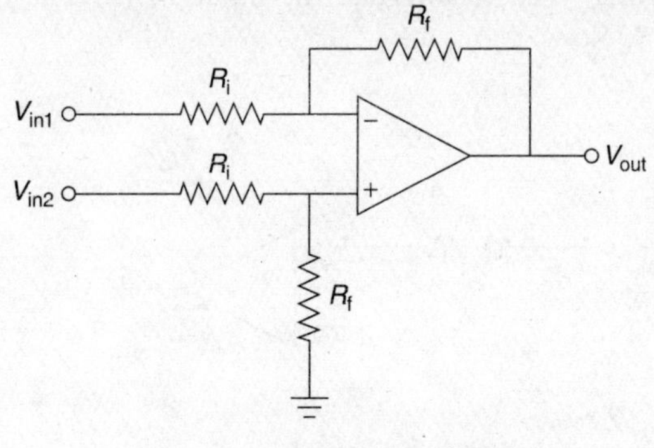

Figure 7 Differential amplifier.

Therefore

$$V_{out} = V_{01} + V_{02} = (V_{in\,2} - V_{in\,1})\frac{R_f}{R_i}$$

15.2.3.4 Adder or Summing Amplifier

This arrangement can be used to add different input voltages (Fig. 8). The voltages V_{in1}, V_{in2}, etc. are connected to inverting input terminal of the OPAMP and the non-inverting input terminal is grounded. Making use of the result obtained for inverting amplifier it can be shown that:

$$V_{out} = -R_f\left(\frac{V_{in1}}{R_{i1}} + \frac{V_{in2}}{R_{i2}} + \frac{V_{in3}}{R_{i3}} + \cdots + \frac{V_{inn}}{R_{in}}\right)$$

If

$$R_{i1} = R_{i2} = R_{i3} = \cdots = R_{in} = R_f$$

Then

$$V_{out} = -(V_{in1} + V_{in2} + V_{in3} + \cdots + V_{in\,n})$$

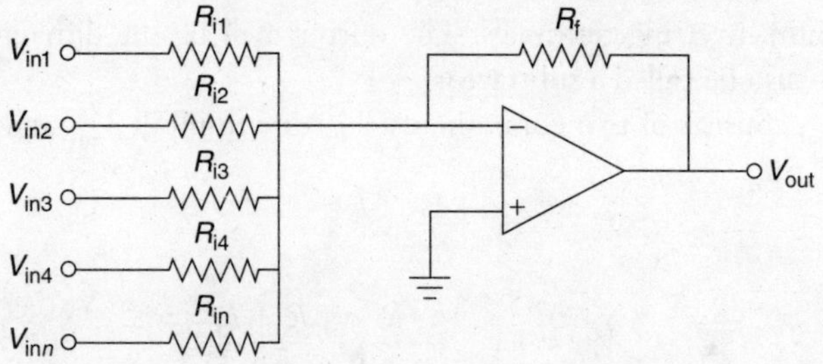

Figure 8 Adder or summing amplifier.

15.2.3.5 Integrator

This is an important OPAMP used to integrate the input signal. The circuit arrangement for an integrator is shown in Fig. 9. Now

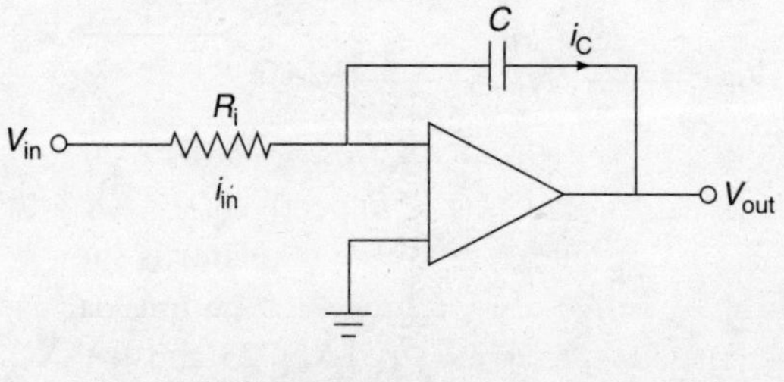

Figure 9 Integrator.

$$i_{in} = i_C = -C\frac{dV_{out}}{dt} = \frac{V_{in}}{R_i}$$

$$V_{in}dt = -CR_i\, dV_{in}$$

Therefore

$$V_{out} = \frac{-1}{CR_i}\int V_{in}\, dt$$

15.2.3.6 Differentiator

An OPAMP differentiator circuit is shown in Fig. 10. Here the output signal is the time derivative of the input signal. Mathematically,

$$V_{out} = -RC\frac{d}{dt}V_{in}$$

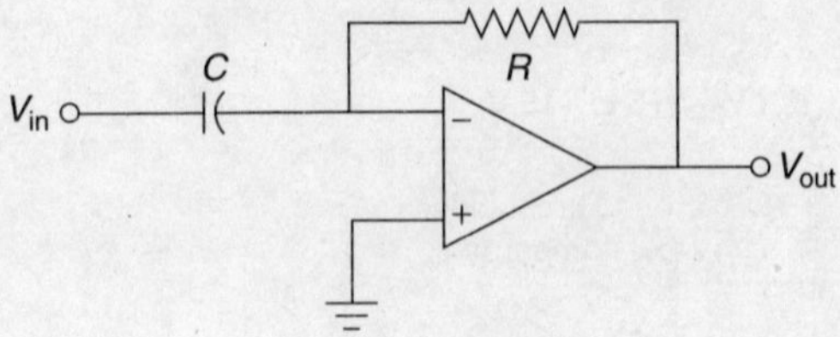

Figure 10 Differentiator.

15.2.3.7 Voltage Regulator

A voltage regulator is an electronic circuit that provides a constant DC voltage independent of the load current and input voltage variations. IC voltage regulators offer the advantages of extremely good regulation, compact size and ease use. Linear IC voltage regulators are classified into the following three types:

1. Fixed positive output voltage regulator.
2. Fixed negative output voltage regulator.
3. Adjustable output voltage regulators.

78 XX series are three-terminal, positive fixed voltage regulators. 7805 IC gives a stable +5 V supply, 7812 IC and 7815 IC provide +12 V and +15 V, respectively. 79 XX series are negative fixed voltage regulators. The circuit arrangement for a monolithic regulator is shown in Fig. 11. Input capacitor C_1 is used to cancel the inductive effect due to long distribution loads and output capacitor C_2 for improving the transient response of the regulator. LM 723 and μA 723 are the adjustable voltage regulators.

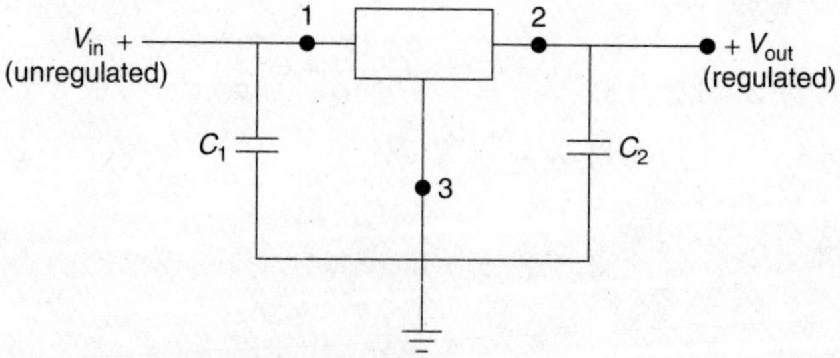

Figure 11 Monolithic voltage regulator.

15.3 IC Timer

Timer circuits play an important part in the control of many industrial and other applications. The IC 555 timer is a versatile IC timer. It can be used as square wave generator, saw tooth wave generator, etc. Accurate time delays can also be obtained from this timer.

15.3.1 Wave from Generator using IC 555

Pin connection and block diagram of IC 555 are shown in Fig. 12. It is clear from this figure that a resistive voltage divider is used to set the comparator voltage level. As the three resistors have equal

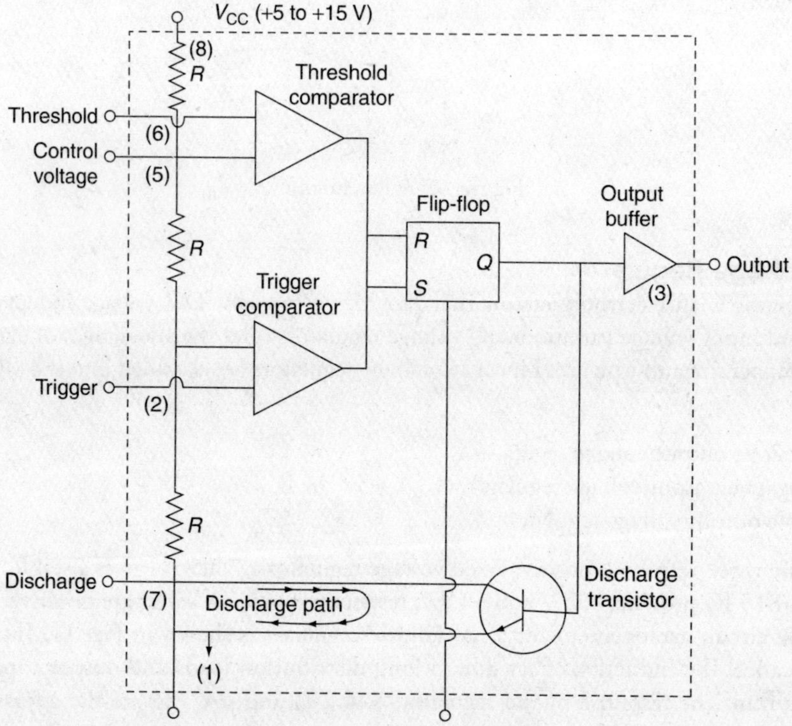

Figure 12 Internal circuit diagram of IC 555 timer.

values (= 5 kW), the upper or threshold comparator has a voltage reference of (2/3) V_{cc}. The lower comparator has a voltage reference of (1/3) V_{cc}. When the trigger voltage goes below (1/3) V_{cc}, the flip-flop sets and the output goes to the high level. The threshold input is fed by on RC network. When the external capacitor voltage increases above (2/3) V_{cc}, the upper comparator resets flip-flop, which in turn switches the output back to its low level. If the output is low, then the discharge transistor turns ON and if an external timing capacitor is connected, then the discharge transistor provides path for it.

Figure 13 shows delay time calculation using an external RC network – one-shot multi-vibrator. It shows the arrangement of IC timer for delay time generation. Time delay $T = 1.1\ RC$, where R and C are the resistance and capacitance of the external RC network.

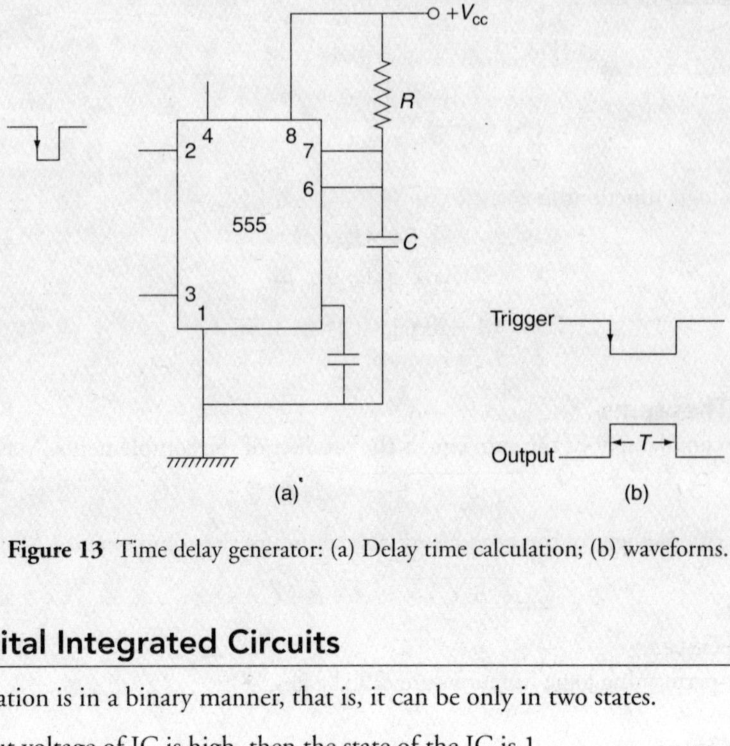

Figure 13 Time delay generator: (a) Delay time calculation; (b) waveforms.

15.4 Digital Integrated Circuits

A digital IC operation is in a binary manner, that is, it can be only in two states.

1. If the output voltage of IC is high, then the state of the IC is 1.
2. If the output is low, then the state of the IC is 0.

This is according to the convention known as positive logic system. There is also a negative logic system where low output voltage corresponds to 1 and high output voltage to 0.

15.4.1 Boolean Algebra

It is the algebra of digital circuits. Here a variable takes only two values, either 0 or 1. This is employed to simplify digital circuits and to minimize the number of devices. Some of the Boolean laws are as follows:

1. **OR Laws**

$$A + 1 = 1$$
$$A + 0 = A$$
$$A + A = A$$
$$A + \bar{A} = 1$$

2. **AND Laws**

$$A \cdot 1 = A$$
$$A \cdot 0 = 0$$
$$A \cdot \bar{A} = 0$$
$$A \cdot A = A$$

3. **Complementation Laws**

$$\bar{\bar{A}} = A$$
$$\bar{I} = 0$$
$$\bar{0} = 1$$

Other important logic functions are:

$$A + AB = A$$
$$A + AB = A + B$$
$$(A + B)\ (A + C) = A + BC$$

DeMorgan's Theorems

Theorem 1: The complement of the sum equals the product of the complements. Mathematically,

$$\overline{A + B + C + \cdots} = \bar{A} \cdot \bar{B} \cdot \bar{C} \cdots$$

Theorem 2: The complement of the product equals the sum of the complements. Mathematically,

$$\overline{A \cdot B \cdot C \cdots} = \bar{A} + \bar{B} + \bar{C} + \cdots$$

15.4.2 Logic Gates

The elements for performing logic functions are called gates.

15.4.2.1 OR Gate

The OR gate performs logical addition, commonly called as OR functions. The OR gate has two or more inputs and only one output. For a two-input OR gate, the symbolic representation and truth table are shown in Fig. 14. Truth table reveals that for an OR gate, the output will be high when both the inputs are high or when any one input is high. The output is 0 when all the inputs are zero. Boolean expression for an OR gate is given by

$$Y = A + B$$

where A, B are inputs and Y is the output.

$Y = A + B$

A	B	$Y = A + B$
0	0	0
1	0	1
0	1	1
1	1	1

(a)

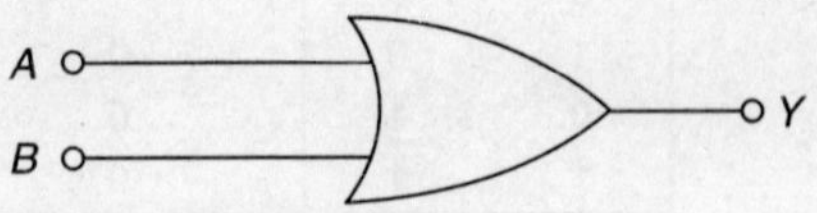

(b)

Figure 14 OR gate: (a) Truth table; (b) symbolic representation.

15.4.2.2 AND Gate

The Boolean expression for an AND gate is

$$Y = A \cdot B$$

The truth table and symbolic representation of AND gate is shown in Fig. 15. The output will be high only when both the inputs are high.

$Y = A \cdot B$

A	B	$Y = A \cdot B$
0	0	0
1	0	0
0	1	0
1	1	1

(a)

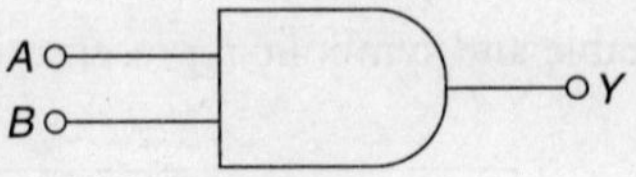

(b)

Figure 15 AND gate: (a) Truth table; (b) symbolic representation.

15.4.2.3 NOT Gate (Invertor)

The NOT gate performs a basic function called inversion or complementation. It is a gate with single input and single output. It is a gate for level changing, that is, 0 to 1 or 1 to 0. The Boolean expression for NOT gate is

$$Y = \overline{A}$$

Figure 16 shows the truth table and symbolic representation of this gate.

$Y = \overline{A}$

A	$Y = \overline{A}$
0	1
1	0

(a)

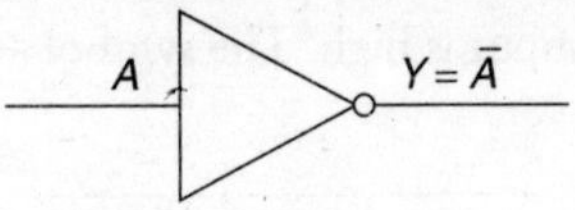

(b)

Figure 16 NOT gate: (a) Truth table; (b) symbolic representation.

15.4.2.4 NOR Gate

It is a series combination of NOT gate and OR gate. It is available as a single unit with many inputs and single output. In this gate, when any one or all the inputs are high, output is low. Only when all the inputs are low, the output is high. The Boolean expression for this gate is

A	B	$Y=\overline{A+B}$
0	0	1
1	0	0
0	1	0
1	1	0

(a)

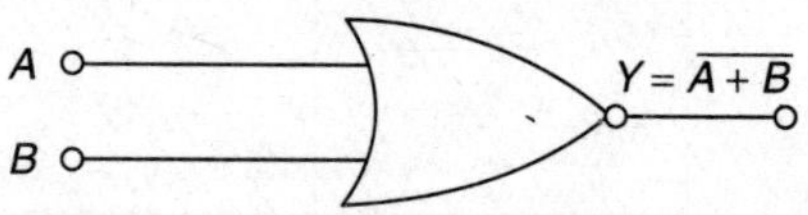

(b)

Figure 17 NOR gate: (a) Truth table; (b) symbolic representation.

$$Y = \overline{A+B}$$

The truth table and symbolic representation are given in Fig. 17.

15.4.2.5 NAND Gate

NAND gate is a combination of NOT and AND gates. It has two or more inputs and only one input. When all the inputs are high, the output is low. If any of the inputs is low, the output is high. The Boolean expression is

$$Y = \overline{A-B}$$

The truth table and symbolic representation is given in Fig. 18.

A	B	$Y=\overline{A-B}$
0	0	1
0	1	1
1	0	1
1	1	0

(a)

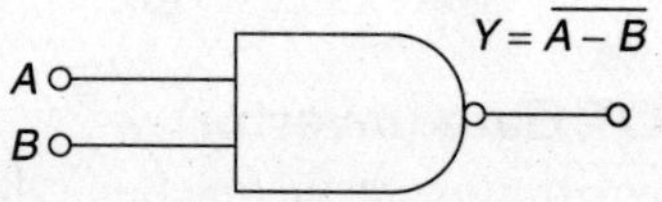

(b)

Figure 18 NAND gate: (a) Truth table; (b) symbolic representation.

15.4.2.6 Exclusive OR (XOR) Gate

The Boolean expression for the gate is given by

The truth table and logical symbol for XOR gate is shown in Fig. 19. The output will be high only if any of the input is high. The symbol $\oplus$ represents XOR operation.

A	B	$Y=A\oplus B$
0	0	0
1	0	1
0	1	1
1	1	0

(a)

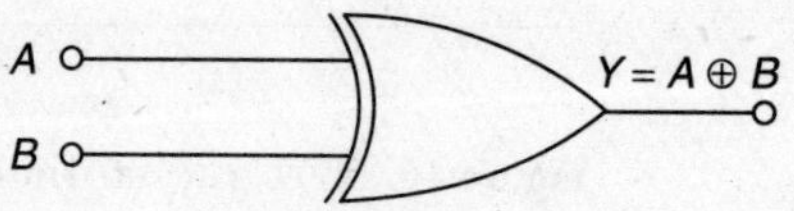

$$Y = \bar{A}B + A\bar{B}$$
$$= A\oplus B$$

(b)

Figure 19 XOR gate: (a) Truth table; (b) symbolic representation.

15.4.2.7 Exclusive NOR (XNOR) Gate

Exclusive NOR gate is the exact inverse of the exclusive OR gate. It is a gate with two inputs and single output. The output is governed by the Boolean expression.

$$Y = AB + \overline{A}\overline{B} = \overline{A \oplus B}$$

Its output is low (0) if the two inputs are at different levels. The truth table and logical symbol are given in Fig. 20.

A	B	$Y = \overline{A \oplus B}$
0	0	1
1	0	0
0	1	0
1	1	1

(a)

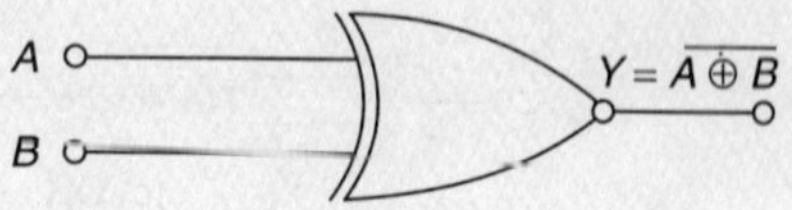

(b)

Figure 20 XNOR gate: (a) Truth table; (b) symbolic representation.

15.4.3 Universal Gates

NAND and NOR gates are known as basic or universal gates as it is possible to synthesize any logic function by using these gates. The logic functions of OR, AND, NOT, and other gates using these universal gates can be realized as shown in Fig. 21.

15.4.4 Flip-Flops

A flip-flop is a storage device used to store a bit of information. It is constructed using logic gates. Flip-flops are also called latches or multivibrators. The two categories of flip-flops based on the occurrence of the events or change of state are as follows:

1. **Synchronous flip-flops:** In the synchronous system, the occurrence of events take place during one of the timing pulses provided by a master oscillator.
2. **Asynchronous flip-flops:** In the asynchronous system, the events occur after the completion of previous event.

The following are the important types of flip-flops:

1. RS flip-flop.
2. D flip-flop.
3. JK flip-flop.

15.4.4.1 RS Flip-Flop

The logic symbol and the truth table for this circuit are given in Fig. 22. The flip-flop has two inputs and two outputs. The two inputs are set (S) and reset (R). The two outputs are Q and its complement $\overline{Q}$.

The construction of RS flip-flop using NAND and NOR gates is shown in Fig. 23. The operation of the circuit is revealed by its truth table. When both the inputs are zero, that is, $R = 0$, $S = 0$, then Q and

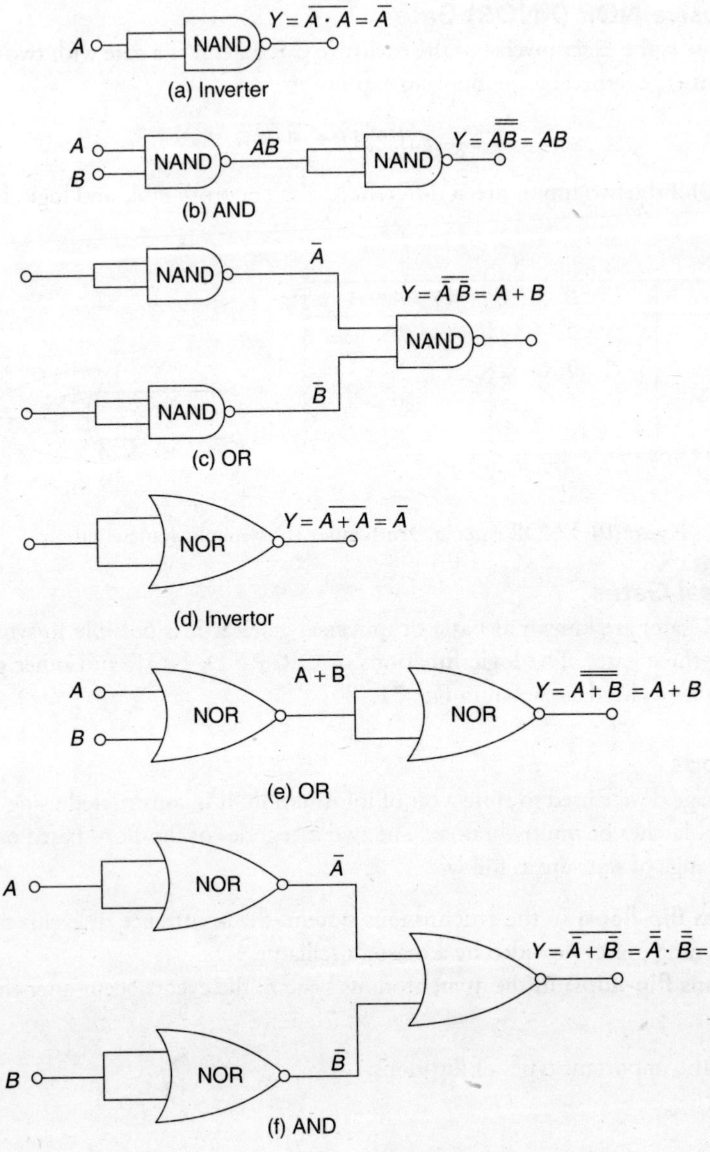

Figure 21 Universal gates: (a) to (c) are using NAND gates; (d) to (f) are using NOR gates.

$\bar{Q}$ remain unchanged. With $R = 0$, set input $S = 1$, the flip-flop output Q becomes 1, $\bar{Q}$ remains at 0, It the reset input $R = 1$ and set input $S = 0$, the output Q flips to 0 and $\bar{Q}$ becomes 1. However, when R and S are simultaneously 1, it produces a racing condition in the flip-flop and the output becomes unpredictable. Therefore, the input condition is not usually permitted.

15.4.4.2 D Flip-Flop

In the case of RS flip-flop, two signals are required to drive the flip-flop. This turns out to be a disadvantage in many applications. This disadvantage is removed in a D flip-flop where only one input is given,

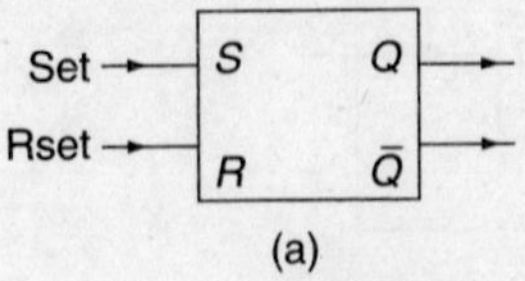

Inputs		Outputs	
S	R	Q	$\bar{Q}$
0	0	Last state	
1	0	1	0
0	1	0	1
1	1	Not allowed	

(b)

Figure 22 RS flip-flop with (a) symbol and (b) truth table.

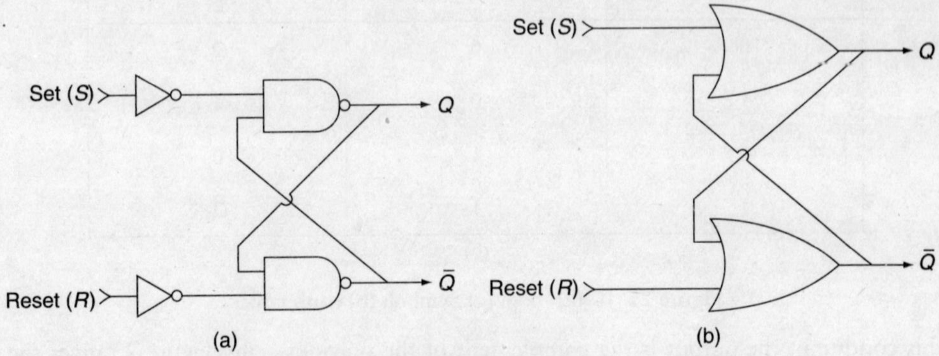

Figure 23 RS flip-flop using (a) NAND gates and (b) NOR gates.

that is, input D. In a level clocked D flip-flop, when D is low, Q is low; when D is high, Q is high. Thus the output follows the input. The symbol and truth table for the D flip-flop are given in Fig. 24. Racing condition obtained in the RS flip-flop when $S = 1$, $R = 1$ is avoided in this D flip-flop, since input R will always be the negation of inputs.

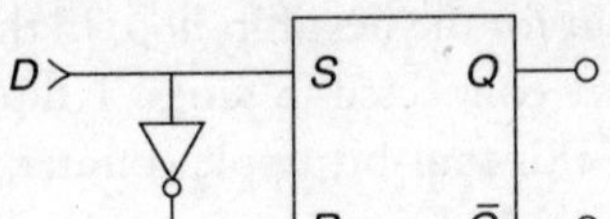

Inputs	Outputs	
D	Q	$\bar{Q}$
0	0	0
1	1	0
0	0	1

(b)

Figure 24 D flip-flop: (a) Symbol and (b) truth table.

15.4.4.3 JK Flip-Flop

The JK flip-flop is much more versatile and popular when compared to RS and D flip-flops. It is an improved version over RS flip-flop, since the racing condition is eliminated and also the inputs $J = 1$ and $K = 1$ are simultaneously permitted.

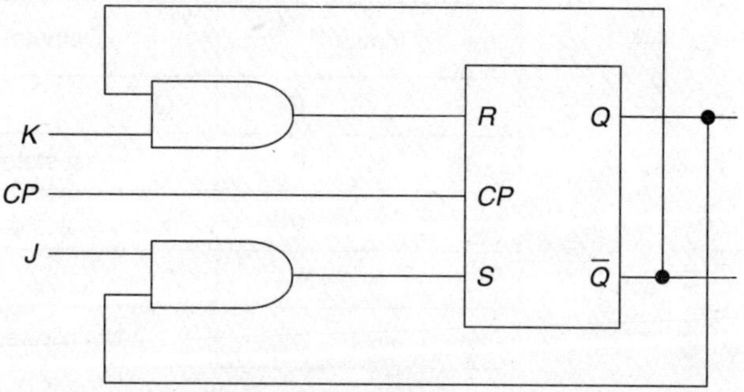

(a)

J	K	Q
0	0	Q_0
1	0	1
0	1	0
1	1	$\overline{Q}_0$

(b)

Figure 25 JK flip-flop: (a) Symbol; (b) truth table.

In this condition, the output is the complement of the previous state. Figure 25 gives the block schematic of basic JK flip-flop along with its truth table. JK flip-flop is widely used in the construction of counters and registers.

15.4.5 Counters

A counter counts the number of clock pulses arriving at the clock input of flip-flops. Because of the frequency of the clock pulses, the counter can also be used as frequency divider. Counters are used that measure frequency and time. There are two types of counters:

1. **Asynchronous counter or serial counter:** Figure 26 shows a four-bit binary ripple counter. The output of each flip-flop serves as the clock input for the next flip-flop. In this type of ripple counter, four negative edge-triggered T flip-flops are connected in series. T flip-flop is obtained if JK inputs are connected to a single input. Using this four-bit ripple counter, clock pulses up to 15 can be counted.

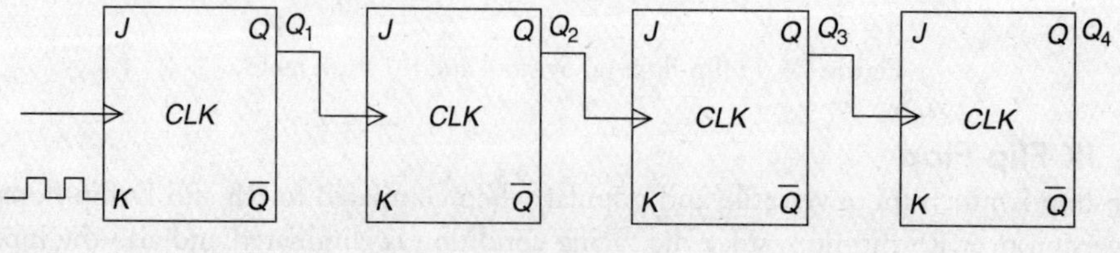

Figure 26 Asynchronous (ripple) counter.

2. **Synchronous counter or parallel counter**: In the case of synchronous counter, all the flip-flops are triggered simultaneously by the clock input pulses.

Summary

In this chapter we have discussed about the following points.

1. Integrated circuit: It is an electronic assembly in which all the devices, components and their interconnections are fabricated on a small silicon chip.
2. Linear ICs: They include operational amplifier, voltage regulators.
3. Digital ICs: These include logic gates, flip-flops, counters.
4. Monolithic ICs: All devices and components are fabricated together on a single piece of silicon material.
5. Operational amplifier (OPAMP): It is a direct-coupled amplifier with higher gain in which feedback is added to control its overall response characteristics.
6. Voltage regulator: It is an electronic circuit that provides a constant DC voltage independent of the total current and input voltage regulations.
7. DeMorgan's Theorems.
8. Logic gates: The elements of performing logic functions are called gates.
9. D flip-flop: Two signals are used to drive the flip-flop.
10. JK flip-flop: It is more advanced since the racing condition is eliminated.
11. Counters: A counter counts the number of clock pulses arriving at the clock input of flip flops.
12. Synchronous counter: All the flip-flops are triggered simultaneously.

Solved Problems

Problem 1

Design an inverting amplifier with a gain of −20 based on an operational amplifier.

Solution

We know

$$A_V = \frac{R_f}{R_i}$$

Therefore

$$-R_f = A_V \cdot R_i$$

$$-R_f = -20 \times 1 \times 10^3$$

$$R_f = 20\ \text{k}\Omega$$

Problem 2

Two voltages +0.6 V and −1.4 V are applied to the two input resistors of a summation amplifier. The respective input resistors are 400 kΩ and 100 kΩ. The feedback resistor is 200 kΩ. Determine the output voltage.

Solution

We have

$$\frac{V_{out}}{R_f} = -\left(\frac{V_A}{R_A} + \frac{V_B}{R_B}\right)$$

$$V_{out} = -200\left(\frac{0.6}{400} + \frac{-1.4}{100}\right)$$

$$= 2.5\text{ V}$$

Problem 3

Design a non-inverting amplifier with a gain 16 based on an operational amplifier. Assume $R_i = 1$ kΩ.

Solution

$$A_V = 1 + \frac{R_f}{R_i}$$

$$16 = 1 + \frac{R_f}{(1 \times 10^3)}$$

$$R_f = (16 - 1) \times (1 \times 10^3)$$

$$R_f = 15\text{ k}\Omega$$

Problem 4

Draw the circuit of gates that would affect the function

$$F = \overline{A + B \cdot C}$$

Simplify this function and hence redraw the circuit that could affect it.

Solution

The gate circuit based on the original function is shown in Fig. 27. Using DeMorgan's theorem

$$F = \overline{A + B \cdot C}$$

$$= \overline{A} \cdot (\overline{B \cdot C})$$

$$= \overline{A} \cdot (\overline{B} + \overline{C})$$

The redrawn circuit is shown in Fig. 28.

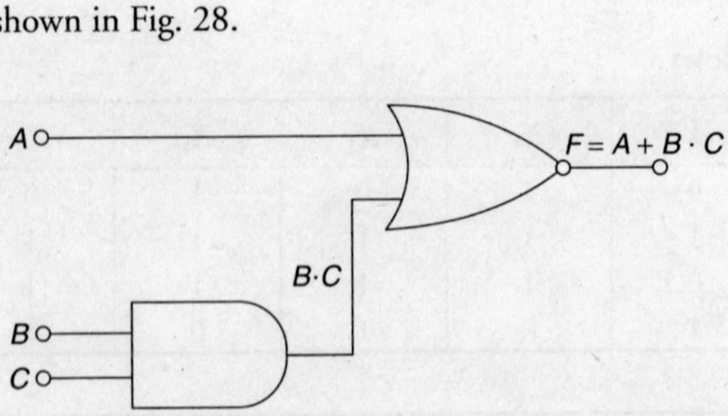

Figure 27 Circuit based on original function.

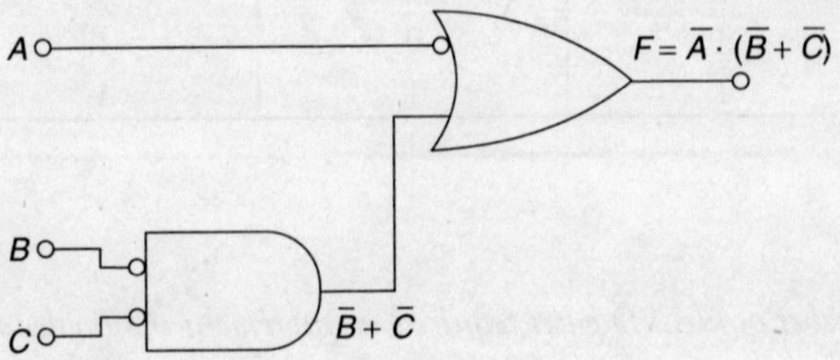

Figure 28 Circuit based on simplified function.

Problem 5

Write the truth table for the logic diagram shown in Fig. 29.

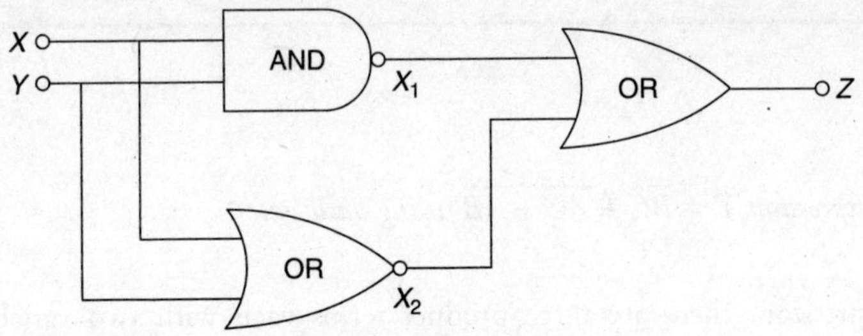

Figure 29 Problem 6.

Solution

The truth table is as follows.

X	Y	X_1	X_2	X_1	X_2	Z
0	0	1	0	1	0	1
0	1	1	1	1	1	1
1	0	1	1	1	1	1
1	1	0	1	0	1	1

Then

X	Y	Z
0	0	1
0	1	1
1	0	1
1	1	1

Problem 6

Find the minimum number of NAND gates required to implement the Boolean function

$$A + A\overline{B} + A\overline{B}C$$

Solution

Solving the given Boolean expression we get

$$\begin{aligned} A + A\overline{B} + A\overline{B}C &= A + A\overline{B}\,(1 + C) \\ &= A + A\overline{B} \\ &= A\,(1 + \overline{B}) \\ &= A \end{aligned}$$

The output is directly taken from A and hence zero NAND gate. Therefore minimum number is zero.

Problem 7

Realize the logic expression $Y = \overline{BC} + \overline{AC} + \overline{AB}$ *using basic gates.*

Solution

In the given expression, there are three product terms each with two variable which can be implemented using three 2-input AND gates, and the product terms can be OR operated together using a 3-input OR gate. The complemented form of individual variable can be obtained by three NOT gates. Thus, the circuit for the given expression is realized as shown in Fig. 30.

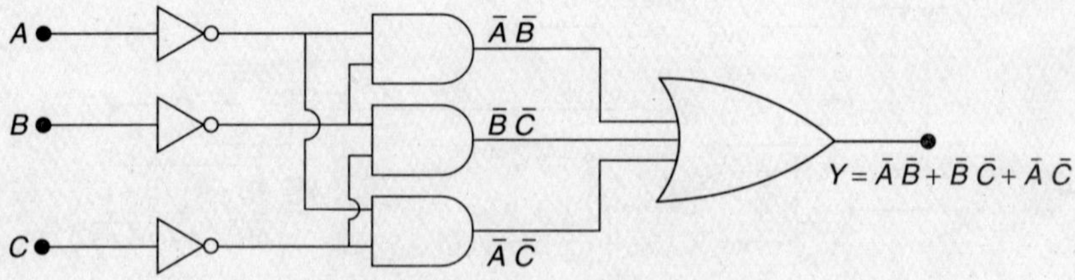

Figure 30 Solution to Problem 7.

Problem 8

Realize the logic expression $Y = (A+B)(\bar{A}+C)(B+D)$ *using basic gates.*

Solution

In the given expression, there are three sum terms which can be implemented using three 2-input OR gates and their outputs are AND operated together by a 3-input AND gate. A NOT gate can be used to obtain the universe of *A*. Now, the realized circuit is shown in Fig. 31.

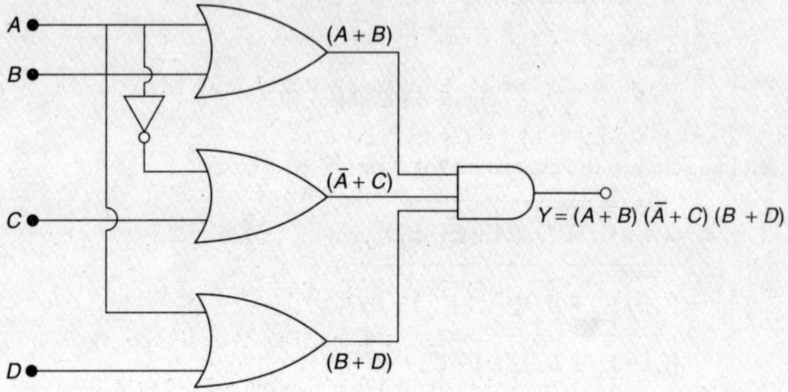

Figure 31 Solution to Problem 8.

Problem 9

Implement $Y = A\bar{B} + A + (B + \bar{C})$ *using NAND gates only.*

Solution

The implementation of the given function in shown in Fig. 32.

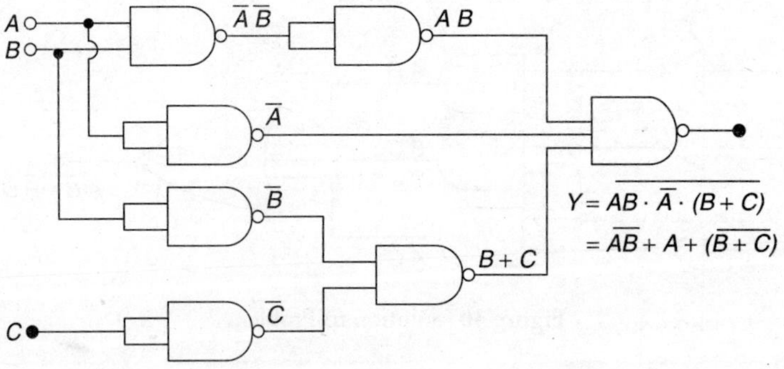

Figure 32 Solution to Problem 9.

Problem 10

Simplify the logic circuit shown in Fig. 33.

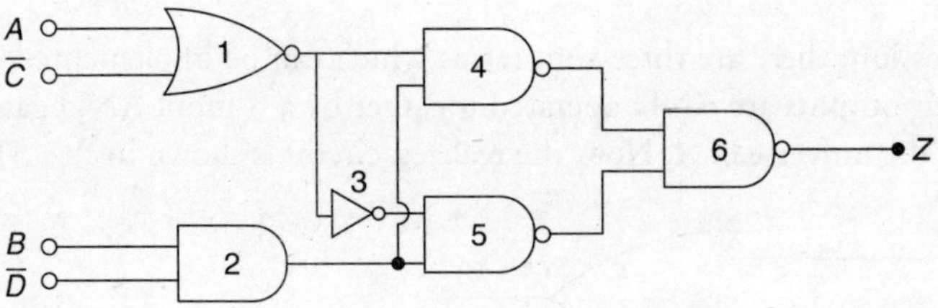

Figure 33 Problem 10.

Solution

From the given logic circuit, the expression for Z can be written as

$$Z=\overline{\overline{\overline{\overline{(A+\overline{C})}\cdot B\overline{D}}}\cdot\overline{\overline{(A+\overline{C})}\cdot B\overline{D}}}$$

$$=\overline{[\overline{\overline{\overline{(A+\overline{\overline{C}})}}}+\overline{B\overline{\overline{D}}}]\cdot[\overline{(A+\overline{\overline{C}})}+\overline{B\overline{\overline{D}}}]}$$

$$=\overline{[(A+\overline{C})+\overline{B\overline{\overline{D}}}\cdot[\overline{(A+\overline{\overline{C}})}+\overline{B\overline{\overline{D}}}]]}$$

$$=\overline{\overline{B\overline{\overline{D}}}+(A+\overline{C})\overline{(A+\overline{\overline{C}})}}\qquad[\because\ (A+B)(A+C)=A+BC]$$

$$=\overline{\overline{B\overline{\overline{D}}}}$$

$$=B\overline{D}\qquad[\because\ A\overline{A}=\overline{O}]$$

Therefore, the above circuit can simplified as shown in Fig. 34.

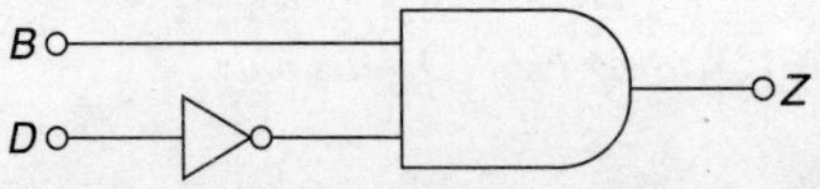

Figure 34 Solution to Problem 11.

Problem 11

Realize $Y = A + BC\bar{D}$ using NAND gates.

Solution

Using NAND gates the given function can be realized as shown in Fig. 35.

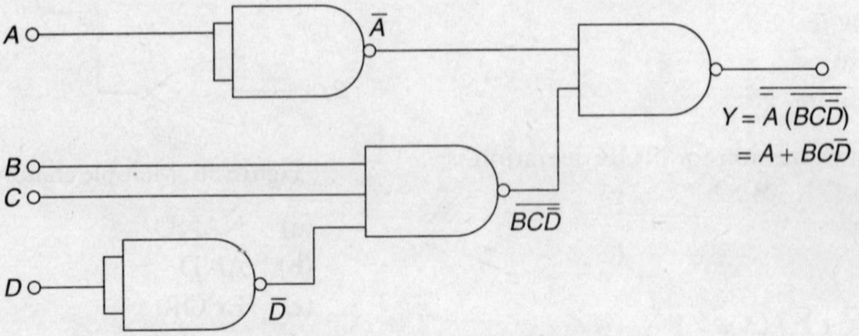

Figure 35 Solution to Problem 12.

Multiple Choice Questions

1. For an AND gate, if one input is High and the other is low, the output is
   (a) high
   (b) low
   (c) zero
   (d) infinite

2. The simplified form of Boolean function $F = (X,Y,Z) = (X+Y+XY)\,(X+Z)$ is
   (a) $X+Y+Z$
   (b) $X+YZ$
   (c) $XY+YZ$
   (d) XYZ

3. For an OR gate, if one input is High, and the other is low, the output is
   (a) high
   (b) low
   (c) zero
   (d) infinite

4. A flip-flop has the number of stable states which is equal to
   (a) 1
   (b) 2
   (c) 31 or 0
   (d) 4

5. The output of two-input NAND gate is high if
   (a) both inputs are low
   (b) one of the input is high and the other one is low
   (c) both inputs are high
   (d) none of these

6. All Boolean expressions can be implemented with
   (a) NAND gates only
   (b) NOR gates only

(c) combinations of NAND gates and NOR gates
(d) any of these

7. For and EX-OR gate, if the two input are the same, the output is
(a) high
(b) low
(c) zero
(d) infinite

8. Boolean expression for NOR operation is
(a) $X+Y$
(b) XY
(c) $\bar{X}\cdot\bar{Y}$
(d) $(\bar{X}+\bar{Y})(X+Y)$

9. Boolean expression for NAND operation is
(a) $\overline{X+Y}$
(b) $\bar{X}+\bar{Y}$
(c) $\bar{X}\cdot\bar{Y}$
(d) $(\bar{Y}+\bar{Y})(X+Y)$

10. The logic performed by the circuit shown in Fig. 36 is

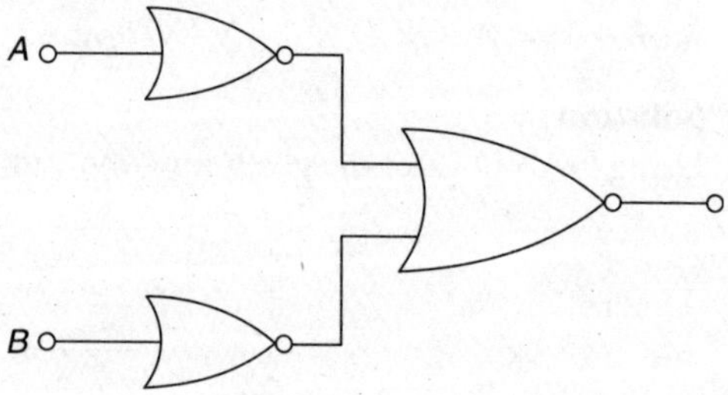

Figure 36 Multiple choice question 10.

(a) NAND
(b) AND
(c) Ex OR
(d) none of these

11. The foundation in which an IC is built is called
(a) insulator
(b) base
(c) wafer
(d) plate

Review Questions

1. What is meant by integrated circuit?
2. Explain the process of epitaxial growth in the fabrication of integrated circuits.
3. Explain the process of diffusion.
4. List the advantages and disadvantages of ICs.
5. How are capacitors fabricated in ICs?
6. Explain the purpose served by SiO_2 layer during different fabrication processes.
7. What is the function of voltage regulator?
8. Write a brief note on 555 IC timer.
9. Show the standard representation of IC voltage regulator.
10. What are universal gates?
11. State DeMorgan's theorems.
12. What are shift registers?
13. Define binary counters.
14. Compare linear and digital ICs.
15. Explain the steps involved in aluminum metallization.
16. How are integrated resistors obtained in monalithitic integrated circuits?
17. Show that the NAND and NOR gates are universal building blocks.
18. Explain the operation of D flip-flops.
19. Mention the different types of flip-flops.
20. Using diode circuit, explain the function of AND and OR logic gates.

Problems

1. Write the Boolean equation at X for the NOT gate. Evaluate the expression if $A=1$.
2. Determine the total number of possible input combinations of a 3-input AND gate. Develop a truth table for this logic gate.
3. Determine the total number of possible input combinations of a 3-input OR gate. Develop the truth table for this logic gate.
4. Sketch the logic circuit for the Boolean expression $(\bar{A}+\bar{B})C$.
5. Sketch the logic circuit for the Boolean expression $(\bar{A}B)+\bar{C}$.
6. Construct a logic circuit that can implement the Boolean expression $X=\bar{A}B+A\bar{B}$.
7. Construct a logic circuit using INVERTER, AND and OR gates for the Boolean expression

$$X=\overline{AB(C+D)}$$

Answers

Multiple Choice Questions

1. (b)
2. (b)
3. (a)
4. (b)
5. (a)
6. (d)
7. (b)
8. (c)
9. (d)
10. (b)
11. (c)

Problems

1. $X=\bar{A}, X=\bar{1}=0$
2. Total number of possible input combinations $=2^3=8$. Truth table is as shown in Table 1.

Table 1 Problem 2

Inputs			*Output*
A	*B*	*C*	*X*
0	0	0	0
0	0	1	0
0	1	0	0
0	1	1	0
1	0	0	0
1	0	1	0
1	1	0	0
1	1	1	1

3. Total number of possible input combinations $=2^3=8$. Truth table is as shown in Table 2.

Table 2 Problem 3

Inputs			*Output*
A	*B*	*C*	*X*
0	0	0	0
0	0	1	1
0	1	0	1
0	1	1	1
1	0	0	1
1	0	1	1
1	1	0	1
1	1	1	1

4. The logic circuit sis given by Fig. 37.

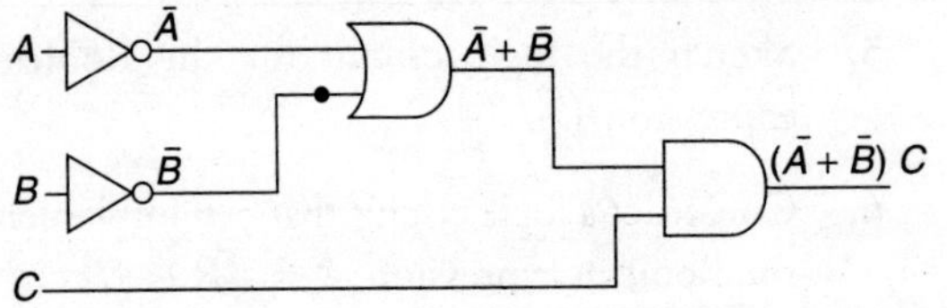

Figure 37 Solution to Problem 4.

5. The logic circuit is given by Fig. 38.

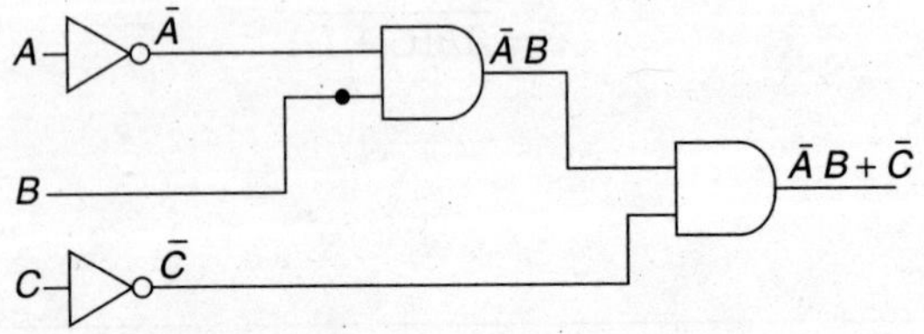

Figure 38 Solution to Problem 5.

6. The logic circuit is given by Fig. 39.

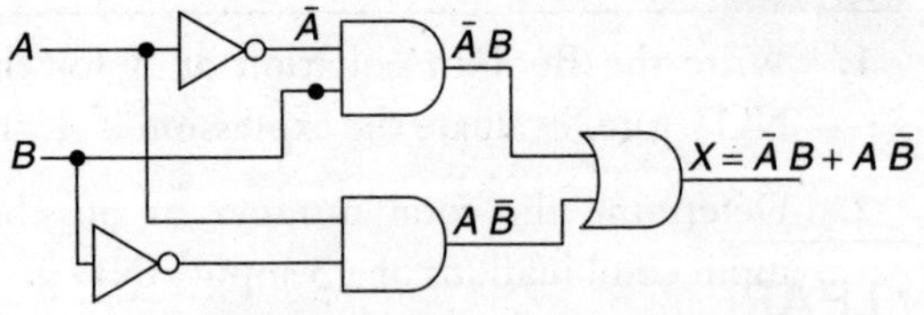

Figure 39 Solution to Problem 6.

7. The logic circuits is given by Fig. 40.

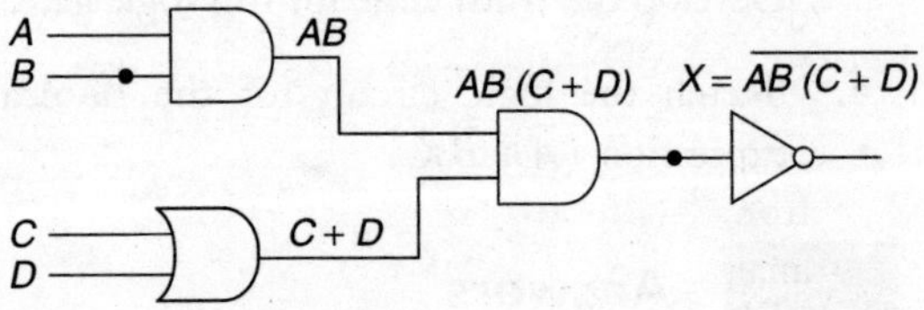

Figure 40 Solution to problem 7.

16 Fundamentals of Computer

LEARNING OBJECTIVES

After reading this chapter, you will be able to:

- Understand the basic concepts of computer devices.
- Study printers and plotters.
- Explain why there is a need to convert signals from analog to digital and from digital to analog.
- Understand how a digital to analog converter circuit operates.
- Explain the concept of workstation main frame computers, distributed and parallel computers.
- Describe the basic architecture of a microprocessor.
- Have an awareness of how microprocessors are controlled.
- Understand the concept of Pentium processors.

Computers have done wonders so far. The need for computer in our day-to-day life is being intensified. Recruitment of personal for any job is based on the individual's computer-handling abilities. Computer predicts the weather and gives warnings. They take pictures of the solar system and send them to earth. Everybody talks about computerization of everything. An electric device is creating a huge impact on the world, and it is high time, everyone acquires atleast some basic knowledge about computers. In this chapter our aim is to get acquainted with basics of computers.

16.1 Input Devices

While working on the computer, data has to be supplied by the user and the results of the processing have to be intimated to the user by the computer. Supplying the required data into computer memory is called *input operation* and getting data out of the memory is called *output operation*. These operations are done by input and output devices, respectively. Input devices allow users and other systems to send data to the computer. Most common input device that is used in modern computers is the keyboard.

A computer keyboard is similar to the typewriter keyboard, but has a few more keys. Each key stroke displays the corresponding character on the visual display unit (VDU). When the Enter Key is pressed, the information typed is transformed into the computer memory. It may be noted that all information is transferred to computer as a stream of binary digits 0 or 1. All characters on the keyboard have equivalent binary codes. One such code is called ASCII (American Standard Code for Information Interchange).

16.2 Central Processing Unit (CPU)

We now discuss the basic principle of CPU. The CPU is the "brain" of the computer – it is the "compute" in computer. Without the CPU, you have no computer. AMD, IBM, Intel, Motorola, SGI and Sun are just a few of the companies that make most of the CPUs used for various kinds of computers including home desktops, office computers, mainframes and supercomputers. Computer CPUs (processors) are composed of thin layers of thousands of transistors. Transistors are tiny, nearly microscopic, bits of material that will block electricity at one voltage (they are non-conductors and "resist" the flow of electricity) and permit electricity to pass through them at different voltage (the material loses its resistance to electricity and becomes a conductor). The ability of these materials (called semi-conductors) to transit from a non-conducting to a conducting state allows them to take two electrical inputs and produce a different output only when one or both inputs are switched ON. A computer CPU is composed of millions (and soon billions) of transistors. Because CPUs are so small, they are often referred to as microprocessors. So, the terms processor, microprocessor and CPU are interchangeable. Modern CPUs are what are called "integrated chips." The idea behind an integrated chip is that several types of components are integrated into a single piece of silicon (a single CPU), such as one or more execution core, arithmetic unit (ALU), logic unit, registers, instruction memory, cache memory and the input/output controller (bus controller).

CPU has a component called arithmetic logic unit (ALU) which performs basic arithmetic operations like addition, subtraction, etc. as well as compares numeric values. The most important point to be noted is that inside the CPU all data are in binary form and the codes for these operations must also be in binary. One important register in ALU is the ACCUMULATOR. As the name implies, it is possible to add a number in a memory location in random access memory (RAM) to the data already existing in ACCUMULATOR. A typical binary instruction for the operation might be as follows:

0	1	0	0	0	1	0	1	0

The basic difference between an ordinary calculator and a computer is the stored program concept used in computers. In calculator, as soon as an instruction is given it is executed. In computer, it is possible to store a sequence of instructions in its main memory, bring them one-by-one to CPU and execute each instruction. The sequence of instructions is stored at a specified area in main memory. The program counter (PC) registers the address of the location where the first instruction is stored. This instruction is first brought to Master Boot Record (MBR) and from MBR to instruction register (IR).

16.3 Output Devices

Any peripheral that receives or displays output from a computer constitutes an output device. Output devices are pieces of computer hardware that allow a computer system to communicate information to a user or another system. This information can be in any form, and includes sound, images and even tactile experiences. Output devices can usually only be used to send data from the computer; items called. Most popular output devices are as follows:

1. Visual display units (VDUs).
2. Printers.
3. Plotters.

A VDU (a monitor or display) is an electronic visual display for computers. Printers can be classified as follows:

1. **Impact or non-impact:** Impact printers work on the principle that the characters are formed by pressing against the paper through a ribbon under the force of an impact. Non-impact printers have images created on paper by techniques like inkjet, heating, electrostatic effects, etc. Multiple copies are possible in impact printing while it is not so in non-impact printing.
2. **Serial, line or page:** Serial printers print one character at a time, line printers print one line at a time and page printers, one page at a time.
3. **Character or matrix:** Character printers have unique letter corresponding to each character, as in typewriters.

Pinters and plotters are discussed in detail in later part of the chapter.

16.4 Memory Devices

16.4.1 Floppy Disk Drive

Floppy disk drive is the most popular auxiliary storage that has now become an internal part of any computer system. A floppy disk drive is a hardware device that reads one of the first types of portable data storage media – floppy diskettes, also known as floppy disks. Floppy disks are read and written by a floppy disk drive or FDD. It was invented by the American Information Technology Company IBM. Floppy disks in 8 inch, 5¼ inch and 3½ inch forms enjoyed nearly three decades as a popular and ubiquitous form of data storage and exchange, from the mid-1970s to the late 1990s. While floppy disk drives still have some limited use, especially with legacy industrial computer equipment, they have now been superseded by USB flash drives, external hard disk drives, optical disks, memory cards and computer networks.

The floppy disk is made of plastic material and is coated with magnetic materials forming circular tracks. The floppy disk rotates and the read/write head of the drive moves radially to the track. The read/write operation is done by changing the direction of the microscopically small magnetic particles on the track.

16.4.2 Hard Disks

There are popularly known as winchester disks and the corresponding drive is known as winchester drive. The magnetic disk is permanently sealed in a casing. The size is very small, but the capacity is very large. The common capacities vary from 20 MB to a few gigabytes. Winchester drives are an essential component of any computer system.

A hard disk drive (HDD) is a non-volatile, random access device for digital data. It features rotating rigid platters on a motor-driven spindle within a protective enclosure. Data are magnetically read and written on the platter by read/write heads that float on a film of air above the platters.

16.4.3 Magnetic Tapes

Magnetic tape is another storage media for storing large volumes of data which are not frequently used. Special tape drives are necessary for using magnetic tapes. Magnetic tapes of different sizes and storage capacities are used. Audio cassette tapes are used with PCs for simple applications like computer games. But this is generally very inconvenient. In mini computers, tapes of width (1/2)″ to 1″ and length of 2400″ are used. Cartridge tapes are also used for backup of memory.

16.5 Memory Terminology

No computer can work without a memory to a store data, programs, instructions, etc. A memory unit in a computer consists of a large number of binary storage cells. The main characteristics of a cell are, it must have two stable states, one to store a "0" and another to store a "1". There are two broad classifications of memory and different technologies are used to build these. They are as follows:

1. **Main memory or primary memory:** This memory always resides inside the computer connected with the CPU. This is relatively large and fast memory. It is used to store programs, data, etc. during the computer operations. It is this memory which is accessed by the CPU whenever necessary. In earlier days, core memory was used as main memory. Now, it has been replaced by the low-cost semiconductor memory.
2. **Secondary memory or auxiliary memory:** Secondary or auxiliary memory is the memory which is not directly accessed by CPU. Whenever the data in this storage is to be used, it will be transferred to main memory and referred. A large volume of data can be stored using different secondary memories available. This also helps to avoid overloading of the main memory.

The memory terminology of a CPU is as follows. It includes new as well as old technologies.

1. **Array:** It is the area of RAM that stores the BITS. The array consists of rows and columns at each intersection that can store a BIT. It is the large rectangular section in centre of die where memory is stored.
2. **Asynchronous cache:** The oldest and slowest type of cache timing is asynchronous cache. Asynchronous means that transfers are not tied to the system clock. A request is sent to the cache, and the cache responds, and this happens independently of what the system clock (on the memory bus) is doing. This is similar to how most system memory works; your typical FPM or EDO memory is also asynchronous (and relatively slow for this reason). As asynchronous cache is not tied to the system clock, it can have problems dealing with faster clock speeds. At slow speeds like 33 MHz, it is capable of 2-1-1-1 timing (which is very good) but at speeds like 60 or 66 MHz, as used in modern Pentium class PCs, it drops down to 3-2-2-2 (which is pretty bad). For this reason, asynchronous cache is commonly found on 486 class motherboards but is not generally used on Pentium or later class machines.
3. **Autofresh:** It is commonly referred to CAS before RAS Refresh or CE before RE Refresh. At internal address counter increments the row each time the memory control indicates the CAS before RES Refresh cycle.
4. **CAS (Column Address Select):** It is a control pin on Dynamic Random Access Memory (DRAM) used to latch the column address. The DRAM has a large array of memory capacitors that are arranged in rows and columns. To read one location in the array, the control circuit first calculates its row number, which it places on the DRAM's address pins. It then toggles the row address select (RAS) pin, causing the DRAM to read the row address. Internally, the DRAM connects the selected row to a bank of amplifiers called *sense amplifiers*, which read the contents of all the capacitors in the row. The control circuit then places the column number of the desired location on the same address pins and toggles the column address select (CAS) pin, causing the DRAM to read the column address. The DRAM uses this to select the output of the sense amplifier

corresponding to the selected column. After a delay called the CAS access time, this output is presented to the outside world on the DRAM's data I/O pin.

5. **CBR:** It implies CAS before RAS. It is a fast refresh technique in which the DRAM keeps track of next row it needs to refresh, thus simplifying what a system would have to do refresh part.

Thus, CPU memory connection shows some disadvantages presented, and so most of consumers go for some different types of memory.

16.5.1 Programmable ROM (PROM)

A programmable read-only memory (PROM) or field programmable read-only memory (FPROM) or one-time programmable non-volatile memory (OTP NVM) is a form of digital memory where the setting of each bit is locked by a fuse or antifuse. Such PROMs are used to store programs permanently. The key difference from a strict ROM is that the programming is applied after the device is constructed. It is a non-volatile memory, that is, once memory is stored it cannot be alternated.

16.5.2 Erasable Programmable ROM

Erasable programmable read-only memory is called EPROM. In this type of memory, the user can not only write information into memory, but also erase the unwanted information by exposing to UV ray after removing it from board.

16.5.3 Electrically Erasable Programmable ROM

Electrically Erasable Programmable Read-Only Memory (EEPROM) is a type of non-volatile memory used in computers and other electronic devices to store small amounts of data that must be saved when power is removed, for example, calibration tables or device configuration. Also the erasing time in EEPROM compared to EPROM is much less.

16.5.4 Flash Memory

Flash memory refers to a particular type of EEPROM. Flash memory is a non-volatile computer storage chip that can be electrically erased and reprogrammed. It is primarily used in memory cards, USB flash drives, MP3 players and solid-state drives for general storage and transfer of data between computers and other digital products. It is a specific type of EEPROM that is erased and programmed in large blocks. In early flash, the entire chip had to be erased at once. Flash memory costs far less than byte-programmable EEPROM and therefore has become the dominant technology wherever a significant amount of non-volatile, solid-state storage is needed. Example applications include PDAs (personal digital assistants), laptop computers, digital audio players, digital cameras and mobile phones. It has also gained popularity in console video game hardware, where it is often used instead of EEPROMs or battery-powered static RAM (SRAM) for game save data.

16.5.5 Semiconductor Memory

Rapid development in electronic field has already made the core memories obsolete. Because of availability of cheap semiconductor memories, the cost of computers have come down. Two types of techniques are used in fabrication of this memory. They are:

1. **Bipolar techniques using transistor-to-transistor logic (TTC) gates:** Though the memory built using the technology is very fast, the number of units integrated in single chip is less. So they are not widely used.
2. **Metal oxide semiconductor:** It is the most technology-efficient method of building main memory. The memory built using the technology is slower compared to the first method. However, this is adopted because more number of units can be integrated together. So the cost is also reduced.

16.6 Mass Storage Data System

The most common form of permanent data storage used in modern computer are archival tapes. This type of data storage was used in very first computers in 1990. In succeeding 55 years, the technology improved remarkably to computer electronic circuit. There are four types of mass data storage data systems:

1. Floppy disk data storage.
2. Magnetic hard disk data storage.
3. Magnetic tape storage.
4. Optical disk data storage.

16.6.1 Floppy Disk Data Storage

The growing market of low-cost microcomputers created the need for low cost memory storage. This led to development of floppy disk. Floppy disks are created out of magnetic oxide coated on polyester substrate over a flexible plastic base. This disk of 7.85 inch diameter is enclosed in an 8 inch square jacket. This gives protection to disk. The envelope has long slit for read/write head across, a hole in centre for mounting the disk drive hub and a hole for index mark sensing as shown in Fig. 1. The floppy is inserted into floppy disk drive and the read/write operations are done similar to that of hard disk.

The data are stored on the tracks. Sectors are used to divide the whole area of disk. Since floppy disks are portable and cheaper, large volumes are sold in market. Computer virus is major threat to floppy disks.

16.6.2 Magnetic Hard Disk Data Storage

Magnetic disks, also called hard disks, provide large storage capabilities with moderate speed. These disks are smooth metal plates coated with a thin film of magnetic material on both sides. A set of such

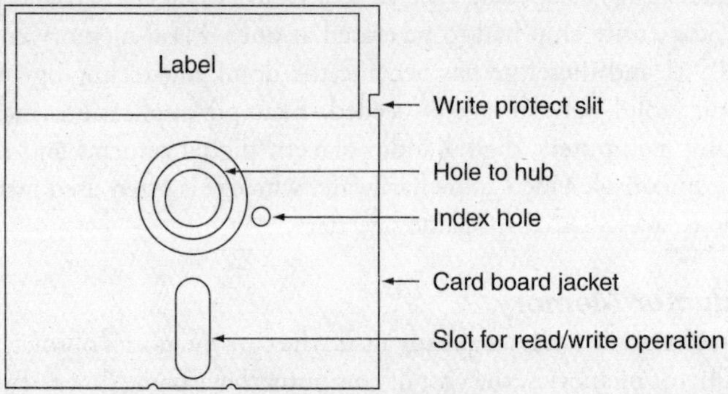

Figure 1 Floppy disk.

disk are piled together in an axis one below the other with some gap. The disk axis is attached with motor so that it can rotate to 360°. The disk space is mounted on disk drive. The drive has a set of read/write heads and which can move in and out in radial direction. The disk is divided into different tracks and sectors (see Fig. 2).

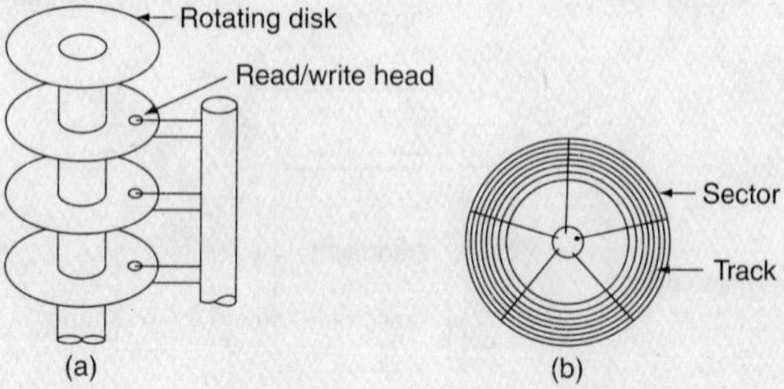

Figure 2 Magnetic hard disk.

To write the data on disk, a current is passed through a coil wound round the head. The current magnetizes the coating on disk surface. The magnetization can be either in clockwise or in anticlockwise direction. These two directions represent a0 and a1, respectively. The read operation is done the same way. When the disk rotates, the magnetized spot below the head induces a current in the coil.

An improved version of hard disk is winchester disk. The read/write heads are light weight. Also the heads are sealed inside the disk pack so that no foreign particle can affect the disk surface.

16.6.3 Magnetic Tape Storage

The magnetic tapes used here are the same as used in audio/video cassettes. They are made up of plastic material coated with a ferromagnetic material. The data are stored in the same way, by magnetism spots or magnetic disks. There are number of lengthwise tracks on the tape.

Writing on or reading from tape is done using tape drive. The drive consists of motors, a tape transport on it and read/write head (refer Fig. 3). Each track will have a read/write head. The movement of tape is a similar to that of film on film projector. It moves from one reel to another over capston and roller and sides over the read/write heads at speed of about 5 m/s.

16.6.4 Optical Disk Data Storage

Optical disks have a great potential for mass storage as ROMs. They are made of transparent glass or plastic and have coat of a thin reflective film of metal. To store data on disk a laser beam is focused at the desired spot. This creates a small hole on the film. Once the data is stored it becomes permanent and we cannot erase it. Presence of a hole is taken as 1 and absence of hole is taken as 0. To read the data from disk, a less-powerful laser beam is passed over the surface. If there is a hole, the amount of beam reflected from disk is altered. This change in reflection is sensed by head and sent to CPU for proceeding.

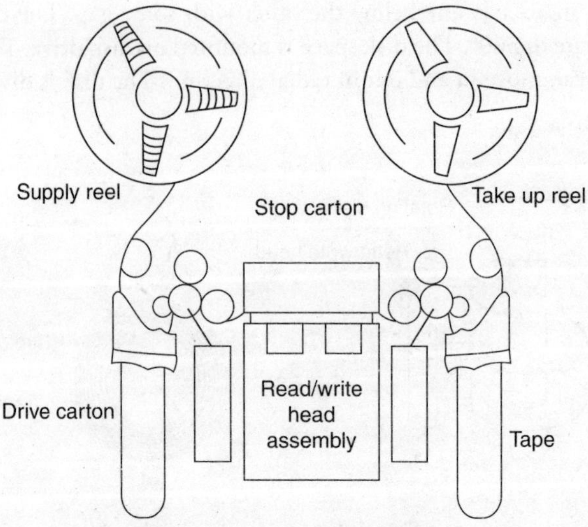

Figure 3 Magnetic tape storage.

Nowadays optical disks come in compact form, called compact disk read-only memory. They are also used as audio storage media.

16.7 Computer Input Devices

The purpose of input devices is to transfer the information from outside world to computer. The instructions and data are entered in computer memory through the input devices such as:

1. Card reader.
2. Paper tape reader.
3. Tele typewriter.
4. Keyboard.
5. Joystick.
6. Mouse.
7. Touch screen technology.
8. Light pen.
9. Magnetic ink character recognition (MICR).
10. Optical character recognition.
11. Bar coding.

16.7.1 Card Reader

Punching card-punching machine-card reader combination was one of early methods to transfer data into computer. The standard punching card earlier was 18.3 cm × 8.3 cm with 80 columns and 12 rows. The first 2 rows are numbered are 12 and 11. The remaining 10 rows are numbered as 0, 1, 2, 3, 4, 5, 6, 7, 8 and 9. When a key is pressed in punching machine, it punches a set of holes along a column.

16.7.2 Paper Tape Reader

The working principle of paper tape reader is similar to a card reader. Here punched card is replaced by continuous strip of paper tape of width ranging from 1/2 inch to 3 inch. Circular holes as punched in paper tape to represent the instruction code.

16.7.3 Tele Typewriter

Tele typewriter is another type of input device which looks similar to an ordinary typewriter. In addition to typing, the tele typewriter transmits the typed data to computer in form of electric pulses. The error in typing a character can be removed by de-pressing suitable keys on keyboard.

16.7.4 Keyboard

Keyboard is standard and a very widely used input device. It consists of electronic switches and is connected to computer through a serial port. When a key is pressed, the closure of corresponding electronic switch generates an electric signal, thus enabling the computer to recognize the key pressed. Various keys present in keyboard include keys to type alphabets.

16.7.5 Joystick

Joystick consists of small vertical levers mounted to base. It is used to steer the cursor on screen. The cursor can be moved by moving the stick.

16.7.6 Mouse

A mouse is a pointing device that functions by detecting two-dimensional motion relative to its supporting surface. Physically, a mouse consists of an object held under one of the user's hands, with one or more buttons. It sometimes features other elements, such as wheels, which allow the user to perform various system-dependent operations, or extra buttons or features that can add more control or dimensional input. The mouse's motion typically translates into the motion of a cursor on a display, which allows for fine control of a graphical user interface.

16.7.7 Touch Screen Technology

A touchscreen is an electronic visual display that can detect the presence and location of a touch within the display area. The term generally refers to touching the display of the device with a finger or hand.

16.7.8 Light Pen

The touch input can also be recorded using light-sensitive pen. It helps in identifying the position of object more accurately. It allows the user to point to displayed objects, or draw on the screen, in a similar way to a touch screen but with greater positional accuracy. It is a computer input device.

16.7.9 Magnetic Ink Character Recognition (MICR)

MICR is a character-recognition system that is widely used in the banking industry for cheque processing. It involves using a stylized font and magnetic ink to print characters in a document. When this document needs to be decoded, it passes through a special machine or equipment, which magnetizes the ink and then translates the magnetic information into characters. If required, the same information in the document can also be read optically through OCR. Over the years, MICR has been proven to be a secure and high-performance mechanism for processing information. The numbers typically found on the bottom of a cheque comprising the cheque number, sort number and account number are usually printed in magnetic ink for use in MICR.

16.7.10 Optical Character Recognition

Optical character recognition (OCR) is the mechanical or electronic translation of scanned images of handwritten, typewritten or printed text into machine-encoded text. It is widely used to convert books and documents into electronic files, to computerize a record-keeping system in an office, or to publish the text on a website. OCR makes it possible to edit the text, search for a word or phrase, store it more compactly, display or print a copy free of scanning artifacts, and apply techniques such as machine translation, text-to-speech and text mining to it. OCR is a field of research in pattern recognition, artificial intelligence and computer vision.

16.7.11 Bar Coding

Bar Coding is a series of parallel vertical lines (bars and space) that can be read by bar code scanners. It is used worldwide as part of product packages, as price tags, carton labels, on invoices, even in credit card bills. When it is read by scanners, a wealth of data is made available to the users. When used with GS1.UCC (Global India One Numbering Uniform Code Council Inc. USA) numbering system, the bar code becomes unique and universal and it can be recognized anywhere in the world. Bar coding is an international concept today. It facilitates unique product identification through using international symbologies/numbering system, promotes brand image and enables timely and accurate capture of product information. This would result in wide ranging benefits including lowering of inventory costs, lower overall supply chain costs and hence reduced costs for Indian products, increasing efficiency of Indian industry and adherence to stringent quality assurance norms through product traceability.

16.8 Printers

Most applications in computing require printed output. Printers are used to take the hard copy of the text documents. Sometimes graphs are also taken. Each printer has a printing head and a drum over which paper scrolls on which the characters are printed.

There are two types of printers:

1. Impact printer
   - Drum printer.
   - Dot matrix printer.
   - Golf ball printer.
2. Non-impact printer
   - Inkjet printer.
   - Laser printer.

16.8.1 Drum Printer

It consists of drum where characters are embossed in rows. The movement of ribbon, movement of printer and printing mechanism are just similar to that of chain printer. When a particular row comes against the hammer, the signals from the buffer activate the hammers of all print positions where that particular character is to be printed. As the drum rotates, one complete line of characters will be printed. A drum printer is illustrated in Fig. 4.

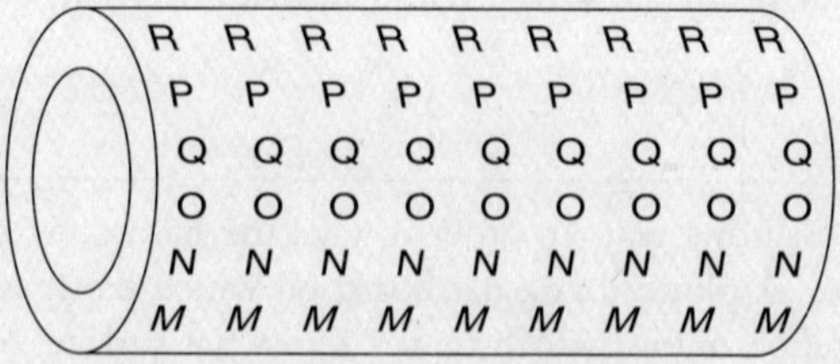

Figure 4 Drum printer.

16.8.2 Dot Matrix Printer

Here the printing head will have a set of pins arranged as a matrix (Fig. 5). The character is formed by selecting appropriate set of pins from this matrix and the paper will be kept on rolling drum on which the head will make impression. Between the paper and head, there will be an inked ribbon. This is used to print a single character. Let us see how the whole line will be printed. The printer will have a temporary storage attached with it, called buffer. CPU will send the line to be printed to this buffer. After printing one character, the head will move one position to right on paper and next character will be printed from here. In this way the head will print up to 80 characters. Now the head is at both position and the printer head will start printing from last character to first character.

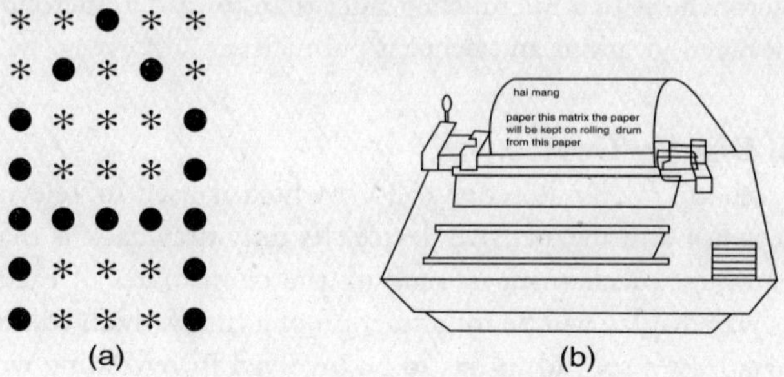

Figure 5 Dot matrix printer.

16.8.3 Golf Ball Printer

Golf ball printer consists of a ball on which the characters are embossed. The desired character can be brought to printer position by moving the ball laterally either up or down. The character is printed where ball strikes against the paper moving on roller, with ribbon in between.

16.8.4 Inkjet Printer

The inkjet printer uses ink cartridges which contain a column of tiny heater. The tiny heater is activated by means of a pulse of current. This causes a drop of ink to ėject onto the paper to form the character. The drawback in printer is that the ink cartridges cost more and also they do not last long.

16.8.5 Laser Printer

Laser printer use laser beam to scan a bit mapped image of the page onto an electrostatic drum. The drum rotates through a developing station, picks up toner and transfers it to paper. The paper then moves to fusing station, while drum is discharged and cleaned.

16.9 Plotters

Printers are best suited for printing text. In order to take the hard copy of graphs, plotters are used. Plotters are of different sizes. A plotter has a flat board on which paper is kept for drawing. There is an arm connected to the stepper motor which moves across flat surface. Multicolor pen are kept in a row on one side of board. According to the program written, a computer keeps up the required color pen and draws graph on the paper plotter. Plotters are used primarily in technical drawing and CAD applications, where they have the advantage of working on very large paper sizes while maintaining high resolution.

16.10 Computer Output on Microfilm

This is often used to store massive data in compact form. The output from the computer, instead of being printed on line printer, is displayed on high resolution cathode ray tube. This is photographed on 35 mm film. A camera is controlled to film successive output pages on the screen of cathode ray tube. A special microfilm reader is used to read the output. In some systems the microfilm is converted into microfiche form, $4'' \times 6''$ sheet of film, that holds 98 frames of $8'' \times 11''$ page images reduced to 24 times. It is easier to microfiche with a microfiche reader than to read a microfilm. Besides this, microfiche is easier to mail between locations and hence is popularity.

16.10.1 Graphical Display Device

VDU is nothing but cathode ray oscilloscope (CRO) which is used in Television. It is most widely used output device. It is fast and inexpensive device. Its only drawback is that it does not deliver a hard copy. Here the simplest mechanism is, sending the coordinates of each pant to be displayed from CPU. But in this way display will be for a fraction of a time, which will not be visible. To make it appear on screen for atleast a second, it has to be involved in refreshing work. To overcome this, the storage tube displays are used. In case the VDU has an extra memory attached to this, it re-fuses the screen.

Mouse, digitizer, light pen and touch technology are all the sophisticated input techniques used with the VDU and the keyboard.

16.10.2 Speech Output Device

Speech output device is one which reads strings of character, stores them and converts them into spoken sentence. Combination of letters is called a phoneme. Electronic chips are designed in such a way that they accept ASCII characters of sentence and combine the sequence of words to phoneme. It is then amplified and the output is from loud speaker device.

16.11 Digital to Analog Converter

Digital to analog conversion is usually abbreviated as D/A conversion. D/A converters are sometimes abbreviated as DACs. The purpose of such conversion is to take a digital signal and convert it to an analog signal.

Typically, a digital signal will be in binary code while an analog signal will be either voltage or current which varies in value over a predetermined range.

We will start by considering the required operation of system – in this case the system is converter with inputs D, C, B and A (Fig. 6). These inputs are likely to be taken from output register of digital system. Details of the binary number system are given.

Being 4-bit system, the input binary numbers can have 16 different values as indicated in Fig. 6. Let us now further assume that output signal voltage has a value numerically equal to that of equivalent binary number.

In the same system, 0110 input would produce on output of 6 mV and so on. There is no need for the relationship to be direct. For instance, we could introduce some proportionality factor such as 7. If this was the case, then 1011 would give as output 77 mV and 0110 would give 42 mV.

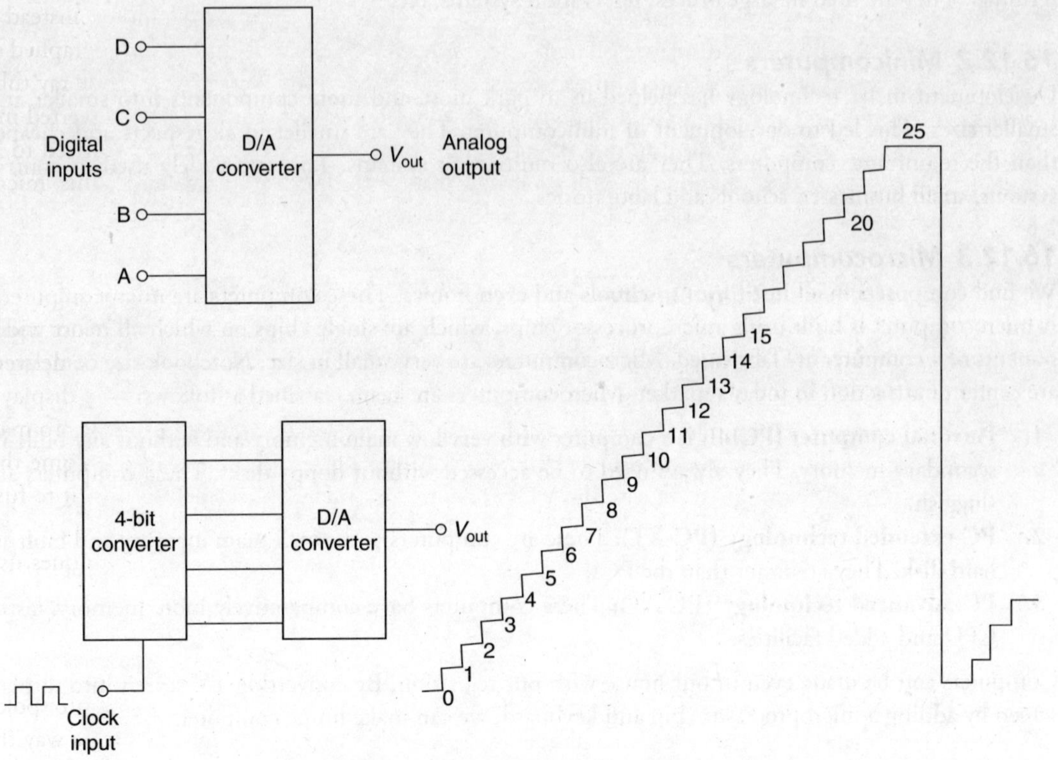

Figure 6 Digital to analog converter.

16.12 Classification of Computers

Computers are classified into the following four types:

1. Mainframe computer.
2. Minicomputer.
3. Microcomputer.
4. Super computer.

16.12.1 Mainframe Computers

In early days, computers were very large, about size of room, and engineers referred to them as mainframe. They were also costly. However, the development in electronics engineering has helped to build powerful, low-cost, small-sized mainframe computers.

More than 10 people can work at a time on a single mainframe computer. It is also called multiaccess or multiuser system. Mainframe computers are now built with large as well as high-speed memory, very sophisticated CPU and I/O devices. They are used in places where large volume of data is to be processed and multiaccess is necessary. They are used to store international databases which can be produced from anywhere in world. They support mass storage devices, fast printers and graphics terminal. They are used in large offices, reservation systems, etc.

16.12.2 Minicomputers

Development in IC technology has helped us to park more and more components into smaller and smaller sizes. This led to development of minicomputer. They are smaller in all respects and cheaper than the mainframe computers. They are also multiaccess systems. They are widely used in control systems, small businesses, schools and laboratories.

16.12.3 Microcomputers

We find computers in all institutions, schools and even homes. These computers are microcomputers. A microcomputer is built using microprocessor chips, which are single chips on which all major components of a computer are fabricated. Microcomputers are very small in size. Notebook size computers are center of attraction in today's market. Microcomputers are again classified as follows:

1. **Personal computer (PC):** It is a computer with very low main memory and without any built-in secondary memory. They always need to be accessed without floppy disks. These computers are sluggish.
2. **PC-extended technology (PC-XT):** These are computers with small main memory and built-in hard disk. They are faster than the PCs.
3. **PC-advanced technology (PC-AT):** These computers have comparatively more memory, faster CPU and added facilities.

Computers can be made even in our house with our television. By converting TV screen into display screen by adding a microprocessor chip and keyboard, we can make home computer.

16.12.4 Super Computers

These computers hold the world record for their computing speed and power. Today's super computers can do their calculation at rate of few billion operations per second. They are relatively specialized in the sense that very few applications need such as high speed. If not for super computers, we would not have enjoyed the spectacular achievements in space technology. Super computer are used for massive scientific and engineering calculations. They are needed for aircraft, missile and spacecraft designing, weather forecasting, seismological studies, etc.

16.12.5 Laptops (PCs)

Laptops are personal computers designed for mobile use. Laptops integrate most of the typical components of a desktop computer including a display, keyboard, a pointing device and speakers into a single unit. A new laptop battery typically stores energy to run the laptop for 3 to 5 hours. Depending upon the computer usage and power management system, Laptops are mostly used in specialized fields such as military and by people such as accountants and sales representatives.

16.12.6 Workstations

Workstations are used by professionals for computation, computer aided design, simulation and graphical visualization. These are also desktop machines. The processing speed is about 10 times higher than PCs. Normally, they have a large color video display unit and large memory unit. Workstations normally use reduced instruction set computer (RISC) processors. Some manufactures of workstations are Silicon Graphics (SIG), Digital Equipment Corporation, Sun Microsystem, IBM and Hewlett Packard. The standard operating systems of workstation are UNIX and its derivative are AIX (IBM), Solarin (SUN) which provide good graphic facilities.

16.12.7 Distributed Computers

Distributed computers consist of number of workstations or PCs interconnected by a communication network. A common mode to use the distributed computers is a client-server configuration. In this mode, specialized services such as large disk resource, numeric-intensive compiling and printing are provided by servers which are used by clients in network. Client will use multiple servers to get the problem solved.

16.12.8 Parallel Computers

A set of computers connected together by a high speed communication network are programmed in such a way that they cooperate to solve a single large problem. Such types of computers are called parallel computers. There are two types:

1. Shared memory parallel computers.
2. Distributed memory parallel computers.

In shared memory computer CPU will share common memory and in distributed memory many independent computers are connected by high-speed network. Here a job is broken into number of parts and allocated to each of the computer in the system.

16.13 Microprocessors

A microprocessor is a single integrated chip in which all important components of computer, ALU, control unit and other circuits, are fabricated together. Microprocessor age began in 1970s when IC technology had advanced enough to make a single chip CPU. Because of low-cost and higher speed qualities of a microprocessor, it is used as controller for almost all the design equipments. Microprocessor chips are very good examples for VLSI chips.

There are varieties of microprocessor chips with different capabilities. The main factor here is number of bits they can handle. For example, microprocessors are made to handle 8-bit, 16-bit or 32-bit operation. An 8-bit chip can handle addition of 8-bit numbers only. To make chip perform 16-bit operations, a special register is needed. We are going to discuss about 8086 microprocessor next.

16.13.1 INTEL 8086 Microprocessor

In 1972, Intel launched the 8008, the first 8-bit microprocessor. It implemented an instruction set designed by Data Print Corporation with programmable CRT terminals in mind that also proved to be of fairly general purpose. The device needed several additional ICs to produce functional computers part due to small 18-pin "memory package" which ruled out the use of separate address bus.

Two years later in 1974, Intel launched the 8086 employing the new 40-pin DIL packages originally developed for calculator IÇs to enable a separate address bus. It had an extended instruction set that was source compatible with the 8008 and also included some 16-bit instructions to make programming easier. The 8086 device, often described as first useful microprocessor, was eventually replaced by depletion load based 8085 which would cope with 5 V power supply instead of three different operating voltages of earlier chips.

16.13.2 Internal Architecture of 8086 Microprocessor

8086 microprocessor is nothing but single chip used CPU in personal computer. The block diagram (Fig. 7) of 8086 shows that it is actually a 8-bit microprocessor capable of handling 16-bit operations with help of some special registers and circuits. Since we have already discussed about the functions of CPU in detail, let us discuss the functions of these special registers.

Apart from control unit and ALU, 8086 microprocessor has following register:

1. Four instruction queue registers which will keep on collecting the instructions from memory so that PU need not wait for the delay in transfer of instructions from memory.
2. Four pairs of general-purposes registers: AH & AL, BH & BL, CH & CL and DH & DL (AH – accumulator high, AL – accumulator low, BH – base register high, BL – base register low, CH – code segment high, CL – code segment low, DH – data segment high, DL – data segment low). They are used by ALU while executing the instructions. These registers independently can hold 8 bits of data; at some time, they can hold 16 bits of data.
3. Processor status word (PSW) is used to identify the status of different registers and signals.
4. Processor registers, viz. ES, CS, SS and DS (CS – code segment register, DS – data segment register, ES – extra segment register, SS – stack segment register) are used the address location when a big program is stored in different memory locations.
5. Pointer registers are SP (stack pointer), BP (base pointer), SI (source index register), DI (density index register) and IP (next instruction pointer register) Among these stack pointer (SP) is an important register. A stack is a special group of storage location in memory. It is organized like the trays arranged one above the other in such a way that, the data stored last is drawn first. So it is called LIFO (last in, first out) memory. SI and DI are used by the control unit to find address of first element. SP is the register always holding the address of top of stack. Instruction pointer is same as program counter discussed in functions of CPU. All the registers are connected together with help of data pathways of buses.

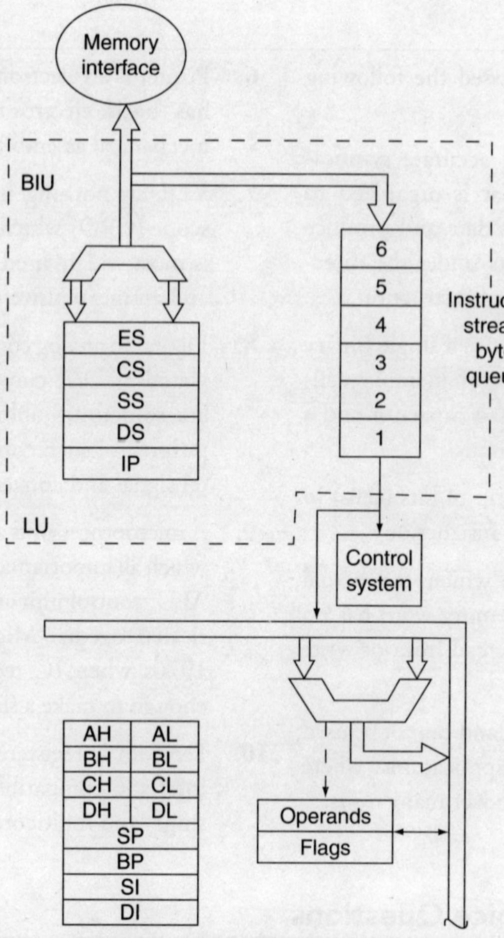

Figure 7 Internal architecture.

16.13.3 Pentium Processors

Pentium is registered brand name of many intel x86 compatible microprocessors, both single and multicore. The name Pentium was derived from the Greek *pente* meaning "five," and the Latin ending *-ium* a name selected after courts had disallowed trademarking of a number of brand names. In 1995, Intel started to employ the registered Pentium trademark also for x86 microprocessor with radically different microarchitectures. The Pentium brand briefly disappeared for Intel's roadmaps, only to remerge in 2007.

Intel introduced the Celeron brand for low-priced microprocessors with the 2006 introduction of "Upper" core 2 brand. There was no plan to use Pentium trademark anymore but Intel developed a line of mid-range dual-core microprocessors with Pentium Dual Core name at request of laptop manufacturers. Pentium brand, thus, lost upper position and was repositioned between the Core 2 and Celeron dual core lines as of 2007.

Summary

In this chapter, we have discussed the following points:

1. Computer is a fast and accurate symbol-manipulating system that is organized to accept, store and process data and produce the desired output results under the direction of a stored program of instruction.
2. Memory cell is used to store a single binary digit (0 or 1). Examples of memory cells include flip-flop, a charged capacitor and a single spot-on magnetic tape.
3. Memory word is the group of bits (cells) in a memory that represent instructions.
4. RAM is a memory in which the actual physical location of a memory word has no effect on how it takes to read from or write into that location.
5. PROMs is very expensive and cannot be used except in high volume applications, where cost would be spread out over many units.
6. Printer is an electromechanical device which has both electronic circuit control and mechanical assemblies.
7. VDU is nothing but cathode ray oscilloscope (CRO) which is used in Television. It is most widely used output device. It is fast and an inexpensive device.
8. Digital to analog conversion is usually abbreviated as D/A conversion. D/A converters are sometimes abbreviated as DACs. The purpose of such conversion is to take a digital signal and convert it to an analog signal.
9. A microprocessor is a single integrated chip in which all important components of computer, ALU, control unit and other circuits, are fabricated together. Microprocessor age began in 1970s when IC technology had advanced enough to make a single chip CPU.
10. Pentium is registered brand name of many intel x86 compatible microprocessors, both single and multicore.

Multiple Choice Questions

1. A device that works in conjunction with a computer but is not a part of it is called
   (a) hardware
   (b) memory
   (c) microprocessor
   (d) peripheral device
2. A device in which computer information and data are stored and relieved at a later time is called
   (a) hardware
   (b) memory
   (c) microprocessor
   (d) peripheral device
3. A single binary digit is called
   (a) byte
   (b) bit
   (c) data
   (d) logic
4. ROM has a
   (a) multiplexer followed by a decoder
   (b) demultiplexer followed by an encoder
   (c) decoder followed by an encoder
   (d) none of the above
5. The basic elements of a microprocessor are
   (a) ALU, memory
   (b) ALU, control unit

(c) ALU, memory, I/O devices
(d) ALU, control unit, memory

6. Which of the following element does not belong to microcomputer system?
(a) Microprocessor
(b) ROM
(c) RAM
(d) All of these

7. Floppy disk is a
(a) flexible storage
(b) permanent storage
(c) ROM
(d) fast storage

8. Permanent data like logarithmic tables are stored in a computer on a
(a) ROM
(b) tape
(c) drum
(d) core

9. Logic shift affects
(a) odd positions
(b) all positions
(c) even positions
(d) zero positions

10. An input port is
(a) tri-state buffer with register
(b) select logic
(c) input device
(d) R/W memory

11. Byte is a synonymous with
(a) band
(b) character
(c) bit
(d) word

12. Mnemonic symbols are used to
(a) denote address
(b) employ hamming code
(c) denote errors
(d) assist human memory

13. Debug is synonymous to
(a) erase
(b) exponent
(c) emulate
(d) trouble shoot

14. After completing the execution, microprocessor returns to
(a) halt state
(b) fetch state
(c) execute state
(d) interrupt state

15. A VSART chip provides
(a) half duplex operation
(b) full duplex operation
(c) duplex operation
(d) full duplex operation but cannot work in asynchronous mode

Review Questions

1. What is a computer?

2. List the major components of a computer.

3. What is the function of control unit?

4. What is the function of ALU?

5. What is the function of input device?

6. What is the function of output device?

7. Mention any four input devices.

8. List the different types of output devices.

9. What are the different types of computers?

10. Define memory device.

11. Mention different types of memories.
12. What is PROM?
13. Define erasable PROM.
14. Define EEPROM.
15. Differentiate between EPROM and EEPROM.
16. Define flash memory.
17. Explain the concept of semiconductor.
18. Explain floppy disk with diagram.
19. What are the different types of data storage systems?
20. Explain magnetic tape storage with diagram.
21. Explain the optical data.
22. Compare tape storage and optical storage.
23. Define printer.
24. What are the different types of printers?
25. Define impact printer.
26. Define non-impact printer.
27. What are the different types of impact printers? Explain them.
28. Write the advantages of printer.
29. List the major companies manufacturing printers.
30. What is ink jet printer?
31. Define drum printer.
32. Define laser printer.
33. Differentiate between laser printer and ink jet printer.
34. Define plotter.
35. Explain details about microfilm and microfiche of a computer.
36. Briefly describe about the graphic display device.
37. What is digital to analog converter?
38. What is the concept of DAC?
39. What is the purpose of laptop?
40. Compare personal computers and laptops.
41. Define microprocessor.
42. Explain the concept of Intel 8086.
43. Explain the architecture of a microprocessor with diagram.
44. What is the purpose of Pentium processors?

Answers

Multiple Choice Questions

1. (d)	**5.** (d)	**9.** (c)	**13.** (d)
2. (b)	**6.** (d)	**10.** (a)	**14.** (b)
3. (b)	**7.** (a)	**11.** (b)	**15.** (b)
4. (c)	**8.** (a)	**12.** (d)	

17 Fundamentals of Communication

LEARNING OBJECTIVES

After reading this chapter, you will be able to:

- Be familiar with the concepts of analog and digital signals.
- Understand the term bandwidth.
- Be able to explain modulation and demodulation.
- Have an understanding of frequency division multiplexing.
- Be aware of basic communication concepts.
- Be able to calculate the theoretical maximum channel capacity.
- Understand the concept, future and goals of ISDN.

Electronic systems are used to transmit information. The process by which this is achieved can be classified as analog or digital. Analog systems use waveforms which are directly proportional to the information being transmitted. Digital systems operate by converting the information into some form of pulse for transmission.

In this chapter, the relative advantages and disadvantages of such systems will be presented. When considering analog systems the concept of bandwidth and modulation will be introduced. By the end of the chapter, you should be familiar with the concepts underlying analog and digital systems and also be aware of the need to develop electronic techniques which can handle one or other system.

17.1 Analog and Digital Signal

Electronic signals fall into two categories: analog and digital. Both types of signal have been used from the earliest days of electrical communication. Analog systems first came to prominence with the telephone, whereas digital systems first appeared when telegraph systems introduced Morse code.

17.1.1 Analog Signal

In analog systems the information or data is given as an electrical signal that varies in direct proportion to the information or data. It follows that the variation must be continuous and between the limits of operation of the system, the variation can have any value from an infinite number of values. Such variation is associated with the production of sound (as is the telephone) is radio receivers or vision in television sets.

An example of an analog signal is shown in Fig. 1. Here we can observe the continuing variation in a complicated pattern, which contrasts with the regulated sinusoidal AC signals.

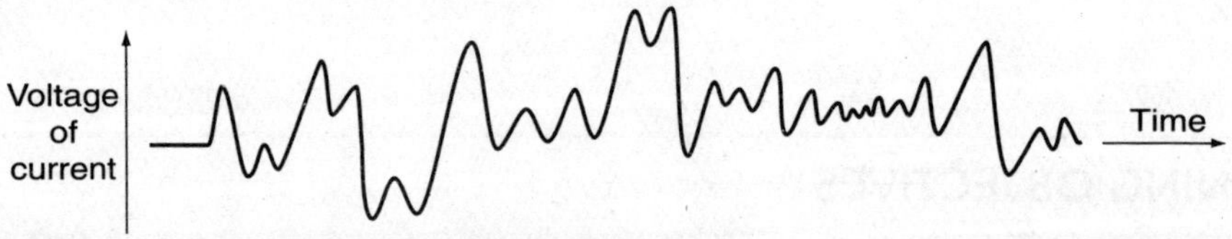

Figure 1 Analog signal.

In practice, analog systems are limited by the linearity of the amplifiers. A linear amplifier should amplify all signals equally. For instance, an amplifier with a gain of 10 should take an input signal of 1 mV and provide an output signal of 10 mV.

Analog signals are a representation of time-varying quantities in a continuous signal. Basically, a time variance is presented in a manner in which some sort of information is passed using various types of methods. These methods can include electrical, mechanical, hydraulic or pneumatic systems. Unlike digital signals, which use a numeric method of transmitting information, analog signals use small fluctuations in the signal itself to pass information.

The analog signals are three types:

1. **Telephone signal:** A Telephone signal transmission device capable of transmitting and receiving dial pulse in a network system which includes a telephone exchange is disclosed.
2. **Video signal:** A video signal is a type of signal which outputs predetermined video and television waveforms, and other signals used to simulate faults in, or aid in parametric measurement of television and video systems.
3. **Radio signal:** Radio signals are a type of electromagnetic radiation with wavelength in the electromagnetic spectrum longer than the infrared light. Like all other electromagnetic waves, they travel at the speed of light.

17.1.2 Digital Signal

Digital signals have one of a limited number of discrete values. In many applications, these are two values which in crude terms could be explained as ON and OFF. These values are usually described as 1 and 0, depending on the presence and absence of the supply voltage or current.

The first form of digital signal in common use was that associated with the Morse code. This uses short and long pulses to make up a code. Examples of this code are shown in Fig. 2. The fact that a code is required can be an obstacle to the application of a digital system.

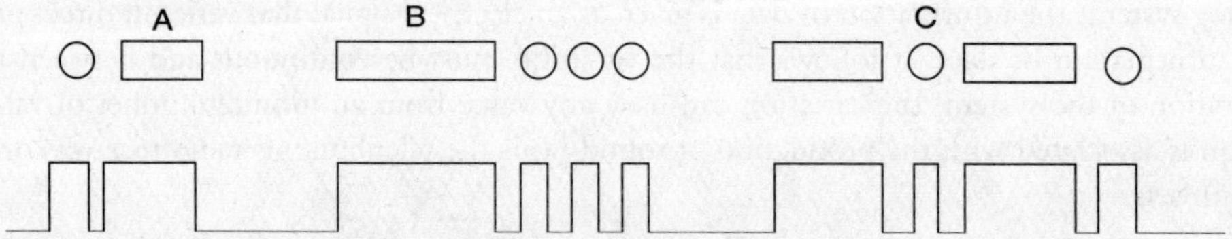

Figure 2 Examples of Morse code.

Let us consider the transmission of a simple pulse over a long time. The input signal is shown in Fig. 3(a) and is seen to be a simple square pulse. Typically the output signal which could result is shown in Fig. 3(b); here we find the pulse has become distorted.

Figure 3 Simple pulse signal: (a) Input signal; (b) output signal.

Following the Morse telegraph systems, most digital systems have progressed to being a series of pulses of identical duration. The period of a pulse is defined as the time between any two similar points of the pulse train. Note

Pulse period symbol: T
Unit: second (s)

The pulse repetition rate or pulse frequency is the number of pulses per second. The term digital signal is used to refer to more than one concept. It can refer to discrete–time signals that have a discrete number of levels (e.g., a sampled and quantified analog signal) or to the continuous–time waveform signals in a digital system.

The digital signals are divided into the following three types:

1. **Radar signal:** Radar signal stands for radio detection and ranging signal. It is an electronic system of collecting distance object by sending electromagnetic waves to them and picking up the waves reflected back (by the distance object).
2. **Binary signal:** Binary signal is a signal that may assume either of two polarities neither of which is zero. A bipolar signal may have a two state non return to zero or a three state non-return zero.
3. **Telecommunication services:** The term telecommunication service means a specified set of user information transfer capabilities provided to a group of users by a telecommunication system and any service provided by a telecommunication provider.

17.2 Data Transmission

Data transmission is the process of transmitting the data from one place to another place in a network. To do this one has to decide how to encode the data to be sent. This will differ according to the type of the data. The various types or data are as follows:

1. Audio data.
2. Text data.
3. Graphical data.
4. Video data.

This data transmission is to be represented very carefully. The data representation can be divided into two categories:

1. **Digital representation:** It means that the information is encoded as a set of binary values, that is, 0 and 1.
2. **Analog representation:** It means that the data will be represented by the variation in continuous physical quantity.

Data transmission is of two types:

1. Line communication.
2. Radio communication.

We will discuss both of them next.

17.2.1 Line Communication

The most common form of the communication is telephone and telex networks. This type of communication is carried out using overhead lines on poles and also by the use of buried cables. The pairs are twisted to avoid cross-talk among subscribers. For long distance calls, repeaters may be used at proper intervals. Such cables are usable upto 500 kHz.

17.2.2 Radio Communication

Radio or wireless communication is the process of sending information from one place and receiving it at another place by means of radio waves. Radio waves are disturbances produced in space by rapidly changing electrical currents flowing through a conductor. Radio waves are electromagnetic waves which travel with the velocity of light which is 186000 miles per second or 3×10^8 m/s.

17.2.3 Propagation of Radio Waves

Radio communication is established between two distant points where radio waves leave a conductor of suitable length called a transmitted antenna are picked up by the receiving end of another conductor of suitable length called the receiving antenna. There are a number of modes or paths by which the radio waves travel from transmitting antenna to receiving antenna. The most important of these modes are

1. ground wave;
2. sky wave;
3. space wave.

17.2.3.1 Ground Wave

A ground wave travels from transmitting antenna to the receiving antenna along the surface of the earth. It is also known as surface wave. The earth being a good conductor of electricity absorbs electrical energy from the ground wave as it glides along the surface of the earth. Energy is also absorbed by intervening objects like trees, buildings, hills, etc. The ground wave, therefore, gets weaker and weaker as it travels along the surface of the earth, till it becomes so weak that it is no longer useful. Ground wave propagation is, therefore, used for local and regional broadcast service.

17.2.3.2 Sky Wave

The transmitting antenna sends out radio waves in all directions. A part of these radio waves travels upward towards the sky and enters a region of the upper atmosphere known as the ionosphere which consists of electrically charged particles called ions. The ionosphere acts like a huge mirror or reflector placed in the sky and sends back to the earth a high frequency radio wave like a ray of light reflected from a mirror. Such a radio wave that is reflected from the ionosphere is called a sky wave.

17.2.3.3 Space Wave

Radio waves above 30 mHz are not reflected by the ionosphere. Radio waves at these frequencies cannot travel more than a few hundred feet along the surface of the earth because of heavy absorption by the earth. These waves travel from the transmitting antenna to the receiving antenna in space at about 15 km above the surface of the earth. This space is known as the troposphere and the waves provide communication within the optical range or the line of right distance only. This range does not normally extend beyond the horizon due to curvature of the earth. Hills, trees and tall buildings also obstruct these waves.

17.3 Modulation

Modulation is basically a process of increasing the frequency of the signal to be transmitted in order to achieve a practical value of antenna dimensions. For the purpose of modulation, we take a high-frequency signal called as carrier signal and vary some characteristics of the carrier signal like amplitude, frequency or phase in accordance with the instantaneous amplitude variation of the message signal. The signal is, thus, said to be modulated. This modulated signal is transmitted from one place to another. On the receiver side, we receiver this signal and retrieve the message in it. This process is called as demodulation. We will discuss this later.

17.3.1 Modulating Signal

The message signal/low-frequency signal which is to be transmitted from one place to other is called as modulating signal (Fig. 4). The high signal used for the purpose of modulation is called as carrier signal (Fig. 5).

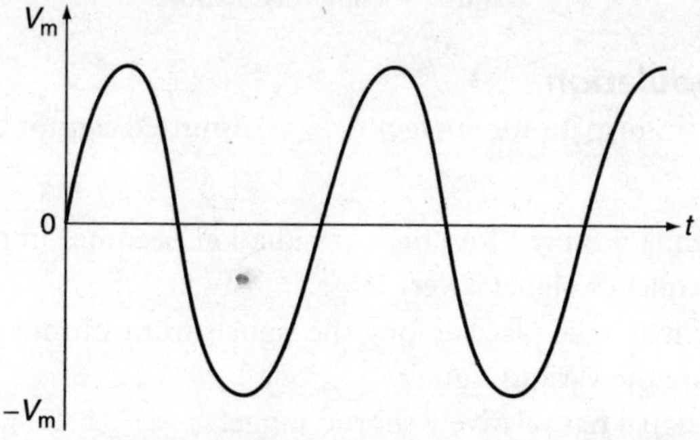

Figure 4 Modulating signal (low frequency).

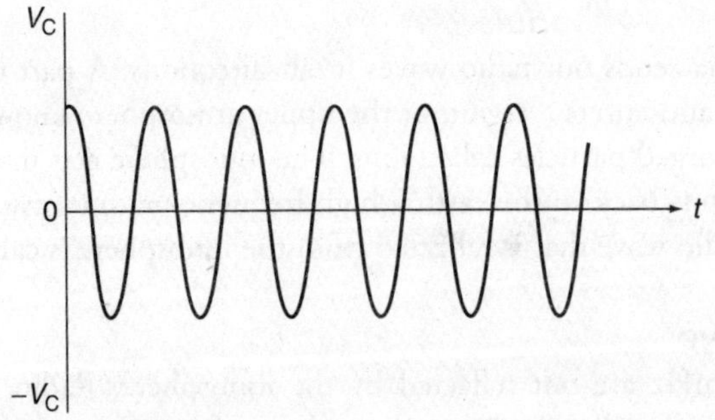

Figure 5 Carrier signal (high frequency).

17.3.2 Classification of Modulation

Modulation can be broadly classified into two types based on the nature of carrier signal.

1. **Sinusoidal modulation:** In sinusoidal modulation, the carrier signal used is a sine signal. The amplitude or frequency of the sinusoidal signal is varied in accordance to the instantaneous amplitude of the modulating signal for the purpose of modulation.
2. **Pulse modulation (Fig. 6):** In pulse modulation, the carrier signal used is a pulse signal. The amplitude, position or width of the pulse is varied in accordance to the instantaneous amplitude variations of the modulating signal for the purpose of modulation.

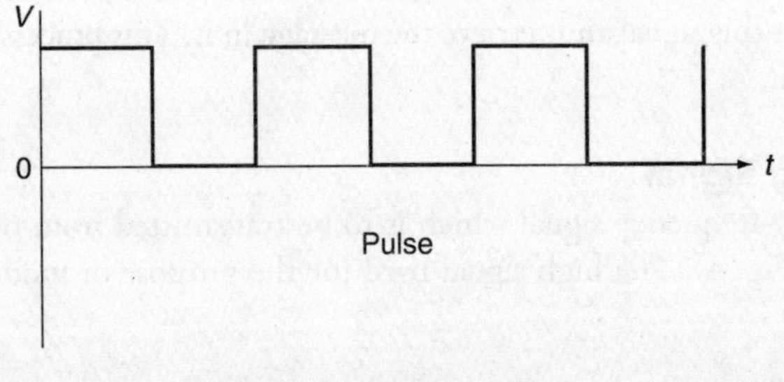

Figure 6 Pulse modulation.

17.3.3 Need for Modulation

The low-frequency signal (information/message) to be transmitted cannot be directly transmitted without modulation because

1. The size of the antenna required for efficient radiation becomes impractically high as the wavelength of the low-frequency signal is very large.
2. Mutual interference may take place among the signals from various transmitters and it becomes impossible to separate the various signals.
3. The low-frequency signal has relatively shorter range.

Thus, it is necessary to modulate the low-frequency signal with the help of a high-frequency carrier wave, before transmitting it through space for large distances.

17.3.4 Continuous Wave Modulation

Continuous wave modulation is a basic form of RF communication. It is essential ON–OFF key of an RF carrier. There are two types of continuous wave modulation, namely,

1. Amplitude modulation.
2. Frequency modulation.

Continuous modulation is also called as sinusoidal modulation.

17.3.4.1 Amplitude Modulation

In amplitude modulation, the amplitude of the carrier is varied in accordance with the instantaneous amplitude of the message signal. The carrier signal is a high-frequency signal. In the process of modulation, the amplitude of the carrier is varied and the variations are proportional to the amplitude of the message signal, that is, whenever the modulating signal reaches maximum amplitude, the amplitude of the carrier is also made higher and whenever the message signal reaches minimum amplitude, the amplitude of the carrier is also minimum (Fig. 7).

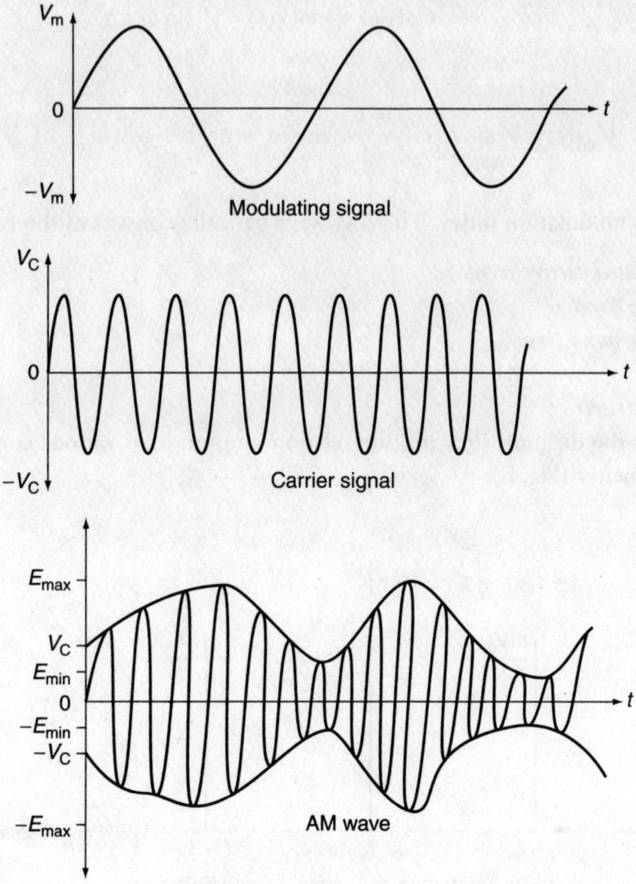

Figure 7 Amplitude modulation wave.

17.3.4.1.1 Modulation Index

In amplitude modulation, we define a parameter called as modulation index which is a measure of depth of modulation. Modulation index may be defined as the ratio of modulating signal amplitude to carrier signal amplitude:

$$m_{\mathrm{a}} = \frac{V_{\mathrm{m}}}{V_{\mathrm{c}}}$$

where m_{a} is the modulation index, V_{m} the modulating signal amplitude, V_{c} the carrier signal amplitude. The value of modulation index normally varies between 0 and 1, that is

$$0 < m_{\mathrm{a}} < 1$$

17.3.4.1.2 Amplitude Modulation (AM) Wave Equation

Mathematically, we can represent an AM wave by means of the following equations:

$$\text{Modulating signal } m(t) = V_{\mathrm{m}} \sin \omega_{\mathrm{m}} t$$

$$\text{Carrier signal } c(t) = V_{\mathrm{c}} \sin \omega_{\mathrm{c}} t$$

Then,

$$V_{\mathrm{AM}}(t) = V_{\mathrm{c}} \sin \omega_{\mathrm{c}} t + \frac{m_{\mathrm{a}} V_{\mathrm{c}}}{2}[\cos(\omega_{\mathrm{c}} - \omega_{\mathrm{m}})t + \cos(\omega_{\mathrm{c}} - \omega_{\mathrm{m}})t] \qquad (17.1)$$

where m_{a} represents modulation index. The AM wave basically consists of three terms:

1. The *unmodulated carrier term.*
2. The *lower side band.*
3. The *upper side band term.*

17.3.4.1.3 Spectrum

Spectrum of AM is the diagram showing the relation amplitude of various component of AM signal with respect to frequency (Fig. 8).

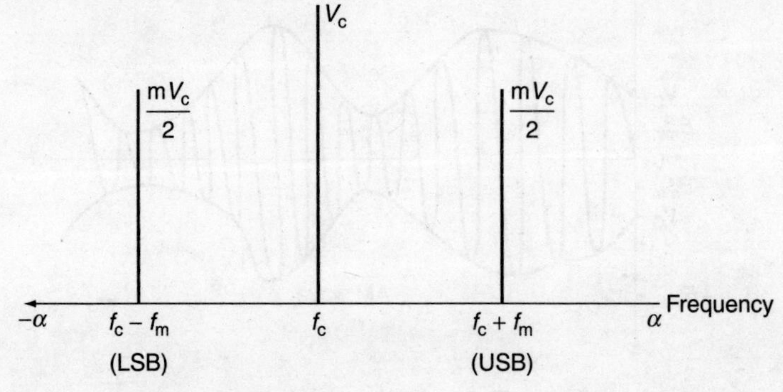

Figure 8 Spectrum of AM. LSB: lower side band; USB: upper side band.

17.3.4.1.4 Bandwidth of AM

Bandwidth is the difference between highest and lowest frequencies in the transmission of AM wave.

$$\text{Bandwidth} = \text{Upper side band frequency} - \text{Lower side band frequency}$$
$$= (f_c + f_m) - (f_c - f_m) = 2f_m$$

This means that the bandwidth of AM is equal to twice the modulating signal frequency.

17.3.4.2 Frequency Modulation

In frequency modulation (FM), the frequency of the carrier signal is varied in accordance to the instantaneous amplitude of the message signal, that is, whenever the amplitude of the message is low, frequency of FM wave is also less. Then the FM wave has a combination of several frequency components (Fig. 9).

While only the frequency of the carrier is varied in accordance to message signal amplitude, the amplitude of FM wave remains constant and is equal to carrier wave.

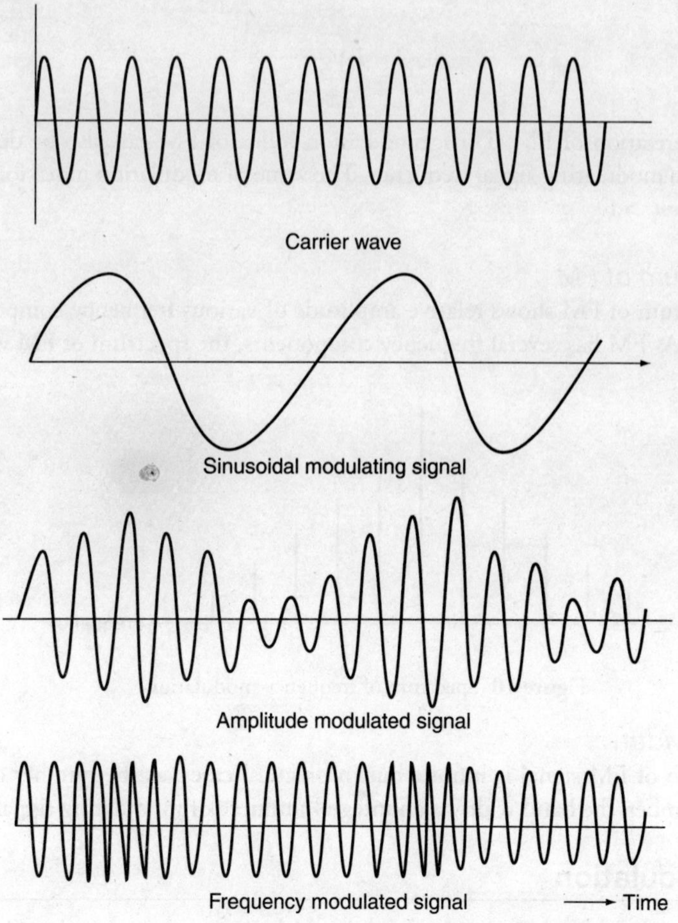

Figure 9 Frequency modulated signal.

17.3.4.2.1 Modulation Index

Modulation index of FM is given by

$$m_f = \frac{k_f V_m}{\omega_m}$$

where k_f is the frequency sensitivity factor, V_m the modulating signal amplitude, ω_m the angular frequency of modulating signal. Now

$$m_f = \frac{k_f V_m}{\omega_m}$$

$$= \frac{k_f V_m}{2\pi f_m}$$

$$m_f = \frac{f_d}{f_m}$$

where

$$f_d = \frac{k_f V_m}{2\pi}$$

is called frequency derivation of FM. Thus, modulation index of FM can also be defined as ratio of frequency deviation to modulating signal frequency. The value of modulation index for FM is normally greater than 1, that is $m_f > 1$.

17.3.4.2.2 Spectrum of FM

Similar to AM, spectrum of FM shows relative amplitude of various frequency components in it with respect to frequency. As FM has several frequency components, the spectrum of FM wave have several side bands (Fig. 10).

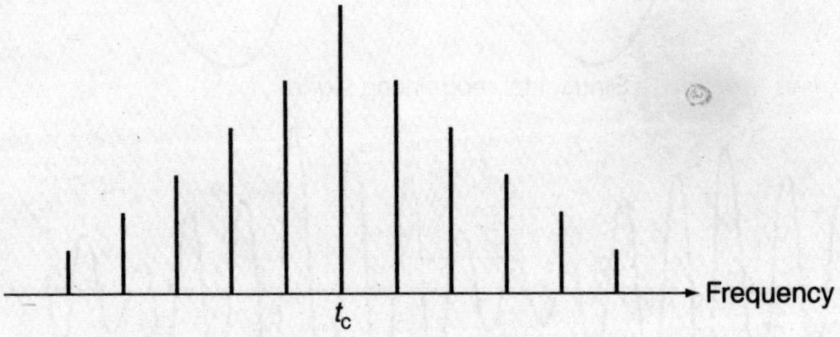

Figure 10 Spectrum of frequency modulation.

17.3.4.2.3 Bandwidth

Ideally the bandwidth of FM signal is infinite but in practical cases, as the number of side bands are limited to a finite number, the band width is an integral multiple of modulating signal frequency.

17.4 Demodulation

n Let us consider the speech transmission system with which we were introduced to filters. After having separated the various channels, it remains to separate the carrier from the original speech signal.

The original signal occupied a band width from a few hertz up to 4 kHz, and it is necessary to remove the carrier from the received signal to regain the original signal. This can be achieved by a reasonably simple network incorporating a rectifier.

The process of separating the original signal from the received signal is called *demodulation.* The modulated carrier is shown in Fig. 11(a) and demodulated signal is shown in Fig. 11(b).

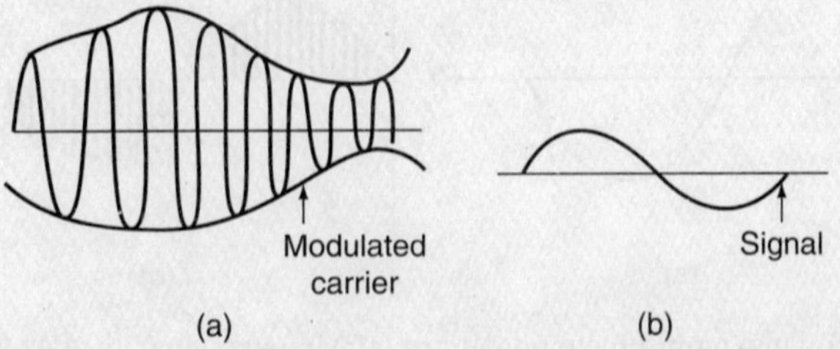

Figure 11 (a) Modulated carrier wave; (b) demodulated signal.

Let us look at an example of the action of demodulation in practice. Figure 12 shows the block system diagram for a simple radio receiver. The aerial picks up the transmitted signal and a bandpass filter is used to select the station to which we wish to listen. The output from the filter is the modulated carrier which generally has to be amplified before the demodulation process can be started. The output from the demodulator leaves the required signal; it then only remains to amplify the signal so that it has sufficient power to drive a loudspeaker.

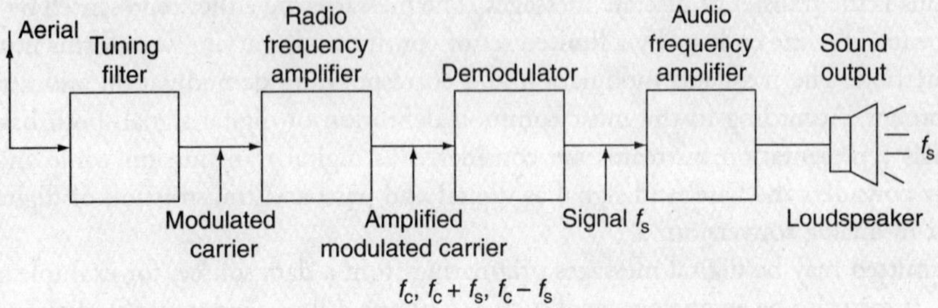

Figure 12 Block system diagram for a simple radio receiver.

17.5 Pulse Modulation

In pulse modulation, a pulse is used as the carrier and amplitude, position or width of the pulse are varied in accordance to the message signal amplitude. Accordingly, pulse modulation may be classified into following types

1. **Pulse amplitude modulation (PAM):** In pulse amplitude modulation (Fig. 13), the amplitude of the pulse is varied in accordance to message signal amplitude. Such a signal is called as pulse amplitude modulated signal.
2. **Pulse position modulation:** In pulse position modulation, the position of the pulse is proportional to the message amplitude. Such a type of modulation is referred to as pulse position modulation.

3. **Pulse width modulation:** In pulse width modulation, the width of the pulse is varied accordance to message signal amplitude. Such a type of modulation is referred to as pulse width modulation.

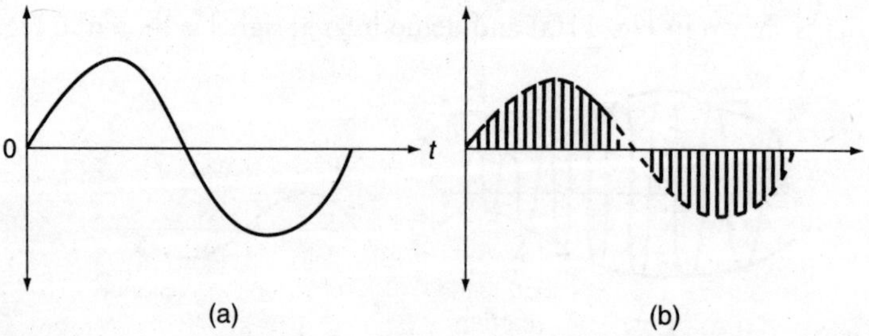

Figure 13 Pulse amplitude modulation: (a) Message signal; (b) PAM signal.

17.6 Digital Data Transmission

Digital data transmission is the physical transfer of data over a point-to-point or point-to-multipoint communication channel. Examples of such channels are copper wire, optical fibers, wireless communication channels, and storage media. The data is represented as an electro-magnetic signal such as an electrical voltage, radiowave, microwave or infrared signal.

While analog communication is the transfer of continuously varying information signal, digital communications is the transfer of discrete messages. The messages are either represented by a sequence of pulses by means of a line code or by a limited set of continuously varying waveforms using a digital modulation method. The passband modulation and corresponding demodulation are carried out by modern equipment. According to the most common definition of digital signal, both baseband and passband signals representation bistreams are considered as digital transmission while an alternative definition only considers the baseband signal as digital and passband transmission of digital data as a form of digital-to-analog conversion.

Data transmitted may be digital messages originating from a data source, for example a computer or a keyboard. It may also be an analog signal such as a phone call or a video signal, digitized into a bit stream.

17.7 Data Communication System

The distance over which data moves within a computer may vary from a thousandth of an inch, as in the case within a single IC chip, to as much as several feet along the backplace of the main circuit board. Over such small distances, digital data may be transmitted as direct, two-level electrical signals over simple copper conductors. Except for the fastest computers, circuit designers are not very concerned about the shape of the conductor or the analog characteristics of signal transmission.

Frequently, however, data must be sent beyond the local circuitry that constitutes a computer. In many cases, the distances involved may be enormous. Unfortunately, as the distance between the source of a message and its destination increases, accurate transmission becomes increasingly difficult.

This results from the electrical distortion of signals travelling through long conductors, and from noise added to the signal as it propagates through a transmission medium. Although some precautions must be taken for data exchange with a computer, the biggest problems occur when data is transferred to devices outside the computer circuitary. In this case, distortion and noise can become so severe that information might get lost.

17.8 MODEMS

In this modern Internet world, every one of us might have come across the device named as MODEM. It is a device used to connect the computer terminal to the communication network such as telephone network. MODEM is an acronym for Modulation DeModulation, which describes exactly what a MODEM does.

At the source or transmitting end, a modem takes the digital signal or binary signal from the computer terminal and converts it into a continuous analog signal that is needed for transmission over the telephone channel. At the destination on receiving end, the modem converts the analog signal coming through the telephone channel into binary signal that is fed into the computer terminal as shown in Fig. 14.

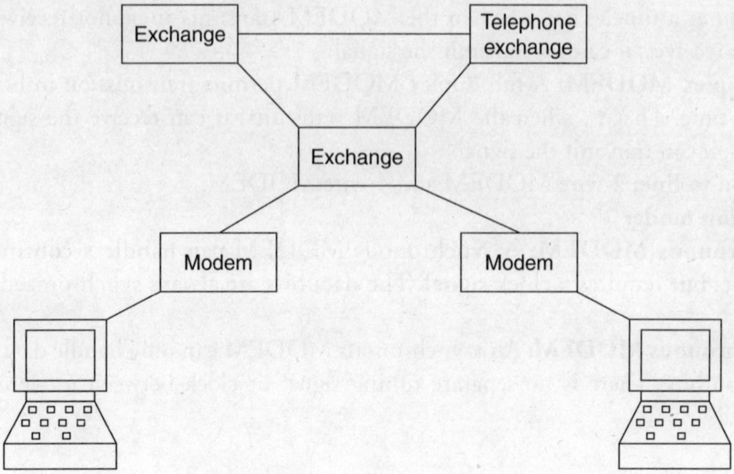

Figure 14 Typical MODEM system.

17.8.1 Working of MODEM

Broadly, a MODEM comprises a transmitter, a receiver and two interfaces (Fig. 15). The digital signal to be transmitted is applied to the transmitter. This transmitter will convert the digital signal to analog signal by means of modulation.

MODEMs use amplitude modulation (AM), frequency modulation (FM), phase modulation (PM) or combination of these to achieve the desired data rates and low error performance over different types of telephone channels.

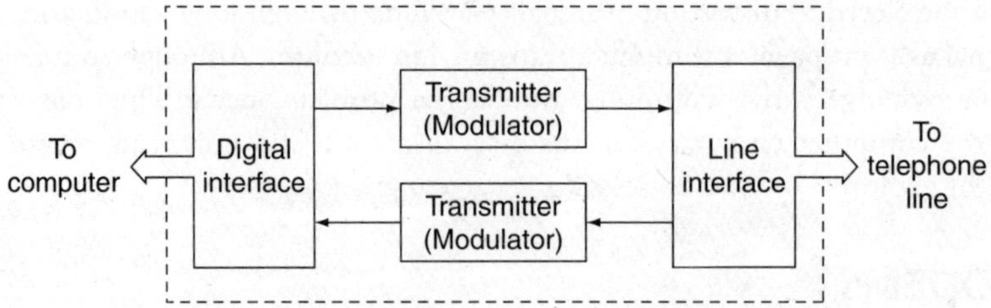

Figure 15 Working of a MODEM.

17.8.2 Different Types of MODEMS

MODEMs can be of different types and they can be classified in a number of ways based on the following features.

1. **Speed:** This is the maximum rate at which the MODEM receives and transmits the data. The various speeds are 28.8 kbps, 38.4 kbps, 56 kbps (kbps – kilo bits per second).
2. **Directional capability:**
   - **Half duplex MODEM:** A half duplex MODEM permits transmission or reception in one direction at a time. That is, when the MODEM transmits it cannot receive the signal and when it receives, it cannot transmit the signal.
   - **Full duplex MODEM:** A full duplex MODEM permits transmission in both directions at a same time. That is, when the MODEM transmits, it can receive the signal and when it receives, it can transmit the signal.
3. **Connection to line:** 2-wire MODEM and 4-wire MODEM.
4. **Transmission mode:**
   - **Synchronous MODEM:** A synchronous MODEM can handle a continuous stream of data bits but requires a clock signal. The data bits are always synchronized with the clock signal.
   - **Asynchronous MODEM:** An asynchronous MODEM can only handle data bytes with start and stop bits. There is no separate timing signal or clock between modem and computer terminal.

17.9 Communication System

The term communication refers to the process of transfer of data/message from one place to other. The message which is being transferred from one place to another may be a voice information, picture information or data in the form of files in a computer. In the process of communication, some information is being transferred from one place to other. In electrical sense, the term communication refers to sending, receiving and processing of information by electrical means. The signal which is being transferred may be called as message, data, information and intelligence. In the process of communication, the signal which may be available in any form, viz. audio/acoustic energy or video signal, is first converted to electrical or electromagnetic form to ease communication. The basic block diagram of a communication system is shown in Fig. 16. The components are as follows:

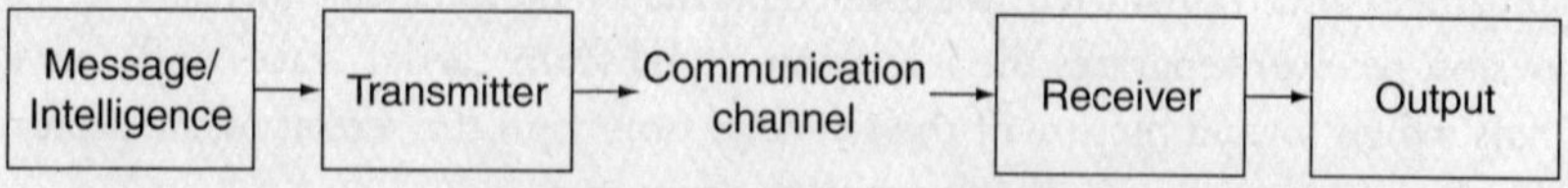

Figure 16 Basic block diagram of a communication system.

1. **Message/Intelligence:** This is the data which has to be communicated and transferred from one place to other. The message may be a human voice or music or some picture information taken from a still camera or motion camera or data files in a computer. Whatever be the nature of available signal to be transferred, we first convert it to an equivalent electrical signal by means of a suitable transducer. A transducer is a device which converts one form of energy to another form of energy. For example, suppose if we want to transfer voice information, the transducer is a microphone which converts the acoustic signal to electric signal. Any data which has to be communicated should be converted to electrical signal.
2. **Transmitter:** This is basically a combination of several blocks within it and the main objective/work of the transmitter is to convert the signal which is available as electrical signal to a form which can be transferred/transmitted through the communication channel. The major blocks of a transmitter section are AF amplifier, modulation, RF amplifier and power amplifier.
3. **Communication channel:** For a data to be transmitted from one place to another, we always require a medium to carry our signal. The channel may be physical medium like a wire or cable or may be free space. Without the help of a medium, data cannot be transmitted from one place to another. An ideal channel serves in just carrying the information between the transmitter and receiver.
4. **Receiver:** The term receiver refers to an arrangement wherein the signal is being converted back to a form suited for the output device. The receiver accepts the signal from the communication channel and retrieves the actual information and gives it to the output device. Normally this consists of RF amplifier, demodulator, AF amplifier and power amplifier.
5. **Output device:** This is the actual system where the information is to be made available. For example, in the case of voice communication, the output device may be a loud speaker, in the case of video information the output device is television and so on.

17.9.1 Radio Communication

Radio or wireless communication is the process of sending information from one place and receiving it in another place by means of radio waves. Radio waves are electromagnetic waves which travel with the velocity of light which is 186000 miles per second or 3×10^8 m/s.

17.9.2 TV Communication System

The word "Television" is a combination of two words "Tele" and "vision" means "seeing" from a "distance." It is a one-way communication system that reproduces fixed or moving images and sound by using electronic circuits.

17.9.2.1 Basic Principle

1. In television, light signals from the object being televised are converted into electrical signals by a television camera and transmitted to distance points by radio carrier waves.
2. The television receiver separates the television signal from carrier waves and converts them into light signals which form a picture of the televised object on the screen of the picture tube.
3. Sound has also to be transmitted along with the picture.
4. Separate carrier waves are used for the transmission of picture signals and sound signal but they are radiated by the same transmitting antenna.
5. The receiving antenna receives both carrier waves but the television receiver converts these signals separately into sound waves which drive a loud speaker and light waves which produce a picture on the screen of the picture tube.
6. For proper display of picture and the reproduction of sound, several controlling signals are also to be transmitted.

17.9.2.2 Television Transmitter

Figure 17 shows the block diagram of black/white TV transmitter. The transmitter simultaneously transmits audio and video signals through a single channel. These signals are separately processed in a TV transmitter. After processing, the two signals are combined using logic transmission through the transmitting antenna in a TV transmitter.

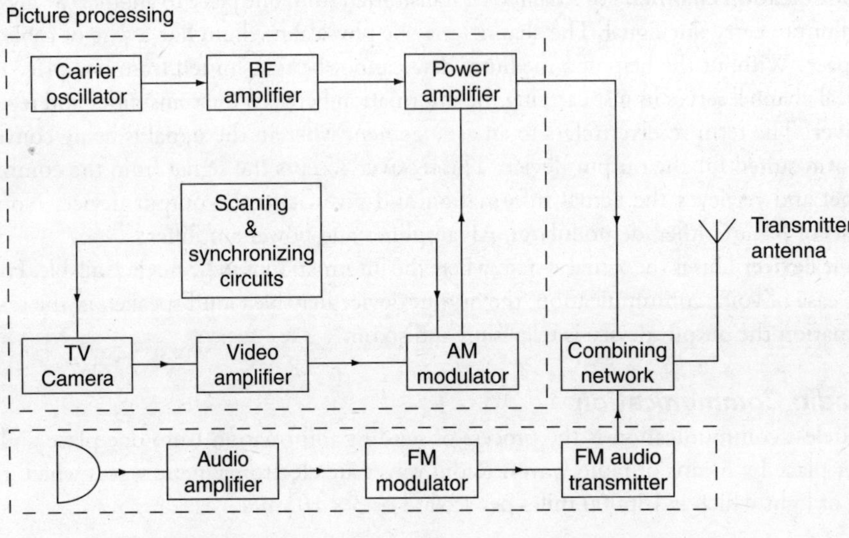

Figure 17 Block diagram of black/white TV transmitter.

17.9.2.3 Picture Transmission

A video signal is transmitted as an amplitude modulated (AM) signal. The TV camera picks up the images and converts them into their corresponding electrical signals. These signals are called video signals.

17.9.2.4 Sound Transmission

Audio signals are transmitted through frequency modulation. The microphone converts the sound signal to its corresponding electrical signal. This signal is called audio signal.

17.9.2.5 Television Receiver

The receiver receives the combined TV signals through the receiving antenna. This antenna receives all the signals transmitted by the various TV stations. The RF tunes and amplifier stage selects the desired signals and rejects all unwanted signals.

17.9.2.6 Scanning Process

Scanning is the process by which the optical image of the televised object formed on the photosensitive plate of the TV camera is broken into series of horizontal lines by an electron beam.

17.9.2.7 Interlocked Scanning

The scanning process in which the scanning electron beam sweeps across each horizontal line in regular succession from top to bottom of the picture is called interlocked or progressive scanning. For continuity of motion, each picture frame is scanned 25 times per second in India.

17.9.2.8 Synchronization

The process by which the horizontal and vertical sweep at the camera and TV receiver are kept in step with each other is known as synchronization.

17.9.2.9 Blanking

On scanning, at the end of each field, the scanning beam quickly returns to start the scanning of another line or field. The path followed by the returning electron beam is called retrace. The retrace is made invisible by a process by as "blanking."

17.9.3 Microwave Communication

Microwave transmission extends from 2 GHz to about 40 GHz. Microwaves travel more or less in straight lines. Long distance transmission can be achieved by using number of microwave repeaters or relay stations between the transmitter and receiver. However, microwave link requires fewer number of repeaters than the coaxial cable system. This may be due to fact that the attenuation of microwaves increases as the square of the distance travelled whereas in a coaxial cable transmission, the attenuation increases logarithmically with distance.

A microwave repeater receives the modulated microwave signal from the previous repeater. The received signal is converted into IF signal of frequency 70 MHz and then amplified. For transmission, these amplified signals are further converted into higher frequency signals. To avoid interference, the frequencies of the received and transmitted signals differ by a few MHz.

The height of the tower used for the microwave link depends on the terrain tower location, etc. Usually the height of the tower is around 25 m. The repeaters make use of 300W DC supply by a battery. The antenna is mounted at the top of the tower. The alignment of the antenna has to be checked during every routine maintenance.

Microwave communication is used for the following:

1. Long haul communication.
2. Transmitting TV signals.
3. Transmitting voice signals.
4. Establishing closed circuit TV.
5. Establishing LAN network.
6. Short distance point-to-point connections between different buildings in a campus.

17.9.4 Satellite Communication

A satellite is a physical object that orbits or revolves around some celestial body. The Earth and other planets are satellites which evolve the sun, and the moon is satellite of the earth.

17.9.4.1 Satellite Orbits

The path along which satellite moves in the space around Earth is called an orbit. A satellite can move in three types of orbits, namely,

1. **Synchronous orbits:** They lie in the equatorial plane of Earth. The synchronous orbit is further classified into:
   - **Geostationary orbit:** An orbit in which a satellite appears stationary relative to any point on Earth is called geostationary orbit. Geostationary satellite remains in a fixed position with respect to a given Earth station and has 24 hours availability time. Three geostationary satellites spaced 120° apart can cover the whole world.
   - **Geosynchronous orbit:** A satellite whose period of revolution equals to period of rotation of Earth about its axis. In this, satellite has the 24 hours non-Equatorial orbit. If a satellite has a time period of 24 hours, then height of the satellite above Earth may be determined to be about 85800 km. The dish antenna at the receiving Earth station receives the down-link signal. This signal is amplified and converted back to original form as at the TV studio.
2. **Polar orbits:** When a satellite rotates in an orbit that takes it over the north and south poles, it is called polar orbit.
3. **Equatorial orbits.** When the satellite rotates in an orbit above the equator, it is called an equatorial orbit.
4. **Inclined orbits:** Any other orbital path besides those discussed above is called an inclined orbit. 100 % of the Earth's surface can be covered with a single satellite in a polar orbit. The satellite is rotating around the Earth in a longitudinal orbit while the Earth is rotating on a latitudinal axis. As a result, every location on Earth lies within the radiation pattern of the satellite twice each day.

These are shown in Fig. 18.

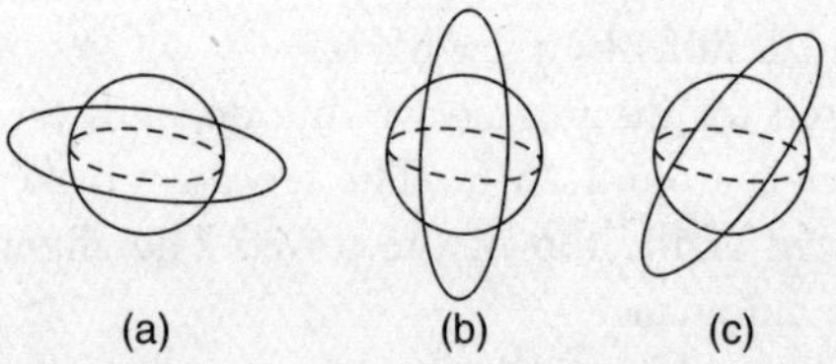

Figure 18 Satellite orbits: (a) Synchronous, (b) polar, (c) inclined.

The other characteristics of orbit are as follows:

1. **Apogee:** It is the farthest point on the orbit of a satellite from the Earth.
2. **Perigee:** It is the nearest point on the orbit of satellite from the Earth.
3. **Inclination:** Inclination of the orbit is determined by the angle it makes with the equator.

Figure 19 illustrates the three points.

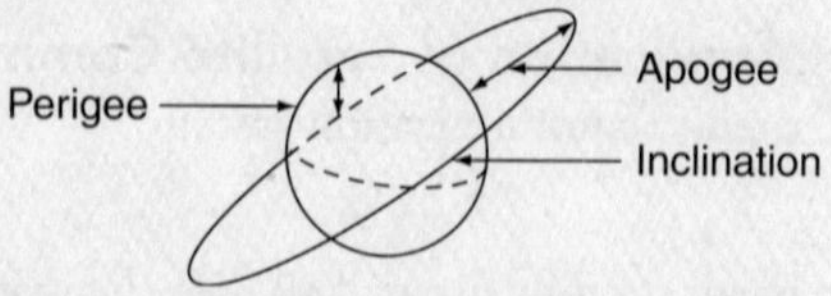

Figure 19 Characteristics of an orbit.

17.9.4.2 Properties of *Satellite Communication*

1. Most communication satellites operate in the microwave frequency spectrum.
2. The microwave spectrum is divided up into frequency bands which have been allocated to satellite as well as other communication services.
3. Most widely used satellite communication band is "c" band. The up-link frequency range is 5.925–6.425 GHz.
4. The bandwidth between the upper and lower sideband frequencies is 500 MHz. This is an incredibly wide band capable of carrying an enormous number of signals.
5. Most new communication satellites will operate in the "ku" band designated as 14/12 GHz. The reason for this shift in frequency is that "c" band is becoming overcrowded.
6. The military uses "x" band (7250–8400 MHz) for its radar and satellite and L band (1530–2700 MHz) is used in marine and aeronautical communications.

17.9.4.3 Satellite Architecture and Organization

All satellite communication system consists of the following two basic parts:

1. Satellite or spacecraft.
2. One or more Earth stations.

Figure 20 illustrates the satellite communication system.

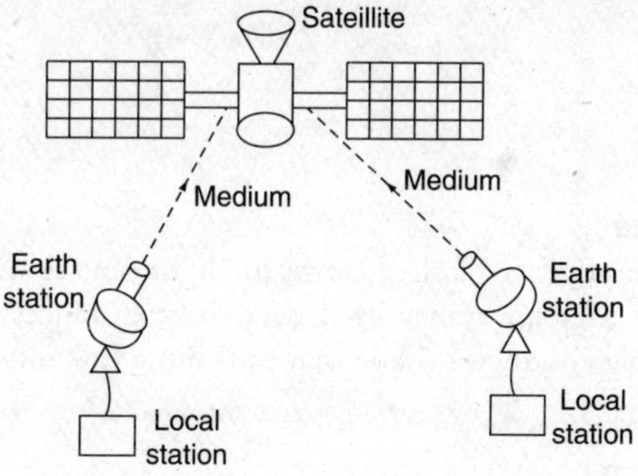

Figure 20 Satellite communication system.

The signal to be transmitted originates from the TV studio and then fed to the Earth station. The Earth station converts this signal into the assigned up-link microwave frequency. The transmitting dish antenna is in direction of the satellite. The receiving dish antenna of the satellite receives up-link signal. The down-link signal is then fed to the transmitter which retransmits it back to Earth through another antenna.

17.9.4.4 Advantages and Application of Satellite Communication System

The following are advantages of satellite communication system:

1. It offers large bandwidth.
2. Global connection among Earth stations situated all over the world is possible.
3. It offers not only point-to-point links but also broadcast capabilities.

The satellite communication systems find application in the following:

1. TV broadcast.
2. Mobile communication.
3. Long distance telephony.
4. Weather forecasting, etc.

17.9.5 Optical Communication System

One of the main limitations of communication system is their restricted information-carrying capabilities. To increase the information capacity of the channel, the bandwidth of the channel must be increased. This reduces available spectrum space.

The information-carrying capability of a radio signal can be increased tremendously if higher carrier frequencies are used. Today the use of microwave frequencies for communication has been greatly expanded. Although the microwave frequency spectrum is not totally occupied at this point, it is well used. A great deal of communication takes place by microwave signal transmitted through satellite relay stations. Satellite can receive and transmit thousands of telephones calls to any place on Earth or can carry key high-speed digital data.

The components of a typical fiber-optical communication system are shown in Fig. 21. They are as follows:

1. Light source.
2. Modulation.
3. Photodetectors.
4. Optical fibers.

17.9.5.1 Optical Source

The digital pulses are then used to flash a powerful light source OFF and ON very rapidly. In simple, low-cost system, they are transmitted over short distances through the light emitting diode (LED). This is a simple conductor diode that puts out a low intensity red light beam. Other colors are also used.

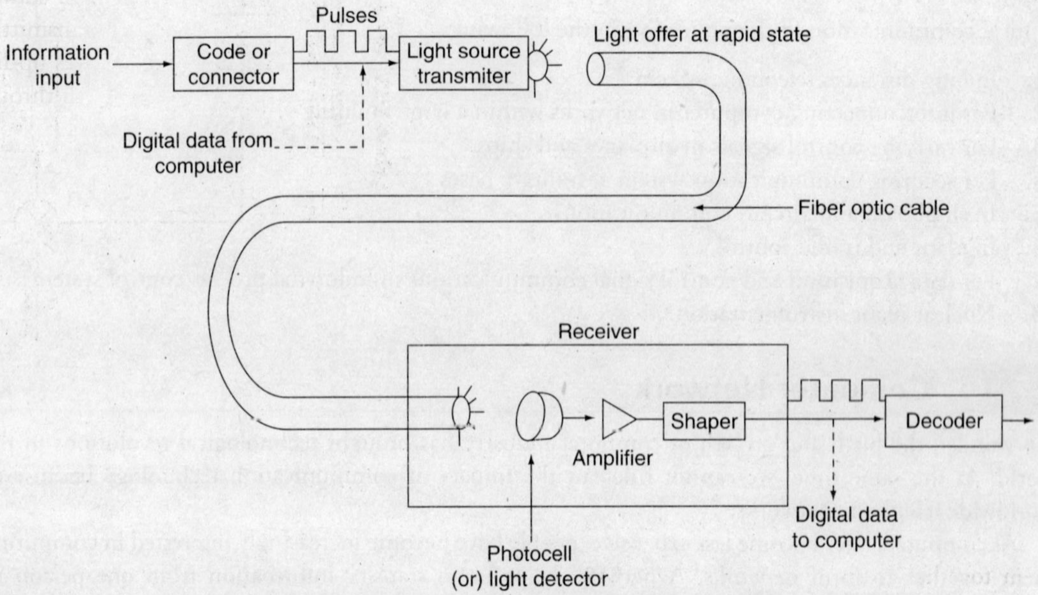

Figure 21 Components of a typical fiber-optical communication system.

Infrared beams like those used in TV remote controls are used in transmission. This is also a semiconductor device that generates an extremely intense single frequency light beam.

17.9.5.2 Optical Destructor

The light beam pulses are then fed into a fiber-optic cable where they are transmitted over long distances. At the received end, a light-sensitive device known as a photocell or light detector is used to detect the light pulses.

In very long transmission systems, repeater units must be used along the way. Since the light is greatly attenuated when it travels over long distances, at some point it may be too weak to be received reliably. To overcome this problem, special relay situations are used to pick up the light beam, convert it back into electrical pulses that are amplified and then retransmit the pulses on another light beam. Several stages of repeaters may be needed to over key long distances. Despite the attenuation problem, the loss is less than that occurs with the electric cables.

17.9.5.3 Advantages and Applications of Optical Communication

The following are advantages of optical communication:

1. Fiber cables are thin, small and weightless.
2. Fiber cables in the range 10^{14} to 10^{15} Hz frequencies.
3. Optical cables require less space and are easy to install.
4. It is virtual impossible to tap information from a fiber cable.
5. The fiber optic systems are cheaper than co-axial cables or microwave links.

Optical communication finds application in the following:

1. In long distances telephone system.
2. For interconnecting computer in networks within a large building.
3. For carrying control signals in airplanes and ships.
4. For securing communication system at military bases.
5. In shipboard and aircraft communications.
6. In plant and traffic control.
7. For data acquisition and control signal communications in industrial process control systems.
8. Nuclear plant instrumentation.

17.10 Computer Network

No wonder, the birth and growth of computer industry has brought technological revolutions in the world. At the same time, we cannot rule out the impact of communication technology because of worldwide telephone networks.

As computers have become less expensive, people have become increasingly interested in computing them together to form networks. A network helps us to transfer information from one person to another, from one place to another and from one continent to another.
There are three types of network.

1. **Local area network (LAN):** It connects the computer system within a building or within 10 km. LAN is a network connecting computers and associated devices, which are used to interchange information. The devices that can be linked are:
   - Key boards, PC terminal and DOS.
   - Printers and plotters.
   - Mass storage devices.
   - Various computer-controlled machines.
2. **Metropolitan area network (MAN):** It connects people across the city.
3. **Wide area network (WAN):** It connects the people across the country.

17.11 Integrated Services Digital Network (ISDN)

Telephony and telegraphy provided the major telecommunication services till the recent past. Even though telephone service was introduced quite later than telegraphy, it was found to be a convenient form of communication and overtook telegraphy. At present there is a well-spread public switch telephone network (PSTN) which is used for local and long distance voice communication.

With development of digital computers, need arose for data communications. It, however, requires a network with better quality and reliability of service as compared to telephony. So telephone networks began to be used for data communication. However, this required that data be converted to voice frequency band used in telephony. Modems were developed for this purpose – modems convert digital data signals to voice frequency band before they are transmitted on the telephone lines, but they also become more complex as the speed of data transmission is increased.

17.11.1 Future of ISDN

ISDN is mostly used in

1. **National ISDN:** Most of the interesting Research and Development is being done in the B-ISDN world of application. Surprisingly though, there is still a lot of standardizing going on with Narrow band ISDN, which amounts to standardizing the standards.
2. **Broadband ISDN:** A broadband ISDN or B-ISDN is a network of large bandwidth to carry date, voice, image and video at a very high data rate. The proposed access rate for the B-ISDN is 150 Mbps. Such high data rate is adequate for image traffic and also allows the interconnection of high speed LANs. This high access rate also allows video broadcast traffic, video conferencing and many potential new applications.

17.11.2 Goals of ISDN

Some of the goals of ISDN are as follows:

1. It is cost-related tariff.
2. It is multipexed support.
3. It is transparent.
4. It is very standard.
5. Separation of competitive function.

17.12 Internet

The Internet is a global system of interconnected networks that use the standard Internet protocol suite to serve billions of users world wide. It is a network of networks that consists of millions of private, public, academic, business and government networks.

Most traditional communication media included telephone music film, etc. The Internet has enabled or accelerated new forms of human interactions through instant messaging, Internet forums and social networking. On-line shopping has boomed both for major retail outlets and small artisans and traders. The commercialization of what was by then on international network in the mid-1990s resulting in its popularization and incorporation into virtually every aspect of modern human life. As of 2009 an estimated quarter of Earth's population used the services of Internet.

Summary

In this chapter, the following points are discussed:

1. Analog data takes continuous values on interval. The most familiar examples of analog signals are the data collected by sensors.
2. A telephone message comprises speech and sound having vowels and consonants.
3. A radar is a device which is used to find out the location of distant objects in terms of range and bearing.
4. Telecommunication systems are used to provide a wide variety of services of community public.

5. Television transmission is meant for general public and occupies a bandwidth of 5.5 MHz to 8 MHz spacing between two adjacent channels.

6. Pulse amplitude modulation is a pulse modulation system in which the signals are sampled at regular intervals and each sample is made proportional to the amplitude of signal at the instant of sampling.

7. A communication satellite is in effect of a microwave relay station. It is used to link two or more round-based microwave transmitter/receiver known as Earth stations or ground stations.

Multiple Choice Questions

1. A modulator is a system to
   (a) separate two frequencies
   (b) impress the information on to a radio frequency carrier
   (c) extract information from the carrier
   (d) amplify the audio frequency signal

2. In AM transmission, the frequency which is not transmitted is
   (a) upper side frequency
   (b) lower side frequency
   (c) audio frequency
   (d) carrier frequency

3. In high-power AM transmission, modulation is done at
   (a) buffer stage
   (b) RF power stage
   (c) oscillator stage
   (d) IF stage

4. The output of a diode detector contains
   (a) modulating signal
   (b) DC voltage
   (c) RF ripple
   (d) all of these

5. The product demodulation essentially is a
   (a) balanced modulator
   (b) mixer
   (c) amplifier
   (d) oscillator

6. The AM broad cast is given by
   (a) 10 kHz to 30 kHz
   (b) 500 kHz to 1500 kHz
   (c) 3 to 30 MHz
   (d) 30 to 300 MHz

7. The function of buffer amplifier in transmitters is to provide
   (a) impedance matching
   (b) frequency stability of oscillator
   (c) amplification of RF signal
   (d) none of these

8. Carrier wave is generated by crystal oscillator because
   (a) it provides high power output
   (b) no buffer stage is needed
   (c) tuned circuit is not required
   (d) it has good stability

9. Typical bandwidth of an FM receiver is
   (a) 20 kHz
   (b) 200 kHz
   (c) 1 kHz
   (d) 200 kHz

10. Telegraphy is
   (a) the transmission of speech through space
   (b) the transmission of picture through space
   (c) the transmission of information using a code
   (d) the detection of objects

Review Questions

1. Define communication.
2. What are the basic types of communication?
3. Define modulation.
4. What are the types of modulation?
5. Define amplitude modulation.
6. Define frequency modulation.
7. What is meant by modulation index for AM wave?
8. What is bandwidth for AM wave?
9. Define modulation index for FM wave.
10. Define pulse amplitude modulation.
11. Define pulse position modulation.
12. Define pulse width modulation.
13. What is mean by sampling?
14. What are the main characteristics of a receiver?
15. What are the three main color TV systems used?
16. What is Geostationary satellite?
17. Write the application of satellite communication?
18. What are the characteristics of an orbit?
19. What are modes of propagation in an optical fiber?
20. How do you define the bandwidth efficiency of digital microwave system?
21. Draw the basic diagram of communication system and explain.
22. Explain the different types of modulation systems.
23. Explain the AM modulation with spectrum.
24. Explain pulse code modulation and show various coding techniques.
25. Explain the propagation of radio waves.
26. Explain with neat block diagram, the principle & working of AM and PM transmitter.
27. What are the characteristics of radio receiver?
28. Explain any one radio receiver with diagram.
29. Explain the working of TV transmitter and receiver.
30. Explain fiber optics communication system.
31. Explain the principle of satellite communication.
32. Explain ISDN.
33. Explain the future of ISDN.
34. Write the goals of ISDN.
35. Write the major application of Internet.
36. Mention the advantage of Internet.
37. Write the principle of operation of Internet.
38. Write the new trends in Internet.
39. Explain any 1 major application of Internet.
40. Write the disadvantages of Internet.

Answers

Multiple Choice Questions

1. (b)
2. (c)
3. (b)
4. (d)
5. (b)
6. (b)
7. (c)
8. (d)
9. (b)
10. (c)